Plant Microbe Interface

Ajit Varma • Swati Tripathi • Ram Prasad
Editors

Plant Microbe Interface

Springer

Editors
Ajit Varma
Amity Institute of Microbial Technology
Amity University Uttar Pradesh
Noida, Uttar Pradesh, India

Swati Tripathi
Amity Institute of Microbial Technology
Amity University Uttar Pradesh
Noida, Uttar Pradesh, India

Ram Prasad
Amity Institute of Microbial Technology
Amity University Uttar Pradesh
Noida, Uttar Pradesh, India

ISBN 978-3-030-19830-5 ISBN 978-3-030-19831-2 (eBook)
https://doi.org/10.1007/978-3-030-19831-2

This Springer imprint is published by the registered company Springer Nature Switzerland AG.
The registered company address is: Gewerbestrasse 11, 6330 Cham, Switzerland

Contents

Chapter 1
Mycorrhizae Resource Allocation in Root Development and Root Morphology

Ibrahim Ortaş, Mazhar Rafique, and Md Toufiq Iqbal

Abstract Plant root systems are influenced by genetics and environmental conditions which are leading to varied root system architectures. Different plant species have diverse root system architectures, and mineral nutrient availability is mainly determined by the root system. Also, the availability of mineral nutrient uptake is played by the role of mycorrhizal fungi. In this chapter, the role of plant root development, root architecture, and mycorrhizal inoculation on mineral nutrition was reviewed. The root development, mainly the physiological, morphological, and molecular responses of plant roots to diverse nutrient uptake in assistance to the mycorrhizal fungi, is one of the hot research areas for plant scientists and plant nutritionists. Keeping in mind the importance of this subject, the present chapter is compiled which covers the importance of nutrient uptake in plant growth and development. Moreover, the importance of roots in nutrient uptake and establishing the symbiotic relationship is essential. Underground relations are set up by the plant roots in coordination with different soil microorganisms. Arbuscular mycorrhizal fungi (AMF) as a major soil organism participate in symbiotic relationship and facilitate the plant in growth and root development. Moreover, it shapes the plant roots for the better cooperation with AMF in nutrient and water uptake facilitation. It may change the root morphology, physiology, and molecular behavior which may vary plant to plant.

I. Ortaş (✉)
Department of Soil Science and Plant Nutrition, University of Cukurova, Adana, Turkey
e-mail: iortas@cu.edu.tr

M. Rafique
Department of Plant Sciences, Quaid-i-Azam University, Islamabad, Pakistan

M. T. Iqbal
Department of Agronomy and Agricultural Extension, University of Rajshahi, Rajshahi, Bangladesh

A. Varma et al. (eds.), *Plant Microbe Interface*,
https://doi.org/10.1007/978-3-030-19831-2_1

1.1 Introduction: Plant Nutrients and Their Requirements

Each plant species has its own root systems. Plant roots are the main source for nutrient uptake. The plant root systems are the basic nutrient absorbing area of the plants. At the same time, plant roots can alter their close environment (in the root-soil interface), called the rhizosphere, by absorbing nutrients and liberating exudate (Shrivastava et al. 2014). This limited area around the root surface (0–20 mm region) is biologically active. The rhizosphere dynamics of any plants, namely, the rhizosphere physics, chemistry, and biology, are completely different than the bulk soil (Marschner 2012; Ortas 1997). Each plant species root structure and rhizosphere differed significantly on plant nutrition acquisition systems. In addition, plant roots develop symbiotic relationship with mycorrhizae fungi and rhizospheric bacteria to improve nutrients uptake (Barea et al. 2005). Nutrient and water uptake, root respiration, root exudate, and pH changes (either increases or decreases) can occur at this thin soil layer.

Mainly most of the plant nutrient uptake mechanisms recently in the late 1970s were recognized by plant and soil chemistry scientists. During 1900 most of the nutrient elements were determined to be special (Epstein 1997; Marschner 1996). All plant species need to have essential plant nutrients. The essentiality was proposed to consider an element as “essential” when it accomplishes three basic principles: (1) in its absence, plants must be incapable of continuing normal growth, (2) its biological functions cannot be carried out by any other element, and (3) it must be directly involved in plant metabolism. Reports of diverse research groups in the early 1970s perfectly illustrated the rhizospheric processes to understand the effects of nutrient availability in root system development. Marschner’s group since the early 1970s clarified several nutrient uptake mechanisms by root and mycorrhizal fungi (Marschner 2012). However, some nutrients which are less mobile in soil, such as P, K, Zn, Cu, Fe, and NH_4-N, are more efficiently taken up (Smith and Read 2010).

1.2 Effect of Nutrient Availability in Root System Development

The evolution of root systems such as root branching, root hair development, and mycorrhizal symbiosis with root processes is still under evaluation process. According to Hemsley and Poole (2004), evolution of root system is a notable process that has led to a progressive transformation from the very simple root systems of early land plants to the diverse and complex root systems of the modern plants. Cruz-Ramirez et al. (2009) indicated the ancient plants did not have to face obstacles to acquire water and obtain nutrients efficiently under moist climatic conditions. In drier environments, the plant root system was developed, availability of nutrients increased, and roots become a vigorous anchor for the establishment in soils with diverse physical and chemical characteristics. Dicotyledonous and monocotyledonous plants have a diverse root system development, and their nutrient and water uptake are determined by root systems. In fibrous root systems, most stem-borne roots develop underground, but in some species, such as maize (*Zea mays*) (Cruz-Ramirez et al. 2009) and sorghum, roots are produced from aboveground structures. Generally, aboveground roots are called aerial roots. It has been reported by Hochholdinger et al. (2004) that prop roots stabilize the main stem and are also capable of branching and taking up mineral nutrients and water.

Lynch (1995) indicated that plant root geometry and morphology are very important for maximizing nutrient uptake because root systems that have higher ratios of surface area to the volume will more effectively explore a larger volume of soil. Several root parameters such as root length, root diameter, root surface area, and root hair density are important for enhanced nutrient uptake by plants. Consequently, changes that arbuscular mycorrhizal fungi (AMF) bring about on root morphology, such as root branching and root elongation, could constitute an additional mechanism by which the fungi enhance P uptake. Adaptive changes of root growth and architecture under P starvation are related to altered carbohydrate distribution between root and shoot, and these changes may be caused by plant hormones (Nacry et al. 2005; Neumann et al. 1999), sugar signaling (Karthikeyan et al. 2007; Vance 2010), and nitric oxide in the case of cluster-root formation in white lupin (Wang et al. 2010).

When soil nutrient concentration is low, root systems of some plant species expanded the capacity by branching thorough this way roots interact with soil organisms and explore the soil. Especially in a low amount of P availability, plant species changed their root branching patterns. Some non-mycorrhizal plants, for example, *Lupinus albus*, a legume plant have special root systems such as cluster roots have a high capacity to uptake P from soil (Marschner et al. 1986; Neumann and Martinoia 2002). The plant species do not establish symbiosis with mycorrhizae in order to get sufficient orthophosphate (Pi) uptake, although their root hairs can contribute with up to 80% of the surface area of the root (Jungk 2001).

Plant root epidermal cells including root hairs (the direct pathway) uptake Pi by leads to lowering of Pi concentrations in the rhizosphere which is called depletion zone. Usually, Pi depletion replacement does not easily keep pace with uptake (Fig. 1.1) (Smith et al. 2011). Plant roots and mycorrhizal fungi that take up P as negatively charged Pi ion ($H_2PO_4^-$) forms (Smith et al. 2011) indicated that P depletion poses an additional problem because the concentration in plant cells is about 1000-fold higher than in the soil solution and the cell membrane has an inside-negative electric potential. According to Bucher (2007) in this case, Pi uptake requires metabolic energy and involves high-affinity transporter proteins in the *Pht1* family. Marschner (1995) indicated that plants had evolved a range of strategies that increase either Pi uptake capacity or availability of Pi in the rhizospheric soil. The most common of these strategies exist for mycorrhizal symbiosis. Some plant species which don't have mycorrhizal symbiosis can produce of dense "cluster roots" that produce organic anions which release Pi from poorly available inorganic

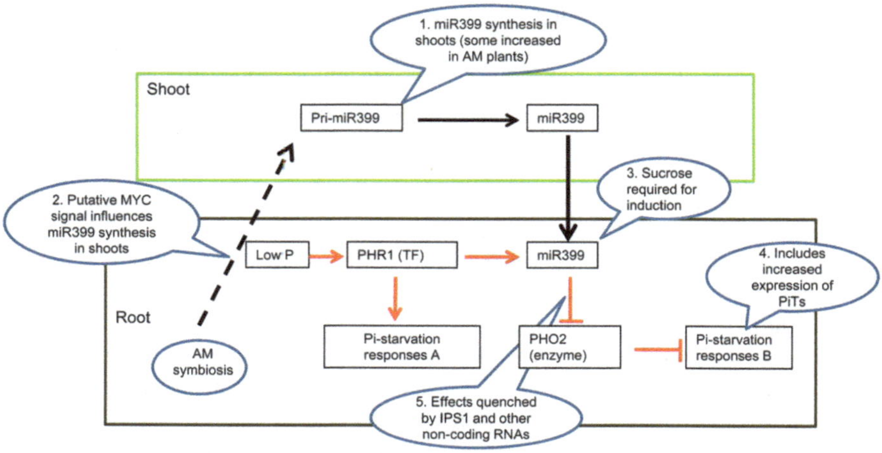

Fig. 1.1 Possible signaling events in AM roots based on studies of Pi starvation in non-mycorrhizal plants and miR399 expression in AM medic. In NM plants, low P increases the activity of the transcription factor (TF) PHR1, which binds to the P1BS element in promoters of several Pi starvation-induced genes (A) and increases their expression. PHR1 also increases the expression of miR399s. miR399s are probably largely synthesized in shoots, where they accumulate more in AM than in NM plants (callout 1); this implies the transfer of (unknown) MYC signals from root to shoot in AM plants (callout 2). miR399s are transferred from shoots to roots. Accumulation in roots is influenced by PHR1 and by such transport from shoots (callout 3). High miR399 levels under low P reduce the activity of the enzyme encoded by PHO2 and hence increase PHO2-dependent Pi-starvation responses, including increased expression of PiTs (callout 4). Effects of miR399s in reducing PHO2 activity can be quenched by noncoding RNAs such as IPS1 (callout 5). PHO2 might then inhibit Pi-starvation responses and reduce the expression of PiTs [modified from Branscheid et al. (2010)]

forms, but these are much less common (Cheng et al. 2011; Lambers et al. 2008, 2010).

Lambers et al. (2008) indicated that in many cases cluster-root formation might be an alternative strategy to AM formation. Phosphorus is involved in several biochemical mechanisms such as the formation of cell membranes, carbohydrate metabolism, protein synthesis, photosynthesis, respiration sugar metabolism, energy storage, and transfer. Schachtman et al. (1998) indicated that since P is a component of key molecules such as nucleic acids, phospholipids, and ATP, consequently, plants cannot grow without a reliable supply of this nutrient. Soil P concentration may be large; however, most of it is not available or less mobile because of the very low solubility of phosphates of iron, ammonium, and calcium, leading the soil solution P concentration of 10 μm or less (Schachtman et al. 1998). The AM hyphae extend plant root system's capacity to explore more nutrient and water in the soil to cope the stress situations (Fig. 1.2) (Manoharan et al. 2010).

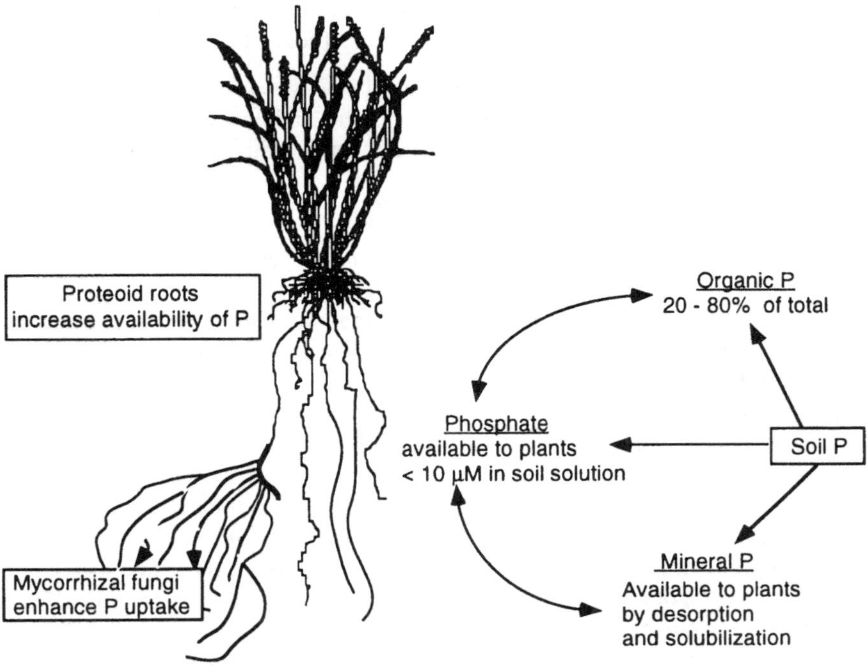

Fig. 1.2 Proteoid roots increase the availability of P. Mycorrhizal fungi enhance P uptake

1.3 Mycorrhizas Alter Root Architecture System

Smith and De Smet (2012) postulated that plant roots involve several varieties of biological processes, such as nutrient uptake, nutrient storage, and mechanical support. In general, root system architecture (RSA) of plants is very dependent on plant species and soil environment (Giri et al. 2018). Root growth is controlled genetically, but it is also influenced by environmental factors (Fageria and Moreira 2011) such as soil quality, mineral nutrient, and soil management. Most tree plants species have short and rare root hairs, and such kinds of plants are dependent on AM colonization. Wu et al. (2010) reported that root architectural alteration in the AM-colonized citrus plant could increase root functioning to explore more water and nutrients under stress conditions. AM mycelia constitute of morphologically different types of hyphae. Relatively coarse and thick-walled hyphae with a diameter between 5 and 20 μm appear to function mainly in nutrient transport and extension of the fungal colony. In general, when plants have a low root-shoot biomass ratio, slow root growth rates, and/or poor root hair development, plant roots demand mycorrhizal symbiosis. If a high amount of total soil P is poorly available, plant demands from mycorrhizal symbiosis and in that case mycorrhizal symbiotic plants uptake more P and grow over the non-mycorrhizal plant (Bolan 1991). Mycorrhizal fungi are important through the following mechanisms in the rehabilitation of decertified ecosystems (Khan et al. 2017; Sharma et al. 2017; Teotia et al. 2017):

- Enhancing establishment and growth of plants by increasing nutrient uptake
- Contributing to the efficient recycling of nutrients and thus to long-term stability
- Stabilizing the soil structure and quality. Since mycorrhizae fungi:
  - Can access greater soil volume
  - Can break molecules down into usable forms
  - Can turn inorganic phosphorus and nitrogen into forms usable by plants

Mycorrhizae also play a key role in soil aggregate formation, and aggregates can keep carbon in soil (Prasad et al. 2017). Recent research suggests that mycorrhizal fungi might be an important component of the soil organic carbon (SOC) pool, in addition to facilitating carbon sequestration by stabilizing soil aggregates. In an ecosystem, the flow of carbon to the soil-mediated by mycorrhizae serves several important functions such as getting nutrients and water from the soil. Measurements of plant carbon allocation to mycorrhizal fungi have been estimated to be 5–20% of total plant carbon uptake (Pearson and Jakobsen 1993), and in some ecosystems, the biomass of mycorrhizal fungi can be comparable to the biomass of fine roots.

The association of roots with AMF is a very widespread strategy by which plants facilitate their acquisition of mineral elements from the soil (Marschner 1995). In addition to element uptake via mycorrhizal mycelia, AMF has also been shown to affect root morphology and functioning, as well as mycorrhizosphere soil properties. This may lead to indirect effects of the AM association on plant nutrient availability and uptake (Smith and Read 2010). With their thin diameter, AM hyphae might be able to access smaller soil pores and better compete with soil microbes for nutrient resources, compared with plant roots. Neumann and George (2010) indicated that like plant root systems, AM hyphae seem to differ considerably in their architecture and physiological activities depending on their genotype. Mycorrhizal hyphae length is also controlled by nutrient level mainly by soil phosphorus levels.

1.4 Mycorrhizae Role on Nutrient Uptake

Mycorrhizal symbioses can increase the spatial availability of P, extending the nutrient absorptive surface by formation of mycorrhizal hyphae. In the symbioses, nutrients are transferred by AMF via their extensive mycorrhizal mycelium to plants, while in return the fungi receive carbon from the plant. The AMF not only influence plant growth through increased uptake of nutrients (P, Zn, and Cu) but may also have no nutritional effects in terms of stabilization of soil aggregates and alleviation of plant stresses caused by biotic and abiotic factors (Smith and Read 2010) .

A primary benefit of AMF is the improved P uptake conferred on symbiotic plants. In low-P soils, mycorrhizal plants usually grow better than non-mycorrhizal plants as a consequence of enhanced direct P uptake of plant roots via the AM

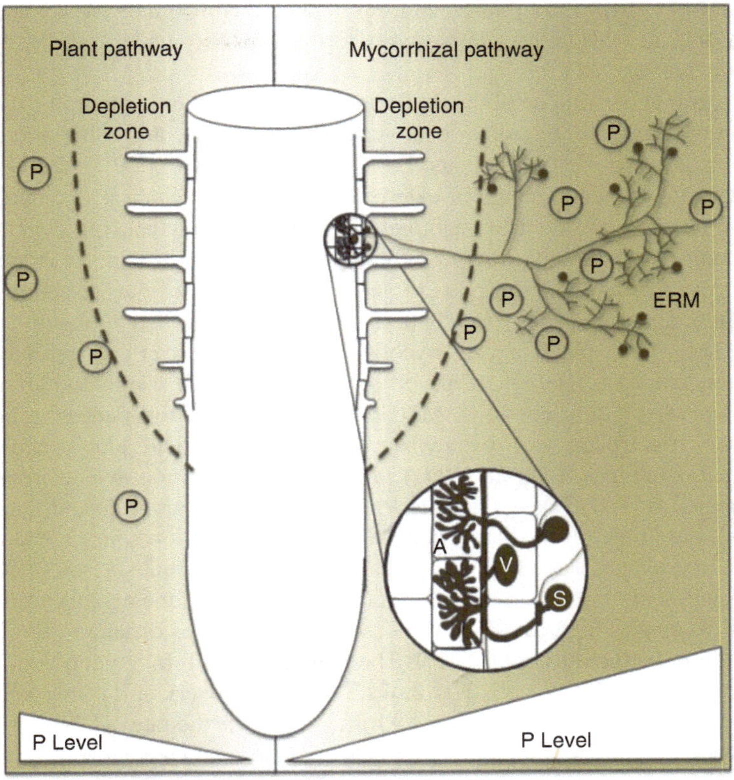

Fig. 1.3 Mycorrhizal effect on nutrient uptake and phosphorus depletion (Bucher 2007; Bücking and Kafle 2015)

pathway. However, plant growth can be suppressed even though the AM pathway contributes greatly to plant P uptake (Smith and Read 2010) . The growth inhibitions might be caused by the downregulation of the direct root P uptake pathway. Recent gene expression study (Feddermann et al. 2010) shows that plants induce a common set of mycorrhiza-induced genes, but there is also variability, indicating that there exists functional diversity in AM symbioses. The differential expression of symbiosis-associated genes among different AM associations is related to the fungal species, plant genotypes, and environmental factors. Therefore, regulation of direct uptake pathways through the epidermis, root hairs, and AM pathways require further investigation (Fig. 1.3) (Bucher 2007; Smith et al. 2010). In the case of mycorrhizae, the area exploited more volume of soil as well. As the plants get more mycorhizzal colonization in roots, it will facilitate in nutrients acquisition and reduced chemical fertilizer requirement (Varma et al. 2017).

Rouphael et al. (2015) reported that AMF symbiosis could induce changes in secondary plant metabolism leading to the enhanced biosynthesis of phytochemicals with health-promoting properties. It has been reported that AMF can secrete phosphatases to hydrolyze phosphate from organic P compounds (Koide and Kabir 2000; Marschner 2011), thus improving crop productivity under low input conditions (Smith et al. 2011).

The extraradical hyphae of the AMF are also important to increase the uptake of ammonium, immobile micronutrients such as Cu and Zn, and other soil-derived mineral cations (K^+, Ca^{2+}, Mg^{2+}, and Fe^{3+}) (Smith and Read 2010). The absences of AMF differ in the total amount of external hyphae length which is the main power for more nutrient uptake. The magnitude of plant growth enhancing effects varies with the nutrient's status of the soil. At the same time, plant rhizosphere mechanisms such as mycorrhizae help the plant to get a high quantity of nutrient mainly P. In most plants, P requirements for growth in the soil are controlled by the mycorrhizal dependences. Also, plant species mycorrhizal dependence can vary with available soil P concentration (Hetrick et al. 1996). The AMF itself cannot stimulate plant growth on very P-deficient soils. Mycorrhizae can utilize soil nutrient efficiently. The AMF affect plant growth only via an increased nutrient supply under well-inoculated conditions. Li et al. (1991) reported that in the non-mycorrhizal plants, the depletion of $NaHCO_3$-extractable P extended about 1 cm into the outer compartment, but in the mycorrhizal plants, a uniform P depletion zone extended up to 11.7 cm (the length of the hyphae compartment) from the root surface. Also in the same experiment, they found that in the outer compartment, the mycorrhizal hyphae length density was high (2.5–7 m cm^{-3} soil) at the various distances (0–11.7 cm) from the root surface (Li et al. 1991). The uptake rate of P by mycorrhizal hyphae was in the range of $3.3–4.3 \times 10^{-15}$ mol s^{-1} cm^{-1}. Recovery of P is low as it is less mobile mineral which moves mainly by diffusion. Since the rate of diffusion of P is very slow (10^{-12} to 10^{-15} m^2 s^{-1}), high plant uptake rates create a zone around the root that is depleted of P (Schachtman et al. 1998).

The external hyphae of AM fungi extend well beyond the depletion zone, accessing supplies of nutrient at a distance and in narrow soil pores. Since the hyphae develop around the root distributed beyond the root area, nutrient uptake is high, and the nutrient depletion zone is expanded. When nutrients are removed from the soil solution more rapidly, a nutrient depletion zone develops and that nutrient can be replaced by diffusion (Li et al. 1997). For a poorly mobile ion such as phosphate and potassium, a sharp and narrow depletion zone develops close to the root. AMF can secrete phosphatases to hydrolyze phosphate from organic P compounds (Koide and Kabir 2000; Marschner 2012) and thus improve crop productivity under low input conditions (i.e., phosphorus deficiency) (Smith and Smith 2011). Douds and Millner (1999) indicated that the extraradical hyphae of the AMF could develop up to 8 cm beyond the root-growing zone and act as extensions of the root system in acquiring nutrients from the soil. The extent of depletion zone makes AM-inoculated plant grows better than a non-mycorrhizal plant. These differences depend on the plant root system, including numbers and extent of root hairs.

1.5 Horticultural Plants Root Development and Mycorrhizae

With the current state of soil technology, inoculation is most feasible for transplanted crops and in areas where soil disturbance has greatly reduced the native inoculum potential. The especially pro-inoculated seedling can get benefit from mycorrhiza to penetrate root and hyphae in the soil. Although for citrus seedling changes in the management of the soil-plant system can be sufficient to optimize the mycorrhizal symbiosis, in horticulture the inoculation of seedlings prior to transplant has given the best results (Ortas 2012).

Roots are surrounded by a matrix of soil organisms, in addition to mycorrhizal fungi, that might influence root function and survivorship in complex ways. The effects of global atmospheric and climatic change on roots might be profound, but are difficult to predict. Some of the expected effects of global change on roots will be mediated indirectly through changes in shoot physiology. Increased C gain under elevated CO_2 might increase root length density, promote shallower root systems by stimulating lateral root production over primary root elongation, increase mycorrhizal colonization, and decrease tissue N concentrations, while at the same time, whole-plant nutrient acquisition is increasing (Pritchard and Rogers 2000; Rogers et al. 1999; Tingey et al. 2000).

Plants are able to respond to P starvation by changing their root architecture, including root morphology, topology, and distribution patterns. Increases in root/shoot ratio, root branching, root elongation, root topsoil foraging, and root hairs are commonly observed in P-deficient plants, while the formation of specialized roots such as cluster roots occurs in a limited number of species (Lynch 1995; Vance 2010). P deficiency has been shown to reduce the growth of primary roots and enhance length and density of root hairs and lateral roots in many plant species (Desnos 2008; López-Bucio et al. 2003). Some plant species, for example, white lupin (*Lupinus albus*), can develop cluster roots with dense and determinative lateral roots, which are covered by a large number of root hairs (Lambers et al. 2006; Vance 2010). Therefore, root architecture plays an important role in maximizing P acquisition because root systems with the higher surface area are able to explore a given volume of soil more effectively (Lynch 1995).

Mycorrhizal fungi using external hypha changed root morphology, increased root absorption, and transport of nutrients to roots (Kungu et al. 2008). The results of Hoshyar et al. (2017) showed that sweet cherry rootstock plantlets inoculated with *Diversispora epigaea* gave the highest leaf area, root diameter, root surface, and phosphorus concentration. The work of Wu et al. (2017b) showed that mycorrhizal species are significantly affecting root morphology (Fig. 1.4).

Biopriming of micropropagated plantlets (horticultural) with AMF helps in the development of a superior and stronger root system (Ponton et al. 1990) by increasing the rooting intensity and surface area of existing roots (Puthur et al. 1998). Colonization of a plant root by AMF can alter the morphology of a root system in a structural, spatial, quantitative, and temporal manner (Atkinson 1992; Atkinson et al.

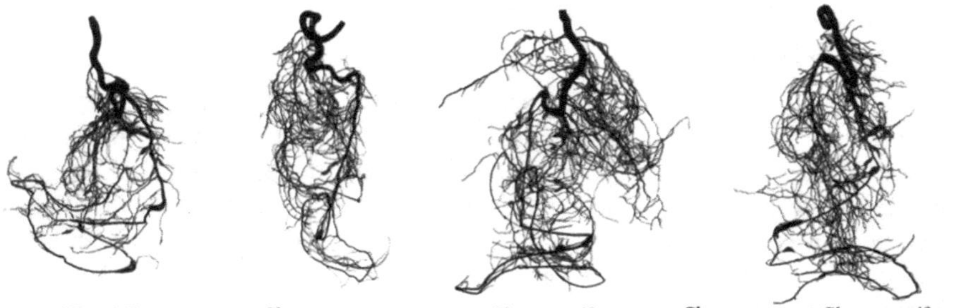

Fig. 1.4 Root system morphology of trifoliate orange (*Poncirus trifoliata*) infected by *Glomus mosseae*, *G. versiforme*, a mixture of *G. mosseae* and *G. versiforme*, and non-AMF (Wu et al. 2017a)

1994; Berta et al. 1993; Norman et al. 1996). The AMF-colonized roots are highly branched, i.e., the root system contains shorter, more branched, adventitious roots of larger diameters and lower specific root lengths (Atkinson et al. 1994; Berta et al. 1993). As a direct consequence, mycorrhizal inoculation stimulates rooting and growth and thereby transplant survival of cuttings and seedlings raised in the nutrient media.

1.5.1 Potato

Potato relies heavily on fertilizer input (Davies et al. 2005) and responds well to mycorrhizal symbiosis. In addition, the potato is a globally plated crop in a wide range of habitats and climate condition. Several reports in controlled environments and field inoculation trials have shown that mycorrhizal inoculants benefit potato production. The first successful study on field inoculation has been reported by Black and Tinker (1977), followed by numerous studies using different potato cultivars and different mycorrhizal inoculants (reviewed by Wu et al. 2013). The results of these studies depended on potato cultivars, inoculants used, phosphorus concentrations, amendment, and tuber size category of interest (Douds Jr et al. 2007). Overall, trial inoculation studies resulted in higher yields and larger tubers than treatments using conventional chemical fertilizers (Douds Jr et al. 2007; Wu et al. 2013; Upadhyaya et al. 2013).

Hijri (2016) analyzed the data set containing 231 field trials conducted by farmers over a 4-year period in Europe and northeast America through an incentive program that was set up by the inoculant industry in 2011 to promote the application of mycorrhizal-based inoculants in agriculture. This data set was unique, in that farmers themselves performed the experiments using their conventional practices under authentic field conditions. In contrast to experiments that are conducted in greenhouses and controlled field trials with many replicates and randomized blocks, the experimental design of this study was simplified in order to encourage farmer involvement. There are no replicates per field trial, but this is compensated by the large sample size. The inoculant used in the present study contains a single strain of one AMF species, *R. irregularis* (synonym *G. irregulare*) DAOM 197198, which exhibits substantial plasticity and an ability to adapt to different environmental conditions. However, *R. irregularis* DAOM 197198 may still be limited to specific agricultural environments, particularly in P-rich soils and under highly stressful conditions. The host plant can influence AM fungal community composition directly, by regulating carbon allocation to roots, by producing secondary metabolites, or by changing the soil environment.

Different mycorrhizal fungi used in the horticultural plants are listed in Table 1.1.

1.6 Role of Mycorrhizae on Field Crops (Mainly Cereal Crop) Root Morphology and Growth

Cereals plants such as maize, wheat, and rice are the main food supply for the majority of world population. Nearly 40% of daily food intake are from cereals. Cereal seed nutrient concentration is very important. It is well known that plant species nutrient acquisition capacity depends on soil and ecological conditions. So, plant root capacity and relation with mycorrhizal fungi are important for sufficient nutrient uptake from soils. It is known that cereal plant has very extensive root systems and long root hairs have high natural ability to take up nutrients from the soil. Mycorrhiza also has a positive effect on various cereal crop root morphologies as well. However, the effect of mycorrhizae on several cereal crops varies from plant to plant. Here, we discussed the effect of mycorrhizae on maize, wheat, and rice plant root morphology.

1.6.1 Maize

During different stages of root development, maize root system consists of different root types (Hochholdinger 2009). In crop species, such as maize, Pi starvation significantly affects the total root length of both primary and lateral roots (He et al. 2003).

Maize is well known as a host plant for spore propagation. There are several maize genotypes, and the genotypes are significantly different in terms of nutrient efficiency through root development. Also, maize plant gives a high response to mycorrhizal inoculation as well. Ortas and Akpinar (2011) showed that several maize genotypes and mycorrhizal inoculation have a diverse response to plant root growth (Fig. 1.5). Compared to the non-inoculated ones, root dry weight tended to be higher with mycorrhizal inoculation.

Table 1.1 Some examples of successful applications of mycorrhization in micropropagation

Name of the plant species	AMF species	References
Actinidia deliciosa (kiwi)	*Glomus* sp. strain E3	Schubert et al. (1992)
Allium cepa	*Gigaspora rosea*, *Glomus mosseae*	Rancillac et al. (1996)
Annona cherimoya	*G. deserticola*	Azcon-Aguilar et al. (1996)
Apple and peach rootstocks	*Glomus* sp. strain A6	Sbrana et al. (1992)
Apple (M9, M26, Golden)	*G. fasciculatum*, *G. mosseae*, *G. intraradices*	Branzanti et al. (1992)
Banana cv. Grade Naine	*G. manihotis*	Rodríguez-Romero et al. (2005)
Banana dwarf cavendish	*Glomus caledonium* and *G. macrocarpum*	Ortas et al. (2017)
Capsicum annuum L. (Chile ancho pepper) cv. San Luis	*G. albidum*, *G. claroides*, *G. diaphanum*	Estrada-Luna and Davies (2003)
Citrus limon L. Burm. "Zagara Bianca"	*G. mosseae* (BEG 116), *Glomus* sp.	Quatrini et al. (2003)
Crab apple cv. Marjatta	*G. claroideum*, *G. fistulosum*	Uosukainen and Vestberg (1997)
Diospyros kaki "Rajo"	*G. intraradices*, *G. mosseae*	Marin et al. (2003)
Fragaria vesca (strawberry)	*G. fistulosum*	Cassells et al. (1996)
Fragaria X ananassa cv. Elvira G. clarum	*G. etunicatum*, *G. intraradices*, *Gi. rosea*, *Gi gigantea*, *G. margarita*, *Scutellospora calospora*, *S. heterogama*, *S. persica*	Taylor and Harrier (2001)
Juglans regia (Walnut)	*G. mosseae*, *G. intraradices*	Dolcet-Sanjuan et al. (1996)
Musa spp. cv. Pacovan *Acaulospora scrobiculata*	*G. clarum*, *G. etunicatum*	Yao et al. (2002)
Musa spp. cv. Grande Naine	*G. proliferum*, *G. versiforme*, *G. intraradices*	Jaizme-Vega et al. (2003)
Musa spp. cv. Grande Naine	*G. intraradices*	Declerck et al. (2002)
Persea americana (avocado)	*G. fasciculatum*	Vidal et al. (1992)
Persea americana	*G. deserticola*, *G. mosseae*	Azcón-Aguilar et al. (1992)
Potato cv. Goldrush	*G. etunicatum*	Yao et al. (2002)
Prunus avium, Syringa japonica	*G. aggregatum*, *G. deserticola*	Arines and Ballester (1992)
Phoenix dactylifera	*G. fasciculatum*, *G. intraradices* (LPA8), *Glomus isolate* (LPA21)	Bouhired et al. (1992)
Pyrus communis	*Glomus* sp.	Rapparini et al. (1996)

(continued)

Table 1.1 (continued)

Name of the plant species	AMF species	References
Rosa hybrida L., cv. New Dawn	*G. intraradices*	Pinior et al. (2005)
Vitis vinifera L.	*G. mosseae, G. manihotis, G. deserticola, Gigaspora gigantea, Acaulospora laevis*	Singh et al. (2003)
Vitis vinifera L.	*G. mosseae, G. manihotis, Scutellospora heterogama, Gigaspora gigantea, Entrophospora colombiana, Acaulospora laevis, A. sorbiculata*	Krishna et al. (2006)

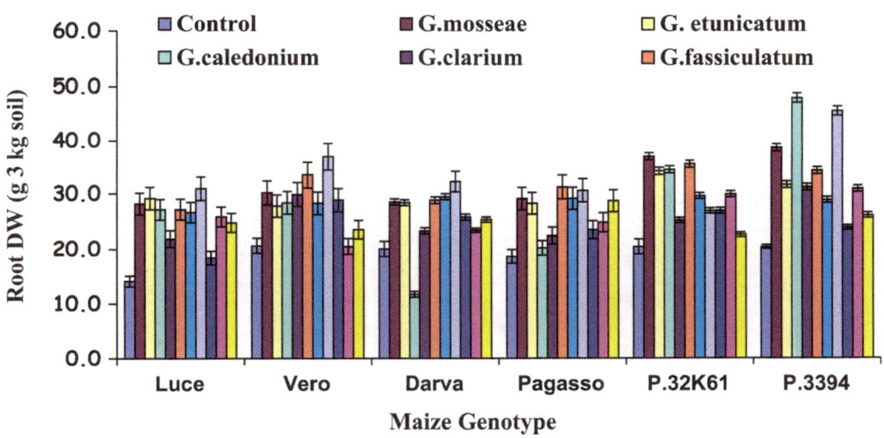

Fig. 1.5 Different maize genotypes response to different mycorrhizae species on root growth

Maize plant root development is also strongly affected by soil ecological conditions such as soil fertility, and soil microbial conditions are very important. Under sterile and non-sterile soil conditions, root attributes significantly affected maize plant growth (Table 1.2) (Ortas 2015). It has been found that under sterile soil conditions, maize plant root length is nearly twice bigger than non-sterile soil conditions. Since with soil sterilization, soil microorganisms especially mycorrhizae spore eliminated plant root growth has become greater. Also, shoot and root ratio was affected under sterile and non-sterile conditions as well. And it is well known that maize plant is responding to mycorrhiza inoculation.

Ortas (2003) reported that indigenous and expartite mycorrhizal fungi strains significantly improved plant growth in maize. In a similar work, maize plant root growth significantly enhanced with mycorrhizal inoculation and also increased with P fertilizer increases under sterile and non-sterilized treatments (Table 1.3). In non-inoculated plants, the root growth was higher in non-sterile treatments; however, in selected mycorrhizal inoculated treatments, the root growth was higher in sterile treatments than non-sterile treatments. With increasing P fertilizer application

Table 1.2 Effect of sterile and non-sterile soil treatments on maize plant parameters, phosphorus, and zinc concentration (Ortas 2015)

Soils treatments		Shoot DW (g pot^{-1})		Root DW (g pot^{-1})		S/R	Root length (m pot^{-1})		Root colonization (%)	
Sultanönü	S	3.87	±0.22c	3.49	±0.27cd	1.11	102	±7c	3	±1d
	NS	3.75	±0.13c	2.96	±0.35d	1.27	60	±5c	11	±3c
Harran	S	4.75	±0.32b	4.18	±0.39ab	1.14	124	±3b	4	±1d
	NS	5.14	±0.39ab	3.74	±0.48bc	1.37	83	±11c	24	±7b
Menzilat	S	5.02	±0.46b	4.63	±0.21a	1.08	222	±12a	3	±1d
	NS	5.74	±0.47a	4.61	±0.02a	1.25	138	±22b	36	±5a

Table 1.3 Effect of mycorrhizal inoculation and P fertilizer on maize root growth under sterile and non-sterile soil conditions

Mycorrhizal species	P Application	Root dry weight (g/plant) Sterile		Non-sterile	
Control	P0	1.25	±0.53ef	0.65	±1.00f
	P1	1.74	±0.60d-e	1.13	±0.52ef
	P2	2.59	±1.01a-e	2.28	±0.69a-e
G. etunicatum	P1	1.99	±0.41d-e	3.53	±0.40a-e
	P2	1.83	±0.93d-e	2.23	±1.87c-e
	P3	3.54	±1.25a-e	4.87	±1.30ab
G. caledonium	P1	2.47	±0.45a-e	4.79	±0.20a-c
	P2	3.72	±2.24a-e	5.49	±4.00a
	P3	4.15	±1.52a-d	5.78	±1.34a
G. mosseae	P1	1.83	±0.84d-e	3.94	±0.84a-d
	P2	3.54	±0.47a-e	4.17	±0.75a-d
	P3	3.97	±0.72a-d	4.71	±1.82a-c

P1 0, P2, 25, and P3 125 mg P_2O_5 kg^{-1}

in both sterile and non-sterile soils, the root dry weight increased. In generally *G. caledonum* inoculation significantly increased maize root growth better than other mycorrhizal species.

1.6.2 Wheat

Wheat inoculated by AMF strengthened root systems in comparison to non-inoculated plants. The AMF-inoculated wheat plant promotes optimization of root morphological characteristics than non-AMF-inoculated wheat plants. This results in longer root length, greater root surface area, smaller root average diameter, and heavier fresh and dry root weights of the wheat plant. Wheat root surface area can also be enhanced through mycorrhizal association (Marschener 1998). Root average diameter was also of great importance in the evaluation of the function of mycorrhizal symbiosis. A glasshouse pot experiment found that root length of the mycorrhizal wheat plant was 674.3 cm plant^{-1}, whereas non-mycorrhizal wheat plant root length was 646.8 cm plant^{-1}. Likewise, root surface area of the mycorrhizal wheat plant was 95.3 cm^2 plant^{-1}, and the non-mycorrhizal wheat plant was 70.9 cm^2 plant^{-1} (Mohammad and Malkawi 2004). Generally, the stronger the root systems are, the smaller the average root diameter should be, which is partly due to the fibrous root system of mycorrhizae-inoculated wheat plant. Thus, AMF has the ability to increase root density of wheat plant.

Wheat spends first into root development during early growth stages and then enters into AMF biomass. The wheat plant also involved in root biomass production before photosynthetic products were used for AM fungal development and AM fungal biomass in roots (Castillo et al. 2012). More than 50% of root segments of wheat were mycorrhizal in a month after seeding. Root colonization patterns of AMF in wheat are highly variable. The root colonization of wheat by AMF is shown in Fig. 1.6. Wheat plant physiology is a major determinant of the levels of colonization.

Wheat plant roots can be classified into several categories according to their ontogenesis and functions. Two root types are distinguished in wheat—the seminal roots (also called primary roots), which develop at the seutellar and epiblast nodes of the embryonic hypocotyls of the germinating caryopsis, and adventitious roots (also called shoot-borne, nodal secondary, or crown roots), which subsequently emerge from the coleoptilar nodes at the base of the apical and tillers. The AMF-inoculated wheat roots extracted water at the fastest rate from the upper soil layers when soil water contents were higher and later extracted water primarily from deeper depths as water in the upper soil layers was depleted. The wheat root dry matter, total root length, and AM colonization were higher under well-watered condition than under dry condition.

Root characteristics and AMF infection of wheat plant influenced by phosphorus (P) supply. Wheat has an extensive root system which makes it responsive to AMF. The AMF infect the root cortex of wheat plants while producing a network of hyphae in the soil. AM symbiosis improves wheat plant growth at vegetative stages through

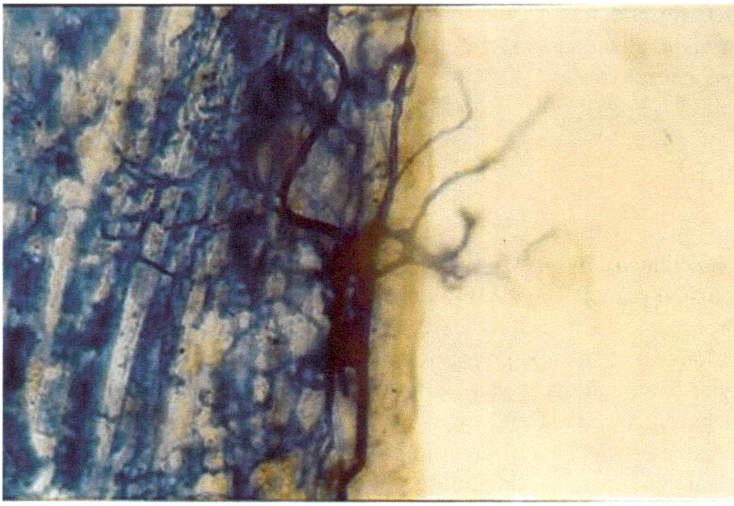

Fig. 1.6 Colonization of wheat root by AMF where mycorrhizal arbuscules and hyphae are visible (Ortas 2000 unpublished pictures)

increasing stomatal conductance, enhancing nutrient use efficiency (NUE), accumulating soluble sugar, and improving ion homeostasis (Zhu et al. 2016). Usually, roots with smaller average diameter result from a high proportion of root hairs, which help roots get into small pores in soil easier and therefore increase soil volume roots are exposed to and finally get chances to absorb more P from the soil solution. The increased absorption and utilization of P element in wheat plants then attribute to bigger root system with heavier root biomass. Mycorrhizal dependence on the symbiosis has a coarser wheat root system and develops fewer root hairs in low fertility soils. Likewise, wheat yield was increased by AMF at low P levels.

Growth rates of mycorrhizal fungi in and around roots probably play a major part in replacement of some mycorrhizal fungi by other types. Wheat varieties (Snowbird and 13NQW1265) responded positively to the inoculation of fungi strain *R. irregularis* (DAOM240442), resulting in a significantly heavier root fresh weight as compared to non-inoculated controls. A comparison between wheat colonized by mycorrhizal strains *F. mosseae*, *F. caledonius*, and *R. irregularis* and non-mycorrhizal controls demonstrated that mycorrhiza inoculation significantly improved root dry weight. Colonization of mycorrhizal strain *R. irregularis* positively benefited wheat root development, represented by a significantly smaller root average diameter than controls and Myke-inoculated wheat. Inoculation of mycorrhizal fungus Myke, without significant differences from non-inoculated control wheat, had no obvious influence on host root average diameter. Colonization with *R. irregularis* fungus contributed to stronger root systems, as evidenced by longer roots, larger root surface area, and smaller root average diameter in *R. irregularis* wheat than in non-mycorrhizal controls. Likewise, AMF strain *R. irregularis* significantly promoted hosts' root growth and development, with longer roots than the non-inoculated controls. Thus, the AMF has genus or even species-specific requirements for successful establishment of the symbiosis. This could be reflected in different accumulation patterns of secondary compounds in roots colonized by different AMF. A small protein, designated Myk15, was found to be strongly induced in wheat (*Triticum aestivum* L.) roots colonized by the AMF *Glomus intraradices*. This protein, which is most abundant in root fractions characterized by strong mycorrhizal colonization, has been characterized using two-dimensional polyacrylamide gel electrophoresis and microsequencing (Fester et al. 2002).

The topology analysis revealed differences in root architecture not detected by any of the other measures of root morphology. This might be due to the fact that wheat plants especially mycorrhizal plants could increase the ability to withstand adversity by delaying protein degradation and maintaining normal metabolism of proteins. In this study, the enzyme activity results showed that AM symbiosis significantly influenced these enzymes to different degrees to respond to the invasion from the environment, which might be the result of a complex interaction between the AMF and plants.

Wheat plant colonized by AMF shows to deplete soil water thoroughly than non-mycorrhizal plants. Because shoot of the wheat plant with AMF usually has larger biomass (more evaporative leaf surface area) than non-AMF wheat plants. Also, the root systems of the wheat plant with AMF inoculation often finely divided and thus have a more absorptive surface area for water and nutrient absorption. The AMF enhances the function of the wheat plant's root hairs and acts as an extension of the root systems allowing the mycorrhizal plants to explore and capture nutrients and water from a larger volume of soil compared to non-AMF plants. Mycorrhizal root colonization increase nutrients absorption, their effective utilization in stress condition, and retaining nutrients for a longer time which ultimately reduces leaching losses. Mycorrhizal wheat plants produce more root dry matter in comparison to non-mycorrhizal plants.

Mycorrhizal wheat plants maintained higher transpiration and shoot water potential than a non-mycorrhizal plant. Wheat plant leaf expansion is more severely reduced when drought affects seminal rather than nodal roots (Volkmar 1997). Wheat plant root signals enhance the fluxes of ions and growth hormones capable of regulating stomatal activity and enzyme biosynthesis, photosynthetic capacity and activity, as well as transpiration. Root tip meristematic activities initiate changes in root distribution, resulting in the synthesis of plant growth regulators, and apparently sense and signal information regarding the soil supply of water and nutrients. The AMF increased the capability of the root system to scavenge water in the drier soil, resulting in less strain to foliage and hence higher stomatal conductance and shoot water potential at particular low soil water content (Duan et al. 1996).

The hyphae of AMF penetrate roots and grow extensively between and within living cortical cells, forming a very large and dynamic interface between symbionts. Root hairs and extrametrical hyphae of vesicular-arbuscular and AMF enlarge the effective absorbing surface area considerably. Root hairs have a diameter of 0.003–0.007 mm, a length of 3–13 mm, and a normal life span of a few days. Wheat roots are usually infected by soilborne vesicular-arbuscular and AMF. The

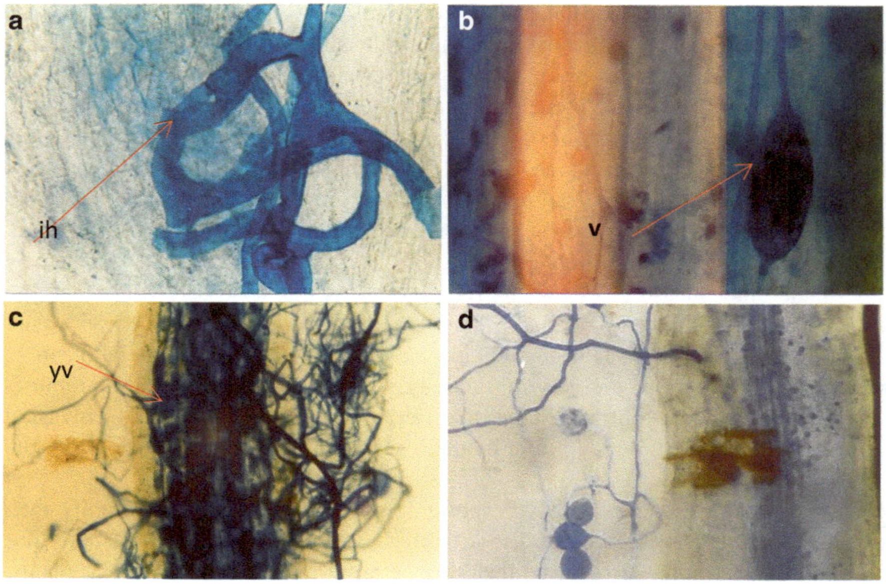

Fig. 1.7 Mycorrhizal structures in wheat roots (*Triticum aestivum*) after artificial inoculation with *G. mosseae*. Roots showed (**a**) intraradical intercellular hyphae (indicated by "ih" and arrow), (**b**) vesicles (indicated by "v" and arrow), (**c**) young vesicles and numerous hyphae (indicated by "yv" and arrow), and (**d**) segments indicated external hyphae and spores. Fungal structures were stained with trypan blue. Bars, 50 lm (Ortas 2000 unpublished pictures)

AMF extrametrical hyphae, which also extend several centimeters away outside of the roots into the soil rhizosphere, are 5–10 times thinner than the root hairs and explore an area around the root, which exceeds the zone of nutrient depletion around uninfected roots. Mycorrhizal structures in wheat roots are shown in Fig. 1.7.

1.6.3 Rice

Rice is mostly cultivated in an anaerobic or flooded condition, where AMF inoculums are reported to decline. The occurrence of AMF colonization of rice roots under the anaerobic condition is still under debate. Anaerobic or flooded condition inhibits root colonization of rice plant due to the anoxic environment (where O_2 is completely absent). The amount of O_2 in anaerobic soil disappears a few hours after flooded condition. When the soil is flooded, water creates a barrier limiting O_2 movement into the soil. The O_2 moves slowly through the water layer and creates a thin surface layer of aerobic soil. Lack of O_2 in the flooded soil causes a shift from aerobic to anaerobic organisms. In this situation, O_2 moves through the stem and roots of water adapted plants like rice via aerenchyma tissue. Thus, the rice root system is affected by an anaerobic condition that resulted in a low percentage of

AMF root colonization. The AMF colonization of rice roots is commonly present at the early growth stages and decline with the age of the plant under anaerobic condition (Ilag et al. 1987). Continuous flooding may exert stress to the rice plant roots and disrupts the morphological and physiological functions of the plant. It decreases root surface area for mycorrhizal colonization. These root changes which become more pronounced over time may explain why in conventional flooded fields, AMF colonization in rice roots diminishes with the age of plant (Lumini et al. 2011; Solaiman and Hirata 1998).

There were several reasons involved to reduce AMF colonization of rice plant under anaerobic condition. It may be possible that due to the oxygen level by itself, the specific chemistry and microenvironment of flooded soils might cause a decrease in AMF colonization. Another reason may be that AMF cannot tolerate microaerophilic condition (Wirsel 2004). Thus, the reduction of AMF under flooded conditions is a consequence not only of a modification of root morphology but also an anatomical change, which leads to the disappearance of the cortical cells that required for intercellular hyphae and arbuscules (Vallino et al. 2014). The whole process on how AMF declines in rice plant under the anaerobic condition is shown in Fig. 1.8.

Rice plant roots form symbiotic associations with AMF under flooded or anaerobic conditions. The AM symbiosis represents the default state of most rice plant root systems and is known to modify root system architecture. This complex root architecture influences AMF colonization. In anaerobic condition, the arbuscular mycorrhizae colonization is regulated by directly influencing rice plant root architecture and anatomy but without the basic AM functionality. The effect of AMF root

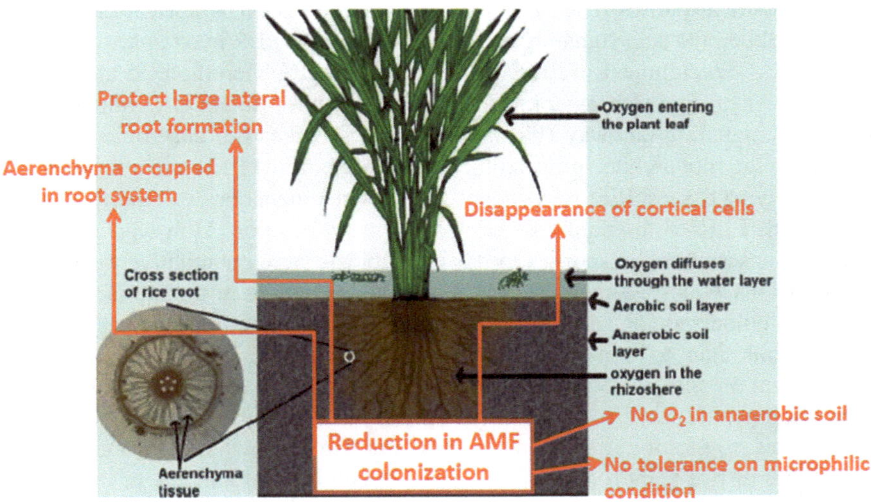

Fig. 1.8 Causes for declining in AMF colonization in rice root system under anaerobic condition

colonization on flooded rice was not symbiotic but rather parasitic under anaerobic condition. Likewise, the effect of AMF on root branching was a consequence of the presence of diffusible signals from the fungus, not of the establishment of the symbiosis (Gutjahr et al. 2009). Similarly, the AMF can grow and colonize rice roots in anaerobic soil while maintaining their signaling properties and functional capacities. The AMF is aerobic microbes, and the occurrence of AM symbiosis in anaerobic condition could be associated with the development of aerenchyma in flooded rice plant roots. The developed aerenchyma allows AMF to obtain atmospheric oxygen under anaerobic condition. It results in a dynamic alteration in the AMF community through different plant physiological changes. This is due to the increase of the aerenchyma tissues, which are not compatible with AMF development.

Rice plant has three types of roots such as (1) crown roots (CR), which emerge from the nodes on the stem and tillers; (2) large lateral roots (LLR), which originate from crown roots and shows indeterminate growth; and (3) fine lateral roots (FLR), which originate both from CR and LLR, whose growth is determinate. The AMF colonization is linked to LLR availability. The LLR has more impact on the success of AM fungal colonization. Rice plant root morphological differences were found between aerobic and anaerobic growth conditions. The branching index (BI) of the rice root apparatus was higher in aerobic condition than the anaerobic condition. The BI is the ratio between the different number of LLR to the length of CR (BI = nLLR/cm CR), which describes the degree of the root apparatus branching (Gutjahr et al. 2009). The AMF stimulates root branching that occurs from a synergic effect between the fungal presence and the positive impact of aerated soil (Maillet et al. 2011). In contrast, under anaerobic conditions, colonization detects in CR system and LLR system but is absent in the FLR system. The root colonization values are also always higher in LLR than CR under anaerobic condition. The LLR and FLR show opposite responses. The LLR support AM colonization, whereas FLR does not because gibberellic acid signaling expressed less than in FLR in comparison to LLR inhibits AM development in FLR. It is presumed that gibberellic acid has a pleiotropic effect on fine root anatomical trails and, in turn, potentially influences the symbiosis signaling pathway (Fiorilli et al. 2015). Likewise, anaerobic condition influences the root architecture leading to a decrease of the LLR and a proliferation of aerenchyma tissues. Rice root system becomes heterogeneously colonized by AM fungi, with LLR preferentially entering into the association. However, root type-specific transcriptional responses to AMF symbiosis were quantitatively more pronounced for CR despite their modest physical engagement in the interaction. Furthermore, colonized CR adopted an expression profile more related to mycorrhizal large lateral than to non-colonized crown roots despite their modest physical engagement in the interaction. Furthermore, colonized crown roots adopted an expression profile more related to mycorrhizal large lateral than to non-colonized crown roots, suggesting fundamental reprogramming of crown root character. Thus, three types of rice plant root involve AMF root colonization directly or indirectly.

1.7 Conclusion

Plant roots are the primary structure for plants which assist in nutrient and water uptake from growth media to the plants. In achieving the proper plant growth, plant roots make the association with the soil microbes such as bacteria and mycorrhizal fungi, depending on the root morphology. They vary from plant to plant, and shortcoming of nutrient and water uptake by the roots is facilitated in association with mycorrhizal fungi. AMF brings the nutrients to the plant roots through a transportation mechanism at the expense of carbon. Moreover, the symbiotic association further shapes the plant roots for a specific plant root-mycorrhizal association. Further studies on root morphology and root plasticity in mycorrhizal inoculated plants are the need of time for better understanding of resource allocation.

References

Arines J, Ballester A (1992) Mycorrhization of micropropagated *Prunus avium* L. plantlets. Micropropagation, root regeneration, and mycorrhizas Joint meeting between COST, Dijon, France, p 45

Atkinson D (1992) Tree root development: the role of models in understanding the consequences of arbuscular endomycorrhizal infection. Agronomie 12:817–820

Atkinson D, Berta G, Hooker J (1994) Impact of mycorrhizal colonisation on root architecture, root longevity and the formation of growth regulators. In: Impact of arbuscular mycorrhizas on sustainable agriculture and natural ecosystems. Springer, Berlin

Azcón-Aguilar C, Barcelo A, Vidal M, De La Vina G (1992) Further studies on the influence of mycorrhizae on growth and development of micropropagated avocado plants. Agronomie 12:837–840

Azcon-Aguilar C, Padilla I, Encina C, Azcon R, Barea J (1996) Mycorrhizal inoculation (Glomus deserticola) enhances plant growth and changes root system morphology in micropropagated *Annona cherimola* Mill. Novel biotechnological approaches to plant production: from sterile root to mycorrhizosphere Joint COST meeting, Pisa, Italy, p 21

Barea JM, Pozo MJ, Azcon R, Azcon-Aguilar C (2005) Microbial co-operation in the rhizosphere. J Exp Bot 56:1761–1778. https://doi.org/10.1093/jxb/eri197

Berta G, Fusconi A, Trotta A (1993) VA mycorrhizal infection and the morphology and function of root systems. Environ Exp Bot 33:159–173

Black R, Tinker P (1977) Interaction between effects of vesicular–arbuscular mycorrhiza and fertiliser phosphorus on yields of potatoes in the field. Nature 267:510–511

Bolan N (1991) A critical review on the role of mycorrhizal fungi in the uptake of phosphorus by plants. Plant Soil 134:189–207

Bouhired L, Gianinazzi S, Gianinazzi-Pearson V (1992) Influence of endomycorrhizal inoculation on the growth of *Phoenix dactylifera*. Micropropagation, root regeneration and mycorrhizas Joint meeting between Cost, Dijon, France, p 53

Branscheid A, Sieh D, Pant BD, May P, Devers EA, Elkrog A, Schauser L, Scheible W-R, Krajinski F (2010) Expression pattern suggests a role of MiR399 in the regulation of the cellular response to local Pi increase during arbuscular mycorrhizal symbiosis. Mol Plant-Microbe Interact 23:915–926

Branzanti B, Gianinazzi-Pearson V, Gianinazzi S (1992) Influence of phosphate fertilization on the growth and nutrient status of micropropagated apple infected with endomycorrhizal fungi during the weaning stage. Agronomie 12:841–845

Bucher M (2007) Functional biology of plant phosphate uptake at root and mycorrhiza interfaces. New Phytol 173:11–26
Bücking H, Kafle A (2015) Role of arbuscular mycorrhizal fungi in the nitrogen uptake of plants: current knowledge and research gaps. Agronomy 5:587
Cassells A, Mark G, Periappuram C (1996) Establishment of arbuscular mycorrhizal fungi in autotrophic strawberry cultures in vitro. Comparison with inoculation of microplants in vivo. Agronomie 16:625–632
Castillo C, Puccio F, Morales D, Borie F, Sieverding E (2012) Early arbuscular mycorrhiza colonization of wheat, barley and oats in Andosols of southern Chile. J Soil Sci Plant Nutr 12:511–524
Cheng L, Bucciarelli B, Liu J, Zinn K, Miller S, Patton-Vogt J, Allan D, Shen J, Vance CP (2011) White lupin cluster root acclimation to phosphorus deficiency and root hair development involve unique glycerophosphodiester phosphodiesterases. Plant Physiol 156:1131–1148
Cruz-Ramirez A, Calderon-Vazquez C, Herrera-Estrella L (2009) Effect of nutrient availability on root system development. https://doi.org/10.1002/9781444310023.ch11
Davies FT, Calderón CM, Huaman Z, Gómez R (2005) Influence of a flavonoid (formononetin) on mycorrhizal activity and potato crop productivity in the highlands of Peru. Sci Hortic 106:318–329
Declerck S, Risede J-M, Delvaux B (2002) Greenhouse response of micropropagated bananas inoculated with in vitro monoxenically produced arbuscular mycorrhizal fungi. Sci Hortic 93:301–309. https://doi.org/10.1016/S0304-4238(01)00347-8
Desnos T (2008) Root branching responses to phosphate and nitrate. Curr Opin Plant Biol 11:82–87
Dolcet-Sanjuan R, Claveria E, Camprubi A, Estaun V, Calvet C (1996) Micropropagation of walnut trees (*Juglans regia* L) and response to arbuscular mycorrhizal inoculation. Agronomie 16:639–645
Douds DD Jr, Nagahashi G, Reider C, Hepperly PR (2007) Inoculation with arbuscular mycorrhizal fungi increases the yield of potatoes in a high P soil. Biol Agric Hortic 25:67–78
Douds DD, Millner P (1999) Biodiversity of arbuscular mycorrhizal fungi in agroecosystems. Agric Ecosyst Environ 74:77–93. https://doi.org/10.1016/s0167-8809(99)00031-6
Duan X, Neuman DS, Reiber JM, Green CD, Saxton AM, Augé RM (1996) Mycorrhizal influence on hydraulic and hormonal factors implicated in the control of stomatal conductance during drought. J Exp Bot 47:1541–1550
Epstein E (1997) The science of composting. CRC Press LLC, Boca Ratón, FL
Estrada-Luna AA, Davies FT (2003) Arbuscular mycorrhizal fungi influence water relations, gas exchange, abscisic acid and growth of micropropagated chile ancho pepper (*Capsicum annuum*) plantlets during acclimatization and post-acclimatization. J Plant Physiol 160:1073–1083
Fageria NK, Moreira A (2011) The role of mineral nutrition on root growth of crop plants. In: Sparks DL (ed) Advances in agronomy, vol 110. Elsevier Academic, San Diego
Feddermann N, Finlay R, Boller T, Elfstrand M (2010) Functional diversity in arbuscular mycorrhiza–the role of gene expression, phosphorous nutrition and symbiotic efficiency. Fungal Ecol 3:1–8
Fester T, Kiess M, Strack D (2002) A mycorrhiza-responsive protein in wheat roots. Mycorrhiza 12:219–222
Fiorilli V, Vallino M, Biselli C, Faccio A, Bagnaresi P, Bonfante P (2015) Host and non-host roots in rice: cellular and molecular approaches reveal differential responses to arbuscular mycorrhizal fungi. Front Plant Sci 6:636
Giri B, Prasad R, Varma A (2018) Root biology. Springer International Publishing (ISBN 978-3-319-75910-4) https://www.springer.com/us/book/9783319759098
Gutjahr C, Casieri L, Paszkowski U (2009) Glomus intraradices induces changes in root system architecture of rice independently of common symbiosis signaling. New Phytol 182:829–837
He Y, Liao H, Yan X (2003) Localized supply of phosphorus induces root morphological and architectural changes of rice in split and stratified soil cultures. Plant Soil 248:247–256
Hemsley AR, Poole I (2004) The evolution of plant physiology. Elsevier, London

Hetrick B, Wilson G, Todd T (1996) Mycorrhizal response in wheat cultivars: relationship to phosphorus. Can J Bot 74:19–25

Hijri M (2016) Analysis of a large dataset of mycorrhiza inoculation field trials on potato shows highly significant increases in yield. Mycorrhiza 26:209–214

Hochholdinger F (2009) The maize root system: morphology, anatomy, and genetics. In: Handbook of maize: its biology. Springer, New York

Hochholdinger F, Woll K, Sauer M, Dembinsky D (2004) Genetic dissection of root formation in maize (*Zea mays*) reveals root-type specific developmental programmes. Ann Bot 93:359–368

Hoshyar Z, Abedi B, Ganji Moghadam E, Davari Nejad G (2017) Effect of arbuscular mycorrhiza on growth and physiological behavior of PHL-C rootstock. J Plant Physiol Breed 7:53–60

Ilag LL, Rosales A, Elazegui F, Mew T (1987) Changes in the population of infective endomycorrhizal fungi in a rice-based cropping system. Plant Soil 103:67–73

Jaizme-Vega M, Rodríguez-Romero A, Hermoso CM, Declerck S (2003) Growth of micropropagated bananas colonized by root-organ culture produced arbuscular mycorrhizal fungi entrapped in Ca-alginate beads. Plant Soil 254:329–335

Jungk A (2001) Root hairs and the acquisition of plant nutrients from soil. J Plant Nutr Soil Sci 164:121–129

Karthikeyan AS, Varadarajan DK, Jain A, Held MA, Carpita NC, Raghothama KG (2007) Phosphate starvation responses are mediated by sugar signaling in Arabidopsis. Planta 225:907–918

Khan MH, Meghvansi M, Prasad K, Siddiqui S, Varma A (2017) Arbuscular mycorrhizal symbiosis and nutrient resource limitation: predicting the linkages and effectiveness of partnership. In: Mycorrhiza-nutrient uptake, biocontrol, ecorestoration. Springer, Cham

Koide RT, Kabir Z (2000) Extraradical hyphae of the mycorrhizal fungus Glomus intraradices can hydrolyse organic phosphate. New Phytol 148:511–517. https://doi.org/10.1046/j.1469-8137.2000.00776.x

Krishna H, Singh S, Minakshi PV, Khawale R, Deshmukh P, Jindal P (2006) Arbuscular-mycorrhizal fungi alleviate transplantation shock in micropropagated grapevine (*Vitis vinifera* L.). J Hortic Sci Biotechnol 81:259–263

Kungu JB, Lasco RD, Dela Cruz LU, Dela Cruz RE, Husain T (2008) Effect of vesicular arbuscular mycorrhiza (VAM) fungi inoculation on coppicing ability and drought resistance of *senna spectabilis*. Pak J Bot 40:2217–2224

Lambers H, Shane MW, Cramer MD, Pearse SJ, Veneklaas EJ (2006) Root structure and functioning for efficient acquisition of phosphorus: matching morphological and physiological traits. Ann Bot 98:693–713

Lambers H, Raven JA, Shaver GR, Smith SE (2008) Plant nutrient-acquisition strategies change with soil age. Trends Ecol Evol 23:95–103. https://doi.org/10.1016/j.tree.2007.10.008

Lambers H, Brundrett MC, Raven JA, Hopper SD (2010) Plant mineral nutrition in ancient landscapes: high plant species diversity on infertile soils is linked to functional diversity for nutritional strategies. Plant Soil 334:11–31. https://doi.org/10.1007/s11104-010-0444-9

Li XL, George E, Marschner H (1991) Phosphorus depletion and pH decrease at the root soil and hyphae soil interfaces of VA mycorrhizal white clover fertilized with ammonium. New Phytol 119:397–404

Li XL, George E, Marschner H, Zhang JL (1997) Phosphorus acquisition from compacted soil by hyphae of a mycorrhizal fungus associated with red clover (*Trifolium pratense*). Can J Bot Revue Canadienne De Botanique 75:723–729

López-Bucio J, Cruz-Ramırez A, Herrera-Estrella L (2003) The role of nutrient availability in regulating root architecture. Curr Opin Plant Biol 6:280–287

Lumini E, Vallino M, Alguacil MM, Romani M, Bianciotto V (2011) Different farming and water regimes in Italian rice fields affect arbuscular mycorrhizal fungal soil communities. Ecol Appl 21:1696–1707

Lynch J (1995) Root architecture and plant productivity. Plant Physiol 109:7

Maillet F, Poinsot V, Andre O, Puech-Pagès V, Haouy A, Gueunier M, Cromer L, Giraudet D, Formey D, Niebel A (2011) Fungal lipochitooligosaccharide symbiotic signals in arbuscular mycorrhiza. Nature 469:58

Manoharan P, Shanmugaiah V, Balasubramanian N, Gomathinayagam S, Sharma MP, Muthuchelian K (2010) Influence of AM fungi on the growth and physiological status of *Erythrina variegata* Linn. Grown under different water stress conditions. Eur J Soil Biol 46:151–156

Marin M, Mari A, Ibarra M, Garcia-Ferriz L (2003) Arbuscular mycorrhizal inoculation of micropropagated persimmon plantlets. J Hortic Sci Biotechnol 78:734–738

Marschener H (1998) Role of root growth, arbuscular mycorrhiza, and root exudates for the efficiency in nutrient acquisition. Field Crop Res 56:203–207

Marschner H (1995) Mineral nutrition of high plants. Academic, London

Marschner H (1996) Mineral nutrient acquisition in nonmycorrhizal and mycorrhizal plants. Phyton Annales Rei Botanicae 36:61–68

Marschner H (2011) Marschner's mineral nutrition of higher plants. Academic, London

Marschner P (2012) Marschner's mineral nutrition of higher plants. Academic, London

Marschner H, Romheld V, Horst WJ, Martin P (1986) Root-induced changes in the rhizosphere – importance for the mineral-nutrition of plants. Zeitschrift Fur Pflanzenernahrung Und Bodenkunde 149:441–456

Mohammad MJ, Malkawi HI (2004) Root, shoot and nutrient acquisition responses of mycorrhizal and nonmycorrhizal wheat to phosphorus application to highly calcareous soils. Asian J Plant Sci 3:363–369

Nacry P, Canivenc G, Muller B, Azmi A, Van Onckelen H, Rossignol M, Doumas P (2005) A role for auxin redistribution in the responses of the root system architecture to phosphate starvation in Arabidopsis. Plant Physiol 138:2061–2074

Neumann E, George E (2010) Nutrient uptake: the arbuscular mycorrhiza fungal symbiosis as a plant nutrient acquisition strategy. In: Arbuscular mycorrhizas: physiology and function. Springer, Amsterdam

Neumann G, Martinoia E (2002) Cluster roots–an underground adaptation for survival in extreme environments. Trends Plant Sci 7:162–167

Neumann G, Massonneau A, Martinoia E, Römheld V (1999) Physiological adaptations to phosphorus deficiency during proteoid root development in white lupin. Planta 208:373–382

Norman J, Atkinson D, Hooker J (1996) Arbuscular mycorrhizal fungal-induced alteration to root architecture in strawberry and induced resistance to the root pathogen Phytophthora fragariae. Plant Soil 185:191–198

Ortas I (1997) Determination of the extent of rhizosphere soil. Commun Soil Sci Plant Anal 28:1767–1776. https://doi.org/10.1080/00103629709369914

Ortas I (2003) Effect of selected mycorrhizal inoculation on phosphorus sustainability in sterile and non-sterile soils in the Harran Plain in South Anatolia. J Plant Nutr 26:1–17. https://doi.org/10.1081/pln-120016494

Ortas I (2012) The effect of mycorrhizal fungal inoculation on plant yield, nutrient uptake and inoculation effectiveness under long-term field conditions. Field Crop Res 125:35–48. https://doi.org/10.1016/j.fcr.2011.08.005

Ortas I (2015) Comparative analyses of Turkey agricultural soils: potential communities of indigenous and exotic mycorrhiza species' effect on maize (*Zea mays* L.) growth and nutrient uptakes. Eur J Soil Biol 69:79–87. https://doi.org/10.1016/j.ejsobi.2015.05.006

Ortas I, Akpinar C (2011) Response of maize genotypes to several mycorrhizal inoculums in terms of plant growth, nutrient uptake and spore production. J Plant Nutr 34:970–987. https://doi.org/10.1080/01904167.2011.555580

Ortas I, Rafique M, Akpinar C, Aka Kacar Y (2017) Growth media and mycorrhizal species effect on acclimatization and nutrient uptake of banana plantlets. Sci Hortic 217:55–60

Pearson J, Jakobsen I (1993) Symbiotic exchange of carbon and phosphorus between cucumber and three arbuscular mycorrhizal fungi. New Phytol 124:481–488

Pinior A, Grunewaldt-Stöcker G, von Alten H, Strasser RJ (2005) Mycorrhizal impact on drought stress tolerance of rose plants probed by chlorophyll a fluorescence, proline content and visual scoring. Mycorrhiza 15:596

Ponton F, Piche Y, Parent S, Caron M (1990) The use of vesicular-arbuscular mycorrhizae in Boston fern production: I. Effects of peat-based mixes. Hortscience 25:183–189

Prasad R, Bhola D, Akdi K, Cruz C, Sairam K, Tuteja N, Varma A (2017) Introduction to mycorrhiza: historical development. In: Mycorrhiza-function, diversity, state of the art. Springer, Cham

Pritchard SG, Rogers HH (2000) Spatial and temporal deployment of crop roots in CO2-enriched environments. New Phytol 147:55–71

Puthur JT, Prasad K, Sharmila P, Saradhi PP (1998) Vesicular arbuscular mycorrhizal fungi improves establishment of micropropagated *Leucaena leucocephala* plantlets. Plant Cell Tissue Organ Cult 53:41

Quatrini P, Gentile M, Carimi F, Pasquale FD, Puglia AM (2003) Effect of native arbuscular mycorrhizal fungi and Glomus mosseae on acclimatization and development of micropropagated *Citrus limon* (L.) Burm. J Hortic Sci Biotechnol 78:39–45

Rancillac M, Cadoux F, Leduc D, Kahane R (1996) Improvement of a protocol to establish in vitro arbuscular mycorrhizal strains with vitro bulbs of onion, *Allium cepa* L. Novel biotechnological approaches to plant production: from sterile root to mycorrhizosphere Joint COST meeting

Rapparini F, Baraldi R, Bertazza G (1996) Growth and carbohydrate status of *Pyrus communis* L plantlets inoculated with Glomus sp. Agronomie 16:653–661

Rodríguez-Romero AS, Guerra MSP, Jaizme-Vega MDC (2005) Effect of arbuscular mycorrhizal fungi and rhizobacteria on banana growth and nutrition. Agron Sustain Dev 25:395–399

Rogers HH, Runion GB, Prior A (1999) Response of plants to elevated atmospheric CO2: root growth, mineral. Carbon dioxide and environmental stress, p 215

Rouphael Y, Franken P, Schneider C, Schwarz D, Giovannetti M, Agnolucci M, Pascale SD, Bonini P, Colla G (2015) Arbuscular mycorrhizal fungi act as biostimulants in horticultural crops. Sci Hortic 196:91–108. https://doi.org/10.1016/j.scienta.2015.09.002

Sbrana C, Vitagliano C, Avio L, Giovanneti M (1992) Influence of vesicular-arbuscular mycorrhizae on transplant stress of micropropagated apple and peach rootstocks. Micropropagation, root regeneration, and mycorrhizas Joint meeting between COST

Schachtman DP, Reid RJ, Ayling SM (1998) Phosphorus uptake by plants: from soil to cell. Plant Physiol 116:447–453

Schubert A, Bodrino C, Gribaudo I (1992) Vesicular-arbuscular mycorrhizal inoculation of kiwifruit (Actinidia deliciosa) micropropagated plants. Agronomie 12:847–850

Sharma S, Sharma AK, Prasad R, Varma A (2017) Arbuscular mycorrhiza: a tool for enhancing crop production. In: Mycorrhiza-nutrient uptake, biocontrol, ecorestoration. Springer, New York

Shrivastava S, Prasad R, Varma A (2014) Anatomy of root from eyes of a microbiologist. In: Root engineering. Springer, Berlin

Singh S, Minakshi G, Khawale R, Patel V, Krishna H, Saxena A (2003) Mycorrhization as an aid for biohardening of in vitro raised Grape (*Vitis vinifera* L.) plantlets. VII International symposium on temperate zone fruits in the tropics and subtropics, p 662

Smith S, De Smet I (2012) Root system architecture: insights from Arabidopsis and cereal crops. Philos Trans R Soc B Biol Sci 367(1595):1441–1452

Smith SE, Read DJ (2010) Mycorrhizal symbiosis. Academic, San Diego, CA

Smith SE, Smith FA (2011) Roles of arbuscular mycorrhizas in plant nutrition and growth: new paradigms from cellular to ecosystem scales. Annu Rev Plant Biol 62:227–250

Smith SE, Facelli E, Pope S, Smith FA (2010) Plant performance in stressful environments: interpreting new and established knowledge of the roles of arbuscular mycorrhizas. Plant Soil 326:3–20

Smith SE, Jakobsen I, Grønlund M, Smith FA (2011) Roles of arbuscular mycorrhizas in plant phosphorus nutrition: interactions between pathways of phosphorus uptake in arbuscular mycorrhizal roots have important implications for understanding and manipulating plant phosphorus acquisition. Plant Physiol 156:1050–1057

Solaiman M, Hirata H (1998) Glomus-wetland rice mycorrhizas influenced by nursery inoculation techniques under high fertility soil conditions. Biol Fertil Soils 27:92–96

Taylor J, Harrier LA (2001) A comparison of development and mineral nutrition of micropropagated Fragaria× ananassa cv. Elvira (strawberry) when colonised by nine species of arbuscular mycorrhizal fungi. Appl Soil Ecol 18:205–215

Teotia P, Kumar M, Prasad R, Kumar V, Tuteja N, Varma A (2017) Mobilization of micronutrients by mycorrhizal fungi. Mycorrhiza-function, diversity, state of the art. Springer, Cham

Tingey DT, Phillips DL, Johnson MG (2000) Elevated CO2 and conifer roots: effects on growth, life span and turnover. New Phytol 147:87–103

Uosukainen M, Vestberg M (1997) Timing of AMF inoculation to microcuttings of crab apple cv. Marjatta. COST Action 821: arbuscular mycorrhizas in sustainable soil-plant systems: report of 1996 activities

Upadhyaya CP, Gururani MA, Prasad R, Varma A (2013) A cell wall extract from *Piriformospora indica* promotes tuberization in potato (*Solanum tuberosum* L.) via enhanced expression of Ca^{+2} signaling pathway and lipoxygenase gene. Appl Biochem Biotechnol 170(4):743–755

Vallino M, Fiorilli V, Bonfante P (2014) Rice flooding negatively impacts root branching and arbuscular mycorrhizal colonization, but not fungal viability. Plant Cell Environ 37:557–572

Vance CP (2010) Quantitative trait loci, epigenetics, sugars, and microRNAs: quaternaries in phosphate acquisition and use. Plant Physiol 154:582–588

Varma A, Prasad R, Tuteja N (2017) Mycorrhiza: nutrient uptake, biocontrol, ecorestoration. Springer International Publishing (ISBN: 978-3-319-68867-1) http://www.springer.com/us/book/9783319688664

Vidal M, Azcón-Aguilar C, Barea J, Pliego-Alfaro F (1992) Mycorrhizal inoculation enhances growth and development of micropropagated plants of avocado. Hortscience 27:785–787

Volkmar K (1997) Water stressed nodal roots of wheat: effects on leaf growth. Funct Plant Biol 24:49–56

Wang B, Tang X, Cheng L, Zhang A, Zhang W, Zhang F, Liu J, Cao Y, Allan D, Vance C (2010) Nitric oxide is involved in phosphorus deficiency-induced cluster-root development and citrate exudation in white lupin. New Phytol 187:1112–1123

Wirsel SG (2004) Homogenous stands of a wetland grass harbour diverse consortia of arbuscular mycorrhizal fungi. FEMS Microbiol Ecol 48:129–138

Wu Q-S, Zou Y-N, He X-H (2010) Contributions of arbuscular mycorrhizal fungi to growth, photosynthesis, root morphology and ionic balance of citrus seedlings under salt stress. Acta Physiol Plant 32:297–304

Wu F, Wang W, Ma Y, Liu Y, Ma X, An L, Feng H (2013) Prospect of beneficial microorganisms applied in potato cultivation for sustainable agriculture. Afr J Microbiol Res 7:2150–2158

Wu Q-S, Srivastava A, Zou Y-N, Malhotra S (2017a) Mycorrhizas in citrus: beyond soil fertility and plant nutrition. Indian J Agric Sci 87:427–443

Wu QS, Sun P, Srivastava AK (2017b) AMF diversity in citrus rhizosphere. Indian J Agric Sci 87:653–656

Yao M, Tweddell R, Desilets H (2002) Effect of two vesicular-arbuscular mycorrhizal fungi on the growth of micropropagated potato plantlets and on the extent of disease caused by Rhizoctonia solani. Mycorrhiza 12:235–242

Zhu X, Song F, Liu S, Liu F (2016) Role of arbuscular mycorrhiza in alleviating salinity stress in wheat (*Triticum aestivum* L.) grown under ambient and elevated CO2. J Agron Crop Sci 202:486–496

Chapter 2
Plant-Mycorrhizal and Plant-Rhizobial Interfaces: Underlying Mechanisms and Their Roles in Sustainable Agroecosystems

Neera Garg, Amrit Bharti, Amrita Sharma, and Shyna Bhalla

Abstract Rhizospheric plant-microbe symbiotic interactions involve numerous microbial populations which have a significant impact on plant growth and productivity. Abiotic stresses are serious threats to agriculture and negatively affect the soil-microbe-plant continuum, which is also responsible for reduced yield. Rhizospheric microbes, especially arbuscular mycorrhizal fungi (AMF) and rhizobia (Rh), are potential economical and eco-friendly resources for counteracting abiotic stresses in plants. These microbial interactions involve the release of signaling molecules, such as Myc factors by AMF and Nod factors by Rh, which initiate communication between these microbes and plants leading to colonization, nodulation, and arbuscule formation. Both these microbes are relatively tolerant to extreme adverse conditions and can improve growth and productivity of stressed plants by improving soil and root system architecture (RSA), nutrient uptake, ion homeostasis, sequestration, and compartmentalization, reducing osmotic and oxidative stress, etc. Moreover, both these symbionts act synergistically and provide various beneficial effects in stressed plants. However, there are a number of gaps in understanding the various steps involved in the establishment of symbioses, the signaling molecules, nutrient exchange through the symbiotic interface and genes involved, as well as the modes of action/mechanism of the AMF and Rh in imparting abiotic stress resistance in plants. This chapter bridges the gap and summarizes the mechanisms adopted by AMF and Rh in imparting stress resistance and enhancing crop productivity for a sustainable agroecosystem.

2.1 Introduction

Rhizospheric plant-microbe symbiotic interactions have the potential to reduce the use of chemical fertilizers in agriculture and have emerged as a central pillar of sustainable agriculture worldwide in the recent decade (Estrada et al. 2013, Bharti

N. Garg (✉) · A. Bharti · A. Sharma · S. Bhalla
Department of Botany, Panjab University, Chandigarh, India

A. Varma et al. (eds.), *Plant Microbe Interface*,
https://doi.org/10.1007/978-3-030-19831-2_2

et al. 2014; Garg et al. 2015; Hashem et al. 2016). Among the innumerable rhizospheric microbial communities, arbuscular mycorrhizal fungi (AMF) and rhizobia (Rh) are the best representatives of two major groups of plant-associated mutualists, both having significant impacts on the growth and productivity of their host plants (Venturi and Keel 2016).

Abiotic stresses such as salt, heavy metal(loid)s, etc. deteriorate soil characteristics such as pH, electrical conductance (EC), architecture, and mineral compositions, which make the soil toxic. Plants inhabiting these toxic soils become prone to declined growth due to insufficient water as well as nutrient uptake, ion imbalance, and altered metabolic responses. The toxic ions also reduce production, density, and germination of spores, hyphal growth, diversity of AMF (Yang et al. 2017), as well as various stages of rhizobial symbiosis (Michiels et al. 1994). Despite these negative relationships, both the microbes never get fully eliminated from soils, indicating extreme adaptability and tolerant nature to varied stress factors as compared to their respective host plants (Bano and Ashfaq 2013). Both mycorrhizal and rhizobial symbioses are known to impart stress resistance in plants by modulating diverse mechanisms and improving their growth and productivity.

The word "mycorrhiza" is formed from two Greek words "myco" meaning fungus and "rhiza" meaning roots (Alizadeh 2011) and is considered as the most primitive symbiont dating back to 450×10^6 years (Bonfante and Genre 2008). AMF belong to the phylum *Glomeromycota*, and their symbiotic associations are widely spread in 80% of the higher plants (Prasad et al. 2017). AMF are obligate symbionts as they depend on reduced carbon (C) of living host plant (4–20% of their photosynthate) and paybacks, including nutrient, water uptake, etc. (Parniske 2008). Establishment of mycorrhizal symbiosis can be divided into three growth phases: (1) asymbiotic hyphal growth phase, in which autonomous hyphae develop following the spore germination; (2) pre-symbiotic growth phase, in which host signal perception by AMF stimulates its hyphal growth and in return release the Myc factors, which are identified by plant potential receptors to induce signaling pathways; and (3) symbiotic phase, where AMF enter plant roots and form intra-radical mycelial (IRM) network, arbuscules, and extra-radical mycelial (ERM) network to recruit nutrients to and from the soil (Smith and Read 2008; Bhandari and Garg 2017). The ERM network envelops and keeps the soil particles compact. The cell walls of the hyphae and spores secrete some glycoproteins (glomalin and glomalin-related proteins) into the soil, which exert a key role in soil aggregation as well as C sequestration, thereby enhancing the fertility of degraded soils (Wu et al. 2012). In addition to improving soil texture, AMF also play roles in imparting plant tolerance to various stresses. The mechanism adopted by AMF in alleviating osmotic and ionic stress is ascribed to the development of ERM, whose hypha delves into the soil and explores the distant areas, successfully access the water-filled pores through their aquaporin (AQP) channels, leading to improved water and nutrient uptake, especially phosphorous (P) (Rajtor and Piotrowska-Seget 2016; Kaur and Garg 2017). Another mechanism includes sequestration of toxic ions in soil and/or in roots which occurs via (1) adsorption of toxic ions to cell wall of spores and hyphae and then their diffusion across the membranes of these organs (Gonzalez-Chavez et al. 2002),

(2) accumulation in ERM, and (3) extracellular chelation (Amir et al. 2014). In order to inactivate absorbed toxic ions, strategy of intracellular compartmentalization has been documented in which toxic ions are translocated into mycorrhizal as well as host cell vacuoles (vacuolization) and vesicles with the help of chelators (Ruiz-Lozano et al. 2012; Amir et al. 2014; Gonzalez-Guerrero et al. 2016).

The term "*Rhizobium*" arrived from the two Ancient Greek words "rhiza" meaning root and "bios" meaning life (Sidahmed 2016). *Rhizobium* and its allies (*Azorhizobium*, *Bradyrhizobium*, *Sinorhizobium*, and *Mesorhizobium*, collectively referred as Rh) are Gram-negative, motile, non-sporulating rods that cause the development of root nodules (which the bacteria inhabit as nitrogen-fixing endosymbionts) in legumes (Santi et al. 2013). The symbiosis between nitrogen (N)-fixing Rh and legumes is considered to have evolved over the past 60 million years (Mohanty et al. 2018) and represents a conventional example of a monospecific partnership (Bonfante 2003; Garg and Manchanda 2007). This process is restricted to a bacterial group (the Rh), from the proteobacterial *Rhizobiaceae* family (Kersters et al. 2006) establishing a symbiotic relationship with plant species belonging to family Fabaceae. Rh extract C from the plant's photosynthates and in mutual exchange supply the plant with valuable organic N, which elevates the respective vigor of the two symbionts (Yamazaki and Hayashi 2015). Rhizobial symbiosis is conceivably split into divergent growth stages: (1) signaling molecules from host plant stimulate Rh to release Nod factors (NFs), which are identified by potential plant receptors, and this activates a well-specified signal cascade; (2) Rh become attached to the host plants and get internalized; and (3) finally, formation of infection thread and other changes take place that trigger a developmental program to give rise to a nodule (Oldroyd 2013). The symbiotic relationship thus established between legumes and Rh plays an important role in improving fertility of degraded soils by enriching it with N and recuperation of agricultural fields due to the capacity of Rh to reduce metal toxicity (bioremediation via various transformation processes) and grow on nutrient-poor soils similar to AMF (Naveed et al. 2015). Rh act as biofertilizers and upregulate N assimilation and iron (Fe) acquisition in plants via biological nitrogen fixation (BNF) which is an ecologically sound and low-cost strategy for enhancing productivity of plants in contaminated soils (Masson-Boivin and Sachs 2018). In addition to N uptake, Rh also regulate the rapid synthesis of various detoxifying agents such as glutathione reductase (GR) (Corticeiro et al. 2006); promote growth-promoting hormones such as auxins, gibberellins, cytokinins, etc. (Reichman 2007); downregulate ACC (1-aminocyclopropane-1-carboxylic acid) synthase as well as upregulate ACC deaminase enzymes (Murset et al. 2012); and facilitate abiotic stress resistance. These mechanisms are controlled by a complex network of signaling cascades which occur during the plant-rhizobial interactions and consequently ensuring stress alleviation (Meena et al. 2017).

Both AMF and Rh live in perfect harmony with each other in the rhizosphere and form a successful "tripartite relationship" due to their functional complementarity in terms of strengthening the nodulation frequency, BNF efficiency, and root colonization of host plant that ultimately result in better nutrient uptake and plant yield

(Meng et al. 2015). Thus, their symbioses become mandatory to enable cultivation of plants in soils contaminated with abiotic stress factors. Mycorrhizal and rhizobial symbioses with plants involve a series of morphophysiological, biochemical, and molecular events that are controlled by various mechanisms. Therefore, knowledge about the relationship of plants with AMF and Rh is of considerable importance for complete elucidation of various mechanisms behind their successful utilization under particular conditions. This review considers the various pre-symbiotic and symbiotic interfaces of both AMF and Rh with plants and explores potentialities toward the management of these resources to enhance the crop productivity under stressed conditions.

2.2 The Biochemical Interface Between Plant-Arbuscular Mycorrhiza Fungi (AMF) and Rhizobia (Rh)

In mycorrhizal and rhizobial symbioses, the symbionts, i.e., plant and microbes, involve recognition and attraction of appropriate partners (Paszkowski 2006) through a chemical dialogue between them in the rhizosphere (Badri et al. 2009) as illustrated in Fig. 2.1. This chemical communication is incessant to form long-standing colonization.

2.2.1 Signaling Cascade at Plant-AMF Interface

The early asymbiotic mycorrhizal growth starts with the development of hyphae that originate from mycorrhizal root portions or resting spores (Koltai and Kapulnik 2009). After spore germination, AMF utilize triacylglycerides and glycogen reserves of spore for the development of short mycelium, because of its inability to uptake C from the soil (Harrison 2005; Leigh et al. 2009). As AMF possess obligate biotrophic nature, therefore, in the absence of host root, development of this short mycelium stops before the depletion of spore reserves and mycelium withdraw their cytoplasm into spore for new germination event and, thus, enter into the dormant stage (Genre et al. 2012). Exploratory mycelial growth pattern alters intensely when hyphae approach a host root and respond to their contiguity (Nasim 2013) which indicates the perception of some signals by AMF from the root exudates (Harrison 2005). Notably, a few molecules exudated from host roots such as auxins (a plant growth hormone), phenolics, flavonoids, and especially strigolactone act as primary signals for establishing mycorrhizal symbiosis (Amballa and Bhumi 2016). Strigolactones are secreted by ATP-binding cassette subtype G (ABCG) transporter PDR1 in hypodermal passage cells, especially during phosphate deficiency (a condition which generates localized rhizospheric gradients to direct root colonization) (Kretzschmar et al. 2012). Strigolactones are well-known stimulants for

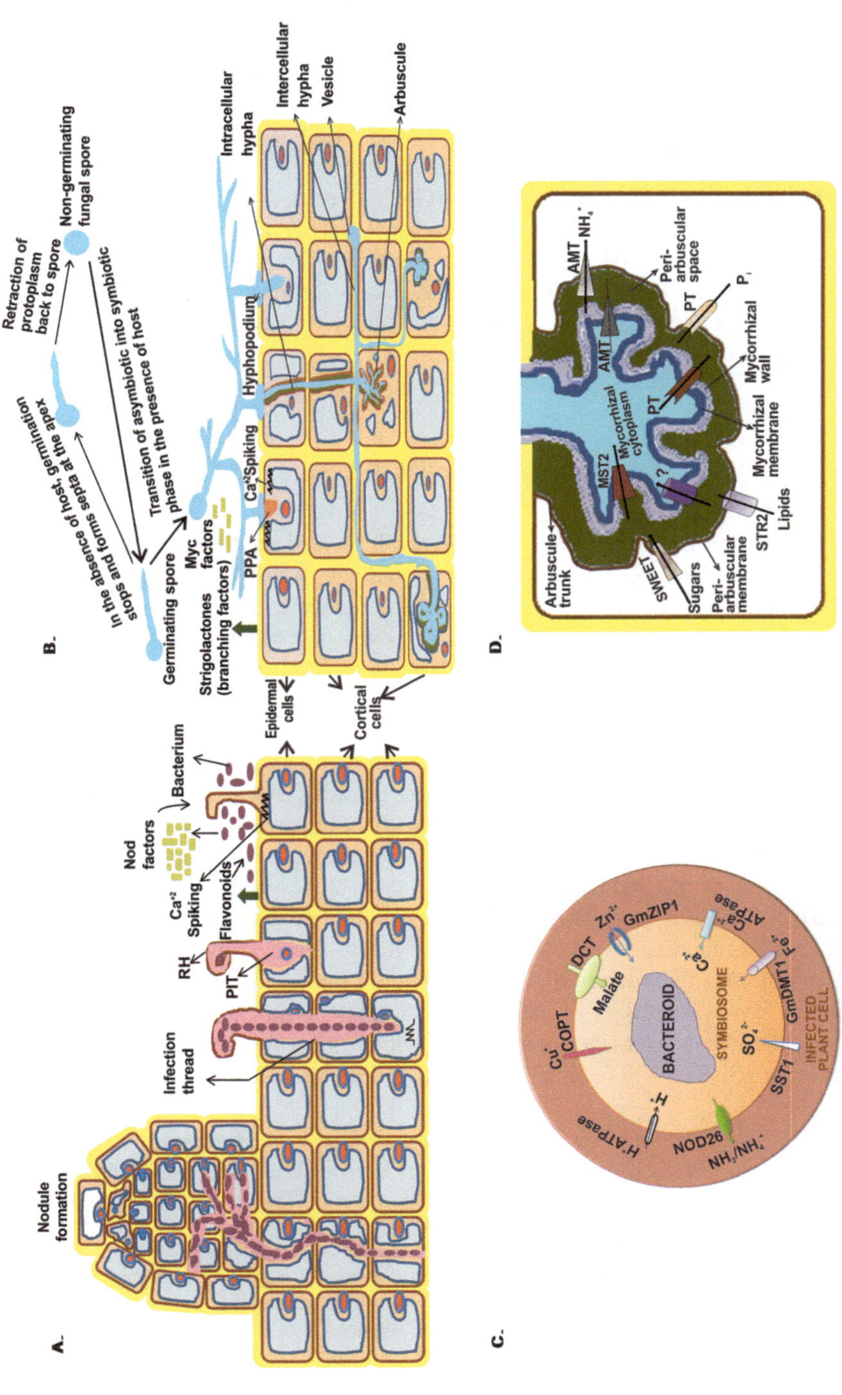

Fig. 2.1 Rhizobial (A) and mycorrhizal (B) symbioses. **(a)** Rhizobia (Rh) stimulate plant to release flavonoids, which in turn stimulate rhizobia to secrete Nod factors that are recognized by plant. Perception of Nod factors leads to calcium (Ca^{2+}) spiking initially in the epidermal cells, but later in cortical cells also. Rh enter the plant root by becoming entrapped in the curled root hair. Infection threads (ITs) are initiated at the curl which allows the rhizobia to invade root cells.

spore germination, hyphal branching, and growth of mycelium toward the host plant (Parniske 2008). Cutin monomers are also recognized as a particular group of plant signaling factors take part in mycorrhizal stimulation (Gobbato et al. 2012). In *Medicago truncatula* mutants, two loci, RAM1 (required for arbuscular mycorrhization)-encode a GRAS domain transcription factor and RAM2-encode an acyltransferase, lead to severely impaired mycorrhizal symbiosis as they are essential for synthesis of cutin monomers (Gobbato et al. 2012; Wang et al. 2012). Strigolactones stimulate AMF to produce molecular signals called Myc factors such as lipochitooligosaccharide (LCO) and short-chain chitin oligomer (Maillet et al. 2011; Waters et al. 2017). *Rhizophagus intraradices* secretes some proteins which enhance the colonization by modifying hormonal signaling pathways in the host plant (Plett and Martin 2015).

2.2.2 Signaling Cascade at Plant-Rh Interface

Rh in the soil are free-living, motile microbes which feed on the dead and decaying organisms. Free-living Rh are unable to fix N and have a different shape from those found in nodules (Datta et al. 2015). They exist as regular cells and appear as straight rods, whereas the N-fixing forms in the nodules exist as irregular cells, the bacteroids that are often club, branched, and Y-shaped (Datta et al. 2015). Rh remain free-living unless the plant and Rh produce a diversity of compounds to attract each other. These compounds perform a deciding role in the commencement of early plant responses together with symbiotic gene activation that leads to mitotic reactivation in cortical cells and formation of preinfection threads (Oldroyd et al. 2011; Santi et al. 2013). The first signaling molecules to be exchanged amidst plant and its bacterial symbiont are plant-generated polycyclic aromatic flavonoids released into the rhizosphere which induce bacterial nod genes (Jones et al. 2007; Venturi and Keel 2016). So far, in vascular plants, more than 4000 diverse flavonoids have been identified (Saxena et al. 2013), among them a few act as stimuli for Rh and are effective in inducing nodulation genes (Garg and Manchanda 2007). Host legumes distinguish from nonhosts by the release of distinct flavonoids from the cells adjoining the zone of root hair emergence, which is the most convenient spot for infection by Rh (Garg and Manchanda 2007). Flavonoids secreted by plants are

---

Fig. 2.1 (continued) Relocation of the nucleus to the infection site predicts the path of ITs. A nodule primordium develops into nodule in cortex, and growing ITs fuse with the nodule tissues. *RH* root hair; *PIT* preinfection thread. (**b**) Plant signaling molecules, especially strigolactones (SL), induce germination of spores as well as branching of hyphae. In reciprocation, arbuscular mycorrhizal fungi (AMF) produce Myc factors that initiate symbiotic signaling pathway leading to calcium (Ca^{2+}) spiking in the epidermal cells. AMF form hyphopodia to penetrate the epidermis. Pre-penetration apparatus (PPA) is formed under the hyphopodium for spatial progress of AMF intrusion and allow hyphal growth in the root cortex. (**c**) Nutrient exchange in rhizobial and (**d**) mycorrhizal symbioses

recognized by a specific protein-NodD released by Rh (Liu et al. 2016), which act as receptor for signal molecules (Wang et al. 2012; Oldroyd 2013). Interaction of Rh with flavonoids turns on NodD, which lead to the transcription of various nod genes (Lindstrom et al. 2002). Nod genes are categorized as common nod genes or as host-specific nod genes (Kamboj et al. 2008). nodA, nodB, and nodC are common nod genes and are found in all Rh strains (Kamboj et al. 2008), whereas nodP, nodQ, and nodH or nodF, nodE, and nodL are the host-specific nod genes (Wais et al. 2002) and vary among different Rh species and dictate the host range (Kamboj et al. 2008). These nod genes then produce lipochitooligosaccharides (LCOs), the derivatives of chitin, which are well-known Nod factors (NFs) (Venturi and Keel 2016). Moreover, several non-flavonoid nod inducers such as stachydrine and trigonelline (alkaloids) have been identified from *Medicago sativa* seedling exudates in 1.1 and 2.3 nmol/seed, respectively, in association with *Sinorhizobium meliloti* (Phillips et al. 1992) and also from roots and root exudates of numerous legumes (Garg and Manchanda 2007). These alkaloids are N-containing compounds, collectively known as betaines (Khan et al. 2017; Webb et al. 2017). The concentrations in which betaines are required (millimolar) for induction of nod genes are much more than those required for flavonoids (micromolar) (Khan et al. 2017).

2.2.3 Common Signaling Symbiotic Pathway (CSSP)

Myc factors, a mixture of sulfated and non-sulfated LCOs, produced by AMF are similar in structure to the NFs produced by Rh (Maillet et al. 2011). A set of legume mutants of LCO-induced signaling were not only unable to undergo rhizobial inter-kingdom communication but also impaired in mycorrhizal symbiosis (Parniske 2008). This strongly suggested that, first, mycorrhizal fungi produce similar LCO-type signaling molecules (named Myc-LCOs) (Maillet et al. 2011) and, second, that most likely during evolution, Rh co-opted the mycorrhizal signaling machinery. The NFs or LCOs produced by Rh are central signal molecules for initiating nodule formation. Unknown plant receptor kinases recognize Myc factors (Mohanty et al. 2018), whereas NFs are recognized and perceived by NFP/NFR5 (Nod factor perception/Nod factor receptor 5) and LYK3/NFR1 (LysM receptor-like kinase 3/Nod factor receptor 1) (Kosuta et al. 2008). These plant receptor kinases at the root epidermis carry lysin motif (LysM) domain (Zipfel and Oldroyd 2017), so that these signals could be transduced by a batch of genes that explicate the signal transduction pathway, the CSSP (common signaling symbiotic pathway) (Mohanty et al. 2018), which is shared by both mycorrhizal and rhizobial symbiosis (Oldroyd 2013). This leads to successful association of both the symbionts (Nanjareddy et al. 2017) and results in nodule and arbuscule formation (Oldroyd 2013). This communication between NFs and plant receptor kinases depends on Nod factor structure, which probably contributes to plant host specificity (Madsen et al. 2010; Maillet et al. 2011; Zipfel and Oldroyd 2017).

The CSSP is composed of a receptor kinase SYMRK (leucine-rich repeat), CASTOR and POLLUX (potassium-permeable channels), nucleoporins (Nup85 and Nup133), CCaMK (a calcium-calmodulin-dependent protein kinase), and CYCLOPS (a nuclear-localized coil-coil protein) (Wang et al. 2017a; Mohanty et al. 2018). A web of GRAS TFs including RAM1 (required for arbuscular mycorrhiza 1) and RAD1 (required for arbuscule development 1) operate in at least few species of AMF (Diedhiou and Diouf 2018). Activation of receptors of cell membrane stimulates oscillations of calcium (Ca^{2+}) in nuclear and perinuclear areas, partially via mevalonate-dependent pathway (Wang et al. 2017a). Ca^{2+} is a crucial secondary messenger at the time of nodulation signaling, and recurrent Ca^{2+} spiking is retained by the transformations in membrane potential. CCaMK decodes the Ca^{2+} signal (Bonfante and Genre 2010) and causes the phosphorylation of its interaction partner CYCLOPS (Wang et al. 2017a). In legumes, the GRAS protein response elements, namely, nodulation signaling pathway 1 (NSP1) and NSP2, including RAM1 (Gobbato et al. 2012) form a complex in an event downstream or parallel to CCAMK as well as CYCLOPS (Nanjareddy et al. 2017; Mohanty et al. 2018) and stimulate the expression of NIN (nodule inception) (Wang et al. 2017a). NIN, as a key transcription factor, is involved in multiple affairs during symbiosis, i.e., proliferation of cortical cells which is achieved by direct activation of NUCLEAR FACTOR-Y(NF-Y) subunit genes, commencement of infection threads, equanimity of nodulation, cytokinin responses, etc., which is necessary for nodule development (Wang et al. 2017a). Various intermediary signaling effectors, such as small noncoding RNA, are also activated by CSSP to regulate gene expression in plants (Diedhiou and Diouf 2018). These molecules encompass siRNA (small interfering RNA) as well as miRNA (microRNA) and play a chief task in majority of developmental processes via regulation of transcription, splicing, RNA-directed DNA methylation, epigenetic functions, and nuclear structures (Ariel et al. 2015) by targeting some TFs such as nodulation signaling pathway 2 (NSP2), NF-YA, AP2/ERF (APETALA 2/ethylene responsive factor), and ARF (Lelandais-Brière et al. 2016).

2.3 Mycorrhizal Symbiosis

Host-symbiont dialogue leads to establishment of the first physical contact between AMF and their partners. Strigolactone gradient in hypodermal passage cells enhances their susceptibility toward hyphal invasion, which induces mycorrhization (Kretzschmar et al. 2012). During establishment of symbiosis, hyphopodia are formed by AMF to penetrate the epidermis (Fig. 2.1b). Hyphopodia are formed particularly on isolated as well as intact cell walls of host plants rather than on any artificial surface, which indicates that it could identify particular cell wall features at the time of its formation (Harrison 2012). During the hyphopodium formation, the host cell prepares itself for the spatial progress of AMF intrusion with the involvement of pre-penetration apparatus (PPA) just underneath the hyphopodium (Genre et al. 2008). Formation of PPA is commenced due to signaling via the CSSP (Harrison 2012). The nucleus inside the epidermal cell migrates near the contact point of hyphopodium, and a dense assemblage of endoplasmic reticulum (ER), microtubules, and actin is formed around the nucleus which collectively constitute PPA (Genre et al. 2005). Thereafter, nucleus shifts toward the opposite direction of the epidermal cell, and a wide cytoplasmic column comprising of PPA links the original position below the hyphopodium with the nucleus (Genre et al. 2008). Studies with FM4-64 labeling indicate that the cell membrane invagination below the contact site of hyphopodium and the penetrating hyphae grow through the cell following the path directed by PPA (Genre et al. 2005). Ca^{2+} spiking has been noticed in epidermal cells just below the hyphopodium as well as in underlying cortical cells, which are linked with the migration of nucleus near the contact point as well as cytoplasmic aggregation (Sieberer et al. 2012). Detailed investigations of outer cortical cells revealed that high-frequency spiking was observed only in penetrated cell, but not detected after the hyphal growth was accomplished across the cell (Sieberer et al. 2012). Studies using electron microscopy and live cell imaging revealed that abundant dictyosomes were present below the hyphopodium as well as around the penetrating hypha. Moreover, EXO-84 (a marker of the exocytotic pathway) and vesicle-associated membrane proteins (VAMP) gather near the hyphopodium and growing hyphal tip (Genre et al. 2012). These observations indicate that in PPA, the important secretory activities are possibly linked with the formation of perifungal membrane as well as matrix that encase the growing hypha in the cell (Harrison 2012). Similar kind of reorganization of ER and migration of nucleus was noticed in the outer cortical cells as well as in the inner cortical cells which prepare the cells for arbuscule development (Genre et al. 2008, 2012; Sieberer et al. 2012). Different morphotypes of AMF such as Arum type and Paris type have different types of size, shape, and development. Arum-type symbiosis involves development of PPA beneath the hyphal contact site, and the nucleus transfers toward the central part of the cell and stays there throughout the arbuscule branching, while in the Paris type, hyphae traverse through an intracellular path across the inner cortical cells. Around five cells in front of the progressing hyphae exhibit a polarized organization with a broad aggregation of ER, which connects the nucleus present in the middle of the cell to the likely contact site and further links to the opposite direction of the cell, from where the hypha will move out of the cell. Time-lapse imaging revealed the role of PPA in directing the pathway of hyphal development where the hyphae move across each cell, occasionally developing large coils within the cell. Such observations indicate that either cell-to-cell communication or a diffusible signal in front of the developing hypha leads to the formation of PPA in Paris type (Genre et al. 2008, 2012; Harrison 2012). During initiation of arbuscule formation, PPA directs the growing hypha toward the center of the inner cortical cell where arbuscule branching begins (Genre et al. 2008). At the time of branching of arbuscule, the ER encases the developing arbuscule, as well as cytoskeletal components, reorganizes, and surrounds its branches (Genre et al. 2008, 2012; Pumplin and Harrison 2009). Numerous Golgi bodies are located next to arbuscule branch nodes (Pumplin and Harrison 2009), and investigations using markers indicate the process of exocytosis at the hyphal tips (Genre et al. 2012). These investigations, regarding rearrangement of organelles and cytoskeleton, suggest that probably the secretory activities are related to the formation of the peri-arbuscular membrane (PAM) and matrix in the peri-arbuscular space (Harrison 2012). In line with this, a number of components of the secretory pathway have been observed to be required for PAM synthesis, such as SNARE proteins and the Exo70I subunit of the exocyst (Luginbuehl et al. 2017). Inside the colonized root cells, specifically arbuscule-containing cells, where PAM encases hyphal branches, a broad contact surface was observed, which is a three-dimensional complex bordered by the plant and fungal membranes. On the basis of live cell imaging of arbuscule and its surrounding membranes, presence of two different PAM domains was observed, one is trunk domain, and another is branch domain (Pumplin and Harrison 2009). The former domain constitutes the bottom of arbuscule consisting of proteins which are also present in the cell membrane, whereas the branch domain encases the fine hyphal branches of arbuscule and harbors a specific set of proteins which mediate the nutrient exchange between the fungus and the plant (Pumplin and Harrison 2009). RNA-silencing approach reveals that two members of the exocytotic vesicle-associated membrane proteins (VAMPs), VAMP721d and VAMP721e, localize on the PAM and are indispensable for the formation of arbuscules (Ivanov et al. 2012). During the mycorrhizal symbiosis, other PAM-localized proteins, such as PT4 and AMT2, are required for phosphate and ammonium transport, respectively (Luginbuehl et al. 2017). The H^+-ATPase HA1 creates a proton gradient across the PAM to energize the nutrient transporters (Luginbuehl et al. 2017). PAM-localized proteins must be secreted and incorporated into the new membrane to attain specialized membrane composition needed for nutrient exchange between the symbionts (Luginbuehl et al. 2017).

2.4 Rhizobial Symbiosis

Rhizobial interaction with plant starts when Rh colonize and attach to the outer surface of the root hair (Fig. 2.1a). The Rh-plant attachment is mediated by rhicadhesin, a specific adhesion protein (Poole et al. 2018), which is present on the surfaces of *Rhizobium* species. The attachment is also mediated by plant-produced lectins (Menendez et al. 2017). The Rh then induce the root hair tips curling, which originates by constant and moderate reorientation of the route of root hair growth (Garg and Manchanda 2007). Within the pocket of curl, Rh become entrapped, where localized degradation of cellulose and other cell wall polysaccharides occurs (Menendez et al. 2017), the inward growth of plasma membrane takes place (Mohanty et al. 2018), and new material is unloaded by plant and Rh (Garg and Manchanda 2007). Simultaneously, cell division activates in the cortical and pericycle cells (McAdam et al. 2018) close to the infection point in front of protoxylem (Kader and Delseny 2011). Ferguson et al. (2010) observed that the actively dividing cortical cells lead to the development of nodules that at maturity may either retain a meristem (indeterminate nodule) or lack it (determinate nodule). Root hair curling occurs only in those root hairs that are susceptible to deformations induced by NFs (Rook et al. 1998). Root hairs of zone II that have nearly completed their growth are most susceptible. However, root hairs of zone I that are actively growing and root hairs of zone III that have already completed their growth become recalcitrant to the impairing activity of NFs (Gage 2004). This deformation of susceptible root hairs starts with isodiametric swelling of root hair tips which occurs due to the disruption of cytoskeleton. This process is preceded by the inception of actively growing other tips that mimics root tips of zone I (Sieberer and Emons 2000). At the root surface, entrapped Rh continue their division to form a community which is called as infection foci, and from these foci, root hair infection threads (ITs) begin to develop (Oldroyd et al. 2011). Interestingly, the tunnellike collection of ER forms in few cells preceding primary infection structure (IT), called as preinfection threads (PITs) (Genre et al. 2008; Lace and Ott 2018). PIT formation is related to enlargement of nuclei, a signal of endoreduplication mediated by decrease in CDK (cyclin-dependent kinase) activities at the G2-M transition (Lace and Ott 2018). The growing IT induces divisions of the cortical cells leading to the formation of nodule primordial. Further development of the primordia demands the mediation of products of specific genes, among which the conserved TFs of legume symbiosis play a key role, which collaborate with DELLA proteins and induce the expression of NIN (Diedhiou and Diouf 2018). The bacteria on reaching the inner cortex are endocytosed by the target cortical cell in an independent, unwalled membrane compartment, the symbiosome, that originates from IT and organizes a niche there (Jones et al. 2007). Within the symbiosome, the bacteria must survive and differentiate into symbiosis-specific form, the bacteroid (Yamaya-Ito et al. 2018), which is specialized for nitrogen fixation and fills the host cells (Ren 2018; Tsyganova et al. 2018). The release of Rh is mediated by local degradation of the cell wall of IT and generation of an infection droplet in nodule cells, which are the regions that still contain plant cell membrane but no cell wall (Ren 2018). From the infection droplets, the individual bacterium is released into the cytoplasm and produce nitrogenase as a result of a physiologic adaptation (Garg and Manchanda 2007). In determinate nodules, the fusion of independent symbiosome occurs, and bacteroids inside the symbiosome divide, resulting in symbiosomes that contain multiple bacteroids which are very much alike in size to free-living bacteria (Ren 2018). However, the individual symbiosomes in case of indeterminate nodules containing bacteroids further divide, resulting in symbiosome with a solitary bacteroid that noticeably expands and loses the ability to restore its free-living state (Ren 2018). A plasmalemma-derived symbiosome membrane (SM) forms a constant envelope around individual bacterium during its release into host cytoplasm and delineates a new space, the symbiosome space (SS), in the middle of SM and the outer bacterial membrane (Pierre et al. 2013). SM serves as a metabolite exchange mediator and also as a physical interface between both symbionts, the functions required for nodule development. SM in full-grown root nodules exhibits an assemblage of proteins that take part in transport, signaling and protein destination, nodule formation and function, metabolic processes, pathogen response, etc. (Garg and Manchanda 2007). The targeted protein and lipid material to the SM impart the symbiosome a biochemical uniqueness (Jones et al. 2007).

2.5 Exchange of Nutrients During Mycorrhizal Symbiosis

The exchange of nutrients between the symbionts is central to the mycorrhizal symbiosis, and this occurs mostly across the extensive interface between the arbuscule and cortical cell (Yang et al. 2012). This large symbiotic interface was provided by highly branched arbuscules for nutrient exchange. P and N are exported from AMF to plant through this symbiotic interface (Bonfante and Genre 2010; Garg and Bharti 2018). Loss of MtPT4 (phosphate transporter from *M. truncatula*) function leads to early degeneration of the arbuscule and loss of symbiosis which indicates a link between symbiotic phosphate delivery and arbuscule life expectancy (Javot et al. 2007). AMF acquire their whole C supply from the host plant which is up to 20% of the C fixed during photosynthesis (Bago et al. 2000). Transfer of hexose to AMF was confirmed through labeling studies coupled with NMR (Pfeffer et al. 1999) as well as with recent research pinpointing the involvement of specific fungal hexose transporters (Helber et al. 2011). It has been recently shown that host-synthesized lipids were transferred directly to AMF, besides the sugar (Luginbuehl et al. 2017). The flux of organic C from host plants to AMF through the symbiotic interface is tightly regulated by both host and AMF (MacLean et al. 2017).

2.5.1 *Transport of Carbon*

It was observed that suppression of root-specific apoplastic invertase activity led to decreased root colonization in *Nicotiana tabaccum* (Schaarschmidt et al. 2007). To supply belowground parts, sucrose is unloaded toward roots and metabolized into monosaccharides (Courty et al. 2015). Enhanced expression of many sucrose transporters (SUTs) in leaves as well as in mycorrhizal roots is associated with the increased sink strength due to root colonization, which is responsible for improved unloading of sucrose from phloem (Doidy et al. 2012) and its export in mycorrhizal cells which further includes sucrose cleavage and strict regulation of sucrose transporters (Courty et al. 2015). In line with it, the sucrose-degrading enzymes, invertases as well as sucrose synthase, get induced in a variety of plant spp. following mycorrhizal inoculation. SWEETs have been found to be involved in export of sucrose as well as monosaccharides; its involvement in the unloading of sucrose from phloem into roots as well as in regulation of sugar export through the symbiotic interface has also been suggested (Manck-Gotzenberger and Requena 2016; Garg and Bharti 2018). SUTs and monosaccharide transporters (MSTs) are induced in the phloem of inoculated roots, suggesting their involvement in mycorrhizal symbiosis (Wang et al. 2017b). In glomeromycotan fungi, GpMST1 (a carbohydrate transporter) is involved in the import of glucose from its host (Schussler et al. 2006). In *Rhizophagus* sp., silencing of a sugar transporter RiMST2 caused not only the reduction in root colonization but also truncated arbuscules (Helber et al. 2011). AMF must transfer the C acquired by IRM from roots to ERM for the growth of its hyphae and formation of spores (Wang et al. 2017b). Besides this, a large fraction of sugars must be transformed into fatty acids, as the lipid is the chief form of C storage in AMF (Becard et al. 1991).

Jiang et al. (2017) found that the heterologous expression of an acyl-ACP thioesterase led to the accretion of C12:0 in AMF when inoculated in *M. truncatula*, which is obstructed in ram2 mutants. The heterodimeric ABC transporters stunted arbuscule (STR) and STR2 (Gutjahr et al. 2012) are located on the PAM and are essential for mycorrhizal symbiosis as well as involved in the supply of lipids to the AMF (Roth and Paszkowski 2017). In vitro substrate inclination of RAM2 for C16:0-CoA and its ability to form C16:02-monoacylglycerol (MAG) indicate that C16:0 is the major received form of fatty acid by AMF from plants (Luginbuehl et al. 2017). Therefore, MAG produced by RAM2 is possibly transported through the heterodimeric ABC transporters STR-STR2 of PAM into the peri-arbuscular space and subsequently to AMF through unknown lipid transporters (Jiang et al. 2017). Depending on the AMF species, spores are made up of lipids up to 95% by weight (Becard et al. 1991) while intra-radical vesicles up to 58% of dry weight (Jabajihare et al. 1984). Moreover, in some ERM regions, lipids also contribute up to 47% of hyphal volume, and *Rhizophagus irregularis*, a fatty acid auxotroph, has lipids with about twofold energy content of carbohydrates per unit weight, indicating import of fatty acid to AMF from host roots as the major portion of their carbon (Wang et al. 2017b).

2.5.2 *Transport of Mineral Nutrients*

There are two pathways for nutrient absorption in mycorrhizal colonized roots, direct and indirect mycorrhizal pathway. In direct pathway, nutrients are absorbed directly via root epidermis and root hairs, while in indirect pathway, mycelium transports nutrients absorbed by ERM from the soil into root cortical cells via symbiotic interfaces (Smith and Read 2008). The import of P by host plant roots from AMF was confirmed by experimentation with ^{32}P- or ^{33}P-labeled substrates in a two-compartment system (Jakobsen et al. 1992). Promotion of root rhizosphere and hence absorbing surface area by AMF leads to the mobilization of sparingly available P; as a result mycorrhizal plants possess higher P influx than the non-mycorrhizal plants. Yang et al. (2012) reported that symbiotic fungi delivered about 70% of the overall P in mycorrhizal inoculated rice.

Expansion of mycelium in the soil determines its efficacy of P uptake (Jakobsen et al. 1992). In various host plants, existence of AMF-inducible phosphate transporters (PTs) has been revealed through expression analysis. PTs, such as StPT3 (*Solanum tuberosum*) and MtPT4 (*M. truncatula*), are suggested to be involved in import of phosphate by the host plant from AMF across the symbiotic interface (Rausch et al. 2001; Javot et al. 2007; Wang et al. 2017b). The transport of P toward plant cells by PTs requires protonation and deprotonation of these transporters accompanied by conformation changes (Karandashov and Bucher 2005). H^+-ATPase of cell membrane is suggested to energize perifungal membranes surrounding the arbuscules (Ramos et al. 2009). H^+-ATPase HA1 in *Oryza sativa* and *M. truncatula* play crucial role in the uptake of P from the symbiotic interface (Wang et al. 2014). PTs localized on the ERM, such as GiPT (*Glomus intraradices*) and GvPT (*Glomus versiforme*), are accountable for the uptake of P from the soil (Harrison and Vanbuuren 1995; Maldonado-Mendoza et al. 2001). Remarkably, GmosPT in *Glomus mosseae* and GigmPT in *Gigaspora margarita* have shown their expression in the ERM and in arbuscule-containing cells (Balestrini et al. 2007; Xie et al. 2016). GigmPT mutant leads to decrease in root colonization, demonstrating that GigmPT involved in P reuptake to AMF from the symbiotic interface is vital for mycorrhizal symbiosis (Wang et al. 2017b). Xie et al. (2016) proposed that during mycorrhizal symbiosis, GigmPT could be involved in phosphate sensing and its reuptake and transport.

Earlier studies with 15N-labeled substrates revealed that root colonization improved the uptake of N by host plants (Hawkins et al. 2000). Glutamine synthetase/glutamine oxoglutarate aminotransferase (GS/GOGAT) cycle assimilates inorganic N, taken up by the AMF, into amino acids and finally forms arginine (Johansen et al. 1996). Among various amino acids present in the ERM, arginine is the most prevalent form which is transported to IRM via intracellular hyphae, where it is converted into ornithine and urea. Urea undergoes hydrolyzation to form ammonia, which is subsequently taken up by host plant through symbiotic interface. Consistent with it, the isotope labeling studies reveal the upregulation of genes related to N assimilation in ERM and breakdown of arginine in IRM (Govindarajulu et al. 2005). But, the mechanism of transfer of arginine from the ERM to IRM is yet to be unfolded. P is primarily translocated from ERM to IRM as negatively charged polyphosphate, and positively charged arginine is supposed to be transported along with it throughout the symbiosis (Parniske 2008).

Many studies (Lopez-Pedrosa et al. 2006; Perez-Tienda et al. 2011; Calabrese et al. 2016) have identified some ammonium (NH_4^+) transporters (AMTs) like GintAMT1, GintAMT2, and GintAMT3 in *Rhizophagus irregularis*. Low NH_4^+ conditions lead to enhanced expression of these AMT genes and uptake of NH_4^+ by the ERM from the soil. Additionally, AMF might obtain N by increasing the rate of decomposition of organic matter (Leigh et al. 2009). Many AMTs such as GmAMT4.1 and ATM2;3 are upregulated in mycorrhizal roots of *Glycine max* and *M. truncatula*, respectively, and show particular expression in cortical cells having arbuscules. Breuillin-Sessoms et al. (2015) reported that such transporters are specifically present on PAMs of branch area than the trunk area, revealing that the arbuscule branches are the active sites for NH_4^+ transfer. In contrast, in *Lotus japonicus*, cloned LjAMT2;2 takes up N in the form of NH_3 from AMF at the symbiotic interface which implies that, besides NH_4^+, NH_3 can be provided at the symbiotic interface by AMF as well as imported by host plant roots (Guether et al. 2009).

2.5.3 Transport of Metal Ions

AMF can enhance potassium (K) uptake in host plant and modulate K deficiency-induced plant responses by averting production of reactive oxygen species (ROS), inducing particular genes and cell membrane K^+/H^+ exchanger. Garcia et al. (2017) have reported that mycorrhization decreased sodium (Na) accretion in shoots of the host plant. The putative sulfate transporter of *M. truncatula*, MtSultr1;2, is upregulated by mycorrhizal symbiosis in sulfate-deficient (Casieri et al. 2012) while LjSultr1;2 in sulfate-sufficient roots of *L. japonicus* (Giovannetti et al. 2014). Faber et al. (1990) observed that uptake of zinc (Zn) was modulated by mycorrhization in *Zea mays*. A putative Zn transporter, GintZnT1, has been isolated from the ERM of *G. intraradices* which is suggested to play an important role in allocating Zn to host plants (Gonzalez-Guerrero et al. 2005). Root colonization downregulates Lemt2 (encodes a metallothionein) and LeNramp1 (encodes a broad range metal ions transporter) in *Lycopersicon esculentum*, suggesting an ameliorative role of AMF under heavy metal stress (Ouziad et al. 2005).

2.6 Exchange of Nutrients During Rhizobial Symbiosis

In rhizobial symbiosis, all exchanges between the host plant and bacteria occur through the SM, and most of them depend on the activity of transport proteins (Fig. 2.1c). The most often discussed exchange is of reduced C and N, in the form of malate and NH_3, respectively (Courty et al. 2015). NH_3 is the product of BNF which contributes about 40 million tonnes of N each year into agricultural systems (Udvardi and Poole 2013; Courty et al. 2015). BNF is powered via plant photosynthesis, and per gram fixation of N consumes approximately 6 g of photosynthetic C (Vance and Heichel 1991). However, Rh depend on the plant for other elements, such as Zn, molybdenum (Mo), sulfur (S), iron (Fe), copper (Cu), etc., that are required for its metabolism (Udvardi and Poole 2013). The fixed N (NH_3) is assimilated with the help of organic C compounds in nodules, which is then received by legumes (Moreau et al. 2002). The controlled dispersal of signaling molecules and metabolites is allowed by the presence of channels and transporters on SM; however, the genes responsible for several of these transport processes are still unknown (Clarke et al. 2014).

2.6.1 Transport of Carbon

C (malate-a dicarboxylate) is translocated through the phloem from source leaves to nodules (Udvardi and Day 1997), which is then transported to the bacteroids through the SM (Clarke et al. 2014). In 1988, the existence of a monovalent dicarboxylate transporter (DCT) was first identified from *G. max* symbiosomes, suggesting malate to be the dominant form transported and then succinate (Udvardi et al. 1988). However, LjSUT4 is described as the first sugar transporter enhanced by nodulation in *L. japonicus* (Flemetakis et al. 2003) and expressed in nodules and roots (Kryvoruchko et al. 2016). SUTs are a group of sucrose-H^+ symporters within the MFS (major facilitator superfamily) (Kryvoruchko et al. 2016). A sugar transporter, DgSTP1, characterized from non-legume *Datisca glomerata* was induced in N-fixing nodules (Schubert et al. 2011), belonging to the SP (sugar porter) family having the highest comparative uptake for glucose than galactose, mannose, and xylose (Kryvoruchko et al. 2016). The increase of DgSTP1 transcripts and proteins at low pH in infected nodule cells suggests the export of glucose toward symbiotic bacteria before the induction of BNF process (Schubert et al. 2011). Another sugar transporter family, called SWEET, was also discovered and MtN3, the nodule-specific gene, identified to function in symbiotic associations (Kryvoruchko et al. 2016).

2.6.2 Transport of Mineral Nutrients

Inorganic N is reduced to NH_3 in bacteroids by the enzyme nitrogenase and is dispersed out from the bacteroids to the SS (acidic), where substantial amount of NH_3 is protonated to NH_4^+ (Pfau 2013). The repression of bacteroid NH_4^+ carrier prevents the reuptake of NH_4^+ by bacteroids during their differentiation (Howitt et al. 1986). SM is triggered by an H^+-ATPase transporting H^+ into the SS (Hwang 2013). This establishes a concentration gradient promoting the discharge of NH_3/NH_4^+ with the help of NH_3 channel Nodulin 26 (NOD26) to the host plant cytoplasm, where NH_3 is rapidly assimilated to glutamine and asparagine (Fortin et al. 1987). NOD26 belongs to MIP/AQP (major intrinsic protein/aquaporin) channel family and predicts the preferred orientation of glutamine synthetase in the cytosol (Masalkar et al. 2010), which would promote NH_4^+ assimilation rapidly (Clarke et al. 2014). In tropical legume nodules, glutamine is transformed into allantoin and allantoic acid, which are transported out of nodules into xylem and translocated to the shoot (Tegeder and Masclaux-Daubresse 2018). Depending on their synthesis and the legume species, these ureides move apoplastically toward the cortical and endodermal cells (nodule vasculature), where the apoplastic flow is blocked by the endodermis (Pelissier et al. 2004). This is essential for their preserved movement back into xylem or phloem for subsequent transportation to the shoot and root, respectively (Tegeder and Masclaux-Daubresse 2018).

2.6.3 Transport of Metal Ions

Calcium ion (Ca^{2+}) is the most important regulator of intracellular processes in plants (Izmailov 2003). It has been suggested that the SM is furnished with Ca^{2+}-ATPase, delivering Ca^{2+} ions into the SS from infected cell cytosol through the SM (Krylova et al. 2017). Ca^{2+}-ATPases identified in the *G. max* SM proteome were demonstrated to be present widely in the tissues endorsing recruitment of a new role and location as part of the symbiosis (Clarke et al. 2014). Ca^{2+} has the potential to regulate symbiosome function, which is shown by the existence of a calcium-dependent protein kinase (CDPK) on the SM and its role in uptake of malate within the symbiosome (Weaver et al. 1991). However, it is yet to be investigated if there is any function of the symbiosome in storing Ca^{2+} in vivo (Clarke et al. 2014). In plants, electrochemical gradient over the plasma membrane is a propulsive force for uptaking majority of nutrients, a large portion of which is produced by the H1-ATPase (Gilroy and Jones 2000).

Iron (Fe) is an innate component of heme and nitrogenase in the plant and the bacteroid, respectively, and thus is essential for symbiosis (Clarke et al. 2014). Consequently, before being allocated to the bacteroid, Fe as Fe^{2+} (ferrous) and Fe^{3+} (ferric) need to pass through the SM (Brear et al. 2013). However, uptake of Fe^{2+} is faster than Fe^{3+} by the symbiosome as well as the bacteroid (Moreau et al. 1998). The only Fe^{2+} transporter specified in nodules is GmDMT1 (*G. max* divalent metal transporter 1) present on the SM in *G. max* (Brear et al. 2013). The SM, like the root PM, has Fe^{3+} reductase activity, though, the protein liable for this activity has not been spotted, yet is anticipated to minimize Fe^{3+} piled up in the SS before its uptake by the bacteroid. Fe^{2+} transfer across the SM may also be mediated by members of other protein families including VIT (vacuolar iron transporter), YSL (yellow stripe-like), NRAMP (natural resistance-associated macrophage protein), and ZIP (zinc resistance transporter, iron-resistance transporter-like proteins) that show intensified execution in nodules and are supposed to function in the transfer of other metals across SM in addition to Fe (Brear et al. 2013). In indeterminate nodules of *M. truncatula*, localization of Fe is not observed at the nodule epidermis (Rodriguez-Haas et al. 2013), which suggests that the main acquisition of Fe by the nodules from the rhizosphere is not by its direct uptake (Brear et al. 2013). The ZIP family transporters can also transport cadmium (Cd), copper (Cu), zinc (Zn), and manganese (Mn) in a diversity of organisms (Hall and Guerinot 2006). Both Fe and Zn can be transported by MtZIP6 (Brear et al. 2013), and the transport is normally directed into the cellular cytosol across the PM and membranes of other organelles (Hall and Guerinot 2006). Therefore, ZIP members could transport Fe into infected cells or out of the SM (Brear et al. 2013). The NRAMP transporters involve transport of general metal ions (including Fe^{2+}) which is operated by a H^+ gradient (Nevo and Nelson 2006). NRAMP proteins in plants can also transport metals other than Fe (Cailliatte et al. 2010) including Mn, Cd, cobalt (Co), and aluminum (Al), but *Arachis hypogaea* AhNRAMP1 gene is presumably concerned with Fe procurement from the soil (Xiong et al. 2012). These NRAMP proteins are also connected with import of Fe into the cellular cytosol. VIT family members are engaged in Fe^{2+} acquisition into the vacuole (Brear et al. 2013). The YSL transporter family forms a marked group in the oligopeptide (OPT) superfamily, which shows little resemblance (< 20%) with other members (Curie et al. 2008). *Zea mays* ZmYS1 is the endowing member of the YSL transporters, which is a symporter coupled to H^+ transport, and its expression is boosted under Fe scarcity (Brear et al. 2013).

In case of Zn, YSL or ZIP families are the candidate transporters mediating apoplastic uptake and indicated as divalent metal (Zn^{2+}) introducer (Curie et al. 2008). GmZIP1 of *G. max* functions in the symbiosis with *Bradyrhizobium japonicum* and localizes to the symbiosome, whereas GmZIP1 mRNA is expressed in nodules (Grotz and Guerinot 2006). Transcriptomic data reveals that the ZIP transporter of *M. truncatula*, MtZIP6, is strongly expressed in Rh-inoculated cells (Limpens et al. 2013). Furthermore, Zn uptake by GmZIP1 into symbiosomes is repressed by antibodies (Grotz and Guerinot 2006). In addition, another ZIP protein has already been associated with BNF as being responsible for preserving Zn homeostasis in the symbiosome in *G. max* (Clarke et al. 2014). YSLs mediate the transport as metal-nicotianamine (NA) complexes (Kramer et al. 2007).

Cu plays a significant role in BNF as many of the key enzymes in the process contain Cu as a cofactor and are also used as a part of Cu-Zn superoxide dismutase in free radical assimilation in the nodules (Gonzalez-Guerrero et al. 2016). In order to survive and satisfy the microaerobic environmental conditions and high-energy demands of the process, bacteroids express a high-affinity Cu-containing cytochrome cbb_3 oxidase (Udvardi and Poole 2013). The host plant has to supply Rh with metallic micronutrients (Senovilla et al. 2018). Therefore, Cu is carried via root vasculature to the apoplast of infected region in indeterminate nodules of *M. truncatula* (Rodriguez-Haas et al. 2013). From here, it is introduced in the cytosol of infected cells by a number of transporters including YSL and COPT (copper transporter) when the substrates are nicotianamine-bound metal, a metal chelator (Zheng et al. 2012), and Cu^+ (Pilon 2011), respectively (Senovilla et al. 2018).

Sulfur (S) is an important constituent of nitrogenase required for reducing N by the metalloclusters and must cross the membranes through the SST1 (symbiotic sulfate transporter 1) present on the SM (Krusell et al. 2005). In *L. japonicus*, SST1 is essential for providing S to bacteria and is required for formation of nodules (Gallardo et al. 2014). Krusell et al. (2005) reported that LjSST1 (*L. japonicus* symbiotic sulfate transporter 1) expression is very important for BNF. A few SST family members can transport molybdate too, besides sulfate (Clarke et al. 2014).

Molybdenum (Mo) is critical for nitrogen-fixing tissues as it is required by nitrogenase in its active center (iron-molybdenum cofactor-FeMoco) (Tejada-Jimenez et al. 2017). The MtMOT1.3 (*M. truncatula* molybdate transporter) has been reported to supply Mo for BNF. MtMOT1.3 belongs to MOT1 (molybdate transporter family), the sole member of the family presenting a nodule-specific expression in *M. truncatula* (Tejada-Jimenez et al. 2017). In *Bradyrhizobium japonicum*, an ABC transport system is involved in carrying molybdate which is encoded by Mod-ABC (Delgado et al. 2006), suggesting that Mo must move across the SM (Clarke et al. 2014). Recently, LjMOT1 (*L. japonicus* molybdenum transporter) has been identified in PM of *Nicotiana tabaccum*, which has a role in Mo uptake from soil and its translocation to the aboveground parts, when expressed in leaves (Gao et al. 2016). In some cases, sulfate transporters could be hypothesized to conciliate molybdate transfer (Tejada-Jimenez et al. 2017). The transfer, in that case, could be accomplished by SST1-like proteins across the SM which are associated with sulfate transportation to symbiosomes (Krusell et al. 2005).

2.7 Role of AMF and Rh Under Various Abiotic Stresses

Abiotic stresses, namely, water deficit, water logging, cold, heat, salinity, heavy metal(loid)s (HM), etc., deteriorate the soil characteristics and make the soil deficient in nutrients, thus negatively affecting the plant-mycorrhizal and plant-rhizobial interactions, ultimately leading to declined growth and productivity of plants. The toxic effects of all the abiotic stresses on both these microbes (AMF and Rh) are deleterious and can reduce synthesis, density, and germination of spores, growth of hyphae, diversity of AMF (Yang et al. 2017), as well as the early stages of rhizobial growth by negatively affecting *nod* gene expression, NF production, and nodule formation (Abd-Alla et al. 2014). However, both the microbes are able to survive, regardless of the presence of high intensity of toxic elements in soil, indicating their extreme flexibility and tolerant nature (Vriezen et al. 2007; Bano and Ashfaq 2013), enabling the plants to cope up with various abiotic stresses.

AMF regulate growth and productivity of plants under stressed conditions either through improving soil and root system architecture, enhancing plant-water relations as well as photosynthesis, maintaining ion equilibrium, downregulating free radical formation, upregulating antioxidant machinery, and accumulating osmolytes or via vacuolization, compartmentalization, as well as sequestration of toxic ions by adapting specific mechanisms (Garg et al. 2017; Varma et al. 2017a, b, c). Similarly, successful implication of rhizobial inoculations has been reported in improving plant growth and concurrently alleviating the extent of toxicity in plants exposed to different kinds of stresses. Various mechanisms employed by Rh in mitigating the effects of stress include solubilization of toxic ions, acidification of rhizosphere, enhancement in root surface area for mineral uptake, and increase in the release of root exudates (Ma et al. 2016). Rh also provide resistance to HM stress by indicating a number of responses such as biosorption (adsorption/absorption), precipitation, bioaccumulation, complexation, and biotransformations (reduction, oxidation), thereby decreasing HM toxicity toward themselves and the plants (Ma et al. 2016). The various roles played by AMF and Rh in bringing abiotic stress tolerance are presented in Fig. 2.2 and discussed in detail as follows.

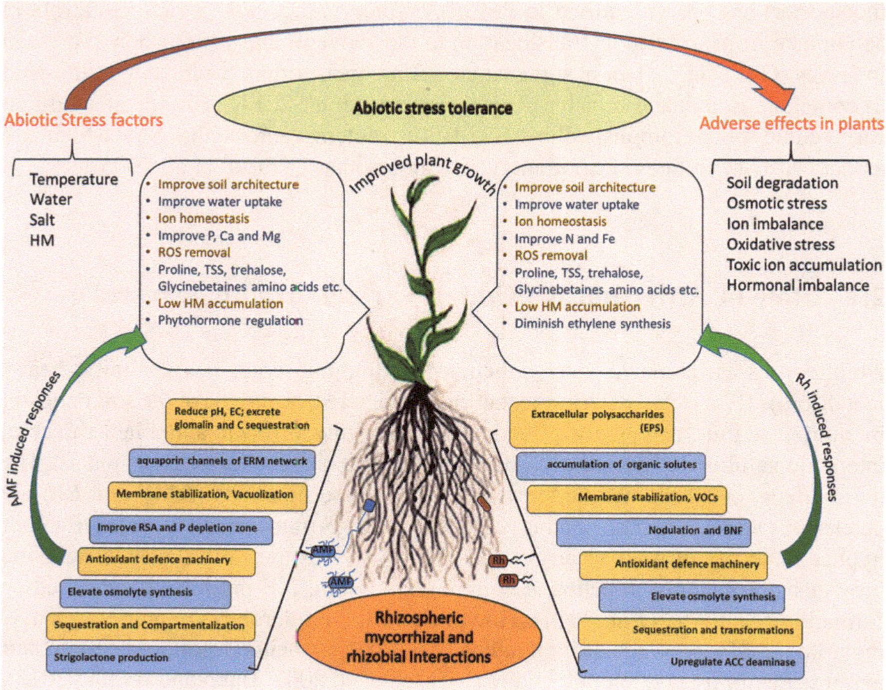

Fig. 2.2 Adverse effects of various abiotic stresses leading to declined growth and productivity, and mechanisms adapted by rhizospheric mycorrhizal and rhizobial interactions causing above ground responses for amelioration of various abiotic stresses. AMF (arbuscular mycorrhizal fungi), Rh (Rhizobia), EC (electrical conductance), C (carbon), ERM (extraradical mycelial), RSA (root system architecture), P (phosphorus), EPS (extracellular polysaccharides), VOCs (volatile organic compounds), BNF (biological nitrogen fixation), ACC (l-aminocyclopropane carboxylic acid), Ca (calcium), Mg (magnesium), TSS (total soluble sugars), HM (heavy metal(loid)s, N (nitrogen), Fe (iron)

2.7.1 Soil and Root System Architecture

Abiotic stresses deteriorate soil characteristics such as pH, electrical conductance (EC), architecture and porosity, as well as nutrient compositions. The nutrient-deficient soil is deleterious to the root system architecture (RSA), and the corresponding root metabolism affects the process of photosynthesis, especially during water-deficit and ionic stresses. The declined photosynthesis leads to alterations in other metabolic responses resulting in reduced vegetative and reproductive yield. Thus, the dynamic modulation of soil as well as RSA under stress conditions is a major challenge for the survival of plants (Chen et al. 2011). AMF have a direct impact on both these attributes as they regulate pH and EC; improve the structure, aggregation of soil, as well as root biomass; and drive the structure of plant communities and productivity (Rillig et al. 2015). Mycorrhizal symbiosis reduces pH as well as EC of the rhizospheric soil by discharging organic acids into the soil and selective uptake of nutrients from the soil (Liu et al. 2016). The extensive, thick ERM network envelops and keeps the soil particles compact, fits into the pores between microaggregates, and builds up stable macroporous aggregates that allow penetration of air and water, thereby preventing soil erosion (Berruti et al. 2014). It has been proposed that glycoproteins (glomalin and glomalin-related proteins) secreted into the soil by AMF could employ a key role in this process. Glomalin is a hydrophobic, iron-rich, heat-stable, long-lasting glycoprotein having chelating properties, assumed to be a homolog of heat shock protein 60 (HSP60) (Ma et al. 2016). It improves structural stability, quality, and water retention properties of soil when exudated in great amount (Rillig et al. 2015). AMF fulfill all their demands from the host plant by exerting a sink demand for carbohydrates, which could result in up to 20% drain of C from the host plant. This indirectly contributes to C sequestration process by storing C in the soils. AMF are efficient to such an extent that they can stabilize soils up to 5 months after the death of the host plant (Soka and Ritchie 2014). Along with C sequestration, AMF play a significant role in the flow of other major nutrients such as P and N (Johnson et al. 2010). This aspect has led to the recognition of the importance of AMF in climate change mitigation.

Rh alleviate the deleterious impacts of abiotic stresses by improving the structure of soil via the secretion of extracellular polysaccharides (EPS) which bind the soil particles (Belimov et al. 2009). A significant improvement in root-adhering soil per root tissue (RAS/RT) ratio has been reported in the rhizosphere of *Helianthus* sp. inoculated with the EPS-producing rhizobial strain YAS34 exposed to drought stress (Alami et al. 2000). Rhizobial inoculations potentially improve RSA in degraded soils by increasing the number of root tips and root surface area, consequently increasing water and nutrient uptake (Egamberdieva and Kucharova 2009).

2.7.2 *Osmotic Balance*

When plant is prone to water, salt, and metal-like abiotic stresses, the first and foremost deleterious impact is the osmotic phase, which induces water deficiency in roots, and consequently, shoot growth is arrested within fraction of time. Osmotic imbalance can occur either due to excess of water (i.e., *flooding/water logging*) or water-deficit (i.e., *drought stress*) conditions. The former produces deleterious condition known as anoxia (lack of O_2) in roots triggering chlorosis, necrosis, and epinasty and ultimately reduction in growth and yield (Bailey-Serres et al. 2010). The latter condition develops due to inadequate supply of water which is essential for normal plant growth, development, and reproducibility, consequently causing dehydration of cells (Zandalinas et al. 2018). To tide over the drought stress conditions, defense mechanisms in plants are directed toward maintaining osmotic balance by adapting number of strategies such as morphological adaptations (increase in root length), stomatal closure (to prevent leaf water loss), regulation of hydraulic conductivity, osmotic adjustment, etc., all aimed for the optimal uptake of water (Farooq et al. 2009). In addition to these mechanisms, plants protect themselves by inducing stress-responsive genes belonging to gene families NHX, DREB, MYB, HKT,

NAC, and WRKY to counteract the devastating effect of osmotic stress (Khan et al. 2017).

Mycorrhizal symbiosis plays key roles in addressing the osmotic balance in stressed plants. Under osmotic stress conditions, relative water content has been found to be more in mycorrhizal plants than non-mycorrhizal plants (Aroca et al. 2011). This is mainly caused by assistance of external hyphae in direct transfer of water to the plant roots. External hypha of AMF extends from an infection point into the soil, undergoes branching and rebranching to establish highly branched ERM network, and links it to the plant roots (Simard et al. 2012). The ERM network serves several functions; the most predominant is the uptake as well as translocation of water and nutrients from soil to roots (Peterson and Massicotte 2004). Water is taken up from the contiguous soil by the hyphal tips (which are hydrophilic) and transferred forward along a hypha to a cortical apoplast either through inner wall layers without the interruption of any membrane in between or cytoplasm (Allen 2007), where it joins the water transport via a root apoplastic pathway (Barzana et al. 2012). As the soil dries out, extracellular hyphae having thin structure (diameter 2–5 μm) delve into soil pores to reach out water source which is inaccessible to root hairs, thereby providing roots more access to available soil water (Khalvati et al. 2005). Wu et al. (2009) proposed bimodal flow of water within the hyphae and indicated that changes in transpiration and hydraulic conductivity are two main factors responsible for hyphal water transport. Aquaporin (AQP) channels of ERM serve as major mechanism for the assisted water flow (Aroca et al. 2009; Kaur and Garg 2017). In this context, first AQP gene (GintAQP1) was cloned from *Rhizophagus intraradices* which could compensate the stress-induced downregulation of AQP of host plant (Aroca et al. 2009). In line with this, increased expression of two AQP genes (*GintAQPF1* and *GintAQPF2*) was observed in root cortical cells of mycorrhizal *Zea mays* exposed to drought stress (Li et al. 2013). In order to ameliorate salt-induced *physiological drought conditions*, AMF inoculations exert changes in root morphology in a structural, spatial, quantitative, and temporal manner and enhance root hydraulic conductivity as well. This not only results in production of better RSA but also allows exploration of a large soil volume, thus enhancing salt tolerance in mycorrhizal plants (Wu et al. 2013). Ruth et al. (2011) reported that about 20% of water absorbed by inoculated roots is through direct and indirect contribution of ERM network. Hence, higher transpiration and photosynthetic rates observed in water-stressed plants are due to the improved water conductance in roots and increased stomatal conductance by mycorrhizal inoculations (Porcel et al. 2012).

Rhizobial strains vary in their response to combat osmotic stress which is evident from the study of Mary et al. (1986) in which 5.5-fold increase in survival of *Sinorhizobium meliloti* strain RCR2011 was observed under physiological drought conditions, while no improvement was observed in *S. meliloti* strain 1.5. Rh use different adaptive mechanisms, such as intracellular accumulation of low molecular weight organic solutes (amino acids such as glutamate and N-acetylglutaminylglutamine, sugars, polyamines, etc.) and beneficial ions such as K, Ca, etc. to combat osmotic stress (Zahran 1999). Rh also regulate water status and stomatal conductance by modulating root hydraulic conductivity and transpiration rate. Besides these, Rh undergo morphological alterations such as changes in cell morphology, modifications in the pattern of EPS, and lipopolysaccharides under osmotic stress. Mayak et al. (2004) observed increased root growth in Rh-inoculated plants exposed to salt stress which improved the water uptake efficiency. The exact mechanism by which EPS bring stress tolerance is not known yet. However, Rinaudo (2004) explored different properties of EPS, e.g., exclusion of toxic compounds and ability to form glasses under water-deficit conditions which probably are the reasons behind their functioning. Another rhizobial adaptation mechanism includes changes in their gene expression in order to resist hyperosmotic stress (Lopez-Raez et al. 2008).

2.7.3 *Ion Homeostasis and Nutrient Uptake*

After osmotic phase, ionic phase arises with time which is caused by the accumulation of toxic ions, predominantly Na^+ and Cl^-, in the cytoplasm. When the rate of inclusion of toxic ions exceeds their rate of exclusion by roots/cellular compartmentalization in vacuoles, they start accumulating in the cytosol and disrupt cellular structures as well as functions (Munns 2002). Major abiotic constraints like salt and HM impede plant growth and development by inducing specific ion toxicity that disturbs the mineral ion homeostasis. Hence, all tolerance mechanisms in plants are directed toward maintaining ion equilibrium.

The mechanisms adapted by mycorrhizal symbiosis for combating salt stress are maintenance of ion homeostasis (by reducing the uptake of Na^+ and Cl^- and enhancing K^+ absorption), translocation of nutrients by providing membrane stability as well as integrity, and intercellular vacuolization of the intruded toxic ions (Garg and Manchanda 2007, 2009). This is accomplished by mycorrhizal symbiosis-mediated upregulation of genes encoding Na^+ and K^+ transporters (phloem-expressed K^+ transporter-ZmAKT2, plasma membrane-localized Na^+/H^+ antiporter-ZmSOS1, and xylem K^+ transporter-ZmSKOR) responsible for maintaining K^+/Na^+ homeostasis in roots as reported in *Zea mays* (Estrada et al. 2013). These transporters favor Na^+ extrusion from the cytoplasm and its sequestration into the vacuoles, leading to enhanced salt tolerance. Other possible tolerance mechanisms include dilution effect where negative impact of toxic ions gets nullified/reduced in specified tissues due to extensive growth of plant (Hashem et al. 2016).

Rhizobial inoculations alter root morphology with extensive rhizosheaths, trap the toxic ions in the EPS matrix, and regulate expression of ion affinity transporters, which boosts K^+ inclusion and Na^+ exclusion via roots and acquisition of both macro- and micronutrients (mineralization) (Lugtenberg et al. 2013). This is evident from a study (Bharti et al. 2014) in which Rh-inoculated salt-stressed plants exhibit ion homeostasis and improved nutrients which could be due to enhanced N fixation and P solubilization. Rh along with some other plant growth-promoting rhizobacteria (*Azospirillum* sp., *Achromobacter* sp., *Serratia* sp., *Aeromonas* sp., *Bacillus* spp.) stimulate yield, growth, and nutrient uptake in different crops such as

Solanum sp. and *Capsicum* sp. under salt stress (Mayak et al. 2004). Pereira et al. (2008) have reported the salt tolerance ability of bacteria to be plasmid-mediated resistance since extra chromosomal genes contribute to salt resistance and get rapidly transferred from salt tolerant to sensitive bacteria.

2.7.4 Redox Balance

Osmotic stress induces stomatal closure due to which CO_2 availability to be fixed by the enzyme Rubisco becomes insufficient, as a result of which, excess of reducing power gets accumulated (Porcel et al. 2012). Such an accumulated reducing power inhibits electron transport from PS II (photosystem II) to PS I (photosystem I) and the excess of electrons thus produced are trapped by oxygen molecules, leading to generation of excessive reactive oxygen species (ROS) such as singlet oxygen ($^1O_2$), superoxides ($O_2^{\bullet -}$), hydroxyl radicals ($^{\bullet}OH$), etc., causing cell damage or death (Miller et al. 2010). Plants possess a complex network of several enzymatic [e.g., superoxide dismutases (SOD), catalases (CAT), and peroxidases (POX)] and nonenzymatic molecules (carotenoids, flavonoids, glutathione, ascorbate, and tocopherols) as a defensive approach to eliminate these ROS (Porcel et al. 2012). However, these molecules get overloaded and are incapable of maintaining an adequate redox balance; the stress thus exerted is called *oxidative stress*.

Mycorrhizal plants exhibit less oxidative damage as compared to non-mycorrhizal plants exposed to osmotic stress, either by higher activities of antioxidant enzymes or by higher levels of antioxidant compounds (Ruiz-Sanchez et al. 2010; Ruiz-Lozano et al. 2012). However, since mycorrhizal plants have the ability to improve their water status and possess higher stomatal conductance than non-mycorrhizal plants, generation of ROS also declined under osmotic stress conditions (Hashem et al. 2018). Numerous reports have confirmed that mycorrhizal symbiosis-mediated enhancement in the activities of antioxidants is allied with alleviation of osmotic stress in plants (Ruiz-Lozano et al. 2001; Ruiz-Sanchez et al. 2010; Hashem et al. 2018). Ruiz-Lozano et al. (2001) confirmed the enhancement in activities of SOD, GR by transcriptomic analysis of the respective genes in *G. max*. By increasing these enzymatic activities, mycorrhizal symbiosis not only lowered the buildup of toxic metabolites, such as malondialdehyde (MDA) and hydrogen peroxide (H_2O_2), but also provided membrane integrity as well as stability and regulated vital metabolic processes in stressed plants (Ruiz-Sanchez et al. 2010). Among other potential ROS scavengers, increased levels of flavonoids have been observed to play a crucial role in reducing oxidative damage, thus alleviating drought stress in mycorrhizal plants (Abbaspour et al. 2012). Several volatile organic compounds, namely, isoprenoids, apocarotenoids, and strigolactones (Lopez-Raez et al. 2008; Walter and Strack 2011), have been characterized in mycorrhizal plants, which may act as a supplementary protective system against various abiotic stresses through direct (ROS scavenging) or indirect (altering ROS signaling) modes of action.

Strong evidences are present which confer the positive roles of ROS and antioxidant machinery in the development of legume-rhizobial symbiosis as well as its functioning. However, the positive role pertains to a certain concentration, and the levels exceeding that concentration are proven to be deleterious for Rh itself (Terpolilli et al. 2016). There are substantial evidences that Rh-inoculated plants can survive under oxidative stress by Rh-mediated modulation of antioxidant enzymes (Wang et al. 2012). Rh inoculations decrease the activities of GR and APX enzymes in leaves of *Lactuca* sp. which are found to increase in response to salt stress (Han and Lee 2005). However, increased activities of SOD, PAL, and CAT are observed in rhizobial-inoculated *Gladiolus* sp. than their uninoculated counterparts under the influence of salt stress (Damodaran et al. 2014). Thus, it is proven that Rh inoculations play significant roles in improving the tolerance capacity of plants toward oxidative stress, but the underlying mechanisms are not clear yet. However, rhizobial strains, host plant, ecological as well as climatic conditions, and type and duration of stress are the factors which might be responsible for such variations in enzymatic activities.

2.7.5 Sequestration and Compartmentalization

Heavy metal(loid)s (HM) like aluminum (Al), arsenic (As), cadmium (Cd), cobalt (Co), chromium (Cr), copper (Cu), iron (Fe), manganese (Mn), nickel (Ni), lead (Pb), antimony (Sb), and zinc (Zn) are present in contaminated soil. Increasing accumulation of HM in agricultural soils results in degradation of soil quality and lowers the productivity of crop plants by affecting diverse metabolic processes (Tamayo et al. 2014).

Extensive researches have been carried out on the tolerance abilities and potential functionalities of AMF in phytoremediation of HM-contaminated soil (Garg et al. 2015; Kaur and Garg 2017). The mechanisms adapted by mycorrhizal symbiosis to provide HM resistance include (1) improvement of rhizospheric soil by lowering pH and EC, (2) restrictive uptake by secreting certain compounds, (3) precipitation in polyphosphate granules, (4) sequestration in soil and/or in roots, (5) accumulation in ERM, (6) extracellular chelation, (7) regulation of gene expression, etc. (Malekzadeh et al. 2011; Amir et al. 2014; Kaur and Garg 2017). Sequestration of metals in ERM involves two steps: metals, firstly, adhere to the chitin-rich cell wall of the hyphae (phytostabilization) and then, secondly, diffuse across the membrane and translocated to different plant tissues (phytoextraction) (Malekzadeh et al. 2011). In order to inactivate absorbed HM, strategy of intracellular compartmentalization of metals by AMF has been documented, in which toxic elements are translocated into fungal vacuoles and also in vesicles by means of chelators such as phytochelatins (PCs), metallothioneins (MTs), glutathione (GSH), etc. where they get deposited away from the plant cytosol (Amir et al. 2014; Gonzalez-Guerrero et al. 2016). Glomalin also exhibits a metal-chelating function (biostabilization) by efficiently sequestering or immobilizing different HM (in vivo and in vitro) at the cell wall of fungal hyphae and in soil (Bano and Ashfaq 2013). Last but not the least, dilution effect is also one possible mechanism of HM tolerance in which harmful effect of metals in specific tissues of plant declines owing to the overall improved plant growth (Hashem et al. 2016). AMF have also found to be beneficial in hyperaccumulation of As in *Pteris vittata* inoculated with AMF which exhibits enhanced frond and leaf surface area (Trotta et al. 2006). Transporters of Cu (*GintABC1*), Fe (MtNRAMP1 and MtNRAMP3), as well as Zn (*GintZnT1*) along with the several putative genes have been identified in *Rhizophagus irregularis* DAOM-197198 through its genome-wide analysis (Gonzalez-Guerrero et al. 2005; Tamayo et al. 2014), but their functions in combating HM stress are yet to be discovered.

Several studies have revealed the role of Rh in diminishing the toxic effects of HM stress on their host plants (Ma et al. 2003; Dary et al. 2010; Pajuelo et al. 2011). One of the most common HM resistance mechanisms in Rh includes extrusion from their cell walls that minimizes the accumulation of the toxic metals to an extent which could not inhibit growth and development of plant. Rh also excrete some proteins that adhere to the metal and store it in the periplasm away from cytoplasm and plasma membrane, which serve as important sites of various reactions and indirectly help phytostabilization (Pajuelo et al. 2011). HM, if absorbed, can be expulsed by Rh through efflux mechanism with the help of ATPases and chemiosmotic ion/H^+ exchangers (Silver and Phung 2005). These mechanisms are complementary to other resistance mechanisms prevailing in Rh such as accumulation and complexation of metal ions inside the cell and biotransformation of toxic forms to less toxic via methylation, precipitation, volatilization, rhizofiltration, and chelation with S-rich ligands like metallothioneins, glutathione, etc. (Gusmao et al. 2006; Dary et al. 2010). Moreover, EPS production and ACC deaminase activity of Rh help in biomass accumulation of HM-affected plants (Ma et al. 2003). These HM tolerance mechanisms are compatible and can act simultaneously.

2.7.6 *Osmolyte Synthesis*

Osmotic adjustment may be defined as "the phenomenon of accumulation of compatible solutes to maintain the cell turgidity within boundaries acceptable for normal cell physiology." It is the first and foremost defense response arising in plants to combat various abiotic stresses, especially cold stress, by increasing synthesis and accumulation of compatible osmolytes such as proline, total soluble sugars (TSS), trehalose, glycine betaine, total free amino acids, etc. (Evelin et al. 2009; Gill and Tuteja 2010). Proline is the major osmolyte synthesized from hydrolysis of proteins in plants which adjusts cytosolic acidity, reduces lipid peroxidation by scavenging ROS, and stabilizes proteins and membranes to tolerate various abiotic stresses (Gill and Tuteja 2010). Consequently, increased synthesis of TSS via upregulated enzymatic activities of amylases and invertases also serves as an important osmolyte in adjusting the osmotic potential of plants under salt stress (Evelin et al. 2009; Garg and Bharti 2018).

Mycorrhizal symbiosis imparts osmotic adjustment by accumulating osmolytes, although the effects show deviation from one plant to another and also show tissue specificity (Sheng et al. 2011). For example, under water-deficit conditions, non-mycorrhizal plants accumulate more proline in shoots than mycorrhizal plants, and on the contrary, in roots, mycorrhizal plants accumulate more proline than non-mycorrhizal plants (Porcel and Ruiz-Lozano 2004; Garg and Baher 2013). Higher accumulation of proline in AMF-inoculated roots takes place through upregulation of Δ1-pyrroline-5-carboxylate synthetase gene (the rate-limiting enzyme-MeP5CS) (Huang et al. 2013), which maintains the water potential gradient favoring water flow into the roots, in order to cope with the low water potential of drying soils. Mycorrhizal inoculation is also effective in enhancing the synthesis of glycine betaine which also contributes to osmotic adjustment and subsequently photosynthesis in stressed plants (Sheng et al. 2011). A positive correlation was found between *Glomus fasciculatum* inoculation and increased total carbohydrate concentrations in *Phragmites australis* plants (Al-Garni 2006). Similar observations have been found in *Rhizophagus intraradices* colonized *G. max* plants (Porcel and Ruiz-Lozano 2004). Increased synthesis of carbohydrates, such as TSS and trehalose, has been observed in mycorrhizal plants through enhanced photosynthesis as well as modulation of carbohydrate and trehalose metabolisms, respectively (Garg and Singh 2018; Garg and Bharti 2018). Ocon et al. (2007) detected moderate transient activities of both the anabolic trehalose-6-phosphate phosphatase and catabolic neutral trehalase in salt-stressed plants, but these were not associated with any transcriptional change leading to trehalose accumulation in *Rhizophagus intraradices*. On the contrary, some reports have revealed that increased accumulation of trehalose in response to stressed conditions leads to membrane stability by forming hydrogen bonds with phosphate groups of membranes and polar headgroups of proteins, keeps them hydrated, and scavenges ROS (Paul et al. 2008).

Rh inoculations produce compatible osmolytes and act in synergism with plants which also exhibit increased levels of osmolytes, thus increasing the osmotic potential within the cells to facilitate plant growth under abiotic stresses (Bharti et al. 2014). But it is difficult to establish whether the main contribution is due to de novo synthesis in plant, i.e., upregulation of proline biosynthesis pathway or by uptake from rhizosphere. Rh-inoculated plants display enhanced proline accumulation when compared with uninoculated plants under drought and salt stress (Bharti et al. 2014). Farooq et al. (2009) revealed that osmolyte accumulation in stressed plants is also facilitated by Rh-mediated modulation of carbohydrate metabolism as well as transport of TSS and directly implicates source-sink relations, photosynthesis, growth rate, and biomass reallocation. Upregulation of amylases activities was observed in Rh-inoculated maize seedlings to alleviate negative effect of drought stress (Bano and Fatima 2009). Biosynthesis of trehalose also imparts osmotolerance to plants inoculated with Rh-stabilizing membranes. Suarez et al. (2008) reported that inoculation of *Phaseolus* sp. with genetically engineered *Rhizobium etli* strain (overexpressing trehalose-6-phosphate synthase) imparts more drought tolerance as compared to the wild-type strain-inoculated counterparts. Such implications have also been observed in *Phaseolus vulgaris* co-inoculated with modified *Rhizobium tropici* and *Paenibacillus polymyxa* strains to overexpress trehalose 6-phosphate gene resulting in increased nodulation, N content, and plant growth. A microarray analysis of nodules of *P. vulgaris* conferring upregulation of stress tolerance genes confirmed that extracellular trehalose effectively induced salt resistance (Figueiredo et al. 2008). Staudinger et al. (2016) have found increased concentrations of pinitol in Rh-inoculated *M. truncatula*, which acts as an osmolyte under drought stress.

2.7.7 Hormonal Signaling

AMF inoculations alter auxins, cytokinins, and strigolactone levels, the major hormones responsible for regulating root architecture (Fusconi 2013). The synthesis and transport of salicylic acid (SA) and jasmonic acid (JA) involve several metabolic processes for plant growth. Hashem et al. (2018) reported mycorrhizal symbiosis-mediated enhancement in the levels of these hormones to enhance stress tolerance in *Cucumis sativus*. In contrast, Medina et al. (2003) reported declined level of SA in mycorrhizal inoculated *Nicotiana tabaccum*. Seed priming with SA and JA plays a crucial role, not only for the plant itself but even for the effective mycorrhizal symbiotic establishment (Song et al. 2013; Garg and Bharti 2018). ABA, the stress hormone, acts as an antitranspirant leading to declined water loss by causing stomatal closure. However, mycorrhizal inoculations block ABA synthesis and thereby regulate root hydraulic properties and rate of transpiration by allowing the stomatal functioning, which in turn enhances photosynthesis (Ruiz-Lozano et al. 2016). All abiotic stresses lead to excessive production of ethylene, which negatively affects the plant growth by inhibiting root elongation as well as root colonization, particularly in P-deprived soils (Zsogon et al. 2008). The enhanced activity of enzyme ACC deaminase in mycorrhizal plants may hydrolyze ACC into α-ketobutyrate and NH_3, which provide nutrients to plants. AMF-treated plants have been found to combat stress by increasing strigolactone production as a result of stimulated AMF development and symbiosis (Aroca et al. 2013).

Rh stop the synthesis of ethylene by the activity of ACC deaminase enzyme and demonstrate normal plant growth. Initial phases of rhizobial symbiosis pose biotic stress in plants which results in excessive accumulation of ACC in the infected roots which is a precursor for ethylene. ACC accumulation and ethylene are the well-established negative regulators of nodulation (Oldroyd et al. 2001). This aspect is confirmed by Ma et al. (2003), who found that ACC deaminase activity (which degrades ACC and inhibits ethylene production) of *Rhizobium leguminosarum* bv. viciae 128C53K enhances nodule genesis in *Pisum sativum* L. cv. Sparkle. Shaharoona et al. (2006) have also found that roots of *Vigna radiata* co-inoculated with *Bradyrhizobium* and rhizobacteria possessing ACC deaminase activity reveal better nodulation as compared to only *Bradyrhizobium*-inoculated roots. The genes *acdS* are responsible for ACC deaminase activity, which is controlled by leucine-responsive regulatory protein (LRP-Li et al. 2001) and is species specific. In *Mesorhizobium loti* MAFF303099, the acdS genes are regulated by the transcriptional control of the N-fixing regulator, nifA (Nukui et al. 2006), while in *Bradyrhizobium japonicumn* USDA110 and *Rhizobium leguminosarum* bv. viciae 128C53K, it is under the regulation of an LRP-like protein and r70 promoter (Ma et al. 2003). Rhizobial symbiosis elicits hormones in systemic tissues of *M. sativa*, which protects the plants from oxidative stress induced by HM stress (Yang et al. 2009). Rh inoculations enhance germination of seeds of *Lactuca* sp. under saline conditions by producing or modifying phytohormones, especially gibberellins (Liu et al. 2018).

2.8 Conclusion and Future Perspectives

Various investigations have established the potential capabilities of AMF and Rh in increasing growth, development, and productivity of stressed plants in order to attain sustainable agroecosystem. However, there are still some unresolved issues, such as identification of signaling molecules, genes, transporters, as well as proteins involved in establishment of symbioses, nutrient exchange through the symbiotic interfaces, and exact mechanism involved in imparting abiotic stress tolerance in plants.

Acknowledgment The authors acknowledge the financial support from Department of Biotechnology (DBT), Ministry of Science and Technology, and University Grants Commission (UGC), Government of India, for carrying out related research.

Conflict of Interest There is no conflict of interest between the authors.

References

Abbaspour H, Saeidi-Sar S, Afshari H, Abdel-Wahhab MA (2012) Tolerance of mycorrhiza infected pistachio (*Pistacia vera* L.) seedling to drought stress under glasshouse conditions. J Plant Physiol 169:704–709

Abd-Alla MH, Issa AA, Ohyama T (2014) Impact of harsh environmental conditions on nodule formation and dinitrogen fixation of legumes. In Adv Biol Ecol of Nitrogen Fixation (InTech) https://doi.org/10.5772/56997

Alami Y, Achouak W, Marol C, Heulin T (2000) Rhizosphere soil aggregation and plant growth promotion of sunflowers by an exopolysaccharide-producing *Rhizobium* sp. Strain isolated from sunflower roots. Appl Environ Microbiol 66:3393–3398

Al-Garni SMS (2006) Increasing NaCl-salt tolerance of a halophytic plant *Phragmites australis* by mycorrhizal symbiosis. Am Eurasian J Agric Environ Sci 1:119–126

Alizadeh O (2011) Mycorrhizal symbiosis. Adv Stud Biol 3(6):273–281

Allen MF (2007) Mycorrhizal fungi: highways for water and nutrients in arid soils. Vadose Zone J 6:291–297

Amballa H, Bhumi NR (2016) Significance of arbuscular mycorrhizal fungi and rhizosphere microflora in plant growth and nutrition. In: Plant-microbe interaction: an approach to sustainable agriculture. Springer, Singapore, pp 417–452
Amir H, Jourand P, Cavaloc Y, Ducousso M (2014) Role of mycorrhizal fungi in the alleviation of heavy metal toxicity in plants. In: Solaiman ZM, Abbott LK, Varma A (eds) Mycorrhizal fungi: use in sustainable agriculture and land restoration. Springer, Berlin, pp 241–258
Ariel F, Romero-Barrios N, Jégu T, Benhamed M, Crespi M (2015) Battles and hijacks: noncoding transcription in plants. Trends Plant Sci 20:362–371
Aroca R, Bago A, Sutka M, Paz JA, Cano C, Amodeo G, Ruiz-Lozano JM (2009) Expression analysis of the first arbuscular mycorrhizal fungi aquaporin described reveals concerted gene expression between salt-stressed and non-stressed mycelium. Mol Plant-Microbe Interact 22:1169–1178
Aroca R, Porcel R, Ruiz-Lozano JM (2011) Regulation of root water uptake under abiotic stress conditions. J Exp Bot 63:43–57
Aroca R, Ruiz-Lozano JM, Zamarreno AM, Paz JA, Garcia-Mina JM, Pozo MJ, Lopez-Raez JA (2013) Arbuscular mycorrhizal symbiosis influences strigolactone production under salinity and alleviates salt stress in lettuce plants. J Plant Physiol 170:47–55
Badri DV, Weir TL, van der Lelie D, Vivanco JM (2009) Rhizosphere chemical dialogues: plant–microbe interactions. Curr Opin Biotechnol 20:642–650
Bago B, Pfeffer PE, Shachar-Hill Y (2000) Carbon metabolism and transport in arbuscular mycorrhizas. Plant Physiol 124:949–957
Bailey-Serres J, Fukao T, Ronald P, Ismail A, Heuer S, Mackill D (2010) Submergence tolerant rice: SUB1's journey from landrace to modern cultivar. Rice 3:138–147
Balestrini R, Gomez-Ariza J, Lanfranco L, Bonfante P (2007) Laser microdissection reveals that transcripts for five plant and one fungal phosphate transporter genes are contemporaneously present in arbusculated cells. Mol Plant-Microbe Interact 20:1055–1062
Bano SA, Ashfaq D (2013) Role of mycorrhiza to reduce heavy metal stress. Nat Sci 5:16–20
Bano A, Fatima M (2009) Salt tolerance in *Zea mays* (L). following inoculation with *Rhizobium* and *Pseudomonas*. Biol Fertil Soils 45:405–413
Barzana G, Aroca R, Paz JA, Chaumont F, Martinez-Ballesta MC, Carvajal M, Ruiz-Lonazo JM (2012) Arbuscular mycorrhizal symbiosis increases relative apoplastic water flow in roots of the host plant under both well-watered and drought stress conditions. Ann Bot 109:1009–1017
Becard G, Doner LW, Rolin DB, Douds DD, Pfeffer PE (1991) Identification and quantification of trehalose in vesicular arbuscular mycorrhizal fungi by in vivo C-13 NMR and HPLC analyses. New Phytol 118:547–552
Belimov AA, Dodd IC, Hontzeas N, Theobald JC, Safronova VI, Davies WJ (2009) Rhizosphere bacteria containing 1-aminocyclopropane-1-carboxylate deaminase increase yield of plants grown in drying soil via both local and systemic hormone signaling. New Phytol 181:413–423
Berruti A, Borriello R, Orgiazzi A, Barbera AC, Lumini E, Bianciotto V (2014) Arbuscular mycorrhizal fungi and their value for ecosystem management. In: Biodiversity-the dynamic balance of the planet. InTech, Rijeka. https://doi.org/10.5772/58231
Bhandari P, Garg N (2017) Dynamics of arbuscular mycorrhizal symbiosis and its role in nutrient acquisition: an overview. In: Mycorrhiza-nutrient uptake, biocontrol, ecorestoration. Springer, Cham, pp 21–43
Bharti N, Barnawal D, Awasthi A, Yadav A, Kalra A (2014) Plant growth promoting rhizobacteria alleviate salinity induced negative effects on growth, oil content and physiological status in *Mentha arvensis*. Acta Physiol Plant 34:45–60
Bonfante P (2003) Plants, mycorrhizal fungi and endobacteria: a dialog among cells and genomes. Biol Bull 204:215–220
Bonfante P, Genre A (2008) Plants and arbuscular mycorrhizal fungi: an evolutionary-developmental perspective. Trends Plant Sci 13:492–498
Bonfante P, Genre A (2010) Mechanisms underlying beneficial plant-fungus interactions in mycorrhizal symbiosis. Nat Commun 1:1–11

Brear EM, Day DA, Smith PMC (2013) Iron: an essential micronutrient for the legume-rhizobium symbiosis. Front Plant Sci 4:1–15

Breuillin-Sessoms F, Floss DS, Gomez SK, Pumplin N, Ding Y, Levesque-Tremblay V, Noar RD, Daniels DA, Bravo A, Eaglesham JB, Benedito VA (2015) Suppression of arbuscule degeneration in *Medicago truncatula* phosphate transporter4 mutants is dependent on the ammonium transporter 2 family prote in AMT2;3. Plant Cell 27:1352–1366

Cailliatte R, Schikora A, Briat JF, Mari S, Curie C (2010) High-affinity manganese uptake by the metal transporter NRAMP1 is essential for *Arabidopsis* growth in low manganese conditions. Plant Cell 22:904–917

Calabrese S, Perez-Tienda J, Ellerbeck M, Arnould C, Chatagnier O, Boller T, Schussler A, Brachmann A, Wipf D, Ferrol N, Courty PE (2016) GintAMT3-a low-affinity ammonium transporter of the arbuscular mycorrhizal *Rhizophagus irregularis*. Front Plant Sci 7:1–14

Casieri L, Gallardo K, Wipf D (2012) Transcriptional response of *Medicago truncatula* sulphate transporters to arbuscular mycorrhizal symbiosis with and without sulphur stress. Planta 235:1431–1447

Chen YL, Dunbabin VM, Postma JA, Diggle AJ, Palta JA, Lynch JP, Siddique KH, Rengel Z (2011) Phenotypic variability and modelling of root structure of wild *Lupinus angustifolius* genotypes. Plant Soil 348:345–364

Clarke VC, Loughlin PC, Day DA, Smith P (2014) Transport processes of the legume symbiosome membrane. Front Plant Sci 5:1–9

Corticeiro SC, Lima AIG, De EM, Figueira AP (2006) The importance of glutathione in oxidative status of *Rhizobium leguminosarum* biovar viciae under Cd exposure. Enzym Microb Technol 40:132–137

Courty PE, Smith P, Koegel S, Redecker D, Wipf D (2015) Inorganic nitrogen uptake and transport in beneficial plant root-microbe interactions. Crit Rev Plant Sci 34:4–16

Curie C, Cassin G, Couch D, Divol F, Higuchi K, Jean ML, mission J, Schikora A, Czernic P, Mari S (2008) Metal movement within the plant: contribution of nicotianamine and yellow stripe 1-like transporters. Ann Bot 103(1):1–11

Damodaran T, Rai RB, Jha SK, Kannan R, Pandey BK, Sah V, Mishra VK, Sharma DK (2014) Rhizosphere and endophytic bacteria for induction of salt tolerance in gladiolus grown in sodic soils. J Plant Interact 9:577–584

Dary M, Chamber-Perez M, Palomares A, Pajuelo E (2010) "In situ" phytostabilisation of heavy metal polluted soils using *Lupinus luteus* inoculated with metal resistant plant-growth promoting rhizobacteria. J Hazard Mater 177:323–330

Datta A, Singh RK, Tabassum S (2015) Isolation, characterization and growth of Rhizobium strains under optimum conditions for effective biofertilizer production. Int J Pharm Sci Rev Res 32:199–208

Delgado MJ, Tresierra-Ayala A, Talbi C, Bedmar EJ (2006) Functional characterization of the Bradyrhizobium japonicum modA and modB genes involved in molybdenum transport. Microbiology 152(1):199–207

Diedhiou I, Diouf D (2018) Transcription factors network in root endosymbiosis establishment and development. World J Microbiol Biotechnol 34(3):1–14

Doidy J, van Tuinen D, Lamotte O, Corneillat M, Alcaraz G, Wipf D (2012) The *Medicago truncatula* sucrose transporter family: characterization and implication of key members in carbon partitioning towards arbuscular mycorrhizal fungi. Mol Plant 5:1346–1358

Egamberdieva D, Kucharova Z (2009) Selection for root colonising bacteria stimulating wheat growth in saline soils. Biol Fertil Soils 45:563–571

Estrada B, Aroca R, Maathuis FJ, Barea JM, Ruiz-Lozano JM (2013) Arbuscular mycorrhizal fungi native from a Mediterranean saline area enhance maize tolerance to salinity through improved ion homeostasis. Plant Cell Environ 36:1771–1782

Evelin H, Kapoor R, Giri B (2009) Arbuscular mycorrhizal fungi in alleviation of salt stress: a review. Ann Bot 104:1263–1280

Faber BA, Zasoski RJ, Burau RG, Uriu K (1990) Zinc uptake by corn as affected by vesicular-arbuscular mycorrhizae. Plant Soil 129:121–130
Farooq M, Wahid A, Kobayashi N, Fujita D, Basra SMA (2009) Plant drought stress: effects, mechanisms and management. In: agronomy for sustainable development. Springer, Dordrecht, pp 153–188
Ferguson BJ, Indrasumunar A, Hayashi S, Lin MH, Lin YH, Reid DE, Gresshoff PM (2010) Molecular analysis of legume nodule development and autoregulation. J Integr Plant Biol 52:61–76
Figueiredo MV, Burity HA, Martinez CR, Chanway CP (2008) Alleviation of drought stress in the common bean (*Phaseolus vulgaris* L.) by co-inoculation with *Paenibacillus polymyxa* and *Rhizobium tropici*. Appl Soil Ecol 40(1):182–188
Flemetakis E, Dimou M, Cotzur D, Efrose RC, Aivalakis G, Colebatch G, Udvardi MK, Katinakis P (2003) A sucrose transporter, Lj SUT4, is up-regulated during *Lotus japonicus* nodule development. J Exp Bot 54:1789–1791
Fortin MG, Morrison NA, Verma DPS (1987) Nodulin-26, a peribacteroid membrane nodulin is expressed independently of the development of the peribacteroid compartment. Nucleic Acids Res 15:813–824
Fusconi A (2013) Regulation of root morphogenesis in arbuscular mycorrhizae: what role do fungal exudates, phosphate, sugars and hormones play in lateral root formation? Ann Bot 113:19–33
Gage DJ (2004) Infection and invasion of roots by symbiotic, nitrogen-fixing rhizobia during nodulation of temperate legumes. Microbiol Mol Biol Rev 68:280–300
Gallardo K, Courty PE, Le Signor C, Wipf D, Vernoud V (2014) Sulfate transporters in the plant's response to drought and salinity: regulation and possible functions. Front Plant Sci 5(580):1–15
Gao JS, Wu FF, Shen ZL, Meng Y, Cai YP, Lin Y (2016) A putative molybdate transporter LjMOT1 is required for molybdenum transport in *Lotus japonicus*. Physiol Plant 158:331–340
Garg N, Singh S (2018) Mycorrhizal inoculations and silicon fortifications improve rhizobial symbiosis, antioxidant defense, trehalose turnover in pigeon pea genotypes under cadmium and zinc stress. Plant Growth Regul 86(1):105–119
Garcia K, Chasman D, Roy S, Ane JM (2017) Physiological responses and gene co-expression network of mycorrhizal roots under K+ deprivation. Plant Physiol 173:1811–1823
Garg N, Baher N (2013) Role of arbuscular mycorrhizal symbiosis in proline biosynthesis and metabolism of *Cicer arietinum* L. (chickpea) genotypes under salt stress. J Plant Growth Regul 32:767–778
Garg N, Bharti A (2018) Salicylic acid improves arbuscular mycorrhizal symbiosis, and chickpea growth and yield by modulating carbohydrate metabolism under salt stress. Mycorrhiza 28 (8):727–746. https://doi.org/10.1007/s00572-018-0856-6
Garg N, Manchanda G (2007) Symbiotic nitrogen fixation in legume nodules: process and signaling: a review. Agron Sustain Dev 27:59–68
Garg N, Manchanda G (2009) Role of arbuscular mycorrhizae in the alleviation of ionic, osmotic and oxidative stresses induced by salinity in *Cajanus cajan* (L.) Millsp. (Pigeonpea). J Agron Crop Sci 195(2):110–123
Garg N, Singla P, Bhandari P (2015) Metal uptake, oxidative metabolism, and mycorrhization in pigeonpea and pea under arsenic and cadmium stress. Turk J Agric For 39:234–250
Garg N, Bhandari P, Kashyap L, Singh S (2017) Arbuscular mycorrhizal symbiosis: a boon for sustainable legume production under salinity and heavy metal stress. In: Aggarwal A, Yadav K (eds) Mycorrhizal fungi. Astral International, New Delhi, pp 247–273
Genre A, Chabaud M, Timmers T, Bonfante P, Barker DG (2005) Arbuscular mycorrhizal fungi elicit a novel intracellular apparatus in *Medicago truncatula* root epidermal cells before infection. Plant Cell 17:3489–3499
Genre A, Chabaud M, Faccio A, Barker DG, Bonfante P (2008) Prepenetration apparatus assembly precedes and predicts the colonization patterns of arbuscular mycorrhizal fungi within the root cortex of both *Medicago truncatula* and *Daucus carota*. Plant Cell 20:1407–1420

Genre A, Ivanov S, Fendrych M, Faccio A, Zarsky V, Bisseling T, Bonfante P (2012) Multiple exocytotic markers accumulate at the sites of perifungal membrane biogenesis in arbuscular mycorrhizas. Plant Cell Physiol 53:244–255

Gill SS, Tuteja N (2010) Reactive oxygen species and antioxidant machinery in abiotic stress tolerance in crop plants. Plant Physiol Biochem 48:909–930

Gilroy S, Jones DL (2000) Through form to function: root hair development and nutrient uptake. Trends Plant Sci 5:56–60

Giovannetti M, Tolosano M, Volpe V, Kopriva S, Bonfante P (2014) Identification and functional characterization of a sulfate transporter induced by both sulfur starvation and mycorrhiza formation in *Lotus japonicus*. New Phytol 204:609–619

Gobbato E, Marsh JF, Vernié T, Wang E, Maillet F, Kim J, Ratet P, Mysore KS (2012) A GRAS-type transcription factor with a specific function in mycorrhizal signaling. Curr Biol 22:2236–2241

Gonzalez-Chavez C, D'haen J, Vangronsveld J, Dodd JC (2002) Copper sorption and accumulation by the extraradical mycelium of different Glomus spp. (arbuscular mycorrhizal fungi) isolated from the same polluted soil. Plant Soil 240(2):287–297

Gonzalez-Guerrero M, Azcon-Aguilar C, Mooney M, Valderas A, MacDiarmid CW, Eide DJ, Ferrol N (2005) Characterization of a *Glomus intraradices* gene encoding a putative Zn transporter of the cation diffusion facilitator family. Fungal Genet Biol 42:130–140

Gonzalez-Guerrero M, Escudero V, Saez A, Tejada-Jimenez M (2016) Transition metal transport in plants and associated endosymbionts: arbuscular mycorrhizal fungi and rhizobia. Front Plant Sci 7:1–21

Govindarajulu M, Pfeffer PE, Jin HR, Abubaker J, Douds DD, Allen JW, Bucking H, Lammers PJ, Shachar-Hill Y (2005) Nitrogen transfer in the arbuscular mycorrhizal symbiosis. Nature 435:819–823

Grotz N, Guerinot ML (2006) Molecular aspects of Cu, Fe and Zn homeostasis in plants. Biochem Biophys Acta Mol Cell Res 1763:595–608

Guether M, Neuhauser B, Balestrini R, Dynowski M, Ludewig U, Bonfante P (2009) A mycorrhizal-specific ammonium transporter from *Lotus japonicus* acquires nitrogen released by arbuscular mycorrhizal fungi. Plant Physiol 150:73–83

Gusmao A, Cacoilo S, Figueira EM (2006) Glutathione-mediated cadmium sequestration in *Rhizobium leguminosarum*. Enzym Microb Technol 39:763–769

Gutjahr C, Radovanovic D, Geoffroy J, Zhang Q, Siegler H, Chiapello M, Casieri L, Guiderdoni E, Kumar CS (2012) The half-size ABC transporters STR1 and STR2 are indispensable for mycorrhizal arbuscule formation in rice. Plant J 69:906–920

Hall BP, Guerinot ML (2006) The role of ZIP family members in iron transport. In: Iron nutrition in plants and rhizospheric microorganisms. Springer, Dordrecht, pp 311–326

Han HS, Lee KD (2005) Plant growth promoting rhizobacteria effect on antioxidant status, photosynthesis, mineral uptake and growth of lettuce under soil salinity. Res J Agric Biol Sci 1:210–215

Harrison MJ (2005) Signaling in the arbuscular mycorrhizal symbiosis. Annu Rev Microbiol 59:19–42

Harrison MJ (2012) Cellular programs for arbuscular mycorrhizal symbiosis. Curr Opin Plant Biol 15:691–698

Harrison MJ, Vanbuuren ML (1995) A phosphate transporter from the mycorrhizal fungus Glomus versiforme. Nature 378:626–629

Hashem A, Abd-Allah EF, Alqarawi AA, Al Huqail AA, Egamberdieva D, Wirth S (2016) Alleviation of cadmium stress in *Solanum lycopersicum* L. by arbuscular mycorrhizal fungi via induction of acquired systemic tolerance. Saudi J Biol Sci 23:272–281

Hashem A, Alqarawi AA, Radhakrishnan R, Al-Arjani ABF, Aldehaish HA, Egamberdieva D, Abd-Allah EF (2018) Arbuscular mycorrhizal fungi regulate the oxidative system, hormones and ionic equilibrium to trigger salt stress tolerance in *Cucumis sativus* L. Saudi J Biol Sci 25:1102–1114. https://doi.org/10.1016/j.sjbs.2018.03.009

Hawkins HJ, Johansen A, George E (2000) Uptake and transport of organic and inorganic nitrogen by arbuscular mycorrhizal fungi. Plant Soil 226:275–285

Helber N, Wippel K, Sauer N, Schaarschmidt S, Hause B, Requena N (2011) A versatile monosaccharide transporter that operates in the arbuscular mycorrhizal fungus *Glomus sp* is crucial for the symbiotic relationship with plants. Plant Cell 23:3812–3823

Howitt SM, Udvardi MK, Day DA, Gresshoff PM (1986) Ammonia transport in free-living and symbiotic *Rhizobium* sp. ANU289. Microbiology 132:257–261

Huang Z, Zhao L, Chen D, Liang M, Liu Z, Shao H, Long X (2013) Salt stress encourages proline accumulation by regulating proline biosynthesis and degradation in Jerusalem artichoke plantlets. PLoS One 8:e62085

Hwang JH (2013) Soybean nodulin 26: a channel for water and ammonia at the symbiotic interface of legumes and nitrogen-fixing rhizobia bacteria. Doctoral dissertations, University of Tennessee, Knoxville

Izmailov SF (2003) Calcium-based interactions of symbiotic partners in legumes: role of peribacteroid membrane. Russ J Plant Physiol 50:553–566

Ivanov S, Fedorova EE, Limpens E, De Mita S, Genre A, Bonfante P, Bisseling T (2012) Rhizobium–legume symbiosis shares an exocytotic pathway required for arbuscule formation. Proc Natl Acad Sci U S A 109:8316–8321

Jabajihare S, Deschene A, Kendrick B (1984) Lipid-content and composition of vesicles of a vesicular-arbuscular mycorrhizal fungus. Mycologia 76:1024–1030

Jakobsen I, Abbott LK, Robson AD (1992) External hyphae of vesicular arbuscular mycorrhizal fungi associated with *Trifolium subterraneum* L 2 Hyphal transport of P-32 over defined distances. New Phytol 120:509–516

Javot H, Penmetsa RV, Terzaghi N, Cook DR, Harrison MJ (2007) A *Medicago truncatula* phosphate transporter indispensable for the arbuscular mycorrhizal symbiosis. Proc Natl Acad Sci U S A 104:1720–1725

Jiang Y, Wang W, Xie Q, Liu N, Liu L, Wang D, Zhang X, Yang C, Chen X, Tang D, Wang E (2017) Plants transfer lipids to sustain colonization by mutualistic mycorrhizal and parasitic fungi. Science 356:1172–1175

Johansen A, Finlay RD, Olsson PA (1996) Nitrogen metabolism of external hyphae of the arbuscular mycorrhizal fungus *Glomus intraradices*. New Phytol 133:705–712

Johnson NC, Wilson GWT, Bowker MA, Wilson JA, Miller RA (2010) Resource limitation is a driver of local adaptation in mycorrhizal symbioses. Proc Natl Acad Sci U S A 107:2093–2098

Jones KM, Kobayashi H, Davies BW, Taga ME, Walker GC (2007) How rhizobial symbionts invade plants: the *Sinorhizobium–Medicago* model. Nat Rev Microbiol 5(8):619

Kader JC, Delseny M (2011) Advances in botanical research, vol 60. Academic, London

Kamboj D, Kumar R, Kumari A, Kundu BS, Pathak D, Sharma PK (2008) Rhizobia, nod factors and nodulation–a review. Agric Rev 29:200–206

Karandashov V, Bucher M (2005) Symbiotic phosphate transport in arbuscular mycorrhizas trends. Plant Sci 10:22–29

Kaur H, Garg N (2017) Zinc-arbuscular mycorrhizal interactions: effect on nutrient pool, enzymatic antioxidants, and osmolyte synthesis in pigeonpea nodules subjected to Cd stress. Commun Soil Sci Plant Anal 48:1684–1700

Kersters K, De Vos P, Gillis M, Swings J, Vandamme P, Stackebrandt E (2006) Introduction to the proteobacteria. In: The prokaryotes. Springer, New York

Khalvati MA, Hu Y, Mozafar A, Schmidhalter U (2005) Quantification of water uptake by arbuscular mycorrhizal hyphae and its significance for leaf growth, water relations, and gas exchange of barley subjected to drought stress. Plant Biol 7:706–712

Khan MR, Khan I, Ibrar Z, Leon J, Naz AA (2017) Drought-responsive genes expressed predominantly in root tissues are enriched with homotypic cis-regulatory clusters in promoters of major cereal crops. Crop J 5(3):195–206

Koltai H, Kapulnik Y (2009) Effect of arbuscular mycorrhizal symbiosis on enhancement of tolerance to abiotic stresses. In: White JF, Torres MS (eds) Defensive mutualism in microbial symbiosis CRC. Taylor Francis, Boca Raton, FL, pp 217–234

Kosuta S, Hazledine S, Sun J, Miwa H, Morris RJ, Downie JA, Oldroyd GE (2008) Differential and chaotic calcium signatures in the symbiosis signaling pathway of legumes. Proc Natl Acad Sci U S A 105:9823–9828

Kramer U, Talke IN, Hanikenne M (2007) Transition metal transport. FEBS Lett 581:2263–2272

Kretzschmar T, Kohlen W, Sasse J, Borghi L, Schlegel M, Bachelier JB, Reinhardt D, Bours R, Bouwmeester HJ, Martinoia E (2012) A petunia ABC protein controls strigolactone-dependent symbiotic signalling and branching. Nature 483:341–344

Krusell L, Krause K, Ott T, Desbrosses G, Krämer U, Sato S, Nakamura Y, Tabata S, James EK, Snadal N, Stougaard J, Kawaguchi M, Miyamoto A, Suganuma N, Udvardi MK (2005) The sulfate transporter SST1 is crucial for symbiotic nitrogen fixation in *Lotus japonicus* root nodules. Plant Cell 17:1625–1636

Krylova VV, Andreev IM, Zartdinova R, Izmailov SF (2017) Ca2+-ATPase in the symbiosome membrane from broad bean root nodules: further evidence for its functioning as ATP-driven Ca2+/H+ exchanger. Acta Physiol Plant 39(11):1–8

Kryvoruchko IS, Sinharoy S, Torres-Jerez I, Sosso D, Pislariu CI, Guan D, Murray JD, Benedito VA, Frommer WB, Udvardi MK (2016) MtSWEET11, a nodule-specific sucrose transporter of *Medicago truncatula* root nodules. Plant Physiol 171(1):554–565. https://doi.org/10.1104/pp.15.01910

Lace B, Ott T (2018) Commonalities and differences in controlling multipartite intracellular infections of legume roots by symbiotic microbes. Plant Cell Physiol 59:661–672

Leigh J, Hodge A, Fitter AH (2009) Arbuscular mycorrhizal fungi can transfer substantial amounts of nitrogen to their host plant from organic material. New Phytol 181:199–207

Lelandais-Brière C, Moreau J, Hartmann C, Crespi M (2016) Noncoding RNAs, emerging regulators in root endosymbioses. Mol Plant-Microbe Interact 29:170–180

Li J, Shah S, Moffatt BA, Glick BR (2001) Isolation and characterization of an unusual 1-aminocyclopropane-1-carboxylic acid (ACC) deaminase gene from *Enterobacter cloacae* UW4. Anton Van Leeuw 80:255–261

Li T, Hu YJ, Hao ZP, Li H, Chen BD (2013) Aquaporin genes GintAQPF1 and GintAQPF2 from *Glomus intraradices* contribute to plant drought tolerance. Plant Signal Behav 8(5):e24030

Limpens E, Moling S, Hooiveld G, Pereira PA, Bisseling T, Becker JD, Küster H (2013) Cell-and tissue-specific transcriptome analyses of Medicago truncatula root nodules. PLoS One 8(5):1–15

Lindstrom K, Terefework Z, Suominen L, Lortet G (2002) Signalling and development of rhizobium: legume symbioses. Biol Environ Proc R Ir Acad 102:61–64

Liu W, Zhang Y, Jiang S, Deng Y, Christie P, Murray PJ, Li X, Zhang J (2016) Arbuscular mycorrhizal fungi in soil and roots respond differently to phosphorus inputs in an intensively managed calcareous agricultural soil. Sci Rep 6:24902

Liu H, Zhang C, Yang J, Yu N, Wang E (2018) Hormone modulation of legume-rhizobial symbiosis. J Integr Plant Biol 60(8):632–648. https://doi.org/10.1111/jipb.12653

Lopez-Pedrosa A, Gonzalez-Guerrero M, Valderas A, Azcon-Aguilar C, Ferrol N (2006) GintAMT1 encodes a functional high-affinity ammonium transporter that is expressed in the extraradical mycelium of *Glomus intraradices*. Fungal Genet Biol 43:102–110

Lopez-Raez JA, Charnikhova T, Gomez-Roldan V, Matusova R, Kohlen W, De Vos R, Verstappen F, Puech-Pages V, Becard G, Mulder P, Bouwmeester H (2008) Tomato strigolactones are derived from carotenoids and their biosynthesis is promoted by phosphate starvation. New Phytol 178:863–874

Luginbuehl LH, Menard GN, Kurup S, Van Erp H, Radhakrishnan GV, Breakspear A, Oldroyd GE, Eastmond PJ (2017) Fatty acids in arbuscular mycorrhizal fungi are synthesized by the host plant. Science 356:1175–1178

Lugtenberg BJ, Malfanova N, Kamilova F, Berg G (2013) Plant growth promotion by microbes. Mol Microbiol Ecol Rhizosphere 1–2:559–573

Ma W, Sebestianova SB, Sebestian J, Burd GI, Guinel FC, Glick BR (2003) Prevalence of 1-aminocyclopropane-1-carboxylate deaminase in *Rhizobium* spp. Anton Van Leeuw 83:285–291

Ma Y, Oliveira RS, Freitas H, Zhang C (2016) Biochemical and molecular mechanisms of plant-microbe-metal interactions: relevance for phytoremediation. Front Plant Sci 7:1–19. https://doi.org/10.3389/fpls.2016.00918

MacLean AM, Bravo A, Harrison MJ (2017) Plant signaling and metabolic pathways enabling arbuscular mycorrhizal symbiosis. Plant Cell 29(10):2319–2335

Madsen LH, Tirichine L, Jurkiewicz A, Sullivan JT, Heckmann AB, Bek AS, Ronson CW, James EK, Stougaard J (2010) The molecular network governing nodule organogenesis and infection in the model legume *Lotus japonicus*. Nat Commun 1:1–12

Maillet F, Poinsot V, André O, Puech-Pagès V, Haouy A, Gueunier M, Cromer L, Giraudet D, Formey D, Niebel A, Martinez EA, Driguez H, Becard G, Denarie J (2011) Fungal lipochitooligosaccharide symbiotic signals in arbuscular mycorrhiza. Nature 469:58–63

Maldonado-Mendoza IE, Dewbre GR, Harrison MJ (2001) A phosphate transporter gene from the extra-radical mycelium of an arbuscular mycorrhizal fungus *Glomus intraradices* is regulated in response to phosphate in the environment. Mol Plant-Microbe Interact 14:1140–1148

Malekzadeh E, Alikhani HA, Savaghebi-Firoozabadi GR, Zarei M (2011) Influence of arbuscular mycorrhizal fungi and an improving growth bacterium on Cd uptake and maize growth in Cd-polluted soils. Span J Agric Res 9:1213–1223

Manck-Gotzenberger J, Requena N (2016) Arbuscular mycorrhiza symbiosis induces a major transcriptional reprogramming of the potato SWEET sugar transporter family. Front Plant Sci 7:487

Mary P, Ochin D, Tailliez R (1986) Growth status of rhizobia in relation to their tolerance to low water activities and desiccation stress. Soil Biol Biochem 18:179–184

Masalkar P, Wallace IS, Hwang JH, Roberts DM (2010) Interaction of cytosolic glutamine synthetase of soybean root nodules with the C-terminal domain of the symbiosome membrane nodulin 26 aquaglyceroporin. J Biol Chem 285:23880–23888

Masson-Boivin C, Sachs JL (2018) Symbiotic nitrogen fixation by rhizobia—the roots of a success story. Curr Opin Plant Biol 44:7–15

Mayak S, Tirosh T, Glick BR (2004) Plant growth-promoting bacteria that confer resistance to water stress in tomatoes and peppers. Plant Sci 166:525–530

McAdam EL, Reid JB, Foo E (2018) Gibberellins promote nodule organogenesis but inhibit the infection stages of nodulation. J Exp Bot 69:2117–2130

Medina MJH, Gagnon H, Piche Y, Ocampo JA, Garrido JMG, Vierheilig H (2003) Root colonization by arbuscular mycorrhizal fungi is affected by the salicylic acid content of the plant. Plant Sci 164:993–998

Meena KK, Sorty AM, Bitla UM, Choudhary K, Gupta P, Pareek A, Singh DP, Prabha R, Sahu PK, Gupta VK, Singh HB, Krishanani KK, Minhas PS (2017) Abiotic stress responses and microbe-mediated mitigation in plants: the omics strategies. Front Plant Sci 8:1–25

Menendez E, Martínez-Hidalgo P, Silva LR, Velazquez E, Mateos PF, Peix A (2017) Recent advances in the active biomolecules involved in rhizobia-legume symbiosis. In: Microbes for legume improvement. Springer, Cham, pp 45–74

Meng L, Zhang A, Wang F, Han X, Wang D, Li S (2015) Arbuscular mycorrhizal fungi and rhizobium facilitate nitrogen uptake and transfer in soybean/maize intercropping system. Front Plant Sci 6:1–10

Michiels J, Verreth C, Vanderleyden J (1994) Effects of temperature stress on bean-nodulating *Rhizobium* strains. Appl Environ Microbiol 60:1206–1212

Miller G, Suzuki N, Ciftci-Yilmaz S, Mittler R (2010) Reactive oxygen species homeostasis and signalling during drought and salinity stresses. Plant Cell Environ 33:453–467

Mohanty SK, Arthikala MK, Nanjareddy K, Lara M (2018) Plant-symbiont interactions: the functional role of expansins. Symbiosis 74:1–10

Moreau S, Day DA, Puppo A (1998) Ferrous iron is transported across the peribacteroid membrane of soybean nodules. Planta 207:83–87

Moreau S, Thomson RM, Kaiser BN, Trevaskis B, Guerinot ML, Udvardi MK, Puppo A, Day DA (2002) GmZIP1 encodes a symbiosis-specific zinc transporter in soybean. J Biol Chem 277:4738–4746

Munns R (2002) Comparative physiology of salt and water stress. Plant Cell Environ 25:239–250

Murset V, Hennecke H, Gabriella P (2012) Disparate role of rhizobial ACC deaminase in root-nodule symbioses. Symbiosis 57:43–50

Nanjareddy K, Arthikala MK, Gómez BM, Blanco L, Lara M (2017) Differentially expressed genes in mycorrhized and nodulated roots of common bean are associated with defense, cell wall architecture, N metabolism, and P metabolism. PLoS One 12(8):e0182328

Nasim G (2013) Host allelopathy and arbuscular mycorrhizal fungi. In: Cheema ZA, Farooq M, Wahid A (eds) Allelopathy. Springer, Berlin, pp 429–450

Naveed M, Mehboob I, Hussain MB, Zahir ZA (2015) Perspectives of rhizobial inoculation for sustainable crop production. In: Arora NK (ed) Plant microbes symbiosis: applied facets. Springer, New Delhi, pp 209–239

Nevo Y, Nelson N (2006) The NRAMP family of metal-ion transporters. Biochim Biophys Acta, Mol Cell Res 1763:609–620

Nukui N, Minamisawa K, Ayabe SI, Aoki T (2006) Expression of the 1-aminocyclopropane-1-carboxylic acid deaminase gene requires symbiotic nitrogen-fixing regulator gene nifA2 in *Mesorhizobium loti* MAFF303099. Appl Environ Microbiol 72:4964–4969

Ocon A, Hampp R, Requena N (2007) Trehalose turnover during abiotic stress in arbuscular mycorrhizal fungi. New Phytol 174:879–891

Oldroyd GE (2013) Speak, friend, and enter: signalling systems that promote beneficial symbiotic associations in plants. Nat Rev Microbiol 11:252–263

Oldroyd GE, Engstrom EM, Long SR (2001) Ethylene inhibits the Nod factor signal transduction pathway of *Medicago truncatula*. Plant Cell 13:1835–1849

Oldroyd GE, Murray JD, Poole PS, Downie JA (2011) The rules of engagement in the legume-rhizobial symbiosis. Annu Rev Genet 45:119–144

Ouziad F, Hildebrandt U, Schmelzer E, Bothe H (2005) Differential gene expressions in arbuscular mycorrhizal-colonized tomato grown under heavy metal stress. J Plant Physiol 162:634–649

Pajuelo E, Rodriguez-Llorente ID, Lafuente A, Caviedes MA (2011) Legume–rhizobium symbioses as a tool for bioremediation of heavy metal polluted soils. In: Biomanagement of metal-contaminated soils. Springer, Dordrecht, pp 95–123

Parniske M (2008) Arbuscular mycorrhiza: the mother of plant root endosymbiosis. Nat Rev Microbiol 6:763–775

Paszkowski U (2006) A journey through signaling in arbuscular mycorrhizal symbioses 2006. New Phytol 172:35–46

Paul MJ, Primavesi LF, Jhurreea D, Zhang Y (2008) Trehalose metabolism and signaling. Annu Rev Plant Biol 59:417–441

Pelissier HC, Frerich A, Desimone M, Schumacher K, Tegeder M (2004) PvUPS1, an allantoin transporter in nodulated roots of French bean. Plant Physiol 134:664–675

Pereira SIA, Lima AIG, Figueirs EMAP (2008) *Rhizobium leguminosarum* isolated from agricultural ecosystems subjected to different climatic influences: the relation between genetic diversity, salt tolerance and nodulation efficiency. In: Liu T-X (ed) Soil ecology research developments. Nova Science, New York, pp 247–263

Perez-Tienda J, Testillano PS, Balestrini R, Fiorilli V, Azcon Aguilar C, Ferrol N (2011) GintAMT2, a new member of the ammonium transporter family in the arbuscular mycorrhizal fungus *Glomus intraradices*. Fungal Genet Biol 48:1044–1055

Peterson RL, Massicotte HB (2004) Exploring structural definitions of mycorrhizas, with emphasis on nutrient-exchange interfaces. Can J Bot 82:1074–1088

Pfau T (2013) Modelling metabolic interactions in the legume-rhizobia symbiosis. Doctoral dissertation, University of Aberdeen, Aberdeen, United Kingdom

Pfeffer PE, Douds DD, Bécard G, Shachar-Hill Y (1999) Carbon uptake and the metabolism and transport of lipids in an arbuscular mycorrhiza. Plant Physiol 120:587–598

Phillips DA, Joseph CM, Maxwell CA (1992) Trigonelline and stachydrine released from alfalfa seeds activate NodD2 protein in *Rhizobium meliloti*. Plant Physiol 99:1526–1531

Pierre O, Engler G, Hopkins J, Brau F, Boncompagni E, Herouart D (2013) Peribacteroid space acidification: a marker of mature bacteroid functioning in *Medicago truncatula* nodules. Plant Cell Environ 36:2059–2070

Pilon M (2011) Moving copper in plants. New Phytol 192(2):305–307

Plett JM, Martin F (2015) Reconsidering mutualistic plant–fungal interactions through the lens of effector biology. Curr Opin Plant Biol 26:45–50

Poole P, Ramachandran V, Terpolilli J (2018) Rhizobia: from saprophytes to endosymbionts. Nat Rev Microbiol 16(5):291–303

Porcel R, Ruiz-Lozano JM (2004) Arbuscular mycorrhizal influence on leaf water potential, solute accumulation, and oxidative stress in soybean plants subjected to drought stress. J Exp Bot 55:1743–1750

Porcel R, Aroca R, Ruiz-Lozano JM (2012) Salinity stress alleviation using arbuscular mycorrhizal fungi. A review. Agron Sustain Dev 32:181–200

Prasad R, Bhola D, Akdi K, Cruz C, Sairam KVSS, Tuteja N, Varma A (2017) Introduction to mycorrhiza: historical development. In: Varma A, Prasad R, Tuteja N (eds) Mycorrhiza. Springer International Publishing AG, Cham, pp 1–7

Pumplin N, Harrison MJ (2009) Live-cell imaging reveals periarbuscular membrane domains and organelle location in *Medicago truncatula* roots during arbuscular mycorrhizal symbiosis. Plant Physiol 15:809–819

Rajtor M, Piotrowska-Seget Z (2016) Prospects for arbuscular mycorrhizal fungi (AMF) to assist in phytoremediation of soil hydrocarbon contaminants. Chemosphere 162:105–116

Ramos AC, Martins MA, Okorokova-Facanha AL, Olivares FL, Okorokov LA, Sepulveda N, Feijo JA, Facanha AR (2009) Arbuscular mycorrhizal fungi induce differential activation of the plasma membrane and vacuolar H+ pumps in maize roots. Mycorrhiza 19:69–80

Rausch C, Daram P, Brunner S, Jansa J, Laloi M, Leggewie G, Amrhein N, Bucher M (2001) A phosphate transporter expressed in arbuscule-containing cells in potato. Nature 414:462–466

Reichman SM (2007) The potential use of legume-rhizobium symbiosis for the remediation of arsenic contaminated sites. Soil Biol Biochem 39:2587–2593

Ren G (2018) The evolution of determinate and indeterminate nodules within the Papilionoideae subfamily. Doctoral dissertation, Wageningen University

Rillig MC, Aguilar-Trigueros CA, Bergmann J, Verbruggen E, Veresoglou SD, Lehmann A (2015) Plant root and mycorrhizal fungal traits for understanding soil aggregation. New Phytol 205:1385–1388

Rinaudo M (2004) Role of substituents on the properties of some polysaccharides. Biomacromolecules 5:1155–1165

Rodriguez-Haas B, Finney L, Vogt S, Gonzalez-Melendi P, Imperial J, González-Guerrero M (2013) Iron distribution through the developmental stages of *Medicago truncatula* nodules. Metallomics 5:1247–1253

Rook MB, Bisseling T, Emons AMC (1998) Lipochito-oligosaccharides re-initiate root hair tip growth in *Vicia sativa* with high calcium and spectrin-like antigen at the tip. Plant J 13:341–350

Roth R, Paszkowski U (2017) Plant carbon nourishment of arbuscular mycorrhizal fungi. Curr Opin Plant Biol 39:50–56

Ruiz-Lozano JM, Collados C, Barea JM, Azcon R (2001) Arbuscular mycorrhizal symbiosis can alleviate drought-induced nodule senescence in soybean plants. New Phytol 151:493–502

Ruiz-Lozano JM, Porcel R, Azcon C, Aroca R (2012) Regulation by arbuscular mycorrhizae of the integrated physiological response to salinity in plants: new challenges in physiological and molecular studies. J Exp Bot 63:4033–4044

Ruiz-Lozano JM, Aroca R, Zamarreno AM, Molina S, Andreo-Jimenez B, Porcel R, Garcia-Mina JM, Ruyter-Spira C, Lopez-Raez JA (2016) Arbuscular mycorrhizal symbiosis induces strigolactone biosynthesis under drought and improves drought tolerance in lettuce and tomato. Plant Cell Environ 39:441–452

Ruiz-Sanchez M, Aroca R, Munoz Y, Polon R, Ruiz-Lozano JM (2010) The arbuscular mycorrhizal symbiosis enhances the photosynthetic efficiency and the antioxidative response of rice plants subjected to drought stress. J Plant Physiol 167:862–869

Ruth B, Khalvati M, Schmidhalter U (2011) Quantification of mycorrhizal water uptake via high-resolution on-line water content sensors. Plant Soil 342:459–468

Santi C, Bogusz D, Franche C (2013) Biological nitrogen fixation in non-legume plants. Ann Bot 111:743–767

Saxena M, Saxena J, Nema R, Singh D, Gupta A (2013) Phytochemistry of medicinal plants. J Pharmacogn Phytochem 1(6):168–182

Schaarschmidt S, Gonzalez MC, Roitsch T, Strack D, Sonnewald U, Hause B (2007) Regulation of arbuscular mycorrhization by carbon: the symbiotic interaction cannot be improved by increased carbon availability accomplished by root specifically enhanced invertase activity. Plant Physiol 143:1827–1840

Schubert M, Koteyeva NK, Wabnitz PW, Santos P, Büttner M, Sauer N, Demchenko K, Pawlowski K (2011) Plasmodesmata distribution and sugar partitioning in nitrogen-fixing root nodules of *Datisca glomerata*. Planta 233:139–152

Schussler A, Martin H, Cohen D, Fitz M, Wipf D (2006) Characterization of a carbohydrate transporter from symbiotic glomeromycotan fungi. Nature 444:933–936

Senovilla M, Castro-Rodríguez R, Abreu I, Escudero V, Kryvoruchko I, Udvardi MK, Imperial J, González-Guerrero M (2018) *Medicago truncatula* copper transporter 1 (MtCOPT1) delivers copper for symbiotic nitrogen fixation. New Phytol 218:696–709

Shaharoona B, Arshad M, Zahir ZA (2006) Effect of plant growth promoting rhizobacteria containing ACC-deaminase on maize (*Zea mays* L.) growth under axenic conditions and on nodulation in mung bean (*Vigna radiata* L.). Lett Appl Microbiol 42:155–159

Sheng M, Tang M, Zhang F, Huang Y (2011) Influence of arbuscular mycorrhiza on organic solutes in maize leaves under salt stress. Mycorrhiza 21:423–430

Sidahmed NEM (2016) The effect of compost, rhizobium inoculation and urea on peanut plant growth. Doctoral dissertation, Sudan University of Science and Technology

Sieberer B, Emons AMC (2000) Cytoarchitecture and pattern of cytoplasmic streaming in root hairs of *Medicago truncatula* during development and deformation by nodulation factors. Protoplasma 214:118–127

Sieberer BJ, Chabaud M, Fournier J, Timmers AC, Barker DG (2012) A switch in Ca2+ spiking signature is concomitant with endosymbiotic microbe entry into cortical root cells of *Medicago truncatula*. Plant J 69:822–830

Silver S, Phung LT (2005) A bacterial view of the periodic table: genes and proteins for toxic inorganic ions. J Ind Microbiol Biotechnol 32:587–605

Simard SW, Beiler KJ, Bingham MA, Deslippe JR, Philip LJ, Teste FP (2012) Mycorrhizal networks: mechanisms, ecology and modelling. Fungal Biol Rev 26:39–60

Smith SE, Read DJ (2008) Mycorrhizal symbiosis, 3rd edn. Academic, New York, p 800

Soka G, Ritchie M (2014) Arbuscular mycorrhizal symbiosis and ecosystem processes: prospects for future research in tropical soils. Open J Ecol 4:11–22

Song YY, Ye M, Li CY, Wang RL, Wei XC, Luo SM, Zeng RS (2013) Priming of anti-herbivore defense in tomato by arbuscular mycorrhizal fungus and involvement of the jasmonate pathway. J Chem Ecol 39:1036–1044

Staudinger C, Mehmeti-Tershani V, Gil-Quintana E, Gonzalez EM, Hofhansl F, Bachmann G, Wienkoop S (2016) Evidence for a rhizobia-induced drought stress response strategy in *Medicago truncatula*. J Proteome 136:202–213

Suarez R, Wong A, Ramirez M, Barraza A, Orozco MDC, Cevallos MA, Lara M, Hernandez G, Iturriaga G (2008) Improvement of drought tolerance and grain yield in common bean by overexpressing trehalose-6-phosphate synthase in rhizobia. Mol Plant Microbe Interact 21:958–966

Tamayo E, Gomez-Gallego T, Azcon-Aguilar C, Ferrol N (2014) Genome-wide analysis of copper, iron and zinc transporters in the arbuscular mycorrhizal fungus *Rhizophagus irregularis*. Front Plant Sci 5:1–13

Tegeder M, Masclaux-Daubresse C (2018) Source and sink mechanisms of nitrogen transport and use. New Phytol 217:35–53

Tejada-Jimenez M, Gil-Diez P, Leon-Mediavilla J, Wen J, Mysore K, Imperial J, Gonzalez-Guerrero M (2017) *Medicago truncatula* MOT1. 3 is a plasma membrane molybdenum transporter required for nitrogenase activity in root nodules. BioRxiv. https://doi.org/10.1101/102517

Terpolilli JJ, Masakapalli SK, Karunakaran R, Webb I, Green R, Watmough NJ, Kruger NJ, Ratcliffe RG, Poole PS (2016) Lipogenesis and redox balance in nitrogen-fixing pea bacteroids. J Bacteriol 198:2864–2875

Trotta A, Falaschi P, Cornara L, Minganti V, Fusconi A, Drava G, Berta G (2006) Arbuscular mycorrhizae increase the arsenic translocation factor in the As hyperaccumulating fern *Pteris vittata* L. Chemosphere 65:74–81

Tsyganova AV, Kitaeva AB, Tsyganov VE (2018) Cell differentiation in nitrogen-fixing nodules hosting symbiosomes. Funct Plant Biol 45(2):47–57

Udvardi MK, Day DA (1997) Metabolite transport across symbiotic membranes of legume nodules. Annu Rev Plant Biol 48:493–523

Udvardi M, Poole PS (2013) Transport and metabolism in legume-rhizobia symbioses. Annu Rev Plant Biol 64:781–805

Udvardi MK, Price GD, Gresshoff PM, Day DA (1988) A dicarboxylate transporter on the peribacteroid membrane of soybean nodules. FEBS Lett 231:36–40

Vance CP, Heichel GH (1991) Carbon in N2 fixation: limitation or exquisite adaptation. Annu Rev Plant Biol 42:373–390

Varma A, Prasad R, Tuteja N (2017a) Mycorrhiza: eco-physiology, secondary metabolites, nanomaterials. Springer International Publishing, Cham. http://www.springer.com/us/book/9783319578484. ISBN: 978-3-319-57849-1

Varma A, Prasad R, Tuteja N (2017b) Mycorrhiza: function, diversity and state-of-art. Springer International Publishing, Berlin. http://www.springer.com/us/book/9783319530635. ISBN: 978-3-319-53064-2

Varma A, Prasad R, Tuteja N (2017c) Mycorrhiza: nutrient uptake, biocontrol, ecorestoration. Springer International Publishing, Cham. http://www.springer.com/us/book/9783319688664. ISBN: 978-3-319-68867-1

Venturi V, Keel C (2016) Signaling in the rhizosphere. Trends Plant Sci 21(3):187–198

Vriezen Jan AC, Bruijn FJDB, Nusslein K (2007) Responses of rhizobia to desiccation in relation to osmotic stress, oxygen, and temperature. Appl Environ Microbiol 73(11):3451–3459

Wais RJ, Keating DH, Long SR (2002) Structure-function analysis of nod factor-induced root hair calcium spiking in Rhizobium-legume symbiosis. Plant Physiol 129:211–224

Walter MH, Strack D (2011) Carotenoids and their cleavage products: biosynthesis and functions. Nat Prod Rep 28:663–692

Wang E, Schornack S, Marsh JF, Gobbato E, Schwessinger B, Eastmond P, Schultz M, Kamoun S, Oldroyd GE (2012) A common signaling process that promotes mycorrhizal and oomycete colonization of plants. Curr Biol 22:2242–2246

Wang E, Yu N, Bano SA, Liu C, Miller AJ, Cousins D, Zhang X, Ratet P, Tadege M, Mysore KS, Downie JA (2014) A H^+-ATPase that energizes nutrient uptake during mycorrhizal symbioses in rice and *Medicago truncatula*. Plant Cell 26:1818–1830

Wang L, Wang L, Zhou Y, Duanmu D (2017a) Use of CRISPR/Cas9 for symbiotic nitrogen fixation research in legumes. In: Progress in molecular biology and translational science, vol 149. Academic, San Diego, pp 187–213

Wang W, Shi J, Xie Q, Jiang Y, Yu N, Wang E (2017b) Nutrient exchange and regulation in arbuscular mycorrhizal symbiosis. Mol Plant 10(9):1147–1158. https://doi.org/10.1016/j.molp.2017.07.012

Waters MT, Gutjahr C, Bennett T, Nelson DC (2017) Strigolactone signaling and evolution. Annu Rev Plant Biol 68:291–322

Webb BA, Karl Compton K, Castañeda Saldaña R, Arapov TD, Keith Ray W, Helm RF, Scharf BE (2017) Sinorhizobium meliloti chemotaxis to quaternary ammonium compounds is mediated by the chemoreceptor McpX. Mol Microbiol 103(2):333–346

Weaver CD, Crombie B, Stacey G, Roberts DM (1991) Calcium-dependent phosphorylation of symbiosome membrane proteins from nitrogen-fixing soybean nodules: evidence for phosphorylation of nodulin-26. Plant Physiol 95:222–227

Wu QS, Levy Y, Zou YN (2009) Arbuscular mycorrhizae and water relations in citrus. In: Tennant P, Benkeblia N (eds) Tree and forestry and science biotechnology, vol 3, pp 105–112

Wu QS, He XH, Zou YN, He KP, Sun YH, Cao MQ (2012) Spatial distribution of glomalin-related soil protein and its relationships with root mycorrhization, soil aggregates, carbohydrates, activity of protease and β-glucosidase in the rhizosphere of *Citrus unshiu*. Soil Biol Biochem 45:181–183

Wu QS, Srivastava AK, Zou YN (2013) AMF-induced tolerance to drought stress in citrus: a review. Sci Hortic 164:77–87

Xie X, Lin H, Peng X, Xu C, Sun Z, Jiang K, Huang A, Wu X, Tang N, Salvioli A, Bonfante P (2016) Arbuscular mycorrhizal symbiosis requires a phosphate transceptor in the *Gigaspora margarita* fungal symbiont. Mol Plant 9:1583–1608

Xiong H, Kobayashi T, Kakei Y, Senoura T, Nakazono M, Takahashi H, Nakanishi H, Shen H, Duan P, Nishizawa NK, Zuo Y (2012) AhNRAMP1 iron transporter is involved in iron acquisition in peanut. J Exp Bot 63:4437–4446

Yamaya-Ito H, Shimoda Y, Hakoyama T, Sato S, Kaneko T, Hossain MS, Shibata S, Kawaguchi M, Hayashi M, Kouchi H, Umehara Y (2018) Loss-of-function of ASPARTIC PEPTIDASE NODULE-INDUCED 1 (APN1) in Lotus japonicus restricts efficient nitrogen-fixing symbiosis with specific *Mesorhizobium loti* strains. Plant J 93:5–16

Yamazaki A, Hayashi M (2015) Building the interaction interfaces: host responses upon infection with microorganisms. Curr Opin Plant Biol 23:132–139

Yang J, Kloepper JW, Ryu CM (2009) Rhizosphere bacteria help plants tolerate abiotic stress. Trends Plant Sci 14:1–4

Yang SY, Grønlund M, Jakobsen I, Grotemeyer MS, Rentsch D, Miyao A, Hirochika H, Kumar CS, Sundaresan V, Salamin N, Catausan S (2012) Non-redundant regulation of rice arbuscular mycorrhizal symbiosis by two members of the PHOSPHATE TRANSPORTER1 gene family. Plant Cell 24:4236–4251

Yang Y, He C, Huang L, Ban Y, Tang M (2017) The effects of arbuscular mycorrhizal fungi on glomalin-related soil protein distribution, aggregate stability and their relationships with soil properties at different soil depths in lead-zinc contaminated area. PLoS One 12(8):1–19

Zahran HH (1999) Rhizobium-legume symbiosis and nitrogen fixation under severe conditions and in an arid climate. Microbiol Mol Biol Rev 63:968–989

Zandalinas SI, Mittler R, Balfagon D, Arbona V, Gomez-Cadenas A (2018) Plant adaptations to the combination of drought and high temperatures. Physiol Plant 162:2–12

Zheng L, Yamaji N, Yokosho K, Ma JF (2012) YSL16 is a phloem-localized transporter of the copper-nicotianamine complex that is responsible for copper distribution in rice. Plant Cell 24:3767–3782

Zipfel C, Oldroyd GE (2017) Plant signalling in symbiosis and immunity. Nature 543(7645):328

Zsogon A, Lambais MR, Benedito VA, Figueira AVDO, Peres LEP (2008) Reduced arbuscular mycorrhizal colonization in tomato ethylene mutants. Sci Agric 65:259–267

Chapter 3
Truffles and Morels: Two Different Evolutionary Strategies of Fungal-Plant Interactions in the *Pezizales*

Francesca Ori, Ian Hall, Carmelo Gianchino, Mirco Iotti, and Alessandra Zambonelli

Abstract Pezizales are a widespread group of fungi, basal to the other filamentous ascomycetes. Most species live in soil as saprobes, in a mycorrhizal relationship with a wide range of plants, or as plant parasites. The lineage Morchellaceae–Discinaceae–Helvellaceae–Tuberaceae includes most of the commercially valuable species in the order. The truffles in the genus Tuber and morels in the genus Morchella arguably command more interest in culinary circles than any other groups of mushrooms. In recent years, the interactions of these fungi with plants have been thoroughly researched although many aspects still need to be clarified. In this chapter, we describe and compare these two groups of mushrooms and take a look at the evidence as to whether there are real trophic differences from those traditionally held and if things are not quite as simple as our forebears would have had us believe. We explore the range of host plants involved in the interactions, the morpho-anatomy of symbiotic structures, the molecular mechanisms of symbiosis, and the influence of other microbial species.

3.1 Introduction

Pezizales are a widespread group of fungi, basal to the other filamentous ascomycetes, and include 1683 species in 16 families (Kirk et al. 2008). Most of the species live in soil as saprobes, in a mycorrhizal relation with a wide range of plants or as plant parasites, although the biotic interactions of many taxa are still to be clarified. Epigeous apothecia are the most common ascomatal form within the *Pezizales*, but several closed

F. Ori · C. Gianchino · M. Iotti (✉)
Department of Life, Health and Environmental Sciences, University of L'Aquila, L'Aquila, Italy
e-mail: mirco.iotti@univaq.it

I. Hall
Truffles and Mushrooms (Consulting) Limited, Dunedin, New Zealand

A. Zambonelli
Department of Agricultural and Food Sciences, DISTAL, University of Bologna, Bologna, Italy

A. Varma et al. (eds.), *Plant Microbe Interface*,
https://doi.org/10.1007/978-3-030-19831-2_3

structures are frequent in some families (Pfister 2015). Among the latter, hypogeous closed ascomata (truffles) are the most frequent and have evolved independently at least 15 times in 6 families from operculate cupuliform lineages (Læssøe and Hansen 2007). Truffle species primarily have an ectomycorrhizal lifestyle which has been considered a precondition for the switch to hypogeous fruiting (Tedersoo et al. 2006).

Any comparison of groups in the *Pezizales* must begin with a brief but concise overview of their taxonomies and nomenclature. Three major lineages are identified within *Pezizales* by Landvik et al. (1997) and Hansen and Pfister (2006): (1) *Pezizaceae-Ascobolaceae*, (2) *Morchellaceae-Discinaceae-Helvellaceae-Tuberaceae* and (3) *Pyronemataceae-Ascodesmidaceae-Glaziellaceae-Sarcosomataceae-Sarcoscyphaceae*. The second lineage includes many of the commercially valuable species in the order. In particular, the truffles in the genus *Tuber* and morels in the genus *Morchella* arguably command more interest in culinary circles than any other groups of mushrooms with prices sometimes reaching eye-watering heights.

All *Tuber* species establish symbiotic relationships with a broad range of host plants and produce subglobose hypogeous ascomata, surrounded by a smooth or ornamented peridium, with a solid gleba usually off-white and pale brown to blackish at maturity and with paler-coloured veins. The ascospores are either ellipsoidal to globose and are generally ornamented with spines or a net. In just one species, *Tuber melosporum*, the ascospores are smooth (Lancellotti et al. 2016). The variable number of ascospores per ascus is typical in the *Pezizales*, and it varies both within and between *Tuber* species. Although the term truffle is applied to all mycorrhizal fungi that produce hypogeous ascomata (Trappe and Claridge 2010), the term "true truffles" is applied only to species belonging to *Tuber* (Bonito and Smith 2016).

The genus *Tuber* diversified in the early Cretaceous (142 Mya), less than 10 Mya after the origin of *Tuberaceae* has been dated (Bonito et al. 2013). It is composed of 11 major clades (Aestivum, Excavatum, Gennadii, Gibbosum, Japonicum, Macrosporum, Maculatum, Melanosporum, Multimaculatum, Puberulum and Rufum), with an estimated number of species between 180 and 220. The main concern on truffle taxonomy is that some species are known only from "environmental" DNA sequences (Bonito and Smith 2016), while cryptic species are common in some lineages (Bonuso et al. 2010; Benucci et al. 2016; Healy et al. 2016). For the European species of *Tuber*, a taxonomy based on morphological features primarily based on those of the ascospores and peridium has been almost adequate to distinguish between species. However, the taxonomy of Chinese species is now rooted firmly in molecular techniques, for example, recent species described by Prof. Li Fan and other Chinese authors (Fan et al. 2017a, b).

The trophic habit in *Morchella* is heterogeneous. A saprobic lifestyle is the most common trophic habit, but a number of interactions with plants are known (Pilz et al. 2007). All members of *Morchella* fruit aboveground and have hollow ascomata with a conical to ovate honeycomb-like cap and elliptical smooth spores contained in 8-spored asci. The terms "true" and "false" are also used for morels to distinguish the members of the genus *Morchella* from those of *Gyromitra*, *Verpa* and *Helvella* (https://www.mushroom-appreciation.com/false-morel.html). The genus *Morchella* has recently been given a thorough reassessment by Du et al. (2012), Kuo et al. (2012) and Richard et al. (2015). In these, modern molecular techniques were used to largely replace the inadequate morphological methods widely employed since *Morchella esculenta* was typified in 1794. The above phylogenetic studies identified 65 species distributed in 2 sister clades Esculenta (the "yellow" morels) and Elata (the "black" morels) that diverged in the early Cretaceous (130 Mya) and 2 other species, *M. rufobrunnea* and *M. anatolica*, forming a basal lineage estimated to have evolved in the late Jurassic (O'Donnell et al. 2011).

Species in *Tuber* and *Morchella* are characterized by high levels of continental endemism and provincialism and are largely restricted to temperate regions of the Northern Hemisphere where fruiting season is limited to spring for morels and throughout the year for true truffles (Richard et al. 2015; Bonito and Smith 2016). As with other *Ascomycota*, morels and truffles have a haploid life cycle, and heterothallism seems to be the unique reproductive mode within the genera (Du et al. 2016; Martin et al. 2010). Ascomata derive from outcrossing of primary homokaryotic mycelia harbouring different mating types where both can act as the maternal partner. However, high inbreeding coefficients have been consistently found in natural populations of both fungal groups (Taschen et al. 2016; Du et al. 2016; Paolocci et al. 2006). The dispersal of *Morchella* spores is by the forcible discharged into the air from asci, which is the classical dispersal mechanism in the *Pezizales*. In contrast, *Tuber* ascospores, which are formed in indehiscent ascomata, are primarily dispersed by wild animals attracted by aromas (Beever and Lebel 2014). Then after ingestion the ornamentation on the outsides of the spores arguably extends the time the spores spend in the gut by becoming tangled between the villi and hence the distance the spores are likely to travel (Zambonelli et al. 2017). Anamorphic states have been reported for both morels (Carris et al. 2015; Masaphy 2010; Alvarado-Castillo et al. 2014) and truffles (Urban et al. 2004; Healy et al. 2013), but the role of conidia in their life cycles is still to be clarified. Some authors suppose that these conidia can act as spermatia allowing the fertilization of a strain of different mating types (Le Tacon et al. 2016).

In this review we describe and compare the two groups and take a look at the evidence as to whether there are real trophic differences from those traditionally held and if things are not quite as simple as our forebears would have had us believe.

3.2 True Truffles

True truffles live in a broad range of forest biomes from the Baltic rim to semiarid areas of North Africa or forested areas of Southern Asia (Lancellotti et al. 2016; Fan et al. 2017a, b). A number of species of *Tuber* have been accidentally introduced into some Southern Hemisphere countries probably on the roots of Northern Hemisphere amenity or plantation forest trees. Others, in particular *Tuber aestivum*, *Tuber borchii* and *Tuber melanosporum*, have been deliberately introduced into almost all countries with temperate climatic zones where their cultivation is now an established industry (Hall et al. 2017).

Although the morels have received considerable attention in the modern scientific literature starting with Delmas (1978), this pales into insignificance when compared to that for *Tuber* spp. To summarize this here would be superfluous, so we direct the reader to some of the books and reviews that have appeared over the past decade (Riousset et al. 2001; Granetti et al. 2005; Reyna 2007; Hall et al. 2007; Olivier et al. 2012). Instead we have distilled from the literature what is particularly pertinent to our comparison.

3.2.1 Interactions with Plants

3.2.1.1 Symbiotic Interactions

Since ancient times, truffles have been the subject of debate over their modes of life. In the 1500s, Andrea Cesalpino (1583) unequivocally allocated truffles to the fungi. Nevertheless, their biology was still far from clear. After observing the connection between various trees and fruiting body production, Frank (1885) was able to identify a symbiotic relationship that he called "mycorrhizal symbiosis". What Frank actually discovered was the ectomycorrhiza, where the fungal hyphae entirely wrap the root tip, induce the loss of root hairs and the formation of an enveloping mantle. He also observed the fungus penetrating between the cortical cells and the formation of the Hartig net, an apoplastic structure where the exchange of nutrients takes place. This structure tends to be superficial in angiosperms but much deeper in gymnosperms, often reaching the endodermis (Smith and Read 2008). Some species of *Tuber* can be distinguished simply in the form of cystidia that are produced. Once the ectomycorrhiza is well developed, the fungal mantle with its emanating hyphae assumes the entire role of water and mineral uptake on which the host plant depends, whereas the roots maintain only the role of nutrient-transport and mechanical support.

The genus *Tuber* has been assumed to be a collection of ectomycorrhizal fungi but recently some species of *Tuber* spp. have been found in association with what were assumed to be non-ectomycorrhizal hosts. *Tuber aestivum*, *Tuber excavatum* and *Tuber maculatum* were found to form orchid mycorrhizas with *Epipactis microphylla*, *Epipactis helleborine* and *Cephalanthera damasonium* (Selosse et al. 2004; Ouanphanivanh et al. 2008). This type of mycorrhiza is characterized by septate fungal hyphae that penetrate cell walls, forming coils in the majority of the cortical cells (Selosse et al. 2004). In addition, *Tuber* spp. was also found to form arbutoid mycorrhizas (Smith and Read 2008). This type of mycorrhiza resembles ectomycorrhizas formed by angiosperms except for the intracellular colonization of the epidermal cells. Using molecular methods, Kennedy et al. (2012) identified the fungal partner of mycorrhizas on *Arbutus menziesii* Pursh as two species of *Tuber*. Later, mycorrhizas of *Tuber borchii* on *Arbutus unedo* were described, from natural truffle grounds, and were later synthetized in the greenhouse (Lancellotti et al. 2014). Arbutoid mycorrhizas formed by *T. borchii* macroscopically appear as typical ectomycorrhizas except for cruciform ramifications. Recently, arbutoid mycorrhizas were also synthetized in pots with *T. aestivum* and *T. melanosporum* (Fig. 3.1 a–b) (Mirco Iotti, unpublished data). As far as we are aware, the association of *Tuber* species with other members of the Ericaceae (ericoid mycorrhizas) have yet to be described. This shows that *Tuber* species can produce at least three types of symbioses that differ both morphologically and functionally (Gryndler 2016).

Fig. 3.1 Ectomycorrhiza (**a**) and arbutoid mycorrhiza (**b**) of *Tuber melanosporum*. Pseudosclerotia of *Morchella* sp. in pure culture viewed under a dissecting (**c**) and optical (**d**) microscope. Ascoma of *Morchella dunalii* in a littoral pine wood (**e**)

3.2.1.2 Interaction with Grasses and Herbaceous Species

The relationships between truffle species and herbaceous plants are rather controversial. *Tuber* species do not form vesicles or arbuscules within root cells like mycorrhizal fungi of *Glomeromycota*, but interactions with the roots of grasses are known. Some deciduous woody plants present areas devoid of herbaceous vegetation around their trunk when a symbiosis is established with some *Tuber* species, such as *Tuber melanosporum*, *Tuber aestivum* and *Tuber indicum* (Streiblová et al. 2012) and *Tuber borchii* (Hall et al. 2007). This circular zone around the host plants, the brûlé, is generally considered as an indication of the presence of truffle although other fungi in the Pezizales such as *Trichophaea* spp. can also produce them (Hall et al. 2007). The nature of the brûlé was hypothesized to be caused by several phenomenon. Firstly, it was explained as the result of competition of *Tuber* mycelium for nutrients and water (Delmas 1983) or as a parasitic symbiosis. This relationship can manifest as root necrosis (Plattner and Hall 1995; Chevalier 2010) or scar formation on host tree's roots (Sourzat 2004). Several authors linked the presence of the brûlé to the production of phytotoxic metabolites by the *Tuber* species (Fasolo-Bonfante et al. 1971; Pacioni 1991; Lanza et al. 2004). *Tuber aestivum* was detected on the root surface of decomposing cell layers of non-ectomycorrhizal hosts (Gryndler et al. 2014). The fungus did not penetrate deep into the root and no visible change to root morphology was evident. However, truffle mycelium in the nonhost roots was one to two orders of magnitude higher than that in the surrounding soil. Recently, a study conducted by Schneider-Maunoury et al. (2018) revealed the presence of *T. melanosporum* DNA on roots of herbaceous plants (e.g., Brassicaceae) growing within the brûlé. This finding was attributed to a colonization of roots by *T. melanosporum* in a manner similar to the endophytic fungi that colonize leaves and stems (Hardoim et al. 2015) without causing morphological modifications. It's not clear what biological benefits that might accrue from these interactions with grasses, but an exchange of carbon cannot be excluded. If the relationship was pathogenic, then presumably it could confer some advantage to the mycorrhizal host plant perhaps from the carbon transfer from the grass to the mycorrhizal host or by supressing root competition. Indeed, truffle species cannot fruit in association with grasses presumably because the carbon balance is not appropriate to food the developing primordia.

A special mention needs to be made for *T. magnatum*, the most valuable and mysterious truffle species. It has been shown beyond doubt that *T. magnatum* can form typical ectomycorrhizas under greenhouse conditions (Mello et al. 2001; Rubini et al. 2001; Boutahir et al. 2013). However, its mycorrhizas are almost impossible to find in the field (Bertini et al. 2006; Leonardi et al. 2013) even though its mycelium is widespread in truffle-producing soils (Zampieri et al. 2010; Iotti et al. 2014). It might be that in *T. magnatum* a saprobic phase persists, while the symbiotic one is transitory or is formed with nonhost species (Riccioni et al. 2016). Bearing in mind *T. magnatum*'s high demand for water (Iotti et al. 2018), it might take refuge within host or nonhost roots as for the ectomycorrhizal desert truffle *Terfezia claveryi* under drought conditions (Navarro-Ródenas et al. 2013; Morte et al. 2000).

3.2.2 Morpho-anatomy of Tuber *Ectomycorrhizas*

Like ectomycorrhizas formed by other ectomycorrhizal fungi, the architecture of the host's cortical cells is highly modified by *Tuber* colonization assuming a "fish-bone" appearance in longitudinal section (Smith and Read 2008; Perotto et al. 2013). Ectomycorrhizas of different *Tuber* species have distinctive species-specific features (Agerer 1991) that vary with the host plant, the age of the mycorrhiza and the growing conditions (Bencivenga and Granetti 1990; Granetti and Bencivenga 1990). The anatomical structure of the mantle and the features of the cystidia (size, ramification type, etc.) are some of the most important characteristics with which to distinguish the ectomycorrhizas formed by different *Tuber* species. All ectomycorrhizas are characterized by a pseudoparenchymatous mantel; most of the species have epidermoid hyphal cells like a jigsaw puzzle, with greater or lesser lobate cells, while only few species show angular cells (Table 3.1) (Zambonelli et al. 2000; Boutahir et al. 2013). As demonstrated differences can even be seen even between strains of the same species such as the degree of cell lobing among five strains of *T. borchii* (Giomaro et al. 2000). Another distinctive feature of *Tuber* ectomycorrhizas is cystidia which arise from the surface of the mantle and differ markedly between species. Their presence is also affected by various factors such as a high soil temperature (Leonardi et al. 2017). Cystidia are found in many *Tuber* species, with the exception of *T. rufum*, where they have never been observed (Iotti et al. 2007). Cystidia shape may be hyphal-like to form a woolly mantle (*T. aestivum*, *T. mesentericum*) or needle-like to form a spiny mantle (*T. magnatum*, *T. borchii*, *T. maculatum*, *T. puberulum*, *T. melanosporum*, *T. indicum*, *T. brumale*) (Zambonelli et al. 2000). Differences in length have also been reported among truffle species, e.g. *T. oligospermum* cystidia are morphologically similar but smaller than those of *T. borchii* (Boutahir et al. 2013). The colour of cystidia may also differ according to the species, being hyaline in *T. borchii*, pale yellow in *T. brumale* and ochre in *T. aestivum* (Zambonelli et al. 2000). Ramifying cystidia are also present in black truffle species such as *T. macrosporum*, *T. melanosporum* and *T. indicum* (Zambonelli et al. 1993; Comandini and Pacioni 1997).

Ectomycorrhizas have also been classified by Agerer (2001) on the basis of the way they explore the surrounding substrate by extra-radical mycelia. The localization of these mycelia may be close to the mycorrhizal mantle, or they may form far-reaching rhizomorphs, resulting in different "exploration types" that represent diverse foraging strategies. *Tuber* species are generally characterized by contact exploration-type ectomycorrhizas with a smooth mantle and only a few emanating hyphae, often in close contact with the surrounding substrates. However, *Tuber* species with long cystidia may be classified as short-distance exploration-type ectomycorrhizas. In any case, rhizomorphs have never been detected in *Tuber* species (Table 3.2).

3.2.3 Plant: Fungus Molecular Interactions

The development of a functional ectomycorrhiza requires the growth of the fungus towards the host plant roots, enveloping the roots, the formation of Hartig's net and establishment of a nutrient exchange system that is the basis of the symbiosis (Martin

Table 3.1 Host families and anatomy of *Tuber* ectomycorrhizas

Tuber sp.	Mantle anatomy	Cystidia		Exploration type	Hosts (family)[a]
		Shape	Colour		
T. magnatum	Type M	Needle like	Hyaline	Short distance	Pinaceae, Betulaceae, Salicaceae, Fagaceae, Malvaceae
T. borchii	Type M (L) Q*	Needle like	Hyaline	Contact	Betulaceae, Cistaceae, Juglandaceae, Fagaceae, Pinaceae, Salicaceae, Malvaceae
T. maculatum	Type M	Needle like	Hyaline	Contact	Pinaceae
T. puberulum	Type M	Needle like	Hyaline	Contact	Pinaceae
T. dryophilum	Type M	Needle like	Hyaline	Contact	
T. rufum	Type M	Never observed	Never observed	Contact	Betulaceae, Fagaceae, Pinaceae
T. excavatum					Pinaceae, Betulaceae, Fagaceae, Salicaceae, Malvaceae
T. aestivum	Type L	Wolly	Ochre	Short distance	Pinaceae, Betulaceae, Juglandaceae, Fagaceae, Cistaceae, Salicaceae, Malvaceae, Ulmaceae
T. indicum	Type M	Needle like	Pale yellow	Short distance (when cystidia are present)	Pinaceae, Fagaceae
T. melanosporum	Type M	Needle like	Pale yellow/ochre	Short distance (when cystidia are present)	Pinaceae, Betulaceae, Fagaceae, Cistaceae, Salicaceae, Malvaceae
T. brumale	Type M	Needle like	Pale yellow	Contact	Pinaceae, Betulaceae, Fagaceae
T. macrosporum	Type M	Needle like	Pale yellow/ochre	Contact	Pinaceae, Betulaceae, Salicaceae, Fagaceae, Malvaceae
T. mesentericum	Type L	Wolly	Ochre	Short distance	Betulaceae, Fagaceae, Pinaceae, Malvaceae
T. fulgens					Betulaceae, Fagaceae, Pinaceae
T. oligospermum	Type M	Needle like with a basal inflation	Hyaline	Contact	Pinaceae, Fagaceae

(continued)

Table 3.1 (continued)

Tuber sp.	Mantle anatomy	Cystidia		Exploration type	Hosts (family)[a]
		Shape	Colour		
T. pseudoexcavatum	Type M	Needle like	Yellowish	Short distance	Fagaceae
T. scruposum					Pinaceae
T. malenconii	Type L	Wolly	Yellowish	Short distance	Fagaceae

Q* on *Pinus* personal observation
[a]References: Agerer (2006), Blaschke (1988), Boutahir et al. (2013), Garcia-Montero et al. (2008), Gryndler (2016), Shamekh et al. (2009), Zambonelli et al. (1995, 2000)

Table 3.2 Hosts and interaction types of morel species in literature

Morchella sp.	Interaction type	Host (families)	References
Esculenta and Elata clades	Endophyte	Poaceae	Baynes et al. (2012)
	Not defined	Betulaceae, Rosaceae, Fabaceae	Philippoussis and Balis (1995)
	Mycorrhiza	Never observed	Kageyama et al. (2008)
	Ectomycorrhiza/ Ectendomycorrhiza	Pinaceae	Dahlstrom et al. (2000)
M. esculenta	Mycelium muff	Cornaceae, Oleaceae	Robert (1865)
	Mycelium muff	Asteraceae	Roze (1883)
	Ectomycorrhizae	Ulmaceae, Rosaceae, Pinaceae	Harbin and Volk (1999)
	Ectendomycorrhiza	*Fraxinus excelsior*	Wipf et al. (1997)
	Mycorrhizae	Pinaceae	Yamada and Katsuya (1995)
	Unpublished data	Rosaceae, Pteridaceae, grass	Lakhanpal et al. (1991)
M. deliciosa	Unpublished data	Rosaceae, Pteridaceae, grass	Lakhanpal et al. (1991)
M. rotunda	Mycelium muff	Herbaceous	Buscot and Roux (1987)
	Ectomycorrhiza	Pinaceae	Buscot and Kottke (1990)
M. crassipes	Endophyte	Poaceae	Yu et al. (2016)
M. elata	Ectomycorrhiza	Ulmaceae, Rosaceae, Pinaceae	Harbin and Volk (1999)
M. conica	Unpublished data	Oleaceae	Goldway et al. (2000)
		Salicaceae	Godbout and Fortin (1985)

et al. 2001). Despite the identification of the symbiotic association types, mechanisms involved in their establishment and maintenance are not completely understood. Little is known about the molecular "talking" between plant and fungus characterizing the pre-contact recognition and the initial phases of ectomycorrhizal structure differentiation. As a reaction to the plant signals, the fungus seems to synthesize molecules that induce changes in root morphology in order to increase the success of the symbiosis (Martin and Hilbert 1991; Peterson and Bonfante 1994). This molecular dialogue between fungus and plant is mediated either by volatile organic compounds (VOCs) or solutes that dissolve in soil water. Specific VOCs are produced during the pre-symbiotic mycelial stage (Splivallo et al. 2007, 2009). A specific spectrum of 29 VOCs was detected by Menotta et al. (2004a) when *T. borchii* and *Tilia americana* grow together in in vitro conditions. It was supposed that some of these VOCs released by the host plant affect mycelial growth and chemotropism of *T. borchii*. Splivallo et al. (2009) demonstrated that ethylene and indole-3-acetic acid (IAA) induce changes such as root shortening and increased branching even before the fungal hyphae reach the rhizoplane. Large quantities of these hormones might also act as potent herbicides that might explain the formation of brûlés (Splivallo et al. 2011). Phenolic VOCs produced by truffles are also a rich source of toxic phytochemicals present in the brûlé (Streiblová et al. 2012). The pre-contact recognition triggers changes in biochemical machinery of *Tuber* mycelium too. Menotta et al. (2004b) revealed an upregulation of 58 *T. borchii* genes before the establishment of the symbiosis. Some of these genes are involved in the apical growth of hyphae towards the host roots, while others, such as glyoxal oxidase and GAS-2 protein-like molecule, are necessary for fungal colonization of the host tissues. Ragnelli et al. (2014) highlighted a higher amount of polyphenols/tannins in the outer cell layers of host roots which lead to their death in the pre-symbiotic phase. The authors considered this phenomenon a form of hypersensitive reaction of the host plants against undesired microorganisms, but it seems pivotal for *Tuber* ectomycorrhiza establishment. For its part, truffle hyphae overexpress a spectrum of genes such as laccases and tyrosinases, which play an important role in degradation of plant cell wall and apoplast colonization (Martin et al. 2010; Zarivi et al. 2011, 2013). The repertory of genes targeting plant cell wall in *Tuber* genome is much poorer than that of phytopathogenic or saprobic fungi but richer than basidiomycetous ectomycorrhizal fungi which do not have enzymes to degrade the middle lamella of host cells (Kohler et al. 2015; Murat et al. 2018). These conditions might suggest that *Tuber* species behave as weak pathogens during the early stages of mycorrhizal formation perhaps reflecting their behaviour in the brûlé later.

When ectomycorrhizas are functional, carbon exchange takes place in the apoplast of the Hartig's net. Ectomycorrhizas represent a strong sink for the photosynthetic sucrose which is hydrolysed by a plant wall-bound invertase (Smith and Read 2008). In contrast to many ectomycorrhizal fungi, *Tuber* species, with some exceptions such as *T. borchii*, can produce a fungal invertase that makes them less dependent on the host plant for carbohydrates (Ceccaroli et al. 2011). The upregulation of all glycogen-synthesizing enzymes in ECM tissue contributes to maintaining the carbon flux towards the fungal cells (Ceccaroli et al. 2011).

Environmental conditions affect the morphology and functioning of plant-fungus interaction. Leonardi et al. (2017) demonstrated that the exposure of seedlings to over 30 °C is detrimental to root colonization. At high temperature the mantle ceases to develop together with the root tip elongation that remains uncolonized, and ectomycorrhizas become senescent, brown and lose their cystidia. The mantle also appears thicker, in response to the production of an additional layer of cells to protect the Hartig's net, and tannin bodies appear between the inner mantle layers. Tannins are also present in stromatic structures formed by black truffles on the root bark surface of host plants (Pargney and Jalade 1995). These fungal proliferations have been explained as microniches in which hyphae can be preserved in old roots (Pargney et al. 2001).

3.2.4 Interaction with Other Microorganisms

Bacteria are considered the third partner in the ectomycorrhizal symbiosis which can be considered as a tripartite symbiosis (Duponnois 2006). Some bacteria, called "mycorrhiza helper bacteria" (Garbaye 1994), can promote mycorrhization by stimulating fungal spore germination, accelerating hyphal extension and facilitating the root system colonization (Tarkka and Frey-Klett 2008). Many bacteria grow specifically in the *Tuber* ectomycorrhizosphere and can be found either on the mantle surface or within the intercell spaces of the outer layers (Schelkle et al. 1996). Culturable *Pseudomonas* spp., actinomycetes and aerobic spore-former bacteria were isolated by Sbrana et al. (2002) from *T. borchii* ectomycorrhizas synthetized in greenhouse, but only few spore-former isolates were able to stimulate the growth of *T. borchii* mycelium in an in vitro dual culture system. On the contrary *Pseudomonas fluorescens* has proven to be able to double the ectomycorrhizal colonization of *T. melanosporum* on *Pinus* seedlings in greenhouse trials (Dominguez et al. 2012, 2015). A total of 183 isolates belonging to 6 different orders (*Actinomycetales*, *Burkholderiales*, *Enterobacteriales*, *Pseudomonadales*, *Rhizobiales* and *Xanthomonadales*) were isolated from ectomycorrhizas of *T. aestivum* by Gryndler and Hršelová (2012). Some of these isolates (*Lysobacter* sp. and *Ensifer* sp.) as well as members of four genera of *Actinobacteria* (*Actinosynnema*, *Allokutzneria*, *Kibdelosporangium* and *Lentzea*) were then found to be significantly positively associated with *T. aestivum* ectomycorrhizas (Gryndler et al. 2013a). On the contrary, members of *Rhizobiales* abound in the fruiting bodies of many truffle species (Barbieri et al. 2016) but are scarce in the *Tuber* ectomycorrhizosphere (Gryndler et al. 2013b). Antony-Babu et al. (2014) also found that bacterial communities associated with *T. melanosporum* ectomycorrhizas were distinct from those living on and within its ascomata and in the surrounding soil. In this study, the ectomycorrhizosphere was significantly enriched with *Actinobacteria* in *Streptomyces* and *Thermoleophilum*. Other studies found that there is a much higher proportion of ECM-inhibiting bacteria in the soil than in the ectomycorrhizosphere (Frey-Klett et al. 2005), and some of them can act selectively against ectomycorrhizal fungal species. *Staphylococcus pasteuri* releases derivatives of aliphatic, aromatic compounds and terpenoids that affect the success of the mycorrhizal colonization by *T. borchii* (Barbieri et al. 2005). This bacterium is able to inhibit the growth of *T. borchii* and *Boletus luridus* mycelia but had no effect on *Hebeloma radicosum* (Barbieri et al. 2005).

Other nonbacterial microorganisms that might selectively live in the *Tuber* ectomycorrhizosphere have been poorly investigated. Zacchi et al. (2003) isolated four strains of *Cryptococcus albidus* from *T. aestivum* ectomycorrhizas with *Quercus pubescens* and *Corylus avellana*. All these isolates did not show extracellular chitinolytic and proteolytic activity as the other *C. albidus* strains obtained from the ascomata but only pectinolytic activity. The differential enzymatic activity of these strains could indicate a "helper" role in degrading pectins of the middle lamella during the apoplast colonization by *T. aestivum* without damaging the fungal hyphae.

3.3 True Morels

Morchella can be found in a broad range of habitats including hardwood and coniferous forests of both ectomycorrhizal and non-ectomycorrhizal hosts, tree fruit orchards, sand dunes, urban parks and gardens (Kuo et al. 2012; Richard et al. 2015). From more than 150 years, mycologists have reported morels in contact with hypogeous plant organs through the base of the stalk or characteristic hyphal structures.

3.3.1 Interactions with Plants: A Facultative Group of Organisms

The nutritional strategy that has evolved in *Morchella* has triggered much debate within the scientific community. In 1865, Eugène Robert was the first to report the interaction between *Morchella* spp. and plants. The author noted that the stalks of *Morchella* spp. were tangled up to roots of *Cornus sanguinea* and *Ligustrum vulgare* and hypothesized a parasitic interaction. A few years later, an ascoma of *Morchella esculenta* was found growing on the end of a *Helianthus tuberosus* tubercle, so Roze (1883) concluded that the fungus was a pathogen. However, in the first half of the twentieth century, the term mycorrhiza was used to define the interaction between morels and trees (Matruchot 1909; Moser 1949). Later Buscot and Roux (1987) were not convinced either way and suggested that *M. rotunda* had either a parasitic or symbiotic behaviour towards different woody and herbaceous plants and its association was not truly mycorrhizal. Similarly, Tedersoo et al. (2006) did not list any *Morchella* in their list of ectomycorrhizal pezizalean fungi. However, Buscot and

Kottke (1990) and Dahlstrom et al. (2000) observed ectomycorrhizas with mantles and Hartig's nets in the wild between species of *M. rotunda* and *Picea abies* and subsequently synthesized these in vitro between *Morchella* spp. and Pinaceae. Wipf et al. (1997) observed intracellular penetration of *M. esculenta* on *Fraxinus excelsior* roots as in ectendomycorrhizal fungi. More recently Harbin and Volk (1999) also synthetized putative mycorrhizal associations in sterile system between *M. esculenta* and *M. elata* and *Ulmus* spp., *Malus sylvestris* and *Picea mariana*. More recently still Baynes et al. (2012) reported *Morchella* spp. to be an endophyte of *Bromus tectorum* in a post-fire woody area, while the interaction between *Morchella crassipes* and sweet corn was considered by Yu et al. (2016) to be more like an ectendomycorrhiza.

In addition to this high heterogeneity of plant-fungus interactions exhibited by the genus, morels can also fructify without a connection with a living host. For example, Ower (1982) obtained mature ascomata on wheat a technique, which was subsequently patented (Ower et al. 1988). Chinese researchers, particularly those in the Sichuan Academy of Agricultural Science, have now developed procedures for the large-scale cultivation of morels in particular *Morchella importuna* and *Morchella sextelata* (Liu et al. 2018; Peng et al. 2015). In some situations, this is without any plants being present and with the inoculum raised in pure culture. Before this, many researchers had labelled this ability of morels to grow independently as a sign they were saprobes (Molliard 1904; Fron 1905; Brock 1951; Robbins and Hervey 1959; Wassom and Holden 1977). However, this ability cannot be considered as conclusive evidence because many ectomycorrhizal fungi and some pathogenic pezizalean fungi can be grown on synthetic culture media (Iotti et al. 2002, 2005; Egger and Paden 1986). Using $\delta^{15}N$ and $\delta^{13}C$ isotopes as indicators of fungal trophic strategy, Hobbie et al. (2001) hypothesized that *Morchella* includes both saprotrophic and mycorrhizal species, while other genera (*Verpa* and *Disciotis*) in the *Morchellaceae* appeared to be only saprotrophic. On the contrary, *Morchella* acomata collected in post-fire environments seem to feed as a saprotroph although they appeared capable of assimilating carbon from organic material between 2 and 22 years old (Hobbie et al. 2016). Probably, the genus is characterized by a great nutritional plasticity which can be manifested in different phases of their life cycles.

Morchella are known to be sporadic and will appear one year and not be seen again for several decades. However, *Morchella* spp. are known to fruit regularly in woody areas particularly after a fire. Post-fire fruiting is usually abundant the year after the fire event, but the factors responsible for this are poorly understood (Pilz et al. 2007; Masaphy and Zabari 2013; Larson et al. 2016). Greene et al. (2010) estimated a yield of 14,900 ascomata ha^{-1} in areas of the Kootenay National Park with 100% tree mortality, while Winder and Keefer (2008) reported 8062 ascomata ha^{-1} in the Rocky Mountain Forest (British Columbia) burned the year before. Fujimura et al. (2005) did not find root tips of Pinaceae colonized by *Morchella* after a fire, and Dahlstrom et al. (2000) speculated that morels might be able to obtain fixed carbon from living plants as weak mycorrhizal formers or parasites or by decomposing tissues of senescent or dead plants. Evidence for the latter comes from their ability to produce extracellular enzymes associated with plant cell wall degradation such as lignin peroxidase, manganese-dependent peroxidase and laccase (Papinutti and Lechner 2008; Kanwal and Reddy 2011), cellulases (Cavazzoni and Manzoni 1994) and laccase-like multicopper oxidase (Kellner et al. 2007). Recently, the analysis of *M. importuna* genome highlighted a large repertoire of plant cell wall-degrading enzymes, indicating the ability to obtain carbon from a large set of plant polysaccharides (Murat et al. 2018).

3.3.2 Morpho-anatomy of Morchella*: Plant Structures*

3.3.2.1 Mycorrhiza-Like Associations

Morchella spp. have been reported to form structures like those generated by mycorrhizal fungi, with hyphae that can remain in the apoplast (Buscot and Kottke 1990; Buscot 1992a; Harbin and Volk 1999) and/or penetrate into the lumens of cortical cells (Wipf et al. 1997; Dahlstrom et al. 2000). Yu et al. (2016) described ectendomycorrhiza-like structures without the formation of a typical Hartig's net. Morel mycelium was found to colonize almost all elongation and maturation zones as far as the Casparian strip and to increase the size of the cortical cells. This interaction stimulated the development of the root system and significantly increased the dry root biomass.

The ectomycorrhizal-like structures have varying morphologies depending on the fungus-plant combination and whether the roots originate from field or in vitro material (Buscot and Kottke 1990; Buscot 1992b; Dahlstrom et al. 2000). Sometimes the mycorrhizas are cottony, felty or shiny with continuous or patchy mantles and collapsed layers of cortical cells. The mantle is cream-coloured and 10–15 μm thick, consisting of an external prosenchyma composed of 2–3-μm-thick hyphae loosely organized with frequently branched hyphae and elongated interhyphal spaces and a compact internal sinenchyma composed of 3–6-μm-thick hyphae. The fungus penetrates between the cortical cells forming an intercellular net of highly branched hyphae, which, in section, appears as an incomplete Hartig's net. These are composed of 4–6 μm thick hyphae usually only penetrating between the cortical cells of the first layer and occasionally reaching those of the one beneath. Portions of root with well-developed mantles have thicker cortices, which increase root diameter and gives a bulbous look to the mycorrhizal roots.

3.3.2.2 Sclerotia

Morels readily form sclerotia both in axenic culture and in the field. These appear as undifferentiated mycelial masses of tightly woven branched hyphae (Pilz et al. 2007). In nature, sclerotia have been described as smooth, irregular in shape (1–5 cm diameter) and clustered in small to large aggregates (Volk and Leonard 1989; Miller et al. 1994). Sclerotia are a nutrient sink, are fed by mycelia and are perennating structures during periods of adverse conditions. They also appear to be essential to ascomata production (Pilz et al. 2007). Sclerotia of *Morchella crassipes* were found by Yu et al. (2016) in roots of sweet corn seedlings colonized by the fungus in pure culture before their death. In pure cultures of *M. importuna* on agar plates, they are composed of hyphae very different to the rapidly growing hyphae which can cross a Petri dish in 2 days (Fig. 3.1c–d).

Sclerotia-like structures associated with the roots of woody and herbaceous plants were described in the 1980s by Buscot and colleagues initially were called "mycelial muffs" (Buscot and Roux 1987; Buscot 1989). Mycelial muffs appeared as compact and spongy hyphal aggregates (max 5 cm in length), encrusted with soil particles, that surrounded coarse roots 0.3–1 cm in diameter. Fungal hyphae colonized the cortical parenchyma as far as the phloem without ever penetrating the xylem. Colonization of cell lumens has also been reported (Buscot 1994). Mycelial muffs were found connected to the ascoma stipe by a cylindrical and fragile downy mass of mycelium about 5 cm in diameter; they branch, become thin as they get deeper and incorporate many soil particles. The authors hypothesized a biennial cycle where mycelial muffs developed during summer and autumn when soil climate was optimal for mycelial growth and differentiation. Then, during winter they became storage and hibernation structures which provide nutrients for developing ascomata in the early spring and then quickly dissipate as the fruiting bodies developed (Buscot 1992b). Mycelial muffs have been found associated with many plant species and have been described by other authors (Lakhanpal et al. 1991; Philippoussis and Balis 1995; Goldway et al. 2000).

3.3.2.3 Radiscisclerotia

The term radiscisclerotia was used by Stefani et al. (2010) for newly recognized vegetative fungal structures which enveloped thin plant roots. They appear as an extension of the ascoma stipe which branch belowground in multiple rootlike solid structures, 5–15 mm in diameter. Radiscisclerotia formed by *M. tomentosa* were white and rigid and composed of straight oversized hyphae. The authors hypothesized that these structures are capable of surviving for several years and capable of support fruiting for 2–3 years. Similar structures were found to connect ascomata of *M. dunensis* to different rhizomatous plants on littoral sand dunes although, in this case, they appeared as sand ropes hardened by hyphae (Snabl et al. 2019).

3.3.3 Interactions with Other Microorganisms

Studies on interactions between morel and other microorganisms are scarce. *Bacillus* sp. was found to promote mycorrhizal formation of *Morchella esculenta* on *Picea abies* and, when associated with morel mycelium, affected the morphology of meristematic root tissues (Buscot 1992a, b). Pion et al. (2013) evaluated the interaction of *Morchella crassipes* mycelium with the endophytic bacterium *Pseudomonas putida*. In vitro coculturing gave benefits to both organisms in a sort of bacteria-fungus symbiosis. Fungal hyphae and sclerotia favoured bacterial dispersal and improved carbon nutrition by the absorption of fungal exudates, while the bacterium increased fungal stress tolerance and represented an alternative food source for sclerotia formation when the medium was depleted. As root endophytes, morels have been shown to reduce the disease severity of *Melampsora* rust on *Populus trichocarpa* (Busby et al. 2016) and *Fusarium verticillioides* on mature ears of *Zea mays* (Yu et al. 2016).

3.4 Conclusion

True truffles (*Tuber* spp.) and morels (*Morchella* spp.) include the most economically important ascomycetous mushrooms, and a few have been successfully cultivated although some aspects of their reproduction strategies and ecology are still to be unravelled.

There is a gap in our knowledge of their biology and soil ecology, and in particular on the complex interrelationships these fungi can establish with symbiotic plants and other soil organisms. Most of the recent studies on truffles and morels have been devoted to their phylogeny and biology, and recently the genomes of some representative species of these groups have been sequenced. This has allowed us to determine the heterothallic nature of them but alas did not allowed us to solve the mysteries of the mating fusion between strains carrying different mating types.

Although the truffles are known to form ectomycorrhizas with the roots of host plants, they are also able to establish endophytic and/or parasitic relationships with herbaceous plants. The interrelationships of morels with plants seem to be more complex and cannot simply be categorized as mutualistic, pathogenic or saprobic.

Truffles and morels life cycle is influenced by other soil microorganisms which share the same niche. It is well known that truffle interact with them in the different phases of their life cycle, from the mycelium to mycorrhiza formation and ascoma maturation, although the studies devoted to better understand the molecular mechanisms of these interactions are scarce. With regard to the morels, this kind of information is inconsistent. A better understanding of these biological and ecological aspects could lead to improved techniques for their cultivation and improved productivity.

References

Agerer R (1991) Characterization of ectomycorrhiza. In: Norris JR, Read DJ, Varma AK (eds) Techniques for the study of mycorrhiza, Methods in microbiology, vol 23. Academic, London, pp 25–73. https://doi.org/10.1016/S0580-9517(08)70172-7

Agerer R (2001) Exploration types of ectomycorrhizae. Mycorrhiza 11:107–114. https://doi.org/10.1007/s005720100108

Agerer R (2006) Fungal relationships and structural identity of their ectomycorrhizae. Mycol Prog 5 (2):67–107. https://doi.org/10.1007/s11557-006-0505-x

Alvarado-Castillo G, Mata G, Sangabriel-Conde W (2014) Understanding the life cycle of morels (*Morchella* spp.). Rev Mex Micol 40:47–50

Antony-Babu S, Deveau A, Van Nostrand JD, Zhou J, Le Tacon F, Robin C, Frey-Klett P, Uroz S (2014) Black truffle-associated bacterial communities during the development and maturation of *Tuber melanosporum* ascocarps and putative functional roles. Environ Microbiol 16:2831–2847. https://doi.org/10.1111/1462-2920.12294

Barbieri E, Gioacchini AM, Zambonelli A, Bertini L, Stocchi V (2005) Determination of microbial volatile organic compounds from against *Tuber borchii* using solid-phase microextraction and gas chromatography/ion trap mass spectrometry. Rapid Commun Mass Spectrom 19:3411–3415. https://doi.org/10.1002/rcm.2209

Barbieri E, Ceccaroli P, Agostini D, Zeppa SD, Gioacchini AM, Stocchi V (2016) Truffle-associated bacteria: extrapolation from diversity to function. In: Zambonelli A, Iotti M, Murat C (eds) True truffle (*Tuber* spp.) in the world, Soil biology, vol 47. Springer, Cham, pp 301–3017. https://doi.org/10.1007/978-3-319-31436-5_18

Baynes M, Newcombe G, Dixon L, Castlebury L, O'Donnell K (2012) A novel plant–fungal mutualism associated with fire. Fungal Biol 116:133–144

Beever R, Lebel T (2014) Truffles of New Zealand: a discussion of bird dispersal characteristics of fruiting bodies. Auckl Bot Soc 69:170–178

Bencivenga M, Granetti B (1990) Biometric valuation of mycorrhizae on *Ostrya carpinifolia* Scop. produced by various truffle species (*Tuber* sp. pl.). In: Bencivenga M, Granetti B (eds) Proceedings of the 2nd international conference on Truffle, 24–27 Nov 1988, Spoleto. Comunità Montana dei Monti Martani, Serano e Subasio, Spoleto, Italy, pp 265–270

Benucci GMN, Gógán Csorbai A, Baciarelli Falini L, Marozzi G, Suriano E, Sitta N, Donnini D (2016) Taxonomy, biology, ecology of *Tuber macrosporum* Vittad. and *Tuber mesentericum* Vittad. In: Zambonelli A, Iotti M, Murat C (eds) True truffle (*Tuber* spp.) in the world, Soil biology, vol 47. Springer International Publishing, Cham, pp 69–86. https://doi.org/10.1007/978-3-319-31436-5_5

Bertini L, Rossi I, Zambonelli A, Amicucci A, Sacchi A, Cecchini M, Gregori G, Stocchi V (2006) Molecular identification of *Tuber magnatum* ectomycorrhizae in the field. Microbiol Res 161:59–64. https://doi.org/10.1016/j.micres.2005.06.003

Blaschke H (1988) Tuber puberulum. In: Agerer R (ed) Colour atlas of ectomycorrhizae. Einhorn, Schwabisch Gmund, p 22

Bonito GM, Smith ME (2016) General systematic position of the truffles: evolutionary theories. In: Zambonelli A, Iotti M, Murat C (eds) True truffle (*Tuber* spp.) in the world, Soil biology, vol 47. Springer International Publishing, Cham, pp 3–18. https://doi.org/10.1007/978-3-319-31436-5_1

Bonito G, Smith ME, Nowak M, Healy RA, Guevara G, Cazares E, Kinoshita A, Nouhra ER, Dominguez LS, Tedersoo L, Murat C, Wang Y, Moreno BA, Pfister DH, Nara K, Zambonelli A, Trappe JM, Vilgalys R (2013) Historical biogeography and diversification of truffles in the Tuberaceae and their newly identified southern hemisphere sister lineage. PLoS One 8(1): e52765. https://doi.org/10.1371/journal.pone.0052765

Bonuso E, Zambonelli A, Bergemann SE, Iotti M, Garbelotto M (2010) Multilocus phylogenetic and coalescent analyses identify two cryptic species in the Italian bianchetto truffle, *Tuber borchii* Vittad. Conserv Genet 11:1453–1466. https://doi.org/10.1007/s10592-009-9972-3

Boutahir S, Iotti M, Piattoni F, Zambonelli A (2013) Morphological and molecular characterization of *Tuber oligospermum* mycorrhizas. Afr J Agric Res 8(29):4081–4087. https://doi.org/10.5897/AJAR2013.7354

Brock TD (1951) Studies on the nutrition of *Morchella esculenta* Fries. Mycologia 43(4):402–422

Busby PE, Ridout M, Newcombe G (2016) Fungal endophytes: modifiers of plant disease. Plant Mol Biol 90:645–655. https://doi.org/10.1007/s11103-015-0412-0

Buscot F (1989) Field observations on growth and development of *Morchella rotunda* and *Mitrophora semilibera* in relation to forest soil temperature. Can J Bot 67:589–593

Buscot F (1992a) Mycorrhizal succession and morel biology. In: Read DJ, Lewis DH, Fitten AH, Alexander IJ (eds) Mycorrhizas in ecosystems. CAB International, Wallingford, Oxon, pp 220–224

Buscot F (1992b) Strategies ecologiques et biologiques des morilles. Cryptogam Mycol 13:171–179

Buscot F (1994) Ectomycorrhizal types and endobacteria associated with ectomycorrhizas of *Morchella elata* (Fr.) Boudier with *Picea abies* (L.) Karst. Mycorrhiza 4:223–232

Buscot F, Kottke I (1990) The association of *Morchella Rotunda* Boundier with roots of *Picea abies* Karst. New Phytol 116:425–430

Buscot F, Roux J (1987) Association between living roots and ascocarps of *Morchella rotunda*. Trans Br Mycol Soc 89:249–252

Carris LM, Peever TL, McCotter SW (2015) Mitospore stages of *Disciotis*, *Gyromitra* and *Morchella* in the inland Pacific Northwest USA. Mycologia 107:729–744. https://doi.org/10.3852/14-207

Cavazzoni V, Manzoni M (1994) Extracellular cellulolytic complex from *Morchella conica*: production and purification. LWT Food Sci Technol 27:73–77

Ceccaroli P, Buffalini M, Saltarelli R, Barbieri E, Polidori E, Ottonello S, Kohler A, Tisserant E, Martin F, Stocchi V (2011) Genomic profiling of carbohydrate metabolism in the ectomycorrhizal fungus *Tuber melanosporum*. New Phytol 189:751–764. https://doi.org/10.1111/j.1469-8137.2010.03520.x

Cesalpino A (1583) De plantis libri XVI. Apud Georgium Marefcottum, Florentiae

Chevalier G (2010) Méthode raisonnée de trufficulture. In: Les nouvelles techniques de culture de la truffe – Actes Colloque Fête internationale de la truffe, 16 January 2009, Sarlat. Service de communication de la ville de Sarlat, Sarlat-en-Périgord, France, pp 55–67

Comandini O, Pacioni G (1997) Mycorrhizae of Asian black truffles, *Tuber himalayense* and *T. indicum*. Mycotaxon 63:77–86

Dahlstrom JL, Smith JE, Weber NS (2000) Mycorrhiza-like interaction by *Morchella* with species of the Pinaceae in pure culture synthesis. Mycorrhiza 9:279–285. https://doi.org/10.1007/PL00009992

Delmas J (1978) The potential cultivation of various edible fungi. In: Chang ST, Hayes WA (eds) The biology and cultivation of edible mushrooms. Academic, New York, pp 699–724

Delmas J (1983) La Truffe et sa Culture, 2nd edn. INRA, Paris

Dominguez JA, Martin A, Anriquez A, Albanesi A (2012) The combined effects of *Pseudomonas fluorescens* and *Tuber melanosporum* on the quality of *Pinus halepensis* seedlings. Mycorrhiza 22(6):429–436. https://doi.org/10.1007/s00572-011-0420-0

Dominguez JA, Medina M, Berrocal-Lobo M Anriquez A, Albanesi A (2015) The combined effects of *Pseudomonas fluorescens* CECT 844 and the black truffle co-inoculation on *Pinus nigra* seedlings. iForest 8:624–630. https://doi.org/10.3832/ifor1334-007

Du XH, Zhao Q, Yang ZL, Hansen K, Taşkin H, Büyükalaca S, Dewsbury D, Moncalvo JM, Douhan GW, Robert VARG, Crous PW, Rehner SA, Rooney AP, Sink S, O'Donnell K (2012) How well do ITS rDNA sequences differentiate species of true morels (*Morchella*)? Mycologia 104(6):1351–1368. https://doi.org/10.3852/12-056

Du XH, Zhao Q, Xia EH, Gao LZ, Richard F, Yang ZL (2016) Mixed-reproductive strategies, competitive mating-type distribution and life cycle of fourteen black morel species. Sci Rep 7:1493. https://doi.org/10.1038/s41598-017-01682-8

Duponnois R (2006) Mycorrhizal helper bacteria: their ecological impact in mycorrhizal symbiosis. In: Mahendra R (ed) Handbook of microbial biofertilizers. Haworth Press, New York, pp 231–250

Egger KN, Paden JW (1986) Pathogenicity of postfire ascomycetes (Pezizales) on seeds and germinants of lodgepole pine. Can J Bot 64(10):2368–2371. https://doi.org/10.1139/b86-312

Fan L, Han L, Zhang PR, Yan XY (2017a) Molecular analysis of Chinese truffles resembling *Tuber californicum* in morphology reveals a rich pattern of species diversity with emphasis on four new species. Mycologia 108:344–353. https://doi.org/10.3852/14-343

Fan L, Zhang PR, Yan XY (2017b) Phylogenetic analyses of Chinese *Tuber* species that resemble *T. borchii* reveal the existence of the new species *T. hubeiense* and *T. wumengense*. Mycologia 108:354–362. https://doi.org/10.3852/14-349

Fasolo-Bonfante P, Fontana A, Montacchini F (1971) Studi sull'ecologia del *Tuber melanosporum*. Demostrazione di un effetto fitotossico. Allionia 17:47–54

Frank AB (1885) Uher die auf Wurzelsymbiose beruhende Ernahrung gewisser Baume durch unterirdische Pilze. Ber Dtsch Bot Ges 3:128–145

Frey-Klett P, Chavatte M, Clausse M, Courrier S, Roux CL, Raaijmakers J, Martinotti MG, Pierrat J, Garbaye J (2005) Ectomycorrhizal symbiosis affects functional diversity of rhizosphere fluorescent pseudomonads. New Phytol 165:317–328. https://doi.org/10.1111/j.1469-8137.2004.01212.x

Fron G (1905) Sur les conditions de developpement du mycelium de la morille. Compt Rend Acad Sci Paris 140:1187–1189

Fujimura KE, Smith JE, Horton TR, Weber NS, Spatafora JW (2005) Pezizalean mycorrhizas and sporocarps in ponderosa pine (*Pinus ponderosa*) after prescribed fires in eastern Oregon, USA. Mycorrhiza 15:79–86

Garbaye J (1994) Helper bacteria: a new dimension to the mycorrhizal symbiosis. New Phytol 128 (2):197–210. https://doi.org/10.1111/j.1469-8137.1994.tb04003.x

Garcia-Montero LG, Di Massimo G, Manjón JL, García-Abril A (2008) New data on ectomycorrhizae and soils of the Chinese truffles *Tuber pseudoexcavatum* and *Tuber indicum*, and their impact on truffle cultivation. Mycorrhiza 19:7–14. https://doi.org/10.1007/s00572-008-0198-x

Giomaro G, Zambonelli A, Cecchini D, Stocchi V, Saffi V (2000) Anatomical and morphological characterization of mycorrhizas of five strains of *Tuber borchii* Vittad. Mycorrhiza 10 (3):107–114. https://doi.org/10.1007/s005720000065

Godbout C, Fortin JA (1985) Synthesized ectomycorrhizae of aspen: fungal genus level of structural characterization. Can J Bot 63:252–262

Goldway M, Rachel A, Goldberg D, Hadar Y, Levanon D (2000) *Morchella conica* exhibiting a long fruiting season. Mycol Res 104:1000–1004

Granetti B, Bencivenga M (1990) Morphological aspects of the mycorrhizae of some *Tuber* species with *Ostrya carpinifolia* Scop. In: Bencivenga M & Granetti B (eds) Proceedings of the 2nd International Conference on Truffle, 24–27 November 1988, Spoleto. Comunità Montana dei Monti Martani, Serano e Subasio, Spoleto, Italy, p 271

Granetti B, De Angelis A, Materozzi G (2005) Umbria terra di tartufi. Regione Umbria, Assessorato Regionale Agricoltura, Foreste. Caccia e Pesca, Perugia

Greene DF, Hesketh M, Pounden E (2010) Emergence of morel (*Morchella*) and pixie cup (*Geopyxis carbonaria*) ascocarps in response to the intensity of forest floor combustion during a wildfire. Mycologia 102:766–773. https://doi.org/10.3852/08-096

Gryndler M (2016) True truffle host diversity. In: Zambonelli A, Iotti M, Murat C (eds) True truffle (*Tuber* spp.) in the world, Soil biology, vol 47. Springer International Publishing, Cham, pp 267–281. https://doi.org/10.1007/978-3-319-31436-5_16

Gryndler M, Hršelová H (2012) Isolation of bacteria from ectomycorrhizae of *Tuber aestivum* Vittad. Acta Mycol 47(2):155–160. https://doi.org/10.5586/am.2012.018

Gryndler M, Soukupová L, Hršelová H, Gryndlerová H, Borovička J, Streiblová E, Jansa J (2013a) *Tuber aestivum*-associative prokaryotes. Environ Microbiol Rep 5:346–352. https://doi.org/10.1111/1758-2229.12014

Gryndler M, Trilčová J, Hršelová H, Streiblová E, Gryndlerová H, Jansa J (2013b) *Tuber aestivum* Vittad. mycelium quantified: advantages and limitations of a qPCR approach. Mycorrhiza 23 (5):341–348. https://doi.org/10.1007/s00572-012-0475-6

Gryndler M, Černá L, Bukovská P, Hršelová H, Jansa J (2014) *Tuber aestivum* association with non-host roots. Mycorrhiza 24(8):603–610. https://doi.org/10.1007/s00572-014-0580-9

Hall IR, Brown GT, Zambonelli A (2007) Taming the truffle. Timber Press, Portland

Hall IR, Fitzpatrick N, Miros P, Zambonelli A (2017) Counter-season cultivation of truffles in the Southern Hemisphere – an update. Ital J Mycol 46:21–36. https://doi.org/10.6092/isnn.2531-7342/XXXX

Hansen K, Pfister DH (2006) Systematics of the Pezizomycetes the operculate discomycetes. Mycologia 98(6):1029–1040. https://doi.org/10.1080/15572536.2006.11832631

Harbin M, Volk TJ (1999) The association of *Morchella* with plant roots. In: Abstract of the XVI international botanical congress, 1–7 Aug 1999. Mycological Society of America, St. Louis, Missouri, p 559

Hardoim PR, Van Overbeek LS, Berg G, Pirttilä AM, Compant S, Campisano A, Döring M, Sessitsch A (2015) The hidden world within plants: ecological and evolutionary considerations for defining functioning of microbial endophytes. Microbiol Mol Biol Rev 79(3):293–320. https://doi.org/10.1128/MMBR.00050-14

Healy RA, Smith ME, Bonito GM, Pfister DH, Ge ZW, Guevara GG, Williams G, Stafford K, Kumar L, Lee T, Hobart C, Trappe J, Vilgalys R, McLaughlin DJ (2013) High diversity and widespread occurrence of mitotic spore mats in ectomycorrhizal Pezizales. Mol Ecol 22 (6):1717–1732. https://doi.org/10.1111/mec.12135

Healy RA, Zurier H, Bonito G, Smith ME, Pfister DH (2016) Mycorrhizal detection of native and non-native truffles in a historic arboretum and the discovery of a new North American species, *Tuber arnoldianum* sp. nov. Mycorrhiza 26(7):781–792. https://doi.org/10.1007/s00572-016-0713-4

Hobbie EA, Weber NS, Trappe JM (2001) Mycorrhizal vs saprotrophic status of fungi: the isotopic evidence. New Phytol 150:601–610. https://doi.org/10.1046/j.1469-8137.2001.00134.x

Hobbie EA, Rice SF, Weber NS, Smith JE (2016) Isotopic evidence indicates saprotrophy in post-fire *Morchella* in Oregon and Alaska. Mycologia 108:638–645. https://doi.org/10.3852/15-281

Iotti M, Amicucci A, Stocchi V, Zambonelli A (2002) Morphological and molecular characterization of mycelia of some *Tuber* species in pure culture. New Phytol 155:499–505. https://doi.org/10.1046/j.1469-8137.2002.00486.x

Iotti M, Barbieri E, Stocchi V, Zambonelli A (2005) Morphological and molecular characterisation of mycelia of ectomycorrhizal fungi in pure culture. Fungal Divers 19:51–68

Iotti M, Amicucci A, Bonito G, Bonuso E, Stocchi V, Zambonelli A (2007) Selection of a set of specific primers for the identification of *Tuber rufum*: a truffle species with high genetic variability. FEMS Microbiol Lett 277(2):223–231. https://doi.org/10.1111/j.6968.2007.00963.x

Iotti M, Leonardi M, Lancellotti E, Salerni E, Oddis M, Leonardi P, Perini C, Pacioni G, Zambonelli A (2014) Spatio-temporal dynamic of *Tuber magnatum* mycelium in natural truffle grounds. PLoS One 9(12):e115921. https://doi.org/10.1371/journal.pone.0115921

Iotti M, Leonardi P, Vitali G, Zambonelli A (2018) Effect of summer soil moisture and temperature on the vertical distribution of *Tuber magnatum* mycelium in soil. Biol Fertil Soils 54:707–716. https://doi.org/10.1007/s00374-018-1296-3

Kageyama SA, Mandyam KG, Jumpponen A (2008) Diversity, function and potential applications of the root-associated endophytes. Mycorrhiza 18:29–57

Kanwal HK, Reddy MS (2011) Effect of carbon, nitrogen sources and inducers on ligninolytic enzyme production by *Morchella crassipes*. World J Microbiol Biotechnol 27:687–691

Kellner H, Luis P, Buscot F (2007) Diversity of laccase-like multicopper oxidase genes in Morchellaceae: identification of genes potentially involved in extracellular activities related to plant litter decay. FEMS Microbiol Ecol 61(1):153–163

Kennedy PG, Smith DP, Horton TR, Molina RJ (2012) *Arbutus menziesii* (Ericaceae) facilitates regeneration dynamics in mixed evergreen forests by promoting mycorrhizal fungal diversity and host connectivity. Am J Bot 99:1691–1701. https://doi.org/10.3732/ajb.1200277

Kirk PM, Cannon PF, Minter DW, Stalpers JA (2008) Dictionary of the fungi. CABI, Wallingford

Kohler A, Kuo A, Nagy LG, Morin E, Barry KW, Buscot F, Canbäck B, Choi C, Cichocki N, Clum A, Colpaert J, Copeland A, Costa MD, Doré J, Floudas D, Gay G, Girlanda M, Henrissat B, Herrmann S, Hess J, H€ogberg N, Johansson T, Khouja HR, LaButti K, Lahrmann U, Levasseur A, Lindquist EA, Lipzen A, Marmeisse R, Martino E, Murat C,

Ngan CY, Nehls U, Plett JM, Pringle A, Ohm RA, Perotto S, Peter M, Riley R, Rineau F, Ruytinx J, Salamov A, Shah F, Sun H, Tarkka M, Tritt A, Veneault-Fourrey C, Zuccaro A, Consortium MGI, Tunlid A, Grigoriev IV, Hibbett DS, Martin F (2015) Convergent losses of decay mechanisms and rapid turnover of symbiosis genes in mycorrhizal mutualists. Nat Genet 47:410–415. https://doi.org/10.1038/ng.3223

Kuo M, Dewsbury DR, O'Donnell K, Carter MC, Rehner SA, Moore JD, Moncalvo JM, Canfield SA, Stephenson SL, Methven AS, Volk TJ (2012) Taxonomic revision of true morels (*Morchella*) in Canada and the United States. Mycologia 104:1159–1177. https://doi.org/10.3852/11-375

Læssøe T, Hansen K (2007) Truffle trouble: what happened to the Tuberales? Mycol Res 111 (9):1075–1099. https://doi.org/10.1016/j.mycres.2007.08.004

Lakhanpal TN, Shad OS, Sagar A (1991) Mycorrhiza: a possible deterrent in artificial cultivation of morels. Curr Sci 60:375–377

Lancellotti E, Iotti M, Zambonelli A, Franceschini A (2014) Characterization of *Tuber borchii* and *Arbutus unedo* mycorrhizas. Mycorrhiza 24(6):481–486. https://doi.org/10.1007/s00572-014-0564-9

Lancellotti E, Iotti M, Zambonelli A, Franceschini A (2016) The Puberulum group sensu lato (whitish truffles). In: Zambonelli A, Iotti M, Murat C (eds) True truffle (*Tuber* spp.) in the world, Soil biology, vol 47. Springer International Publishing, Cham, pp 105–124

Landvik S, Egger K, Schumacher T (1997) Toward a subordinal classification of the Pezizales (Ascomycota): phylogenetic analyses SSU rDNA sequences. Nord J Bot 17:403–418. https://doi.org/10.1111/j.1756-1051.1997.tb00337.x

Lanza B, Oweczarek M, De Marco A, Raglione M (2004) Evaluation of phytotoxicity and genotoxicity of substances produced by *Tuber aestivum* and distributed in the soil using *Vicia faba* root micronucleus test. Fresenius Environ Bull 13:1410–1414

Larson AJ, Cansler CA, Cowdery SG, Hiebert S, Furniss TJ, Swanson ME, Lutz JA (2016) Post-fire morel (*Morchella*) mushroom abundance, spatial structure, and harvest sustainability. For Ecol Manag 377:16–25

Le Tacon F, Rubini A, Murat C, Riccioni C, Robin C, Belfiori B, Zeller B, De la Varga H, Akroume E, Deveau A, Martin F, Paolocci F (2016) Certainties and uncertainties about the life cycle of the Périgord black truffle (*Tuber melanosporum* Vittad.). Ann For Sci 73 (1):105–117. https://doi.org/10.1007/s13595-015-0461-1

Leonardi M, Iotti M, Oddis M, Lalli G, Pacioni G, Leonardi P, Maccherini S, Perini C, Salerni E, Zambonelli A (2013) Assessment of ectomycorrhizal fungal communities in the natural habitats of *Tuber magnatum* (Ascomycota, Pezizales). Mycorrhiza 23(5):349–358. https://doi.org/10.1007/s00572-012-0474-7

Leonardi P, Iotti M, Donati Zeppa S, Lancellotti E, Amicucci A, Zambonelli A (2017) Morphological and functional changes in mycelium and mycorrhizas of Tuber borchii due to heat stress. Fungal Ecol 29:20–29

Liu Q, Ma H, Zhang Y, Dong C (2018) Artificial cultivation of true morels: current state, issues and perspectives. Crit Rev Biotechnol 38:259–271

Martin FM, Hilbert JL (1991) Morphological, biochemical and molecular changes during ectomycorrhiza development. Experientia 47:321–331. https://doi.org/10.1007/BF01972073

Martin F, Duplessis S, Ditengou F, Lagrange H, Voiblet C, Lapeyrie F (2001) Developmental cross talking in the ectomycorrhizal symbiosis: signals and communication genes. New Phytol 151:145–154. https://doi.org/10.1046/j.1469-8137.2001.00169.x

Martin F, Kohler A, Murat C, Balestrini R, Coutinho PM, Jaillon O, Montanini B, Morin E, Noel B, Percudani R, Porcel B, Rubini A, Amicucci A, Amselem J, Anthouard V, Arcioni S, Artiguenave F, Aury JM, Ballario P, Bolchi A, Brenna A, Brun A, Buée M, Cantarel B, Chevalier G, Couloux A, Da Silva C, Denoeud F, Duplessis S, Ghignone S, Hilselberger B, Iotti M, Mello M, Miranda M, Pacioni G, Quesneville H, Riccioni C, Ruotolo R, Splivallo R, Stocchi V, Tisserant E, Viscomi AR, Zambonelli A, Zampieri E, Henrissat B, Lebrun MH, Paolocci F, Bonfante P, Ottonello S, Wincker P (2010) Perigord black truffle genome uncovers

evolutionary origins and mechanisms of symbiosis. Nature 464(7291):1033–1038. https://doi.org/10.1038/nature08867
Masaphy S (2010) Biotechnology of morel mushrooms: successful fruiting body formation and development in a soilless system. Biotechnol Lett 32:1523–1527. https://doi.org/10.1007/s10529-010-0328-3
Masaphy S, Zabari L (2013) Observations on post-fire black morel ascocarp development in an Israeli burnt forest site and their preferred micro-sites. Fungal Ecol 6:316–318
Matruchot L (1909) La culture des champignons comestibles, Paris, pp 418–659
Mello A, Fontana A, Meotto F, Comandini O, Bonfante P (2001) Molecular and morphological characterization of *T. magnatum* mycorrhizas in a long-term survey. Microbiol Res 155 (4):279–284. https://doi.org/10.1016/S0944-5013(01)80005-7
Menotta M, Amicucci A, Sisti D (2004a) Differential gene expression during pre-symbiotic interaction between *Tuber borchii* Vittad. and *Tilia americana* L. Curr Genet 46:158–165. https://doi.org/10.1007/s00294-004-0518-4
Menotta M, Gioacchini AM, Amicucci A, Buffalini M, Sisti D, Stocchi V (2004b) Headspace solid-phase microextraction with gas chromatography and mass spectrometry in the investigation of volatile organic compounds in an ectomycorrhizae synthesis system. Rapid Commun Mass Spectrom 18:206–210. https://doi.org/10.1002/rcm.1314
Miller SL, Torres P, McClean TM (1994) Persistence of basidiospores and sclerotia of ectomycorrhizal fungi and *Morchella* in soil. Mycologia 86:89–95
Molliard M (1904) Mycelium et forme conidienne de la morille. CR Hebd Acad Sci 138:516–517
Morte A, Lovisolo C, Schubert A (2000) Effect of drought stress on growth and water relations of the mycorrhizal association *Helianthemum almeriense-Terfezia claveryi*. Mycorrhiza 10:115–119. https://doi.org/10.1007/s005720000066
Moser M (1949) Über das Massenauftreten von Formen der Gattung *Morchella* auf Waldbrandflächen Syndowia. Ann Mycol Ser 2:174–195
Murat C, Payen T, Noel B, Kuo A, Morin E, Chen J, Kohler A, Krizsán K, Balestrini R, Da Silva C, Montanini B, Hainaut M, Levati E, Barry KW, Belfiori B, Cichocki N, Clum A, Dockter RB, Fauchery L, Guy J, Iotti M, Le Tacon F, Lindquist EA, Lipzen A, Malagnac F, Mello A, Molinier V, Miyauchi S, Poulain J, Riccioni C, Rubini A, Sitrit Y, Splivallo R, Traeger S, Wang M, Žifčáková L, Wipf D, Zambonelli Z, Paolocci F, Nowrousian M, Ottonello S, Baldrian P, Spatafora JW, Henrissat B, Nagy LG, Aury JM, Wincker P, Grigoriev IV, Bonfante P, Martin FM (2018) Pezizomycetes genomes reveal the molecular basis of ectomycorrhizal truffle lifestyle. Nat Ecol Evol 2:1956–1965. https://doi.org/10.1038/s41559-018-0710-4
Navarro-Ródenas A, Bárzana G, Nicolás E, Carra A, Schubert A, Morte A (2013) Expression analysis of aquaporins from desert truffle mycorrhizal symbiosis reveals a fine-tuned regulation under drought. Mol Plant Microb Interact 26(9):1068–1078. https://doi.org/10.1094/MPMI-07-12-0178-R
O'Donnell K, Rooney AP, Mills GL, Kuo M, Weber NS, Rehner SA (2011) Phylogeny and historical biogeography of true morels (*Morchella*) reveals an early Cretaceous origin and high continental endemism and provincialism in the Holarctic. Fungal Genet Biol 48:252–265. https://doi.org/10.1016/j.fgb.2010.09.006
Olivier JM, Savignac JC, Sourzat P (2012) Truffe et trufficulture. Fanlac, Périgueux
Ouanphanivanh N, Merényi Z, Orczán AK, Bratek Z, Szigeti Z, Illyés Z (2008) Could orchids indicate truffle habitats? Mycorrhizal association between orchids and truffles. Acta Biol Szegediensis 52:229–232
Ower R (1982) Notes on the development of the morel ascocarp: *Morchella esculenta*. Mycologia 74:142–144
Ower RD, Mills GL, Malachowski JA (1988) Cultivation of *Morchella*. US Patent No. US4866878A
Pacioni G (1991) Effects of *Tuber* metabolites on the rhizospheric environment. Mycol Res 95:1355–1358. https://doi.org/10.1016/S0953-7562(09)80384-5

Paolocci F, Rubini A, Riccioni C, Arcioni S (2006) Reevaluation of the life cycle of *Tuber magnatum*. Appl Environ Microbiol 72(4):2390–2393. https://doi.org/10.1128/AEM.72.4.2390-2393.2006

Papinutti L, Lechner B (2008) Influence of the carbon source on the growth and lignocellulolytic enzyme production by *Morchella esculenta* strains. J Ind Microbiol Biotechnol 35:1715–1721

Pargney JC, Jalade M (1995) Cytological study of fungal structures forming stromas on roots of truffle mycorrhizal plantlets. Bull Acad Soc Lorraines Sci 34:27–44

Pargney JC, Chevalier G, Dupré C, Genet P, Jalade M (2001) Etude des stromas fongiques se développant sur les racines des plants mycorhizés par la truffe. In: Actes Vème Congrès International "Science et Culture de la Truffe", 4–6 Mars 1999, Aix-en-Provence. Fédération Française des Trufficulteurs, Paris, France, pp 167–172

Peng W, Chen Y, Tan H, Tang J, Gan B (2015) Artificial cultivation of morels is blooming in Sichuan. Bulletin of the world society for mushroom biology and mushroom products. www.wsmbmp.org/Bol13/5.html

Perotto S, Angelini P, Bianciotto V, Bonfante P, Girlanda M, Kull T, Mello A, Pecoraro L, Perini C, Persiani AM, Saitta A, Sarrocco S, Vannacci G, Venanzoni R, Venturella G, Selosse MA (2013) Interactions of fungi with other organisms. Plant Biosyst 147:208–218. https://doi.org/10.1080/11263504.2012.753136

Peterson RL, Bonfante P (1994) Comparative structure of vesicular arbuscular mycorrhizas and ectomycorrhizas. Plant Soil 159:79–88

Pfister DH (2015) 2 pezizomycotina: pezizomycetes, orbiliomycetes. In: McLaughlin D, Spatafora J (eds) Syst evol, The mycota, vol 7B. Springer, Berlin. https://doi.org/10.1007/978-3-662-46011-5_2

Philippoussis A, Balis C (1995) Studies on the morphogenesis of sclerotia and subterranean mycelial network of ascocarps in *Morchella* species. Sci Cultivation Edible Fungi 2:847–855

Pilz D, McLain R, Alexander S, Villarreal-Ruiz L, Berch S, Wurtz TL, Parks CG, McFarlane E, Baker B, Molina R, Smith JE (2007) Ecology and management of morels harvested from the forests of western North America. General Technical Report PNW-GTR-710, USDA, Forest Service, PNW 161

Pion M, Spangenberg JE, Simon A, Bindschedler S, Flury C, Chatelain A, Bshary R, Job D, Junier P (2013) Bacterial farming by the fungus *Morchella crassipes*. Proc R Soc Lond [Biol] 280 (1773):2013–2242

Plattner I, Hall IR (1995) Parasitism of non-host plants by the mycorrhizal fungus *Tuber melanosporum*. Mycol Res 99(11):1367–1370. https://doi.org/10.1016/S0953-7562(09)81223-9

Ragnelli AM, Aimola P, Maione M, Zarivi O, Leonardi M, Pacioni G (2014) The cell death phenomenon during *Tuber* ectomycorrhiza morphogenesis. Plant Biosyst 148:473–482

Reyna S (2007) Truficultura Fundamentos y técnicas. Mundi-Prensa, Madrid

Riccioni C, Rubini A, Belfiori B, Gregori G, Paolocci F (2016) *Tuber magnatum*: the special one. What makes it so different from the other *Tuber* spp.? In: Zambonelli A, Iotti M, Murat C (eds) True truffle (*Tuber* spp.) in the world, Soil biology, vol 47. Springer International Publishing, Cham, pp 87–103

Richard F, Bellanger JM, Clowez P, Hansen K, O'Donnell K, Urban A, Sauve M, Courtecuisse R, Moreau PA (2015) True morels (*Morchella*, Pezizales) of Europe and North America: evolutionary relationships inferred from multilocus data and a unified taxonomy. Mycologia 107 (2):359–382. https://doi.org/10.3852/14-166

Riousset G, Riousset L, Chevalier G, Bardet MC (2001) Truffes d'Europe et de Chine. INRA, Paris

Robbins WJ, Hervey A (1959) Wood extract and growth of Morchella. Mycologia 51(3):356–363

Robert ME (1865) Relation entre la famille des oleïnées et les morilles. Bull Soc Bot Fr 12 (5):244–246

Roze ME (1883) Le parasitisme du *Morchella esculenta* Pers. sur l'*Helianthus tuberosus* L. Bull Soc Bot Fr 30:139–143

Rubini A, Paolocci F, Granetti B, Arcioni S (2001) Morphological characterization of molecular typed *Tuber magnatum* ectomycorrhizae. Mycorrhiza 11:179–185. https://doi.org/10.1007/s005720100116

Sbrana C, Agnolucci M, Bedini S, Lepera A, Toffanin A, Giovannetti M, Nuti MP (2002) Diversity of culturable bacterial populations associated to Tuber borchii ectomycorrhizas and their activity on T. borchii mycelial growth. FEMS Microbiol Lett 211:195–201. https://doi.org/10.1111/j.1574-6968.2002.tb11224.x

Schelkle M, Ursic M, Farquhar M, Peterson RL (1996) The use of laser scanning confocal microscopy to characterize mycorrhizas of *Pinus strobus* L. and to localize associated bacteria. Mycorrhiza 6:431–440. https://doi.org/10.1007/s005720050143

Schneider-Maunoury L, Leclercq S, Clément C, Covès H, Lambourdièere J, Sauve M, Richard F, Selosse MA, Taschen E (2018) Is *Tuber melanosporum* colonizing the roots of herbaceous, non-ectomycorrhizal plants? Fungal Ecol 31:59–68

Selosse MA, Faccio A, Scappaticci G, Bonfante P (2004) Chlorophyllous and achlorophyllous specimens of *Epipactis microphylla* (Neottieae, Orchidaceae) are associated with ectomycorrhizal septomycetes, including truffles. Microb Ecol 47:416–426. https://doi.org/10.1007/s00248-003-2034-3

Shamekh S, Donnini D, Zambonelli A, Leisola M (2009) Wild Finnish truffles. Acta Bot Yunnanica 31:69–71

Smith SE, Read DJ (2008) Mycorrhizal symbiosis, 3rd edn. Academic, Amsterdam

Snabl M, Guidori U, Gianchino C, Iotti M, Zambonelli A (2019) Morels on the sand dunes of the Emilia-Romagna coast (Northwestern Adriatic Sea, Italy). Ital J Mycol 48:16–25. https://doi.org/10.6092/issn.2531-7342/9374

Sourzat P (2004) Questions d'écologie appliquées a la trufficulture. Lycée professionnel agricole de Cahors-Le Montat, Impression 43029, Le Montat

Splivallo R, Novero M, Bertea CM, Bossi S, Bonfante P (2007) Truffle volatiles inhibit growth and induce an oxidative burst in *Arabidopsis thaliana*. New Phytol 175(3):417–424. https://doi.org/10.1111/j.1469-8137.2007.02141.x

Splivallo R, Fischer U, Gobel C, Feussner I, Karlovsky P (2009) Truffles regulate plant root morphogenesis via the production of auxin and ethylene. Plant Physiol 150:2018–2029

Splivallo R, Ottonello S, Mello A, Karlovsky P (2011) Truffle volatiles: from chemical ecology to aroma biosynthesis. New Phytol 189:688–699

Stefani FOP, Sokolski S, Wurtz TL, Piché Y, Hamelin RC, Fortin JA, Bérubé JA (2010) Morchella tomentosa: a unique belowground structure and a new clade of morels. Mycologia 102(5):1082–1088

Streiblová E, Gryndlerová H, Gryndler M (2012) Truffle brûlé: an efficient fungal life strategy. FEMS Microbiol Ecol 80(1):1–8. https://doi.org/10.1111/j.1574-6941.2011.01283.x

Tarkka MT, Frey-Klett P (2008) Mycorrhiza helper bacteria. In: Varma A (ed) Mycorrhiza. Springer, Berlin, pp 113–132

Taschen E, Rousset F, Sauve M, Benoit L, Dubois M, Richard F, Selosse M (2016) How the truffle got its mate: insights from genetic structure in spontaneous and planted Mediterranean populations of *Tuber melanosporum*. Mol Ecol 25:5611–5627. https://doi.org/10.1111/mec.13864

Tedersoo L, Hansen K, Brian AP, Kjøller R (2006) Molecular and morphological diversity of pezizalean ectomycorrhiza. New Phytol 170(3):581–596. https://doi.org/10.1111/j.1469-8137.2006.01678.x

Trappe JM, Claridge A (2010) The hidden life of truffles: not just for gourmands, truffles play essential roles in the health of ecosystems. Sci Am 302(4):78–84. https://doi.org/10.1038/scientificamerican0410-78

Urban A, Neuner-Plattner I, Krisai-Greilhuber I, Haselwandter K (2004) Molecular studies on terricolous microfungi reveal novel anamorphs of two Tuber species. Mycol Res 108:749–758. https://doi.org/10.1017/S0953756204000553

Volk TJ, Leonard TJ (1989) Physiological and environmental studies of sclerotium formation and maturation in isolates of *Morchella crassipes*. Appl Environ Microbiol 55:3095–3100

Wassom JJ, Holden DJ (1977) The use of plant tissue culture techniques for studying the growth of morel [mushroom, *Morchella esculenta*]. Proc S D Acad Sci 56:197–206

Winder RS, Keefer ME (2008) Ecology of the 2004 morel harvest in the Rocky Mountain forest. D BC Botany 86:1152–1167

Wipf D, Koschinsky S, Clowez P, Munch JC, Botton B, Buscot F (1997) Recent advances in ecology and systematics of morels. Cryptogam Mycol 18:95–109

Yamada A, Katsuya K (1995) Mycorrhizal association of isolates from sporocarps and ectomycorrhizas with *Pinus densiflora* seedlings. Mycoscience 36:315–323

Yu D, Bu F, Hou J, Kang Y, Yu Z (2016) A morel improved growth and suppressed *Fusarium* infection in sweet corn. World J Microbiol Biotechnol 32:192

Zacchi L, Vaughan-Martini A, Angelini P (2003) Yeast distribution in a truffle-field ecosystem. Ann Microbiol 53:275–282

Zambonelli A, Salomoni S, Pisi A (1993) Caratterizzazione anatomo-morfologica delle micorrize di *Tuber* spp. su *Quercus pubescens* Willd. Micol Ital 22:73–90

Zambonelli A, Salomoni S, Pisi A (1995) Caratterizzazione anatomo-morfologica delle micorrize di *Tuber borchii*, *Tuber aestivum*, *Tuber mesentericum*, *Tuber brumale*, *Tuber melanosporum* su *Pinus pinea*. Micol Ital 24:119–137

Zambonelli A, Rivetti C, Percudani R, Ottonello S (2000) TuberKey: a DELTA-based tool for the description and interactive identification of truffles. Mycotaxon 74:57–76

Zambonelli A, Ori F, Hall I (2017) Mycophagy and spore dispersal by vertebrates. In: Dighton A, White JF (eds) The fungal community: its organization and role in the ecosystem, 4th edn. Tylor and Francis/ACR Press, Boca Raton, FL

Zampieri E, Murat C, Cagnasso M, Bonfante P, Mello A (2010) Soil analysis reveals the presence of an extended mycelial network in a *Tuber magnatum* truffle ground. FEMS Microbiol Ecol 71 (1):43–49. https://doi.org/10.1111/j.1574-6941.2009.00783.x

Zarivi O, Bonfigli A, Colafarina S, Aimola PP, Ragnelli AM, Pacioni G, Miranda M (2011) Tyrosinase expression during black truffle development: from free living mycelium to ripe fruitbody. Phytochemistry 72(18):2317–2324. https://doi.org/10.1016/j.phytochem.2011.08.025

Zarivi O, Bonfigli A, Colafarina S, Aimola P, Ragnelli AM, Miranda M, Pacioni G (2013) Transcriptional, biochemical and histochemical investigation on laccase expression during *Tuber melanosporum* Vittad. development. Phytochemistry 87:23–29. https://doi.org/10.1016/j.phytochem.2012.11.019

Chapter 4
AM Fungi and *Trichoderma* Interaction for Biological Control of Soilborne Plant Pathogen *Fusarium oxysporum*

Khirood Doley, Mahesh Borde, and Mohan Kulkarni

Abstract *Fusarium oxysporum* is an important soilborne destructive plant pathogen that has an effect on several plant species worldwide. The suggested practice for their effective control was the integration of several management practices, but it remains elusive till now. Since it is important to develop disease-resistant and high-yielding crops due to the increase in food demand with minimum utilization of natural resources, it is necessary that the prerequisite employment methodology be of biological origin and that the candidature for the role of biological control agents implies antagonists in various plant–microbe interactions such as arbuscular mycorrhizal fungi and *Trichoderma* spp. This review proposes a framework that might be helpful in the use of AM fungi and *Trichoderma* spp. for their effective biocontrol of several plant pathogens and insights into the mechanisms involved. Also, a relationship between arbuscular mycorrhizal fungi or *Trichoderma* spp. and the host plant is being emphasized upon for improved health and growth for production in present agricultural systems. Therefore, this review focuses on some approaches aimed at the biocontrol of *F. oxysporum* and biotechnological advancement involved in it for paving insights for future research.

4.1 Introduction

Fusarium oxysporum is a soilborne plant fungal pathogen which causes a variety of vascular wilt diseases in plant species through the roots. This species is well documented among soilborne fungal pathogens as it is ranked fifth for being important scientifically and economically (Dean et al. 2012; Li-Jun et al. 2013). *F. oxysporum* is considered to be a general constituent of fungal communities in the rhizosphere of plants (Fravel et al. 2003). It is the most significant plant pathogen which affects diverse host plants (Saremi and Saremi 2013; Mostert et al. 2017; Li

K. Doley · M. Borde
Department of Botany, Savitribai Phule Pune University, Pune, India

M. Kulkarni (✉)
Division of Biochemistry, Department of Chemistry, Savitribai Phule Pune University, Pune, India

A. Varma et al. (eds.), *Plant Microbe Interface*,
https://doi.org/10.1007/978-3-030-19831-2_4

et al. 2018) and is supposed to be endemic in tropical and subtropical countries (O'Donnell et al. 2008). *Fusarium* is a cosmopolitan genus of filamentous ascomycete fungi (*Sordariomycetes*: *Hypocreales*: *Nectriaceae*) with a remarkably broad host range that may include various toxin-producing plant pathogens which are of agricultural significance. Together, *Fusarium* diseases may include blights, rots, wilts and cankers of many horticultural, field, ornamental and forest crops in both agricultural and natural ecosystems (Li et al. 2013).

The following paragraph introduces *F. oxysporum* Schlecht as an example that has gained global significance scientifically and for food security (crop production). As the *F. oxysporum* spp. are considered as soilborne ascomycete pathogens and can cause vascular wilt on several plant species, they compromise plant health and its production (Dean et al. 2012). Also, clinically *Fusarium* sp. infections have been reported to have a virtual 100% death rate among persistent neutropenic patient (Nucci and Anaissie 2007). All isolates of *F. oxysporum* are considered saprophytic as they survive in the soil as organic matter. Furthermore, some isolates are pathogenic to plant species, and other isolates do not invade the root tissues or cause diseases (Alabouvette et al. 1993).

Generally, the life cycle of *F. oxysporum* begins with spore germ tube in the soil, or the mycelium, penetrating through lesions, thereby breaking through the root tips of the probable host plant (Nelson et al. 1997). The developing mycelium enters the xylem vessels through their pits after traversing the root cortex and travels upwards through the plant towards the stem and crown (Aboul-Soud et al. 2004). During disease development, the pathogen continues to grow in the adjoining parenchyma cells in the vascular system, resulting in the production of enormous quantities of conidia (Jimenez and Ricardo 2017). There are three types of asexual spores such as microconidia, macroconidia and chlamydospores produced by *F. oxysporum* (Agrios 1988).

F. oxysporum lacks a sexual cycle in nature, and the means by which its latest pathogenic ancestry has become known is intangible. *F. oxysporum* has a two-partite genome which consists of a core part that is orthologous to other *Fusarium* species and a highly dynamic lineage-specific part (Ma et al. 2010). In tomato, the pathogenic *F. oxysporum* f. sp. *lycopersici* strain 4287, the lineage-specific (LS) part that consists of genomic regions where genes and transposons related to pathogenicity indicated horizontal transfer.

Since *F. oxysporum* is soilborne, this pathogen is difficult to eradicate, and using fungicides does not seem economic or entirely effectual in eradicating *Fusarium* wilt (Sharma et al. 2010). Therefore, for its prevention, several integrated disease management practices need to be functional over traditional ones. Since the present agricultural practices involve the use of various chemical-based fertilizers and pesticides for the improvement of crop yields and the damages incurred upon by several pathogens, it has also brought several other harmful disadvantages to the well-being of humans and animals. Therefore, using organisms that are of biological origin and that do not impose a burden on natural resources will be much more beneficial (Alabouvette et al. 2009) for managing *F. oxysporum*. Earlier, the European directive 2009/128/CE encouraged decreased use of chemical-based pesticides by substitution of chemical means for the effective management of pests and diseases in agricultural practices that are biological, physical or mechanical in nature (EU 2009). Therefore, only biocontrol microbes have been proved to have enormous potentials. Several plant-associated microbial populations have been studied extensively for disease suppression over the last few years. Many biocontrol agents have already shown their potentials in this regard, which include *Bacillus*, *Pseudomonas*, *Trichoderma* spp. and AM fungi for their role in providing pathogen resistance against *Fusarium* wilt in several plants (Cangelosi et al. 2017; Gadag and Krishnaraj 2017; Komy et al. 2015; Rao et al. 2015). Also, effective employment of more than one antagonist is being compared over single isolates which can prove to be superior for diverse environmental niches (Probst et al. 2011).

Therefore, the efficiency of applying AM fungi and *Trichoderma* spp. as biological control agents involves better prospective for sustainable agricultural practices in the near future.

4.2 Arbuscular Mycorrhizal Fungi

Perhaps the most prevalent and certainly significant mutualism between plants and fungi may be referred to as arbuscular mycorrhizal (AM) symbiosis, formed between the majority of land plants and members of the ancient and diverse phylum of fungi known as the *Glomeromycota* (Wang and Qiu 2006; Prasad et al. 2017). It is the most common plant symbiosis known as it involves more than 80% vascular plants. Phylogenetic analysis reports that the symbiosis signalling genes are present in the genomes of the algae which are closest relatives to the land plants and that the function of the encoded proteins is conserved, which indicates that these plant ancestors were preadapted for symbiosis (Delaux et al. 2014). In this mutuality, there is a bidirectional flow of matter between the symbiotic partners, where the fungi obtain mineral nutrients such as phosphorus (P) and provide carbon (C) for the plant (Ferrol et al. 2002; Trepanier et al. 2005). There is evidence that during land colonization by plants whose root systems were not well developed at that time, these fungi assisted plants for extraction of nutrients and water from the soil (Lambais 2006). AM fungi are classified into five broad classes according to their morphological characteristics, viz. arbuscular, arbutoid, ectoericoid, monotropoid and orchid. The AM fungi are extensively identified by the morphology of spores (Robinson-Boyer et al. 2009). During the development of AM colonization, signalling occurs between the two symbionts followed by hyphopodium formation at the root surface and an opening occurs through the epidermal cell layer into the cortex of the root (Wang et al. 2017). Strigolactones are hypothesized to induce germination of AM fungal spores and hyphal growth towards host roots (Waters et al. 2017). Subsequently, when inside the root, the fungus may form highly branched structures known as arbuscules which are the site for nutrient exchange between the fungus and the host plant and therefore are very vital in this mutualism. In addition, there have been reports that not only sugar is transferred by the host plant to AM fungi but also lipids are transported as genes encoding the fatty acid synthase I subunit are found to be absent in mycorrhizal fungi (Keymer and Gutjahr 2018). The AM fungal populations are highest when there is a high diversity of plants such as in tropical rainforests or temperate grasslands where they find more favourable conditions to capitalize on most potential host plants (Smith and Read 2002). As the global environment is changing considerably, one of the factors leading to this may be urbanization. Therefore, urban area as a whole has unique physical, chemical and biological characteristics which need to be studied currently. One such study of novel ecosystems of urban green roof by Chaudhary et al. (2018) has demonstrated the presence of viable and physiologically active propagules of AM fungi on urban green roofs from a variety of roof ages, heights and soil depths extending the presence of AM fungi from below ground to above ground. The study also suggested a new ecological horizon for AM fungi in uninoculated urban green roofs that showed the presence of viable AM fungal propagules between 1 and 8 years after installation with the potential to form new mycorrhizal symbioses.

4.3 Plant–Mycorrhizal Interaction

4.3.1 Role of AMF in Disease Management

Several investigations have shown that when AM fungi are associated with plants, they may confer various adjustments under several pathogen attacks. Many studies have investigated the beneficial role of AM fungi to demonstrate their possible role in the inhibition of plant pathogens in several plant species (Table 4.1). The AM fungi (*Glomus intraradices*) when co-inoculated with rhizobacteria in tomato plants showed various beneficial effects such as increase in growth (fresh and dry root weight), phosphorus content and significantly decreased harmful effects of pathogen in various combinations of both antagonists (Akkopru and Demir 2005). Many investigations have been carried out in the last few years on the role played by AM fungi in disease resistance against several pathogens of plants (Utkhede 2006; Pozo and Azcon-Aguilar 2007; Veerabhadraswamy and Garampalli 2011; Arabi et al. 2013; Eke et al. 2016). Moreover, a study by Zimmermann et al. (2016) has shown abundance of AM fungi *Gigaspora margarita* stimulated by *F. oxysporum* f. sp. *strigae* (Fos) while suppressing pathogen *Striga hermonthica* in maize rhizosphere. Furthermore, it has been observed that when AM fungi are co-inoculated with other biological species such as rhizobacteria, they are very efficient in the management of diseases (Pusztahelyi et al. 2017). Also, when AM fungi are co-inoculated with another species of different biological origin such as *Trichoderma* spp., they have proven to be beneficial, for instance, a study by Omomowo et al. (2018) revealed that when arbuscular mycorrhizal fungi and *Trichoderma* spp. were co-inoculated in cowpea plant for the inhibition of powdery mildew disease and for enhancing its growth, significant reduction in disease severity and incidence was observed along with a significant increase in the overall

Table 4.1 Some examples of AM fungi as biocontrol agent for *F. oxysporum*

Plant species	Disease	Pathogen	References
Chick pea	*Fusarium wilt*	*F. oxysporum* f. cp. *ciceris*	Mahajan et al. (2018)
Tomato	*Fusarium wilt*	*F. oxysporum* f. sp. *lycopersici*	Komy et al. (2015)
Tomato	*Fusarium wilt*	*F. oxysporum* f. sp. *lycopersici*	Raman et al. (2001)
Strawberry	*Fusarium wilt*	*F. oxysporum* f. sp. *fragariae*	Matsubara et al. (2004)
Tomato	*Fusarium wilt*	*F. oxysporum* f. sp. *lycopersici*	Hage-Ahmed et al. (2013)
Cucumber	*Fusarium wilt*	*F. oxysporum*	Wang et al. (2012)
Cucumber	*Fusarium wilt*	*F. oxysporum* f. sp. *radicis-cucumerinum*	Elwakil et al. (2013)
Tomato	*Fusarium wilt*	*F. oxysporum* f. sp. *lycopersici*	Akkopru and Demir (2005)
Tomato	*Fusarium wilt*	*F. oxysporum* f. sp. *lycopersici*	Singh et al. (2010)
Asparagus	*Fusarium wilt*	*F. oxysporum* f. sp. *asparagi*	Matsubara et al. (2001)
Banana	*Panama wilt*	*Fusarium oxysporum* f. sp. *cubense*	Mohandas et al. (2010)
Tomato	*Fusarium wilt*	*F. oxysporum* f. sp. *radicis-lycopersici*	Caron et al. (1986)
Chickpea	*Fusarium wilt*	*F. oxysporum* f. sp. *ciceris*	Garcia-Limones et al. (2002)
Tomato	*Fusarium wilt*	*F. oxysporum* f. sp. *lycopersici*	Scheffknecht et al. (2006)
Tomato	*Fusarium wilt*	*F. oxysporum* f. sp. *lycopersici*	Scheffknecht et al. (2007)
Date palm	Bayoud	*F. oxysporum* f. sp. *albedinis*	Jaiti et al. (2008)
Tomato	*Fusarium wilt*	*F. oxysporum* f. sp. *lycopersici*	Diedhiou et al. (2003)
Cucumber	*Fusarium wilt*	*F. oxysporum*	Hao et al. (2005)
Tomato	*Fusarium wilt*	*F. oxysporum* f. sp. *lycopersici*	Ozgonen et al. (1999)
Persian buttercup	*Fusarium wilt*	*F. oxysporum* f. sp. *ranunculi*	Cangelosi et al. (2017)
Tomato	*Fusarium wilt*	*F. oxysporum* f. sp. *lycopersici*	Ren et al. (2010)
Melon	*Fusarium wilt*	*F. oxysporum*	Martinez-Medina et al. (2009)
Strawberry	*Fusarium wilt*	*F. oxysporum* f. sp. *fragariae*	Li et al. (2010)
Pepper		*F. oxysporum*	Oyetunji and Salami (2011)

(continued)

Table 4.1 (continued)

Plant species	Disease	Pathogen	References
	Fusarium wilt		
Tomato	*Fusarium* wilt	*F. oxysporum* f. sp. *lycopersici*	Gadag and Krishnaraj (2017)
Tomato	*Fusarium* wilt	*F. oxysporum* f. sp. radicis-lycopersici	Utkhede (2006)
Tomato	*Fusarium* wilt	*F. oxysporum*	Al-Hmoud and Al-Momany (2015)

plant growth such as plant height, root length, shoot fresh weight, number of leaves, root dry weight and leaf area when compared to single inoculation of either mycorrhizal or *Trichoderma*. However, in the arbuscular mycorrhizal fungi (*Glomus versiforme*), mutagenesis was induced for strain improvement under a UV lamp at 254 nm wavelength at a distance of 30 cm to the plates for different time intervals (30, 60 and 90 min). The mycorrhizal fungi along with *Trichoderma* spp. and *Pseudomonas fluorescens* were inoculated while planting banana against pathogen *F. oxysporum* f. sp. *cubense*. The results showed significant improvement in plant height and girth along with a decrease in Panama wilt in banana plants (Mohandas et al. 2010). Even flower crops (*Ranunculus asiaticus*) demonstrated protective effects by their association with AM fungi (Cangelosi et al. 2017). The AM fungi (*G. fasciculatum*), *T. viride* and *Pseudomonas fluorescens* were employed in different combinations for the biocontrol of *Fusarium* wilt in tomato plants. When all three antagonists were applied together, it reduced disease suppression by 94% when compared to other treatments. The higher level of phosphorus uptake was facilitated by AM fungi and *T. viride* assisted in the induction of antioxidant enzymes, viz. catalase and peroxidase (Tayal et al. 2011). In a novel method implemented by Elwakil et al. (2013), the seeds of cucumber plant were presoaked with antioxidant formulations along with AM fungal spores against pathogen *F. oxysporum*. The results indicated significant control of the pathogen, and the overall health, rate of photosynthesis and AM fungal colonization were found to be upregulated. Some argued that if AM fungi are applied in consortia relative to single mycorrhizal species rather than employing single mycorrhizal species, they would increase growth parameters and significantly reduce disease incidence and severity due to *F. solani* on common bean seedlings (Eke et al. 2016). Furthermore, when colonization of mycorrhizal species was observed, it showed lower colonization when *T. harzianum* was present which might be due to the competition among them. Similarly, *Trichoderma* spp. (*T. harzianum*, *T. virens* and *T. viride*) and the mycorrhizal consortium (*Funneliformis mosseae*, *Glomus cerebriforme*, *Rhizophagus irregularis*) were evaluated for their role in growth promotion by inoculating them on pigeon pea. It not only enhanced disease suppression caused by *F. udum* but also helped in the upregulation of several growth parameters (Dehariya et al. 2015). Several plants that are inoculated with AM fungi confer resistance to the host;

however, elicitation of defence response is not generated against AM fungi association by the host which still needs further investigation (Schmitz and Harrison 2014). Moreover, how the AM fungi enable survival of the host plant during a pathogen attack and the molecular mechanism involved in mycorrhiza-induced resistance (MIR) need further clarification (Basu et al. 2018). Most importantly, a contradictory study by Svenningsen et al. (2018) demonstrated suppressive activity incurred upon AM fungi by the surrounding microbiota. AM fungi are well known to colonize a large number of plants, and their nutritional acquirement is provided by extra-radical mycelium (ERM) which extends in the soil where other soil microbial communities are present. The ERM activity was found to vary significantly which suggested AMF suppression, and this suppression occurred due to the presence of *Acidobacteria*. Furthermore, even though AM fungal inoculants are being commercially manufactured worldwide by various manufacturers singly or in combinations with other biocontrol species (Gianinazzi and Vosatka 2004), there are several constraints in employing them for actual use in the field. The reasons may include being economically not feasible, low acceptance among growers, labour inducing and low performance in local conditions (Mishra et al. 2018). Moreover, AM fungi are obligate symbionts which means they would not survive on their own, and making pure cultures separately and then manufacturing them on a large scale would be a major constraint in AM inoculum production. However, if AM fungi can be utilized as biofertilizers instead of traditional chemical fertilizers, then it would be more suitable to sustain our agricultural systems in the future (Berruti et al. 2016). Thus, as discussed above, identifying an effective combination of AM fungi might be valuable for transposition of this knowledge not only to increase crops production but also in pathogen management.

4.3.2 Role in Plant Nutrition

Phosphorus is considered to be one of the most important macronutrients required for plant metabolism and growth. Phosphate is a crucial component of macromolecules such as sugar phosphates, phospholipids and nucleotides. As phosphate is present in an immobile inorganic (Pi) form and is made available by the symbiotic association of AM fungi with the host plants, phosphorus is scavenged efficiently from the soil and is made available to the roots, thereby providing phosphate nutrition which is an indirect pathway. In this, an AM fungus prevails over the depletion zone due to reduction in uptake of Pi due to its rapid absorption (Smith and Read 2008). For this limited availability of phosphorus, the host plant then explores the soil with the help of AM fungi (Hinsinger et al. 2011). As far as uptake of phosphorus is concerned in plants whether associated with AM fungi or not, the foremost course for phosphorus uptake preferably will be that of the AM uptake pathway only (Jakobsen 1999). However, if phosphorus is adequately present in the soil, then AM association is discouraged by the host plant by repressive genes (Breuillin et al. 2010). The ability of AM fungi for the uptake of Pi in broad soil pH range is facilitated by proton- and sodium-coupled transporters (H+/Pi and Na+/Pi) (Johri et al. 2015). The mutual site for exchange of nutrients that takes place between the host plant and the fungi is considered to be the symbiosome although research is needed to shed more light on the structural composition of the peri-arbuscular membrane that can explain the symbiotic interaction (Basu et al. 2018). If the soil consists of AM fungi, then phosphorus levels are upregulated in the host plant (Halder and Ray 2006; Tawaraya et al. 2007). In addition, if phosphate is utilized for soil cultivation as in the present time, then it will not be economically feasible to extract and is expected to be exhausted very soon (Sawers et al. 2007). Also, the AM fungi associations can decrease the risk of nutrient loss by enhancing nutrient immobilization when compared to non-mycorrhizal plants or help in alteration of soil nutrients and water cycling processes in ways that favour the retention of nutrients in the soil (Cavagnaro et al. 2015).

Thus, the prospects of employing AM fungi seem to have more implications in the future.

The growth of plants is affected by the diminution of nitrogen from the soil, and AM fungi facilitate the uptake of nitrogen from the soil and make it available to the plants. They also make resources available to increase the usage of different forms of nitrogen (Basu et al. 2018). In addition, AM fungi-inoculated plants can uptake nitrogen directly and transfer it to the host roots (Auge 2001; Andrade et al. 2015). However, the role of AM fungi in nitrogen acquirement in the host plant is inconsistent (Reynolds et al. 2005). While the molecular mechanism of nitrogen uptake is scarce, fungal glutamine synthase and nitrate reductase genes in mycorrhizal fungi explain the incorporation of mineral forms of nitrogen (Tian et al. 2010; Veresoglou et al. 2012). As far as preference is concerned between nitrate ($NO3^-$) and ammonium ($NH4^+$) assimilation in microbes, they prefer direct assimilation of $NH4^+$ due to high energy demands for the reduction of $NO3^-$ into $NH4^+$ (Courty et al. 2015), despite nitrate being the most important source of nitrogen in soil. In addition, the AM fungi-colonized root only induces plant nitrate transporters (Willmann et al. 2014). Even zinc nutrition in AM fungal associations has been reported, but has not been investigated so far. Watts-Williams and Cavagnaro (2018) recently demonstrated that when modern barley was inoculated with AM fungi (*Rhizophagus irregularis*), the roots showed significant upregulation of zinc concentration from zinc-deficient soil by several *ZIP* transporter genes. However, an increase in plant biomass was not observed, but an improvement in zinc concentration of barley grains seemed a better prospective. Lastly, besides assisting the host plant in nutrient acquirement, it also helps in improving tolerance in several crop plants to several abiotic and biotic stresses (Wu et al. 2013; Sikes 2010).

4.3.3 Role in Photosynthesis

When AM fungi colonize the host plant, they induce Calvin-Melvin cycle which in turn results in an increased rate of photosynthesis. Sheng et al. (2008) demonstrated that when the maize plants were grown under salt stress condition, the plants were significantly able to grow along with an overall increase in water status, chlorophyll concentration, gas exchange and chlorophyll fluorescence. The upregulation in the process of photosynthesis may be attributed to the improvement in nutritional status due to the colonization by AM fungi to the host plant (Dong et al. 2008). Moreover, higher concentrations of chlorophyll have been reported in AM fungi-associated plants, and increased chlorophyll level helps in increasing photosynthetic rates (Davies et al. 1993; Mathur and Vyas 1995). The level of sugar content in AM fungi-associated plants was found to be increased (Sheng et al. 2011).

4.3.4 Role in the Production of Phytohormones

Mycorrhization of plants by AM fungi often results in more vigorous growth of plants in terms of their overall growth which is expected to be under the control of phytohormones, and such hormones may be jasmonic acid (JA), salicylic acid (SA), ethylene and abscisic acid (ABA) which also play significant roles as signalling compounds in biotic or abiotic interactions (Garcia-Garrido and Ocampo 2002; Nair et al. 2015; Pozo et al. 2015). The more vigorous the plant, the better the resistance of pathogens (Harrier and Watson 2004). Abscisic acid (ABA) was found to be considerably upregulated in plants that are associated with AM fungi establishment (Ludwig-Muller 2010). Also, ABA production in response to an AM fungus has been shown to contribute in abiotic salt stress tolerance in *Sesbania cannabina* seedlings (Ren et al. 2018). Generally, ABA converts environmental signals in gene expressions (Suzuki et al. 2016). It is also related to a higher rate of photosynthesis in plants (Zorb et al. 2013). The upregulation of auxin hormone after AMF colonization have shown to increase high level of lateral root formations which constitute preferential penetration sites for AM fungi's hyphae which help in disease resistance by closing the infection cycle (Ludwig-Muller and Guther 2007). Also, the role and increase of IAA (indole acetic acid) and cytokinin level were observed during AM fungi inoculations with the host plants (Barker and Tagu 2000; Meixner et al. 2005). Also, mycorrhizal fungi were investigated for their role in the production of auxin-, gibberellin- and cytokinin-like substances that stimulated the growth of plants (Barea and Azcon-Aguilar 1982). The ethylene signalling pathway is found to be involved in potato plants inoculated with AM fungi (*Rhizophagus irregularis* MUCL 41833) by upregulation in the expression of ethylene response factor (EFR3) in mycorrhizal potato plants against *Rhizoctonia solani* (Velivelli et al. 2015). Strigolactones are one of the most important root exudates that assist during initial establishment of AM colonization towards roots of host plants (Akiyama et al. 2005); hence, in response to this, the plant's immune system locally responds by activation of salicylic acid-dependent defences (Gallou et al. 2011). Jasmonic acid is one of the most studied phytohormones in mycorrhiza associated with plants. The AM fungal colonization in tomato plant has been shown to be strongly controlled by the jasmonate signalling pathway. Moreover, jasmonic acid has been shown to contribute to the susceptibility of tomato plants to infection by AM fungi and play a regulatory role in the development of mycorrhizal colonization (Herrera-Medina et al. 2008). Also, the significant jasmonate levels in AM fungi-associated plants have demonstrated reorganization and alterations in the cytoskeleton, biosynthesis of flavonoids and overall improvement in the health of plants (Hause et al. 2007). Cytokinins (CK) comprise phytohormones that assist in the regulation of several fundamental aspects in terms of overall plant development (Kieber and Schaller 2014). Cosme et al. (2016) used transgenic tobacco (*Nicotiana tabacum*) to demonstrate CK regulation in nutrient exchange during AM symbiosis where the level of shoot CK showed a positive impact on AM fungal development in roots and on the root transcript level of an AM-responsive phosphate transporter gene (NtPT4). Foo et al. (2013) in his investigation showed significant upregulations in the level of gibberellin-deficient *na-1* mutants (dwarf) compared with wild-type pea plants when they were associated with mycorrhizal fungi.

4.3.5 AM Fungi-Induced Defence Signalling Pathways

Defence-related compounds such as cytosolic calcium, fatty acids, jasmonic acid, reactive oxygen species (ROS), salicylic acid and ethylene get activated when plants interact with microbes through highly complex molecular signalling by cascades of reactions. The AM fungi's myc factor (myc) leads to further induction of G-protein alterations, MAPK and ROS generation. The phenylpropanoid pathway led to the production of several aromatic secondary metabolites, flavonol synthase such as flavonols which proved to be beneficial for host plants in providing defence against pathogens (Wuyts et al. 2006). Khan et al. (2010) illustrated AM fungi-induced defence responses in plants where the myc factor from AM fungi triggers cytoplasmic Ca^{2+} which leads to the induction of ROS generation and G-protein, and MAPK alterations as well. Also, ROS induces lipoxygenase leading to jasmonic acid biosynthesis, resulting in the phosphorylation of enzymes such as superoxide dismutase, peroxidase, catalase and ascorbate peroxidase which play an important role in ROS metabolism through G-protein and MAPK. MAPK and G-protein trigger defence genes in plant also. Thus, when a pathogen invades, defence-related genes are triggered that encode proteins and subsequently attack and inhibit the pathogen. Furthermore, antioxidant enzymes along with ROS act constitutively on the pathogenic site, and a hypersensitivity reaction is initiated which lastly leads pathogenic cells towards apoptosis. During transcriptome analysis, it was observed that the upregulated expression of ROS scavenging-related genes, viz. thioredoxin, glutaredoxin and glutathione peroxidase (GPX), was related to atrazine (herbicide) stress tolerance in *Glomus mosseae/Medicago sativa* (Song et al. 2015). In addition, the plant utilizes one of the most common responses under stressful condition which is the generation of reactive oxygen species (ROS) as the major signalling agent and activates various adaptive defence mechanisms (Sewelam et al. 2016). Thus, the plants have mechanisms for protection from the ROS by the production of various antioxidant enzymes (Schmitt et al. 2014).

4.3.6 Role in Enzyme Activities

Plants under pathogen attack generally respond by the induction of defence-related marker enzyme activity, namely peroxidase (POX), polyphenol oxidase (PPO), β-1,3-glucanase, chitinase, phenols and phenolics (Elwakil et al. 2013; Thakker et al. 2013). Also, the plant adjusts itself during pathogen invasion by the biosynthesis of compounds such as alkaloids, flavonoids or phenolic acids. The phenolic acids may act as a signalling molecule during resistance to plant pathogens (Ruelas et al. 2006). Lopez-Raez et al. (2010) reported that upon mycorrhization, the major phenolic acids are also altered in tomato roots in addition to the altered hormonal and transcriptional profiles which can be attributed to their regulation in plant defence mechanisms during symbiosis. The PAL activity through the phenylpropanoid pathway and phenolics along with flavonoid activity has been found to be altered in mycorrhiza-inoculated bean plants against *Fusarium solani* (Eke et al. 2016). Zeriouh et al. (2017) have reported that phenolic extract (PEOL) from oleaster (*Olea europaea* var. *sylvestris*) leaves lead to the production of reactive oxygen species (ROS), resulting in the induction of stress in the endoplasmic reticulum by PEOL and finally apoptosis. The upregulation in peroxidase is one of the most well-known biochemical activities in pathogenic plants. Against pathogen infections like *F. oxysporum*, peroxidase's involvement is considered to be one of the important defence compounds that are activated in plants (Morkunas and Gemerek 2007). Peroxidases have an important function in plant defence mechanisms which is the ability of oxidizing key metabolites such as phenolics in plants or pathogens (Mohammadi and Kazemi 2002). Moreover, the AM fungi have been reported to regulate catalase and peroxidase enzyme activities (Blee and Anderson 2000). However, in a study by Kumar et al. (2009), the AM fungi (*Piriformospora indica*) inoculations significantly decreased the antioxidant enzyme activities, viz. catalase (CAT), glutathione reductase (GR), glutathione S-transferase (GST) and superoxide dismutase (SOD), in the roots of maize plant which suggests that the presence of AMF (*P. indica*) might have assisted the maize plant to overcome the disease load of pathogen *Fusarium verticillioides*. The polyphenol oxidase enzyme is considered to take part in the induction of defence resistance against pathogens by oxidation of polyphenols into quinines which are antimicrobial compounds, and it also causes cell wall lignifications to inhibit further infection (Mayer 2006). In addition, the mycorrhizal tomato plant has shown increased polyphenol activities in leaves and roots as compared to non-mycorrhizal plants, suggesting that AM colonization alleviates wilt disease by enhancement in plant resistance (Ren et al. 2010). The role of phenolic compounds as defence-related enzymes has been studied by Abdel-Fattah et al. (2011), who found that the antimicrobial phenolic compounds were increased when the common bean was inoculated with AM fungi against *Rhizoctonia soloni* Kuhn and resulted in the enhancement against it.

4.4 Induction of Systemic Resistance

The contemporary agricultural system has put great emphasis on the utilization of chemicals for the management of disease in plants which is hazardous for humans and the environment. Therefore, the phenomenon of induced systemic biological control could be a better alternative that could minimize the mass scale of chemical-based pesticides or insecticides. Investigations have already suggested AM fungi-mediated systemic resistance in the biocontrol of several pathogens (Pozo and Azcon-Aguilar 2007; Pineda et al. 2010; Castellanos-Morales et al. 2011; Jung et al. 2012; Vos et al. 2014). Systemic acquired resistance (SAR), which is also the induced resistance in plants, is expressed during pathogen invasion, and it assists in resistance to several pathogen attacks. The infected tissue shows accumulation and synthesis of endogenous salicylic acid and is often characterized by pathogenesis-related (PR) proteins such as chitinases, and β-1,3-glucanases which catalyse the hydrolysis of chitin and β-1,3-glucanases that enable direct degradation of microbial cell wall components by antimicrobial activity (van Loon 1997; Xu et al. 2009). Furthermore, new insights are shedding more light on the interaction of AM fungi and pathogen with respect to resilience in the host plant. The plants recognizes pathogens by their pattern recognition receptors (PRRs) such as chitin in *Arabidopsis* (Cao et al. 2014) is a highly conserved component of fungi's cell wall is recognized by LysM-RLK *At* LYK5 (LysM-containing receptor-like kinase 5) which helps in recognizing pathogen-associated molecular patterns (PAMPs). Plant PRRs are perceived to be receptor-like kinases (RLKs) or receptor-like proteins (RLPs) which sense any extracellular threat that leads to the activation of intracellular immune signalling (Macho and Zipfel 2014). It is apparent that during induction of systemic plant defence responses when a plant starts to have an association with AM fungi, the MAMPs (microbe-associated molecular patterns) recognize AM fungi to be pathogenic as MAMPs are conserved between beneficial and pathogenic fungi (Zamioudis and Pieterse 2012). After recognition of MAMP by plant PRRs, MAMP activated MTI (MAMP-triggered immunity response) directs the formation of defence line against the pathogen to inhibit its advancement (Jones and Dangl 2006). In the roots of wheat plants, the AM symbiosis led to significant upregulation of some proteins such as three acidic endochitinases, elicitor-responsive gene 3, hypersensitive-induced reaction protein 3 and the cysteine-rich receptor-like protein kinases, Crk 25 and Crk6 that were found to be involved in defence-related activities in the wheat plant (Fiorilli et al. 2018). This study further confirmed the systemic response by proteomic analysis where chitinase, Germin-like genes and proteins (GLP), phenylalanine ammonia lyase and PR protein were found to be upregulated. Appropriate activation of these molecular events may be fundamental for providing immunity in host plants.

4.5 Application of Mycorrhizal Genes

In the induction of AM fungi associations with host plants, several defence-related genes have been reported which confer defence-related activities such as anti-fungal activities, antimicrobial, and direct the production of phenolic substance and phyto-alexins, viz. *pl 176*, *Pal*, *PR-1a*, *TC104515* (Ruiz-Lozano et al. 1999; Blilou et al. 2000; Conrath et al. 2006; Liu et al. 2007). Generally, for the maintenance of a safe ion concentration, chloride channel proteins play a significant role contained by the plant cells. For the *Fusarium* wilt suppression, 14-3-3-like protein, which is believed to mediate during innate immunity, was found to be significantly increased in the resistant genotype in pea plants together with an increase of chloride channel protein which was attributed to have provided resistance against pathogen *F. oxysporum* (De Angeli et al. 2007; Castillejo et al. 2015). Sun et al. (2018) have characterized three genes, viz. *Fm201*, *Ri14-3-3* and *RiBMH2*, in the AM fungi, namely *Funneliformis mosseae* and *Rhizophagus irregularis*, that encode for 14-3-3-like proteins perceived to be involved during arbuscule formation and abiotic stress. Fiorilli et al. (2018) also demonstrated the expression of 29 more novel genes in AM fungi-associated plant leaves which were pathogenesis-related protein *PR-1*, three barley mildew resistance locus O (*MLO*) genes and putative homologues of *resistance to Pseudomonas syringae* pv. maculicola 1 (*RPM1*) protein that were supposed to be associated with those expressed during biotic stress. Hence, the prospects of genes from AM fungi are of major significance for crop production and protection for future agricultural systems.

4.6 The Fungus *Trichoderma*

Trichoderma spp. are ubiquitous, often a major component of the soil mycoflora, organic matter and occur as saprophytes in rhizospheric ecosystems of almost all climatic zones, viz. agricultural land, forest, marsh and desert soils of Antarctic, tundra or tropical regions (Druzhinina et al. 2011). Also, they are well established in air, plant biomass, water and in the surrounding areas of almost all plant species (Mukherjee et al. 2013a). The fungus genus *Trichoderma* that belongs to Division—Ascomycota, Subdivision—Pezizomycotina, Class—Sordariomycetes, Order—Hypocreales, Family—*Hypocreaceae* has been well acknowledged since the 1920s for its capability to function as biological control agents (BCAs) against plant pathogens (Samuels 1996). Moreover, *Trichoderma* spp. have been in existence since 1865 (Bisby 1939). The strains of *Trichoderma* spp. were found to colonize plant roots of both monocots and dicots (Harman and Shoresh 2007). It is well known for its ability to control plant pathogens for the sustainability of modern day agricultural practice systems with less implication on natural resources. Weindling (1932) firstly investigated the potential application of *Trichoderma* for the control of plant pathogenic fungi *Rhizoctonia solani*. Since then, it has been reviewed many times for its ability as biological control agent for many plant diseases (Verma et al. 2007; Bhagat and Pan 2008; Van Wees et al. 2008; Saravanakumar et al. 2016; Amira et al. 2017; Abhiram and Masih 2018; Mahajan et al. 2018). The Table 4.2 illustrates biocontrol efficacy of *Trichoderma* spp. against *F. oxysporum* in various plants. *Trichoderma* is often characterized by rapidly growing colonies with repeatedly branched conidiophores along with green or hyaline conidia. The primary branch of the conidiophore generally produces secondary branches which may also give rise to tertiary branches. In the end, the majority of single phialides are found to be structurally simple (Rifai 1969). When *Trichoderma* has favourable growth conditions *in vitro*, it readily grows very rapidly, and the colonies may appear greenish, with a reddish tinge, white, blue-green, yellow-green or colourless (Błaszczyk et al. 2014; De Hoog et al. 2000). *Trichoderma* spp. have generally been considered for various beneficial effects apart from providing disease resistance for example as a plant growth promoter, inoculating *T. viride* in cotton seeds resulting in increased germination (Shanmugaiah et al. 2009). Toghueo et al. (2018) showed that *Trichoderma* spp. not only sporulated on the growth of the pathogenic *Fusarium solani* but also helped in the promotion of seed germination in bean plant which led to significant biocontrol of plant pathogens and promoted plant growth as well. Also, when *Trichoderma* spp. colonize plant roots, they not only are advantageous for root growth and development or crop productivity but also assist in the uptake and use of nutrients. Current investigation on *Trichoderma* has revealed its qualities of being opportunistic, avirulent plant symbionts, over and above being parasites of other fungi, which makes it a perfect candidate for its employment as an inhibitor or inducer of resistance against plant pathogens, systemic tolerance against biotic as well as abiotic stresses together with enhancement of plant growth, maintenance of soil, nutrient uptake and increased seed germination (Howell et al. 2000; Escande et al. 2002; Waghunde et al. 2016; Zeilinger et al. 2016; Mendoza-Mendoza et al. 2018). Qualhato et al. (2013) reported that there was a positive correlation between the amounts of secreted cell wall degrading enzymes by *Trichoderma* strains and their ability to control plant pathogenic fungi.

In addition, *Trichoderma* produces and/or releases a variety of metabolites or factors responsible for disease control in many plants via amylase, antibiotics, anthraquinones, aryl-β-glucosidase, β-1,3-glucanase, β-glucosidase, chitinase, endochitinase, endoglucanase, enzymes, extracellular proteases, *N*-acetylglucosaminidase, protease, total cellulase, terpenoids and non-volatile and volatile antibiotics (De Marco et al. 2003; Score and Palfreyman 1994; Hanson 2000; Roco and Perez 2001; Kredics et al. 2005; Komy et al. 2015; Saravanakumar et al. 2016). Even though certain strains of *Trichoderma* spp. are reported to be systemic endophytes, whether they are symbiotic, endophytic or pathogenic still needs further investigation (Druzhinina et al. 2011; Mendoza-Mendoza et al. 2018).

At present formulations of *Trichoderma* spp. are doing well as bio-fungicides in accounting for more than 60% of the registered bio-fungicides worldwide (Verma et al. 2007), and if there are developments for them to be used as a nano-encapsulated product, then it will put forward novel strategies for biocontrol in an

Table 4.2 *Trichoderma* spp. and biocontrol of *F. oxysporum* in various plants

Strain	Pathogens	Mechanism	Disease	References
T. asperellum	*Fusarium oxysporum*	Antibiosis, mycoparasitism and competition for nutrients	Wilt of tomato	Cotxarrera et al. (2002)
T. harzianum	*Fusarium oxysporum f.* sp. *radicis-lycopersici*	Enzymes, antibiotics and anti-fungal properties	Common plant fungal diseases	Grondona et al. (1997)
T. asperellum	*Fusarium oxysporum*	Secondary metabolites	*Fusarium* wilt	Saravanakumar et al. (2016)
T. asperellum	*Fusarium oxysporum f. sp. lycopersici*	Enzymes	*Fusarium* wilt of tomato	Komy et al. (2015)
T. viride	*Fusarium oxysporum* and *Alternaria alternate*	PO, PPO, PAL, catalase, total phenols and antioxidants	Pigeon pea, moong bean	Rao et al. (2015)
T. hamatum	*F. oxysporum* f. sp. *lentis*	Mycoparasitism, antibiosis	Vascular wilt disease of lentil	El-Hasan et al. (2012)
T. harzianum	*F. oxysporum* f. sp. *radicis-cucumerinum*	Defence-related genes	Cucumber	Alizadeh et al. (2013)
T. harzianum	*F. oxysporum* f. sp. *lycopersici*	Competition	*Solanum melongena* L.	Balaji and Ahir (2011)
T. reesei	*F. oxysporum* f. sp. *ciceris*	Lysis	Vegetables	Mukherjee (1997)
T. harzianum	*F. oxysporum* f. sp. *lycopersici*	Competition	Tomato	Marzano et al. (2013)
T. harzianum	*F. oxysporum* f. sp. *ciceris*	Competition	Chickpea	Kumar et al. (2014)
T. harzianum, *T. viride*	*F. oxysporum* f. sp. *ciceri*	Competition	*Fusarium* wilt	Patole et al. (2017)
Trichoderma spp.	*F. oxysporum*	Competition	Wilting	Alabouvette et al. (2009)
T. asperellum	*F. oxysporum* f. sp. *carthami*	Competition	*Fusarium* wilt	Waghmare and Kurundkar (2011)
T. viride	*F. oxysporum*, *F. oxysporum* f. sp. *gladioli*	Competition	Corm rot	Mir et al. (2011)
T. harzianum	*F. oxysporum* f. sp. *ciceri*	Enzymes	Wilt	Jayalakshmi et al. (2009)
Trichoderma spp.	*F. oxysporum* f. sp. *melonis*	Competition for nutrients	Melon wilt	Gava and Pinto (2016)

environment friendly way (Sharma et al. 2017). In India, approximately 250 products exist for applications in the field, but this represents only a fractional percent-share of *Trichoderma* spp.-based bio-fungicides which is not comparable to the huge chemical-based fungicide market (Singh et al. 2009). The main constraints of these bio-fungicides are that they act slowly and environmental factors influence them, thereby restricting their efficacy under field conditions. Therefore, designing new strains is a must by 'genetic manipulation' for making them more effective than the native ones. Therefore, the mechanism of interactions of *Trichoderma* spp. with abiotic as well as biotic factors by getting hold of molecular mechanisms could be more successful (Mukherjee et al. 2012).

4.7 Role in Plant Growth

Rhizospheric microbial populations may prove to be beneficial as they include many properties which might be decisive of the growth of the plant as a whole which not only helps in sustaining plant growth but also elevates defence against pathogens in plants (Daguerre et al. 2014; Giurgiu et al. 2018). Hence, the microbial population near the region of the roots affects root architecture by enhancing root biomass and increases the development of root hair (Harman et al. 2004). The *T. harzianum qid74* gene that encodes for cellular protection was found to be increased in *Trichoderma* high-density oligonucleotides (HDO) in the roots of tomato plants, which altered the architecture of the root and in so doing resulted in an increased absorptive surface that helped in increased efficacy of nutrient utilization for the plant (Samolski et al. 2012). *Trichoderma* spp. when treated on seeds not only increased the yield in spring wheat but also decreased root rot severity caused by *Fusarium* spp. in Canada. However, it was suggested that trials should be conducted at multiple locations and over different time periods for the actual determination of *Trichoderma* spp. as a biocontrol agent (Xue et al. 2017). *Trichoderma* spp. were suggested as plant growth-promoting fungi for their role in the production of phytohormones, phosphates-solubilizing enzymes and siderophores (Doni et al. 2014). Volatile compounds such as 6-pentyl-2H-pyran-2-one, auxins for root enhancement or branching, secondary metabolites and small peptides play a significant role not only in plant growth and development but also in providing innate immunity in plants (Contreras-Cornejo et al. 2009; Lopez-Bucio et al. 2015). Moreover, auxins have been known to play an important role in the development and reproduction of plants by the auxin pathway (Kazan 2013). However, currently how *Trichoderma*–root interaction occurs is still being investigated (Mendoza-Mendoza et al. 2018). Enhancement in the number of leaves, tillers and plant height in rice by *Trichoderma* spp. has been reported (Doni et al. 2014). Kiriga et al. (2018) reported that *Trichoderma* isolates can improve pineapple root mass growth through fresh root enhancement. Al-Hazmi and Javeed (2016) reported that *Trichoderma* spp. can improve tomato plant development as compared to control ones. Furthermore, these enhanced morphological modifications are possible in plants as *Trichoderma* produces plant growth regulators, stimulates and increases the surface of roots for better absorption of nutrients and enables several other plant growth-promoting activities, viz. ability to solubilize Ca3(PO4)2 and to produce cellulases, siderophores, IAA, proteases and chitinases (Djonovic et al. 2007; Li et al. 2018).

4.8 Biocontrol Genes of *Trichoderma* and Their Role

The study of genomics has characterized and quantified important genes in *Trichoderma* spp., which may be helpful in applying them in plant growth activities and as biocontrol agents (BCAs). The genome sequencing of *Trichoderma* spp. has provided significant information from bioinformatical analysis to comprehend their character for sustainable agricultural systems and it provides opportunity for the development of biological approaches. Several isolated, cloned and characterized genes for biocontrol include protease, chitinase, glucanase, tubulins, proteinase, xylanase, monooxygenase, galacturonase, cell adhesion proteins and stress tolerant genes which have a distinctive role in the biocontrol mechanism, viz. cell wall degradation, hyphal growth, stress tolerance and parasitic activity (Sharma et al. 2011). Presently, the genome sequences of *Trichoderma* spp. are as follows: *Trichoderma atroviride*, *Trichoderma longibrachiatum*, *Trichoderma reesei*, *Trichoderma asperellum*, *Trichoderma citrinoviride*, *Trichoderma harzianum* and *Trichoderma virens* are accessible (Sharma et al. 2011). Genes in *Trichoderma* spp. such as *ech42*, gene-encoding cellobiohydrolase (cbh1) and endochitinase *ech42* were reported for the suppression of red rot in sugarcane and stem rot caused by *S. sclerotiorum*, respectively (Singh et al. 2016; Vinodkumar et al. 2017). Cellulase genes such as *Thph1* and *Thph2* from *T. harzianum* have been reported to control foliar disease of *Fusarium* stalk rot in maize by the activation of Jasmonic acid (JA), ethylene (ET), systemic acquired resistance (SAR), salicylic acid (SA), 2,4-dihydroxy-7-methoxy-2H-1,4-benzoxazin-3(4H)-one (DIMBOA) and plant innate immunity (PII)-related genes (Saravanakumar et al. 2018). Further research is needed for greater understanding on disease resistance provided by *Trichoderma* spp. when they are under threat from pathogens. So, many researchers have started the use of genomic and proteomic analysis to enclose the changes that occur in *Trichoderma*, plant and pathogens when they form a network with each other.

4.9 Induced Resistance Against Biotic and Abiotic Stress

During the life cycle of a plant, it is susceptible to various kinds of stress which may be biotic or abiotic. The immune system of a plant recognizes any penetration by any unfamiliar substance, but *Trichoderma* spp. have established themselves with a plant immune system (Zamioudis and Pieterse 2012). Therefore, plants have developed a wide variety of strategies for their resistance (Ponce de Leon and Montesano 2013).

Hence, the resistance in plants can be in the form of passive, non-host, physical or chemical barriers, rapid active defences and delayed active defences. In the case of rapid active defence, various responses such as cell wall reinforcement, initial oxidative outburst, hypersensitive response or changes in membrane function are involved which ultimately lead to programmed cell death. In delayed active defences, injury is repaired and suppression of pathogen, pathogenesis-related (PR) gene expression and systemic acquired resistance (SAR) occurs (Rahman et al. 2018). Induced systemic resistance is considered to be one of the most important mechanisms of the biocontrol effects of *Trichoderma* (Harman 2006). When plants interact with other microorganisms, resistance is supposed to be induced by *Trichoderma* spp. by defence-related signalling molecules that include salicylic acid (SA), jasmonic acid (JA) and ethylene (ET), in beneficial plant–microbe interactions (Hermosa et al. 2012). The immune system of the plant does accumulate mRNA that leads to the expression of the lipoxygenase pathway gene encoding hydroxyperoxidase lyase (HPL) and the phenylpropanoid pathway gene encoding phenylalanine ammonia lyase (PAL) for the biocontrol of plant pathogens. Hence, the above-mentioned investigations studied the implications of beneficial microorganisms for secondary metabolism stimulation, and thus help in improving the bioactive substance content of the plants (Lopez-Bucio et al. 2015).

4.10 *Trichoderma*–Pathogen Interaction

As the ecological niche of *Trichoderma* spp. is widespread and diverse, they often compete with other organisms for nutrients and space, thereby promoting plant growth and simultaneously providing resistance due to pathogen attacks (Druzhinina et al. 2011; Mukherjee et al. 2013b). The following are the modes of action employed for the biocontrol of plant pathogens.

4.11 Mycoparasitism

In the progression of mycoparasitism which is a biocontrol mechanism, there is a direct interaction between *Trichoderma* spp. and the pathogen where the pathogen is killed by the *Trichoderma* spp. (Troian et al. 2014). The process of mycoparasitism involves several events, viz. sensing, detection of host, hyphae from *Trichoderma* being directed towards the pathogen leading to appressorium formation, coiling, secretions of many fungitoxic hydrolytic enzymes (β-glucosidase, endochitinases, mannosidases and proteases) and subsequently lysis of various pathogens (Harman et al. 2004). Oligomers are released as a result of hydrolyzation due to these hydrolytic enzymes, and this activates genes that are required for mycoparasitism (Vinale et al. 2008). Finally, *Trichoderma* spp. penetrate the lumen of the host pathogen, incorporating and metabolizing the protoplasmic contents (Suarez et al.

2007). *T. harzianum* has been shown to be actively associated with fungal mycelia through scanning electron microscopy (SEM) where enzymatic lysis of mycelia filaments has occurred (Braun et al. 2018). *T. asperellum* (NVTA2) significantly prevented not only mycelial growth of *S. sclerotiorum* but also sclerotial production as well against stem rot pathogen (Vinodkumar et al. 2017). Despite *Trichoderma* spp. being categorized for their mycoparasitic characteristics, they have been found to be dependent on the host in response to pathogens where variations in hyphae winding and secreted proteins have been observed (Monteiro et al. 2010). In mycoparasitism, several proteins, metabolites and no less than 20–30 genes have been reported which are involved directly. Potential mycoparasitic proteins secreted including glycoside hydrolases, chitinase, mutanase, α-1,3-glucanase, α-1,2-mannosidase, carboxylic hydrolase ester, carbohydrate-binding module family 13, glucan 1,3-β-glucosidase, α-galactosidase and neutral protease 2 have been demonstrated for the biological control of *Guignardia citricarpa* by *T. atroviride* (Lima et al. 2016). A study by Denga et al. (2018) has recently shown that P6281 (protease) from *T. harzianum* which is upregulated during mycoparasitism that could degrade and destroy the mycelial cell wall and cause outer layer detachment of the cell wall could be helpful in understanding the exact mechanism of mycoparasitism. In addition, the functions of various glucanases and chitinases are well investigated from *Trichoderma* spp. that involve mycoparasitism using gene-for-gene experiments or genetic exploitations to understand this complex process (Daguerre et al. 2014; Hirpara and Gajera 2018). However, the exact mechanism involved in the case of mycoparasitism still needs further studies.

4.12 Antibiosis

In the mechanism of antibiosis, secondary metabolites or antibiotics produced by a microorganism are considered helpful for the inhibition of plant pathogens; hence, mutants are being developed for enhancement in metabolite production (Rey et al. 2000). The antagonist possesses a potential role in inhibiting or killing the beleaguered microorganism. Roles of antibiotics from *Trichoderma* spp., such as gliovirin and gliotoxin, against fungal phytopathogens have been reported (Howell 2006; Mukherjee et al. 2012). Thus, in order to survive, *Trichoderma* spp. are known to produce various antibiotic metabolites which prove to be toxic towards plant pathogens such as lignoren sesquiterpenoid, trichoviridin and isocyanides, trichodermol, terpenes, trichodermin, harzianolide, T39 butenolide or extracellular hydrolytic enzymes (Brewer et al. 1982; Berg et al. 2004; Eziashi et al. 2006; Vinale et al. 2008; Andrabi et al. 2011). *T. atroviride* was shown to express an array of genes involved in the production of secondary metabolites, GH16 β-glucanases, various proteases and small secreted cysteine-rich proteins. Also, *T. virens* was found to express genes that are required for the biosynthesis of gliotoxin, the respective precursors and also glutathione, which is necessary for gliotoxin biosynthesis (Atanasova et al. 2013). In the case of *Trichoderma* spp. isolated and selected from the biodiversity hotspot of India, *Trichoderma* spp. were found to exhibit various activities, viz. protease, chitinase, β-1,3-glucanase and cellulase, and production of volatile and non-volatile compounds was also assayed, which significantly reduced mycelia growth of *Pythium aphanidermatum* that was causing damping-off disease of beans (*Phaseolus vulgaris* L.) (Kamala and Indira 2011). The *ech33* genes that encode chitinases are believed to be the most effective and are being isolated from *Trichoderma* spp. and transferred to plants to make them resistant to several fungal plant pathogens. Therefore, investigators are looking for the presence of *ech33* gene and cloning them for future use (Sharma and Bhat 2012). *T. hamatum* has been well documented for its ability to produce extracellular chitinases (*N*-acetyl-β-D-glucosaminidase), which help in the biocontrol of pathogens (Harman et al. 2004). Furthermore, the role of peptaibols from *Trichoderma* spp. is also considered to be important in stimulation of self-defensive activities in plants to suppress plant pathogens (Viterbo et al. 2007; Sharma et al. 2017).

4.13 Competition

Trichoderma strains share an ecological niche with several microorganisms present nearby in the rhizosphere; hence, it needs to eradicate other microorganisms due to the presence of limiting nutrients. The reason for this may be attributed not only to the ability of *Trichoderma* strains to produce antibiotics or extracellular enzymes but also to their competitive ability with the fungal pathogen for space and nutrients (Howell 2003). Also, *Trichoderma* spp. have the ability to make nutrients available for plants to use them. Moreover, filamentous fungi require essential elements like carbon and iron for their feasibility, and *Trichoderma* has been found to compete for carbon with other fungi such as *F. oxysporum* (Alabouvette et al. 2009). The increase in the competitive ability of *Trichoderma* spp. may be attributed to the presence of ATP-binding cassettes (ABC) transporters which help in reducing several toxic effects, thereby making them resistant to toxins or antimicrobial substances released by other microorganisms or plants. The capability of making immobile nutrients available and their exploitation makes one more competitive than the other soil microbes. Likewise, in the investigation of Atanasova et al. (2013), *T. reesei* was observed to express mainly genes for nutrient acquisition, and it attempted to compete for nutrients with the other fungus by more rapid acquisition of nutrients. In *Trichoderma*, protein production was suggested when it competes with other root colonizers (Brotman et al. 2008). It has been observed that when filamentous fungi are iron deficient, they start secreting siderophores known as low molecular weight ferric iron-specific chelators. Therefore, *Trichoderma* spp. release these siderophores for iron uptake and alongside cause biocontrol of other fungus (Benitez et al. 2004). *T. asperellum* (CHF 78) has been reported to secrete two kinds of siderophores that may have caused efficient reduction in disease severity of *Fusarium* wilt of tomato (Li et al. 2018).

4.14 *Trichoderma*-Pathogen Signalling

Whenever there is infection in plants, they detect possible detrimental invaders directly or indirectly through cellular perturbation, and cell surface-localized pattern recognition receptors (PRRs) activate pattern-triggered immunity (PTI) that recognizes conserved MAMPs (Microbe-Associated Molecular Patterns), which helps in differentiating various pathogen classes and host-derived DAMPs (damage-associated molecular patterns) (Bohm et al. 2014). Cell surface pattern recognition receptors (PRRs) recognize the fungal cell wall which consists of β-glucans and chitin components (Ranf 2017). In the same way, PRRs act in response to microbial hydrolytic activities which release cell wall oligomers or cuticular fragments on plant tissues known as damage-associated molecular patterns (DAMPs). In this way, MAMP-triggered immunity (MTI) is activated which is a result of PRR stimulation, leading to signal transduction. Due to this, various events take place such as cell wall reinforcement by the formation of lignin and callose and phytohormone accumulation such as jasmonic acid (JA), salicylic acid (SA) and ethylene (ET), whose signal is systemically transmitted. In addition, the accumulation of pathogenesis-related (PR) proteins such as chitinases and glucanases helps in degrading the cell walls of potential fungal or oomycete invaders. Secondary metabolites such as phytoalexins also play significant roles (Pieterse et al. 2009; Mendoza-Mendoza et al. 2018). In nature, the largest protein family of G-protein-coupled receptors (GPCR) which is known as seven-transmembrane receptors involves firstly in sensing of prey such as Gpr1, and then a cascade of signal transduction events occurs through Gα, Gβ and Gγ subunits coupled to cAMP and MAPK (mitogen-activated protein kinase) when the ligand binds with receptors thereby being active in mycoparasitism and biocontrol (Gutkind 1998; Schmoll 2008; Druzhinina et al. 2011). In addition, most other filamentous fungi, *Trichoderma* spp. have three MAPK cascades which consist of MAPKKK, MAPKK and MAPK pathways observed during mycoparasitism and biocontrol (Schmoll 2008). In maize, for the biocontrol of pathogens, expressions of plant innate immunity related genes (MAPK5α, MAPK6, and MAPK20) were shown to be increased by the *Trichoderma* spp. (Saravanakumar et al. 2018). *Trichoderma* spp. are also known for the production of non-peptaibiotic secondary metabolites. The volatile secondary metabolite 6-pentyl-α-pyrone (6PAP) produced from *Trichoderma* spp. was found to suppress the *Fusarium* toxin *fusaric* acid (FA) during interaction between pathogen *Fusarium moniliforme* and *Trichoderma* spp. (El-Hasan et al. 2008). The production and release of cell wall degrading enzymes (CWDEs) is considered one of the most important chemical armouries deployed by *Trichoderma* spp. for the inhibition of pathogens. The involvement of extracellular cell wall degrading enzymes (chitinase and β-glucanase) of the antagonistic *Trichoderma* spp. was found to be significant in the reduction of the pathogen *Fusarium oxysporum* f. sp. *lycopersici* (FOL) in tomato against tomato wilt (El-Komy et al. 2015). Also, the increase in the content of Si in roots of tomato has been significantly correlated with disease suppression (Huang et al. 2011). The method of inoculation with *Trichoderma* spp. on *Hypericum perforatum* seedlings against the pathogen *F. oxysporum* has been shown to be responsible for the successful biocontrol and twofold increase in foliar biomass and better development of the roots than the plants inoculated with the pathogen (Giurgiu et al. 2018).

4.15 Future Prospects

Using biological antagonists for the management of pathogens not only maintains natural resources but also provides an attractive alternative for hazardous synthetic chemicals. Genomic sequences of potential biological antagonists should be developed and molecular exploration in AM fungal and *Trichoderma* spp. interactions with plants and the underlying mechanisms involved will be helpful in the development of strategies to employ them as efficient biocontrol agents as well as crop improvers. It is also suggested in this review that compatible strains of biological antagonists be developed with other bacterial or fungal antagonists, and their inoculations will be especially promising in future strategies which will have more significant implications for crop improvement as well as for biological control of plant pathogens. Furthermore, user friendly, labour exclusive, effective at local natural level and uncomplicated economic formulations of biological antagonists should be developed, so that they can be employed by agricultural practitioners simply for sustainable agricultural practices in the near future.

Conflicts of Interest The authors declare no competing and conflict of interest.

References

Abdel-Fattah GM, El-Haddad SA, Hafez EE, Rashad YM (2011) Induction of defense responses in common bean plants by arbuscular mycorrhizal fungi. Microbiol Res 166:268–281

Abhiram P, Masih H (2018) *In vitro* Antagonism of *Trichoderma viride* against *Fusarium oxysporum* strains. J Pharmacogn Phytochem 7:2816–2819

Aboul-Soud MA, Yun BW, Harrier LA, Loake GJ (2004) Transformation of *Fusarium oxysporum* by particle bombardment and characterisation of the resulting transformants expressing a GFP transgene. Mycopathologia 158:475–482

Agrios GN (1988) Plant pathology, 3rd edn. Academic, New York, p 803

Akiyama K, Matsuzaki K, Hayashi H (2005) Plant sesquiterpenes induce hyphal branching in arbuscular mycorrhizal fungi. Nature 435:824–827

Akkopru A, Demir S (2005) Biological control of Fusarium wilt in tomato caused by *Fusarium oxysporum* f. sp. *lycopersici* by AMF *Glomus intraradices* and some rhizobacteria. J Phytopathol 153:544–550

Alabouvette C, Lemanceau P, Steinberg C (1993) Recent advances in the biological control of Fusarium wilts. Pestic Sci 37:365–373

Alabouvette C, Olivain C, Migheli Q, Steinberg C (2009) Microbiological control of soil-borne phytopathogenic fungi with special emphasis on wilt-inducing *Fusarium oxysporum*. New Phytol 184:529–544

Al-Hazmi AS, Javeed MT (2016) Effects of different inoculum densities of *Trichoderma harzianum* and *Trichoderma viride* against *Meloidogyne javanica* on tomato. Saudi J Biol Sci 23:288–292

Al-Hmoud G, Al-Momany A (2015) Effect of four mycorrhizal products on *Fusarium* root rot on different vegetable crops. J Plant Pathol Microb 6:255

Alizadeh H, Behboudi K, Ahmadzadeh M, Javan-Nikkhah M, Zamioudis C, Pieterse CM, Bakker PA (2013) Induced systemic resistance in cucumber and *Arabidopsis thaliana* by the combination of *Trichoderma harzianum* Tr6 and *Pseudomonas* sp. Ps14. Biol Control 65:14–23

Amira MB, Lopeza D, Mohamedc AT, Khouajab A, Chaard H, Fumanala B, Gousset-Duponta A, Bonhommee L, Labela P, Goupila P, Ribeiroa S, Pujade-Renauda V, Juliena JL, Auguing D, Venissea JS (2017) Beneficial effect of *Trichoderma harzianum* strain Ths97 in bio-controlling *Fusarium solani* causal agent of root rot disease in olive trees. Biol Control 110:70–78

Andrabi M, Vaid A, Razdan VK (2011) Evaluation of different measures to control wilt causing pathogens in chickpea. J Plant Prot Res 51:55–59

Andrade SAL, Domingues AP, Mazzafera P (2015) Photosynthesis is induced in rice plants that associate with arbuscular mycorrhizal fungi and are grown under arsenate and arsenite stress. Chemosphere 134:141–149

Arabi MIE, Kanacri S, Ayoubi Z, Jawhar M (2013) Mycorrhizal application as a biocontrol agent against common root rot of barley. Res Biotechnol 4:07–12

Atanasova L, Crom SL, Gruber S, Coulpier F, Seidl-Seiboth V, Kubicek CP, Druzhinina IS (2013) Comparative transcriptomics reveals different strategies of *Trichoderma* mycoparasitism. BMC Genomics 14:121

Auge RM (2001) Water relations, drought and vesicular-arbuscular mycorrhizal symbiosis. Mycorrhiza 11:3–42

Balaji LP, Ahir RR (2011) Evaluation of plant extracts and biocontrol agents against leaf spot disease of brinjal. Indian Phytopath 64:378–380

Barea JM, Azcon-Aguilar C (1982) Production of plant growth-regulating substances by the vesicular-arbuscular mycorrhizal fungus *Glomus mosseae*. Appl Environ Microbiol 43:810–813

Barker S, Tagu D (2000) The roles of auxins and cytokinins in mycorrhizal symbioses. J Plant Growth Regul 19:144–154

Basu S, Rabara RC, Negi S (2018) AMF: the future prospect for sustainable agriculture. Physiol Mol Plant Pathol 102:36–45

Benitez T, Rincon AM, Limon MC, Codon AC (2004) Biocontrol mechanism of *Trichoderma* strains. Int Microbiol 7:249–260

Berg A, Wangun H, Kemami V, Nkengfack AE, Schlegel B (2004) Lignoren, a new sesquiterpenoid metabolite from *Trichoderma lignorum* HKI 0257. J Basic Microbiol 44:317–319

Berruti A, Lumini E, Balestrini R, Bianciotto V (2016) Arbuscular mycorrhizal fungi as natural biofertilizers: let's benefit from past successes. Front Microbiol 6:1–13

Bhagat S, Pan S (2008) Variability in production of extra cellular hydrolytic enzymes by *Trichoderma* spp. and induction of disease resistance in gram (*Ciccer arietinum*). J Biol Conserv 22:57–66

Bisby GR (1939) *Trichoderma viride* Pers. ex. Fries, and notes on *Hypocrea*. Trans Br Mycol Soc 33:149–168

Błaszczyk L, Siwulski M, Sobieralski K, Lisiecka J, Jędryczka M (2014) *Trichoderma* spp. – application and prospects for use in organic farming and industry. J Plant Prot Res 54:309–317

Blee KA, Anderson AJ (2000) Defense responses in plants to arbuscular mycorrhizal fungi. In: Podila GK, Douds DD Jr (eds) Current advances in mycorrhizae research. American Phytopathological Society Press, St. Paul, MN, pp 27–44

Blilou I, Ocampo JA, Garcia-Garrido JM (2000) Induction of Ltp (Lipid transfer protein) and Pal (Phenylalanine ammonia-lyase) gene expression in rice roots colonized by the arbuscular mycorrhizal fungus *Glomus mosseae*. J Exp Bot 51:1969–1977

Bohm H, Albert I, Fan L, Reinhard A, Nurnberger T (2014) Immune receptor complexes at the plant cell surface. Curr Opin Plant Biol 20:47–54

Braun H, Woitsch L, Hetzer B, Geisen R, Zange B, Schmidt-Heydt M (2018) *Trichoderma harzianum*: inhibition of mycotoxin producing fungi and toxin biosynthesis. Int J Food Microbiol 280:10–16

Breuillin F, Schramm J, Hajirezaei M, Ahkami A, Favre P, Druege U, Hause B, Bucher M, Kretzschmar T, Bossolini E, Kuhlemeier C, Martinoia E, Franken P, Scholz U, Reinhardt D (2010) Phosphate systemically inhibits development of arbuscular mycorrhiza in *Petunia hybrida* and represses genes involved in mycorrhizal functioning. Plant J 64:1002–1017

Brewer D, Taylor A, Keeping JW, Taha AA, Thaller V (1982) Production of experimental quantities of isocyanide metabolites of *Trichoderma hamatum*. Can J Microbiol 28:1252–1260

Brotman Y, Briff E, Viterbo A, Chet I (2008) Role of swollenin, an expansin-like protein from *Trichoderma*, in plant root colonization. Plant Physiol 147:779–789

Cangelosi B, Curir P, Beruto M, Monroy F, Borriello R (2017) Protective effects of arbuscular mycorrhizae against *Fusarium Oxysporum* F. Sp. *Ranunculi* in *Ranunculus Asiaticus* cultivations for flower crop. J Hortic Sci Res 1:36–41

Cao Y, Liang Y, Tanaka K, Nguyen CT, Jedrzejczak RP, Joachimiak A, Stacey G (2014) The kinase LYK5 is a major chitin receptor in Arabidopsis and forms a chitin-induced complex with related kinase CERK1. elife 3:e03766

Caron M, Fortin JA, Richard C (1986) Effect of inoculation sequence on the interaction between *Glomus intraradices* and *Fusarium oxysporum* f. sp. *radicis-lycopersici* in tomatoes. Can J Plant Pathol 8:12–16

Castellanos-Morales V, Keiser C, Cárdenas-Navarro R, Grausgruber H, Glauninger J, García-Garrido JM, Steinkellner S, Sampedro I, Hage-Ahmed K, Illana A, Ocampo JA, Vierheilig H (2011) The bioprotective effect of AM root colonization against the soil-borne fungal pathogen *Gaeumannomyces graminis* var. Tritici in barley depends on the barley variety. Soil Biol Biochem 43:831–834

Castillejo MA, Bani M, Rubiales D (2015) Understanding pea resistance mechanisms in response to *Fusarium oxysporum* through proteomic analysis. Phytochemistry 115:44–58

Cavagnaro TR, Bender SF, Asghari HR, van der Heijden MGA (2015) The role of arbuscular mycorrhizas in reducing soil nutrient loss. Trends Plant Sci 20:283–290

Chaudhary VB, Sandall EL, Lazarski MV (2018) Urban mycorrhizas: predicting arbuscular mycorrhizal abundance in green roofs. Fungal Ecol:1–8

Conrath U, Beckers GJM, Flors V, García-Agustín P, Jakab G, Mauch F, Newman MA, Pieterse CMJ, Poinssot B, Pozo MJ, Pugin A, Schaffrath U, Ton J, Wendehenne D, Zimmerli L, Mauch-Mani B (2006) Priming: getting ready for battle. Mol Plant-Microbe Interact 19:1062–1071

Contreras-Cornejo HA, Macías-Rodríguez L, Cortés-Penagos C (2009) *Trichoderma virens*, a plant beneficial fungus enhances biomass production and promotes lateral root growth through an auxin dependent mechanism in *Arabidopsis*. Plant Physiol 149:1579–1592

Cosme M, Ramireddy E, Franken P, Schmülling T, Wurst S (2016) Shoot- and root-borne cytokinin influences arbuscular mycorrhizal symbiosis. Mycorrhiza 26:709–720

Cotxarrera L, Trillas-Gay MI, Steinberg C, Alabouvette C (2002) Use of sewage sludge compost and *Trichoderma asperellum* isolates to suppress fusarium wilt of tomato. Soil Biol Biochem 34:467–476

Courty PE, Smith P, Koegel S, Redeker D, Wipf D (2015) Inorganic nitrogen uptake and transport in beneficial plant root–microbe interactions. Crit Rev Plant Sci 34:4–16

Daguerre Y, Siegel K, Edel-Hermann V, Steinberg C (2014) Fungal proteins and genes associated with biocontrol mechanisms of soil-borne pathogens: a review. Fungal Biol Rev 28:97–125

Davies FT, Potter JR, Linderman RG (1993) Drought resistance of mycorrhizal pepper plants independent of leaf P-concentration – response in gas exchange and water relations. Physiol Plant 87:45–53

De Angeli A, Thomine S, Frachisse JM, Ephritikhine G, Gambale F, Barbier-Brygoo H (2007) Anion channels and transporters in plant cell membranes. FEBS Lett 581:2367–2374

De Hoog GS, Guarro J, Gene J, Figueras MJ (2000) Atlas of Clinical Fungi, vol 1, 2nd edn. Centraalbureau voor Schimmelcultures, Utrecht

De Marco JL, Valadares-Inglis MC, Felix CR (2003) Production of hydrolytic enzymes by *Trichoderma* isolates with antagonistic activity against *Crinipellis perniciosa*, the causal agent of witches' broom of cocoa. Braz J Microbiol 34:33–38

Dean R, Van Kan JA, Pretorius ZA, Hammond-Kosack KE, Di Pietro A, Spanu PD, Rudd JJ, Dickman M, Kahmann R, Ellis J, Foster GD (2012) The top 10 fungal pathogens in molecular plant pathology. Mol Plant Pathol 13:414–430

Dehariya K, Shukla A, Sheikh IA, Vyas D (2015) *Trichoderma* and arbuscular mycorrhizal fungi based biocontrol of *Fusarium udum* Butler and their growth promotion effects on pigeon pea. J Agric Sci Technol 17:505–517

Delaux PM, Varala K, Edger PP, Coruzzi GM, Pires JC, Ane JM (2014) Comparative phylogenomics uncovers the impact of symbiotic associations on host genome evolution. PLoS Genet 10:e1004487

Denga JJ, Huanga WQ, Lia ZW, Lua DL, Zhangb Y, Luo XC (2018) Biocontrol activity of recombinant aspartic protease from *Trichoderma harzianum* against pathogenic fungi. Enzym Microb Technol 112:35–42

Diedhiou PM, Hallmann J, Oerke EC, Dehne HW (2003) Effects of arbuscular mycorrhizal fungi and non-pathogenic *Fusarium oxysporum* on *Meloidogyne incognita* infestation of tomato. Mycorrhiza 13:199–204

Djonovic S, Vittone G, Mendoza-Herrera A, Kenerley CM (2007) Enhanced biocontrol activity of *Trichoderma virens* transformants constitutively coexpressing β-1, 3-and β-1, 6-glucanase genes. Mol Plant Pathol 8:469–480

Dong Y, Zhu YG, Smith FA, Wang Y, Chen B (2008) Arbuscular mycorrhiza enhanced arsenic resistance of both white clover (*Trifolium repens* L.) and ryegrass (*Lolium perenne* L.) plants in an arsenic contaminated soil. Environ Pollut 15:174–181

Doni F, Isahak A, Zain CRCM, Ariffin SM, Mohamad WNW, Yusoff WMW (2014) Formulation of *Trichoderma* sp. *SL2* inoculants using different carriers for soil treatment in rice seedling growth. Springerplus 3:1–5

Druzhinina IS, Seidl-Seiboth V, Herrera-Estrella A, Horwitz BA, Kenerley CM, Monte E, Mukherjee PK, Zeilinger S, Grigoriev IV, Kubicek CP (2011) *Trichoderma*: the genomics of opportunistic success. Nat Rev Microbiol 9:749–759

Eke P, Chatue GC, Wakam LN, Kouipou RMT, Fokou PVT, Boyom FF (2016) Mycorrhiza consortia suppress the fusarium root rot (*Fusarium solani* f. sp. *Phaseoli*) in common bean (*Phaseolus vulgaris* L.). Biol Control 103:240–250

El-Hasan A, Walker F, Buchenauer H (2008) *Trichoderma harzianum* and its metabolite 6-Pentyl-alpha-pyrone suppress fusaric acid produced by *Fusarium moniliforme*. J Phytopathol 156:79–87

El-Hasan SA, Gowen SR, Pembroke B (2012) Use of *Trichoderma hamatum* for biocontrol of lentil vascular wilt disease: efficacy, mechanisms of interaction and future prospects. J Plant Protect Res 53:12–26

El-Komy MH, Saleh AA, Eranthodi A, Molan YY (2015) Characterization of novel Trichoderma asperellum isolates to select effective biocontrol agents against tomato Fusarium wilt. Plant Pathol J 31:50–60

Elwakil MA, Baka ZA, Soliman HM, Sadek MS (2013) A modern tactic for reducing the biotic stress on cucumber plants caused by *Fusarium oxysporum*. Plant Pathol J 12:26–31

Escande AR, Laich FS, Pedraza MV (2002) Field testing of honey-bee dispersed *Trichoderma* spp. to manage sunflower head rot (*Sclerotinia sclerotiorum*). Plant Pathol 51:346–351

European Union (2009) Directive 2009/128/EC of the European Parliament and of the Council establishing a framework for Community action to achieve the sustainable use of pesticides. Off J Eur U. http://eur-lex.europa.eu/search.html?qid=1456841738609&text=2009/128&scope=EURLEX&type=quick&lang=en. Accessed Aug 2018

Eziashi EI, Uma NU, Adekunle AA, Airede CE (2006) Effect of metabolites produced by *Trichoderma* species against *Ceratocystis paradoxa* in culture medium. Afr J Biotechnol 5:703–706

Ferrol N, Barea JM, Azcon-Aguilar C (2002) Mechanisms of nutrient transport across interfaces in arbuscular mycorrhizas. Plant Soil 244:231–237

Fiorilli V, Vannini C, Ortolani F, Garcia-Seco D, Chiapello M, Novero M, Domingo G, Terzi V, Morcia C, Bagnaresi P, Moulin L, Bracale M, Bonfante P (2018) Omics approaches revealed how arbuscular mycorrhizal symbiosis enhances yield and resistance to leaf pathogen in wheat. Sci Rep 8:1–18

Foo E, Ross JJ, Jones WT, Reid JB (2013) Plant hormones in arbuscular mycorrhizal symbioses: an emerging role for gibberellins. Ann Bot 111:769–779

Fravel D, Olivain C, Alabouvette C (2003) *Fusarium oxysporum* and its biocontrol. New Phytol 157:493–502

Gadag AS, Krishnaraj PU (2017) Effect of Actinobacteria and *Glomus fasiculatum* against *Fusarium oxysporum* f. sp. *lycopersici* in Tomato Plant. Int J Curr Microbiol App Sci 6:488–498

Gallou A, Mosquera HPL, Cranenbrouck S, Suaarez JP, Declerck S (2011) Mycorrhiza induced resistance in potato plantlets challenged by *Phytophthora infestans*. Physiol Mol Plant Pathol 76:20–26

Garcia-Garrido JM, Ocampo JA (2002) Regulation of the plant defence response in arbuscular mycorrhizal symbiosis. J Exp Bot 53:1377–1386

Garcia-Limones C, Hervas A, Navas-Cortes JA, Jimenez-Diaz RM, Tena M (2002) Induction of antioxidant enzyme system and other oxidative stress markers associated with compatible and incompatible interactions between chickpea (*Cicer arietinum* L.) and *Fusarium oxysporum* f. sp. *ciceris*. Physiol Mol Plant Pathol 61:325–337

Gava CAT, Pinto JM (2016) Biocontrol of melon wilt caused by *Fusarium oxysporum* Schlect f. sp. *melonis* using seed treatment with *Trichoderma* spp. and liquid compost. Biol Control 97:13–20

Gianinazzi S, Vosatka M (2004) Inoculum of arbuscular mycorrhizal fungi for production systems: science meets business. Can J Bot 82:1264–1271

Giurgiu R, Dumitraş A, Morar G, Scheewe P, Schröder F (2018) A study on the biological control of *Fusarium oxysporum* using *Trichoderma* spp., on soil and rockwool substrates in controlled environment. Not Bot Horti Agrobot Cluj-Na 46:260–269

Grondona I, Hermosa R, Tejada M, Gomis MD, Mateos PF, Bridge PD, Monte E, Garcia-Acha I (1997) Physiological and biochemical characterization of *Trichoderma harzianum*, a biological control agent against soilborne fungal plant pathogens. Appl Environ Microbiol 63:3189–3198

Gutkind JS (1998) Cell growth control by G protein-coupled receptors: from signal transduction to signal integration. Oncogene 17:1331–1342

Hage-Ahmed K, Krammer J, Steinkellner S (2013) The intercropping partner affects arbuscular mycorrhizal fungi and *Fusarium oxysporum* f. sp. *Lycopersici* interactions in tomato. Mycorrhiza 23:543–550

Halder S, Ray MB (2006) Effect of VAM soil containing *Glomus fasciculatum* on growth of *Withania somnifera* Dun. Asian J Exp Sci 20:261–268

Hanson LE (2000) Reduction of *Verticillium* wilt symptoms in cotton following seed treatment with *Trichoderma virens*. J Cotton Sci 4:224–231

Hao Z, Christie P, Qin L, Wang C, Li X (2005) Control of *Fusarium Wilt* of cucumber seedlings by inoculation with an arbuscular mycorrhical fungus. J Plant Nutr 28:1961–1974

Harman GE (2006) Overview of mechanisms and uses of *Trichoderma* spp. Phytopathology 96:190–194

Harman GE, Shoresh M (2007) The mechanisms and applications of symbiotic opportunistic plant symbiont. In: Vurro M, Gressel J (eds) Novel biotechnologies for biocontrol agent enhancement and management. NATO security through science series. Springer, Dordrecht

Harman GE, Howell CR, Viterbo A, Chet I, Lorito M (2004) Trichoderma species—opportunistic, avirulent plant symbionts. Nat Rev Microbiol 2(1):43–56

Harrier LA, Watson CA (2004) The potential role of arbuscular mycorrhizal (AM) fungi in the bioprotection of plants against soil borne pathogens in organic and/or other sustainable farming systems. Pest Manag Sci 60:149–157

Hause B, Mrosk C, Isayenkov S, Strack D (2007) Jasmonates in arbuscular mycorrhizal interactions. Phytochemistry 8:101–110

Hermosa R, Viterbo A, Chet I, Monte E (2012) Plant-beneficial effects of *Trichoderma* and of its genes. Microbiology 158:17–25

Herrera-Medina MJ, Tamayo MI, Vierheilig H, Ocampo JA, Garcia-Garrido JM (2008) The jasmonic acid signalling pathway restricts the development of the arbuscular mycorrhizal association in tomato. J Plant Growth Regul 27:221–230

Hinsinger P, Betencourt E, Bernard L, Brauman A, Plassard C, Shen J, Tang X, Zhang F (2011) P for two, sharing a scarce resource: soil phosphorus acquisition in the rhizosphere of intercropped species. Plant Physiol 156:1078–1086

Hirpara DG, Gajera HP (2018) Molecular heterozygosity and genetic exploitations of *Trichoderma* inter-fusants enhancing tolerance to fungicides and mycoparasitism against *Sclerotium rolfsii* Sacc. Infect Genet Evol 66:26–36

Howell CR (2003) Mechanisms employed by *Trichoderma* species in the biological control of plant diseases: the history and evolution of current concepts. Plant Dis 87:4–10

Howell CR (2006) Understanding the mechanisms employed by *Trichoderma virens* to effect biological control of cotton diseases. Phytopathology 96:178–180

Howell CR, Hanson EL, Stipanovic RD, Puckhaber LS (2000) Induction of terpenoid synthesis in cotton roots and control of *Rhizoctonia solani* by seed treatment with *Trichoderma virens*. Phytopathology 90:248–252

Huang C, Roberts PD, Datnoff LE (2011) Silicon suppresses *Fusarium* crown and root rot of tomato. J Phytopathol 159:546–554

Jaiti F, Meddich A, El-Hadrami I (2008) Effectiveness of arbuscular mycorrhizal fungi in the protection of date palm (*Phoenix dactylifera* L.) against bayoud disease. Physiol Mol Plant Pathol 71:166–173

Jakobsen I (1999) Transport of phosphorus and carbon in arbuscular mycorrhizas. In: Varma A, Hock B (eds) Mycorrhiza. Springer, Berlin

Jayalakshmi SK, Raju S, Usha Rani S, Benagi VI, Sreeramulu K (2009) *Trichoderma harzianum* L1 as a potential source for lytic enzymes and elicitor of defense responses in chickpea (*Cicer arietinum* L.) against wilt disease caused by *Fusarium oxysporum* f. sp. *ciceri*. Aust J Crop Sci 3:44–52

Jimenez C, Ricardo D (2017) Soybean root rot caused by *Fusarium oxysporum* and *Fusarium graminearum*: interactions with biotic and abiotic factors. Graduate theses and dissertations. PhD dissertation, Iowa State University, Ames, IA

Johri AK, Oelmuller R, Dua M, Yadav V, Kumar M, Tuteja N, Varma A, Bonfante P, Persson BL, Stroud RM (2015) Fungal association and utilization of phosphate by plants: success, limitations, and future prospects. Front Microbiol 6:1–13

Jones JDG, Dangl JL (2006) The plant immune system. Nature 444:323–329

Jung SC, Martinez-Medina A, Lopez-Raez JA, Pozo MJ (2012) Mycorrhiza-induced resistance and priming of plant defenses. J Chem Ecol 38:651–664

Kamala T, Indira S (2011) Evaluation of indigenous *Trichoderma* isolates from Manipur as biocontrol agent against *Pythium aphanidermatum* on common beans. 3 Biotech 1:217–225

Kazan K (2013) Auxin and the integration of environmental signals into plant root development. Ann Bot 112:1655–1665

Keymer A, Gutjahr C (2018) Cross-kingdom lipid transfer in arbuscular mycorrhiza symbiosis and beyond. Curr Opin Plant Biol 44:137–144

Khan MH, Meghvansi MK, Panwar V, Gogoi HK, Singh L (2010) Arbuscular mycorrhizal fungi-induced signalling in plant defence against phytopathogens. J Phytol 2:53–69

Kieber JJ, Schaller GE (2014) Cytokinins. In: The arabidopsis book. The American Society of Plant Biologists, p e0168

Kiriga AW, Haukelandb S, Kariukia GM, Coynec DL, Beek NV (2018) Effect of *Trichoderma* spp. and *Purpureocillium lilacinum* on *Meloidogyne javanica* in commercial pineapple production in Kenya. Biol Control 119:27–32

Komy MHE, Saleh AA, Eranthodi A, Molan YY (2015) Characterization of novel *Trichoderma asperellum* isolates to select effective biocontrol agents against tomato *Fusarium* wilt. Plant Pathol J 31:50–60

Kredics L, Antal Z, Szekeres A, Hatvani L, Manczinger L, Vágvölgyi C, Nagy E (2005) Extracellular proteases of *Trichoderma* species: a review. Acta Microbiol Immunol Hung 52:169–184

Kumar M, Yadav V, Tuteja N, Johri AK (2009) Antioxidant enzyme activities in maize plants colonized with *Piriformospora indica*. Microbiology 155:780–790

Kumar V, Shahid M, Singh A, Srivastava M, Mishra A, Srivastava YK, Pandey S, Shrarma A (2014) Effect of biopriming with biocontrol agents *Trichoderma harzianum* (Th.Azad) and *Trichoderma viride* (01pp) on chickpea Genotype (Radhey). J Plant Pathol Microb 5:1–4

Lambais MR (2006) Unraveling the signaling and signal transduction mechanisms controlling arbuscular mycorrhiza development. Sci Agric 63:405–413

Li Y, Yanagi A, Miyawaki Y, Okada T, Matsubara Y (2010) Disease tolerance and changes in antioxidative abilities in mycorrhizal strawberry plants. J Jpn Soc Hortic Sci 79:174–178

Li Q-R, Tan P, Yiang Y-L, Hyde KD, Mckenzie EHC, Bahkali AH, Kang C, Wang Y (2013) A novel *Trichoderma* species isolated from soil in Guizhou, *T. guizhouense*. Mycol Prog 12:167–172

Li YT, Hwang SG, Huang YM, Huang CH (2018) Effects of *Trichoderma asperellum* on nutrient uptake and *Fusarium* wilt of tomato. Crop Prot 110:275–282

Li-Jun M, Geiser DM, Proctor RH, Rooney AP, O'Donnell K, Trail F, Gardiner DM, Manners JM, Kazan K (2013) *Fusarium* Pathogenomics. Annu Rev Microbiol 67:399–416

Lima FB, Felix C, Osorio N, Alves A, Vitorino R, Domingues P, Correia A, RTdS R, Esteves AC (2016) Secretome analysis of *Trichoderma atroviride* T17 biocontrol of *Guignardia citricarpa*. Biol Control 99:38–46

Liu J, Maldonado-Mendoza I, Lopez-Meyer M, Cheung F, Town CD, Harrison MJ (2007) Arbuscular mycorrhizal symbiosis is accompanied by local and systemic alterations in gene expression and an increase in disease resistance in the shoots. Plant J 50:529–544

Lopez-Bucio J, Pelagio-Flores R, Herrera-Estrella A (2015) *Trichoderma* as biostimulant: exploiting the multilevel properties of a plant beneficial fungus. Sci Hortic 196:109–123

Lopez-Raez JA, Flors V, Garcia JM, Pozo MJ (2010) AM symbiosis alters phenolic acid content in tomato roots. Plant Signal Behav 5:1–3

Ludwig-Muller J (2010) Hormonal responses in host plants triggered by arbuscular mycorrhizal fungi. In: Koltai H, Kapulnik Y (eds) Arbuscular mycorrhizas: physiology and function. Springer, Dordrecht, pp 169–190

Ludwig-Muller J, Guther M (2007) Auxins as signals in arbuscular mycorrhiza formation. Plant Signal Behav 2:194–196

Ma LJ, van der Does HC, Borkovich KA, Coleman JJ, Daboussi MJ, Di Pietro A, Dufresne M, Freitag M, Grabherr M, Henrissat B, Houterman PM, Kang S, Shim WB, Woloshuk C, Xie X, Xu JR, Antoniw J, Baker SE, Bluhm BH, Breakspear A, Brown DW, Butchko RA, Chapman S, Coulson R, Coutinho PM, Danchin EG, Diener A, Gale LR, Gardiner DM, Goff S, Hammond-Kosack KE, Hilburn K, Hua-Van A, Jonkers W, Kazan K, Kodira CD, Koehrsen M, Kumar L, Lee YH, Li L, Manners JM, Miranda-Saavedra D, Mukherjee M, Park G, Park J, Park SY, Proctor RH, Regev A, Ruiz-Roldan MC, Sain D, Sakthikumar S, Sykes S, Schwartz DC, Turgeon BG, Wapinski I, Yoder O, Young S, Zeng Q, Zhou S, Galagan J, Cuomo CA, Kistler HC, Rep M (2010) Comparative genomics reveals mobile pathogenicity chromosomes in *Fusarium*. Nature 464:367–373

Macho AP, Zipfel C (2014) Plant PRRs and the activation of innate immune signaling. Mol Cell 54:263–273

Mahajan K, Sharma JK, Dhage A (2018) Evaluation of *Trichoderma* sp. against Fusarium wilt of chickpea caused by *Fusarium oxysporum* f. sp. *ciceris* under in vitro condition. Int J Curr Microbiol Appl Sci 7:595–599

Martinez-Medina A, Pascual JA, Lloret E, Roldán A (2009) Interactions between arbuscular mycorrhizal fungi and *Trichoderma harzianum* and their effects on Fusarium wilt in melon plants grown in seedling nurseries. J Sci Food Agric 89:1843–1850

Marzano M, Gallo A, Altomare C (2013) Improvement of biocontrol efficacy of *Trichoderma harzianum* vs. *Fusarium oxysporum* f. sp. *lycopersici* through UV-induced tolerance to fusaric acid. Biol Control 67:397–408

Mathur N, Vyas A (1995) Influence of VAM on net photosynthesis and transpiration of *Ziziphus mauritiana*. J Plant Physiol 147:328–330

Matsubara Y, Ohba N, Fukui H (2001) Effect of arbuscular mycorrhizal fungus infection on the incidence of *Fusarium* root rot in asparagus seedlings. J Jpn Soc Hortic Sci 70:202–206

Matsubara Y, Hirano I, Sassa D, Koshikawa K (2004) Increased tolerance to fusarium wilt in mycorrhizal strawberry plants raised by capillary watering methods. Environ Contam Biol 42:185–191

Mayer AM (2006) Polyphenol oxidases in plants and fungi: going places? A review. Phytochemistry 67:2318–2331

Meixner C, Ludwig-Muller J, Miersch O, Gresshoff P, Staehelin C, Vierheilig H (2005) Lack of mycorrhizal autoregulation and phytohormonal changes in the super nodulating soybean mutant nts1007. Planta 222:709–715

Mendoza-Mendoza A, Zaid R, Lawry R, Hermosa R, Monte E, Horwitz BA, Mukherjee PK (2018) Molecular dialogues between *Trichoderma* and roots: role of the fungal secretome. Fungal Biol Rev 32:62–85

Mir GH, Devi LS, Ahamd S, Kumar VM, Williams P (2011) Antagonistic potential of native isolates of *Trichoderma viride* on corm rot pathogen complex of saffron (*Crocus sativas*) in Kashmir. Plant Pathol J 10:73–78

Mishra V, Ellouze W, Howard RJ (2018) Utility of arbuscular mycorrhizal fungi for improved production and disease mitigation in organic and hydroponic greenhouse crops. J Hortic 5:1–10

Mohammadi M, Kazemi H (2002) Changes in peroxidase and polyphenol oxidase activities in susceptible and resistant wheat heads inoculated with *Fusarium graminearum* and induced resistance. Plant Sci 162:491–498

Mohandas S, Manjula R, Rawal RD, Lakshmikantha HC, Saikat C, Ramachandra YL (2010) Evaluation of arbuscular mycorrhiza and other biocontrol agents in managing *Fusarium oxysporum* f. sp. *cubense* infection in banana cv. Neypoovan. Biocontrol Sci Technol 20:165–181

Monteiro VN, Silva RN, Steindorf AS, Costa FT, Noronha EF, Ricart CAO, Sousa MV, Vainstein MH, Ulhoa CJ (2010) New insights in *Trichoderma harzianum* antagonism of fungal plant pathogens by secreted protein analysis. Curr Microbiol 61:298–305

Morkunas I, Gemerek J (2007) The possible involvement of peroxidase in defense of yellow lupin embryo axes against *Fusarium oxysporum*. J Plant Physiol 164:497–506

Mostert D, Molina AB, Daniells J, Fourie G, Hermanto C, Chao C-P, Fabregar E, Sinohin VG, Masdek N, Thangavelu R, Li C, Yi G, Mostert L, Viljoen A (2017) The distribution and host range of the banana Fusarium wilt fungus, *Fusarium oxysporum* f. sp. *cubense*, in Asia. PLoS One 12:e0181630

Mukherjee PK (1997) *Trichoderma* sp. as a microbial suppressive agent of *Sclerotium rolfsii* on vegetables. World J Microbiol Biotechnol 13:497–499

Mukherjee M, Mukherjee PK, Horwitz BA, Zachow C, Berg G, Zeilinger S (2012) *Trichoderma*–plant–pathogen interactions: advances in genetics of biological control. Indian J Microbiol 52:522–529

Mukherjee PK, Horwitz BA, Singh US, Mukherjee M, Schmoll M (2013a) *Trichoderma* in agriculture, industry and medicine: an overview. In: Mukherjee PK, Horwitz BA, Singh US,

Mukherjee M, Schmoll M (eds) *Trichoderma:* biology and applications. CABI, Nosworthy, Way, Wallingford, OX, pp 1–9
Mukherjee PK, Horwitz BA, Herrera-Estrella A, Schmoll M, Kenerley CM (2013b) *Trichoderma* research in the genome era. Annu Rev Phytopathol 51:105–129
Nair A, Kolet SP, Thulsiram HV, Bhargava S (2015) Systemic jasmonic acid modulation in mycorrhizal tomato plants and its role in induced resistance against Alternaria alternata. Plant Biol (Stuttg) 17:625–631
Nelson BD, Hansen JM, Windels CE, Helms TC (1997) Reaction of soybean cultivars to isolates of *Fusarium solani* from the Red River Valley. Plant Dis 81:664–668
Nucci M, Anaissie E (2007) *Fusarium* infections in immunocompromised patients. Clin Microbiol Rev 20:695–704
O'Donnell K, Sutton DA, Fothergill A, McCarthy D, Rinaldi MG, Brandt ME, Zhang N, Geiser DM (2008) Molecular phylogenetic diversity, multilocus haplotype nomenclature, and in vitro antifungal resistance within the *Fusarium solani* species complex. J Clin Microbiol 46:2477–2490
Omomowo IO, Fadiji AE, Omomowo OI (2018) Assessment of bio-efficacy of *Glomus versiforme* and *Trichoderma harzianum* in inhibiting powdery mildew disease and enhancing the growth of cowpea. Ann Agric Sci 63:9–17
Oyetunji OJ, Salami AO (2011) Study on the control of Fusarium wilts in the stems of mycorrhizal and trichodermal inoculated pepper (*Capsicum annum* L.). J Appl Biosci 45:3071–3080
Ozgonen H, Bicici M, Erkilic A (1999) The effect of salicylic acid and endomycorrhizal fungus *G. intraradices* on plant development of tomato and fusarium wilt caused by *Fusarium oxysporum* f. sp. *lycopersici*. Turk J Agric For 25:25–29
Patole SP, Dhore SB, Pradhan RS, Shankara K (2017) *In vitro* evaluation of *Trichoderma viride* and *Trichoderma harzianum* against *Fusarium* wilt of Chickpea. Int J Pure App Biosci 5:460–464
Pieterse CM, Leon-Reyes A, Van der Ent S, Van Wees SC (2009) Networking by small-molecule hormones in plant immunity. Nat Chem Biol 5:308–316
Pineda A, Zheng SJ, van Loon JJ, Pieterse CM, Dicke M (2010) Helping plants to deal with insects: the role of beneficial soil-borne microbes. Trends Plant Sci 15:507–514
Ponce de Leon I, Montesano M (2013) Activation of defense mechanisms against pathogens in mosses and flowering plants. Int J Mol Sci 14:3178–3200
Pozo MJ, Azcon-Aguilar C (2007) Unraveling mycorrhiza-induced resistance. Curr Opin Plant Biol 10:393–398
Pozo MJ, Lopez-Raez JA, Azcon-Aguilar C, García-Garrido JM (2015) Phytohormones as integrators of environmental signals in the regulation of mycorrhizal symbioses. New Phytol 205:1431–1436
Prasad R, Bhola D, Akdi K, Cruz C, Sairam KVSS, Tuteja N, Varma A (2017) Introduction to mycorrhiza: historical development. In: Varma A, Prasad R, Tuteja N (eds) Mycorrhiza. Springer International Publishing AG, Cham, pp 1–7
Probst C, Bandyopadhyay R, Price LE, Cotty PJ (2011) Identification of atoxigenic *Aspergillus flavus* isolates to reduce aflatoxin contamination of maize in Kenya. Plant Dis 95:212–218
Pusztahelyi T, Holb IJ, Pocsi I (2017) Plant-fungal interactions: special secondary metabolites of the biotrophic, necrotrophic, and other specific interactions. In: Merillon JM, Ramawat KG (eds) Fungal metabolites. Springer International Publishing, Cham, pp 133–190
Qualhato TF, Lopes FAC, Steindorff AS, Brandão RS, Jesuino RSA, Ulhoa CJ (2013) Mycoparasitism studies of Trichoderma species against three phytopathogenic fungi: evaluation of antagonism and hydrolytic enzyme production. Biotechnol Lett 35(9):1461–1468
Rahman SFSA, Singh E, Pieterse CMJ, Schenk PM (2018) Emerging microbial biocontrol strategies for plant pathogens. Plant Sci 267:102–111
Raman N, Gnanaguru M, Srinivasan V (2001) Biological control of *Fusarium wilt* of tomato by VA mycorrhizal fungus *Glomus fasciculatum*. Bulletin-OILB-SROP 24:33–36
Ranf S (2017) Sensing of molecular patterns through cell surface immune receptors. Curr Opin Plant Biol 38:68–77

Rao GS, Reddy NNR, Surekha C (2015) Induction of plant systemic resistance in legumes *Cajanus cajan, Vigna radiata, Vigna mungo* against plant pathogens *Fusarium oxysporum* and *Alternaria alternata* – a *Trichoderma viride* mediated reprogramming of plant defense mechanism. Int J Recent Sci Res 6:4270–4280

Ren L, Lou Y, Sakamoto K, Inubushi K, Amemiya Y, Shen Q, Xu G (2010) Effects of Arbuscular mycorrhizal colonization on microbial community in rhizosphere soil and *Fusarium* wilt disease in tomato. Commun Soil Sci Plant Anal 41:1399–1410

Ren CG, Kong CC, Xie ZH (2018) Role of abscisic acid in strigolactone induced salt stress tolerance in arbuscular mycorrhizal *Sesbania cannabina* seedlings. BMC Plant Biol 18:74

Rey M, Delgado J, Rincon A, Carmen L, Benitez T, Perez E (2000) Improvement of *Trichoderma* strains for biocontrol. Rev Iberoam Micol 17:31–36

Reynolds HL, Hartley AE, Vogelsang KM, Bever JD, Schultz PA (2005) Arbuscular mycorrhizal fungi do not enhance nitrogen acquisition and growth of old-field perennials under low nitrogen supply in glasshouse culture. New Phytol 167:869–880

Rifai MA (1969) A revision of genus *Trichoderma*. Commonow Mycol Inst Mycol:116–156

Robinson-Boyer L, Grzyb I, Jeffries P (2009) Shifting the balance from qualitative to quantitative analysis of arbuscular mycorrhizal communities in field soils. Fungal Ecol 2:1–9

Roco A, Perez LM (2001) *In vitro* biocontrol activity of *Trichoderma harzianum* on *Alternaria alternata* in the presence of growth regulators. Electron J Biotechnol 4:68–73

Ruelas C, Tiznado-Hernandez ME, Sanchez-Estrada A, Robles-Burgueno MR, Troncoso-Rojas R (2006) Changes in phenolic acid content during *Alternaria alternata* infection in tomato fruit. J Phytopathol 154:236–244

Ruiz-Lozano JM, Roussel H, Gianinazzi S, Gianinazzi-Pearson V (1999) Defense genes are differentially induced by a mycorrhizal fungus and *Rhizobium* sp. in wild-type and symbiosis-defective pea genotypes. Mol Plant-Microbe Interact 12:976–984

Samolski I, Rincon AM, Pinzon LM, Viterbo A, Monte E (2012) The qid74 gene from *Trichoderma harzianum* has a role in root architecture and plant bio-fertilization. Microbiology 158:129–138

Samuels GJ (1996) *Trichoderma*: a review of biology and systematic of the genus. Mycol Res 100:923–935

Saravanakumar K, Yu C, Dou K, Wang M, Li Y, Chen J (2016) Synergistic effect of *Trichoderma*-derived antifungal metabolites and cell wall degrading enzymes on enhanced biocontrol of *Fusarium oxysporum* f. sp. *Cucumerinum*. Biol Control 94:37–46

Saravanakumar K, Dou K, Lu Z, Wang X, Li Y, Chen J (2018) Enhanced biocontrol activity of cellulase from *Trichoderma harzianum* against *Fusarium graminearum* through activation of defense-related genes in maize. Physiol Mol Plant Pathol 103:130–136

Saremi H, Saremi H (2013) Isolation of the most common *Fusarium* species and the effect of soil solarisation on main pathogenic species in different climatic zones of Iran. Eur J Plant Pathol 137:585–596

Sawers RJH, Gutjahr C, Paszkowski U (2007) Cereal mycorrhiza: an ancient symbiosis in modern agriculture. Trends Plant Sci 13:93–97

Scheffknecht S, Mammerler R, Steinkellner S, Vierheilig H (2006) Root exudates of mycorrhizal tomato plants exhibit a different effect on microconidia germination of *Fusarium oxysporum* f. sp. *lycopersici* than root exudates from non-mycorrhizal tomato plants. Mycorrhiza 16:365–370

Scheffknecht S, St-Arnaud M, Khaosaad T, Steinkellner S, Vierheilig H (2007) An altered root exudation pattern through mycorrhization affecting microconidia germination of the highly specialized tomato pathogen *Fusarium oxysporum* f. sp. *lycopersici* (*Fol*) is not tomato specific but also occurs in *Fol* non-host plants. Can J Bot 85:347–351

Schmitt FJ, Renger G, Friedrich T, Kreslavksi VD, Zharmukhadmedov SK, Los DA, Kuznetsov VV, Allakhverdiev SI (2014) Reactive oxygen species: re-evaluation of generation, monitoring and role in stress-signaling in phototrophic organisms. Biochim Biophys Acta 1837:835–848

Schmitz AM, Harrison MJ (2014) Signaling events during initiation of arbuscular mycorrhizal symbiosis. J Integr Plant Biol 56:250–261

Schmoll M (2008) The information highways of a biotechnological workhorse-signal transduction in *Hypocrea jecorina*. BMC Genomics 9:430
Score AJ, Palfreyman JW (1994) Biological control of the dry rot fungus *Serpula lacrymans* by *Trichoderma* species: the effects of complex and synthetic media on interaction and hyphal extension rates. Int Biodeterior Biodegrad 33:115–128
Sewelam N, Kazan K, Schenk PM (2016) Global plant stress signaling: reactive oxygen species at the cross-road. Front Plant Sci 7:1–21
Shanmugaiah V, Balasubramanian N, Gomathinayagam S, Manoharan PT, Rajendran A (2009) Effect of single application of *Trichoderma viride* and *Pseudomonas fluorescens* on growth promotion in cotton plants. Afr J Agric Res 4:1220–1225
Sharma R, Bhat S (2012) Molecular screening of the presence of *ech33* from *Trichoderma* isolate of India. Wudpecker J Agric Res 1:429–432
Sharma A, Rathour R, Plaha P, Katoch V, Khalsa GS, Patial V, Singh Y, Pathania NK (2010) Induction of Fusarium wilt (*Fusarium oxysporum* f. sp. *pisi*) resistance in garden pea using induced mutagenesis and in vitro selection techniques. Euphytica 173:345–356
Sharma P, Ponnusamy VK, Ramesh R, Saravanan K, Deep S, Sharma M, Mahesh S, Dinesh S (2011) Biocontrol genes from *Trichoderma* species: a review. Afr J Biotechnol 10:19898–19907
Sharma V, Salwan R, Sharma PN (2017) The comparative mechanistic aspects of *Trichoderma* and probiotics: scope for future research. Physiol Mol Plant Pathol 100:84–96
Sheng M, Tang M, Chen H, Yang B, Zhang F, Huang Y (2008) Influence of arbuscular mycorrhizae on photosynthesis and water status of maize plants under salt stress. Mycorrhiza 18:287–296
Sheng M, Tang M, Zhang F, Huang Y (2011) Influence of arbuscular mycorrhiza on organic solutes in maize leaves under salt stress. Mycorrhiza 21:423–430
Sikes BA (2010) When do arbuscular mycorrhizal fungi protect plant roots from pathogens? Plant Signal Behav 5:763–765
Singh HB, Singh BN, Singh SP, Singh SR, Sarma BK (2009) Biological control of plant diseases: current status and future prospects. In: Johri JK (ed) Recent advances in biopesticides: biotechnological applications. New India Publishing Agency, New Delhi, p 322
Singh PK, Mishra M, Vyas D (2010) Effect of root exudates of mycorrhizal tomato plants on microconidia germination of *Fusarium oxysporum* f. sp. *lycopersici* than root exudates from non-mycorrhizal tomato plants. Arch Phytopathol Plant Protect 43:1495–1503
Singh R, Maurya S, Upadhyay RS (2016) The improvement of competitive saprophytic capabilities of *Trichoderma* species through the use of chemical mutagens. Braz J Microbiol 47:10–17
Smith SE, Read DJ (2002) Mycorrhizal symbiosis. Academic, London
Smith SE, Read DJ (2008) Mycorrhizal symbiosis. Academic, San Diego, CA
Song Y, Chen D, Lu K, Sun Z, Zeng R (2015) Enhanced tomato disease resistance primed by arbuscular mycorrhizal fungus. Front Plant Sci 6:1–13
Suarez MB, Vizcaíno JA, Llobell A, Monte E (2007) Characterization of genes encoding novel peptidases in the biocontrol fungus *Trichoderma harzianum* CECT 2413 using the TrichoEST functional genomics approach. Curr Genet 51:331–342
Sun Z, Song J, Xin X, Xie X, Zhao B (2018) Arbuscular mycorrhizal fungal 14-3-3 proteins are involved in arbuscule formation and responses to abiotic stresses during AM symbiosis. Front Microbiol 9:1–17
Suzuki N, Bassil E, Hamilton JS, Inupakutika MA, Zandalinas SI, Tripathy D, Luo Y, Dion E, Fukui G, Kumazaki A, Nakano R, Rivero RM, Verbeck GF, Azad RK, Blumwald E, Mittler R (2016) ABA is required for plant acclimation to a combination of salt and heat stress. PLoS One 11:e0147625
Svenningsen NB, Watts-Williams SJ, Joner EJ, Battini F, Efthymiou A, Cruz-Paredes C, Nybroe O, Jakobsen I (2018) Suppression of the activity of arbuscular mycorrhizal fungi by the soil microbiota. ISME J 12:1296–1307
Tawaraya K, Turjaman M, Ekamawanti HA (2007) Effect of arbuscular mycorrhiza colonization on nitrogen and phosphorus uptake and growth of *Aloe vera* L. HortSci 42:1737–1739

Tayal P, Kapoor R, Bhatnagar AK (2011) Functional synergism among *Glomus fasciculatum*, *Trichoderma viride* and *Pseudomonas fluorescens* on *Fusarium wilt* in tomato. J Plant Pathol 93:745–750

Thakker JN, Patel S, Dhandhukia PC (2013) Induction of defense-related enzymes in banana plants: effect of live and dead pathogenic strain of *Fusarium oxysporum* f. sp. *cubense*. ISRN Biotechnol. doi:https://doi.org/10.5402/2013/601303

Tian C, Kasiborski B, Koul R, Lammers PJ, Bucking H, Shachar-Hill Y (2010) Regulation of the nitrogen transfer pathway in the arbuscular mycorrhizal symbiosis: gene characterization and the coordination of expression with nitrogen flux. Plant Physiol 153:1175–1187

Toghueo RMK, Eke P, Zabalgogeazcoa I, de Aldana BRV, Nana LW, Boyom FF (2018) Bio control and growth enhancement potential of two endophytic *Trichoderma* spp. from *Terminalia catappa* against the causative agent of Common Bean Root Rot (*Fusarium solani*). Biol Control 96:8–20

Trepanier M, Becard G, Moutoglis P, Willemot C, Gagne S, Avis TJ, Rioux JA (2005) Dependence of arbuscular-mycorrhizal fungi on their plant host for palmitic acid synthesis. Appl Environ Microbiol 71:5341–5347

Troian RF, Steindorff AS, Ramada MHS, Arruda W, Ulhoa CJ (2014) Mycoparasitism studies of *Trichoderma harzianum* against *Sclerotinia sclerotiorum*: evaluation of antagonism and expression of cell wall-degrading enzymes genes. Biotechnol Lett 36:2095–2101

Utkhede R (2006) Increased growth and yield of hydroponically grown greenhouse tomato plants inoculated with arbuscular mycorrhizal fungi and *Fusarium oxysporum* f. sp. radicis-lycopersici. BioControl 51:393–400

van Loon LC (1997) Induced resistance in plants and the role of pathogenesis-related proteins. Eur J Plant Pathol 103:753–765

Van Wees SCM, der Ent SV, Pieterse CMJ (2008) Plant immune responses triggered by beneficial microbes. Curr Opin Plant Biol 11:443–448

Veerabhadraswamy AL, Garampalli RH (2011) Effect of arbuscular mycorrhizal fungi in the management of black bundle disease of maize caused by *Cephalosporium acremonium*. Sci Res Repot 1:96–100

Velivelli SL, Lojan P, Cranenbrouck S, de Boulois HD, Suarez JP, Declerck S, Franco J, Prestwich BD (2015) The induction of Ethylene response factor 3 (ERF3) in potato as a result of co-inoculation with Pseudomonas sp. R41805 and *Rhizophagus irregularis* MUCL 41833-a possible role in plant defense. Plant Signal Behav 10:e988076

Veresoglou SF, Chen B, Rillig MC (2012) Arbuscular mycorrhiza and soil nitrogen cycling. Soil Biol Biochem 46:53–62

Verma M, Brar S, Tyagi R, Surampalli R, Valero J (2007) Antagonistic fungi, *Trichoderma* spp.: panoply of biological control. Biochem Eng J 37:1–20

Vinale F, Sivasithamparam K, Ghisalberti EL, Marra R, Woo SL, Lorito M (2008) *Trichoderma*-plant-pathogen interactions. Soil Biol Biochem 40:1–10

Vinodkumar S, Indumathi T, Nakkeeran S (2017) *Trichoderma asperellum* (NVTA2) as a potential antagonist for the management of stem rot in carnation under protected cultivation. Biol Control 113:58–64

Viterbo A, Wiest A, Brotman Y, Chet I, Kenerley C (2007) The 18mer peptaibols from *Trichoderma virens* elicit plant defence responses. Mol Plant Pathol 8:737–746

Vos CM, Yang Y, De Coninck B, Cammue BPA (2014) Fungal (-like) biocontrol organisms in tomato disease control. Biol Control 74:65–81

Waghmare SJ, Kurundkar BP (2011) Efficacy of local isolates of *Trichoderma* spp against *Fusarium oxysporum sp. carthami* causing wilt of safflower. Adv Plant Sci 24:37–38

Waghunde RR, Shelake RM, Sabalpara AN (2016) *Trichoderma*: a significant fungus for agriculture and environment. Afr J Agric Res 11:1952–1965

Wang B, Qiu YL (2006) Phylogenetic distribution and evolution of mycorrhizas in land plants. Mycorrhiza 16:299–363

Wang C, Li X, Song F (2012) Protecting cucumber from *Fusarium wilt* with arbuscular mycorrhizal fungi. Commun Soil Sci Plant Anal 43:2851–2864

Wang W, Shi J, Xie Q, Jiang Y, Yu N, Wang E (2017) Nutrient exchange and regulation in arbuscular mycorrhizal symbiosis. Mol Plant 10:1147–1158

Waters MT, Gutjahr C, Bennett T, Nelson DC (2017) Strigolactone signaling and evolution. Annu Rev Plant Biol 68:291–322

Watts-Williams SJ, Cavagnaro TR (2018) Arbuscular mycorrhizal fungi increase grain zinc concentration and modify the expression of root ZIP transporter genes in a modern barley (*Hordeum vulgare*) cultivar. Plant Sci 274:163–170

Weindling R (1932) *Trichoderma lignorum* as a parasitic of other fungi. Phytopathology 22:837–845

Willmann A, Thomfohrde S, Haensch R, Nehls U (2014) The poplar NRT2 gene family of high affinity nitrate importers: impact of nitrogen nutrition and ecto-mycorrhiza formation. Environ Exp Bot 108:79–88

Wu QS, Srivastava AK, Zou YN (2013) AMF-induced tolerance to drought stress in citrus: a review. Sci Hortic (Amst) 164:77–87

Wuyts N, Swennen R, De Waele D (2006) Effects of plant phenylpropanoid pathway products and selected terpenoids and alkaloids on the behaviour of the plant-parasitic nematodes *Radopholus similis*, *Pratylenchus penetrans* and *Meloidogyne incognita*. Nematology 8:89–101

Xu M, Dong J, Wang H, Huang L (2009) Complementary action of jasmonic acid on salicylic acid in mediating fungal elicitor-induced flavonol glycoside accumulation of Ginkgo biloba cells. Plant Cell Environ 32:960–967

Xue AG, Guo W, Chen Y, Siddiqui I, Marchand G, Liu J, Ren C (2017) Effect of seed treatment with novel strains of *Trichoderma* spp. on establishment and yield of spring wheat. Crop Prot 96:97–102

Zamioudis C, Pieterse CMJ (2012) Modulation of host immunity by beneficial microbes. Mol Plant-Microbe Interact 25:139–150

Zeilinger S, Gruber S, Bansal R, Mukherjee PK (2016) Secondary metabolism in *Trichoderma*-chemistry meets genomics. Fungal Biol Rev 30:74–90

Zeriouh W, Nani A, Belarbi M, Dumont A, de Rosny C, Aboura I, Ghanemi FZ, Murtaza B, Patoli D, Thomas C, Apetch L, Rebe C, Delmas D, Khan NA, Ghiringhelli F, Rialland M, Hichami A (2017) Phenolic extract from oleaster (*Olea europaea* var. *Sylvestris*) leaves reduces colon cancer growth and induces caspase-dependent apoptosis in colon cancer cells via the mitochondrial apoptotic pathway. PLoS One 12:e0170823

Zimmermann J, Musyoki MK, Cadisch G, Rasche F (2016) Biocontrol agent *Fusarium oxysporum* f. sp. *strigae* has no adverse effect on indigenous total fungal communities and specific AMF taxa in contrasting maize rhizospheres. Fungal Ecol 23:1–10

Zorb C, Geilfus CM, Mühling KH, Ludwig-Muller J (2013) The influence of salt stress on ABA and auxin concentrations in two maize cultivars differing in salt resistance. J Plant Physiol 170:220–224

Chapter 5
Management of Soil-Borne Diseases of Grain Legumes Through Broad-Spectrum Actinomycetes Having Plant Growth-Promoting and Biocontrol Traits

Subramaniam Gopalakrishnan and Vadlamudi Srinivas

Abstract Chickpea (*Cicer arietinum* L.) and pigeonpea (*Cajanus cajan* L.) are the two important grain legumes grown extensively in the semiarid tropics (SAT) of the world, where soils are poor in nutrients and receive inadequate/erratic rainfall. SAT regions are commonly found in Africa, Australia, and South Asia. Chickpea and pigeonpea suffer from about 38 pathogens that cause soil-borne diseases including wilt, collar rot, dry root rot, damping off, stem canker, and *Ascochyta*/*Phytophthora* blight, and of which three of them, wilt, collar rot, and dry root rot, are important in SAT regions. Management of these soil-borne diseases are hard, as no one control measure is completely effective. Advanced/delayed sowing date, solarization of soil, and use of fungicides are some of the control measures usually employed for these diseases but with little success. The use of disease-resistant cultivar is the best efficient and economical control measure, but it is not available for most of the soil-borne diseases. Biocontrol of soil-borne plant pathogens has been managed using antagonistic actinobacteria, bacteria, and fungi. Actinobacterial strains of *Streptomyces*, *Amycolatopsis*, *Micromonospora*, *Frankia*, and *Nocardia* were reported to exert effective control on soil-borne pathogens and help the host plants to mobilize and acquire macro- and micronutrients. Such novel actinomycetes with wide range of plant growth-promoting (PGP) and antagonistic traits need to be exploited for sustainable agriculture. This chapter gives a comprehensive analysis of important soil-borne diseases of chickpea and pigeonpea and how broad-spectrum actinomycetes, particularly *Streptomyces* spp., could be exploited for managing them.

S. Gopalakrishnan (✉) · V. Srinivas
International Crops Research Institute for the Semi-Arid Tropics (ICRISAT), Patancheru, Telangana, India
e-mail: s.gopalakrishnan@cgiar.org

A. Varma et al. (eds.), *Plant Microbe Interface*,
https://doi.org/10.1007/978-3-030-19831-2_5

5.1 Introduction: Soil-Borne Pathogens of Chickpea

Chickpea affected by more than 170 plant pathogens including bacteria, fungi, nematodes, mycoplasmas, and viruses, whereas only 38 of these, associated with 19 genera of fungi, cause soil-borne diseases, such as wilt, collar rot, dry root rot, canker, damping-off, and blight. The three most important of them, wilt, collar rot, and dry root rot, are briefed here.

5.1.1 Wilt

The causal organism of wilt in chickpea is *Fusarium oxysporum* Schl. emend. Snyd. and Hans. f. sp. *ciceri* (Padwick; FOC). Other species of *Fusarium* also reported to cause wilt and produce toxins (Gopalakrishnan et al. 2005; Gopalakrishnan and Strange 2005). Wilt is the third most important disease of chickpea throughout the world and is reported in about 32 countries (Dubey et al. 2007). It is a serious problem of chickpea between the latitudes 30°N and 30°S of the equator, where the season is warm and dry (semiarid), but common in all growing areas of the world (Dubey et al. 2007). Wilt causes total yield loss (up to 100%) in chickpea under favorable conditions (Landa et al. 2004). FOC is a facultative saprophyte and thus can survive in soil and/or crop residues for 6 years as chlamydospores. The typical symptoms of wilt include drooping of leaves, petiole, and rachis and browning of xylem vessels (Gopalakrishnan et al. 2005). Pigeonpea, lentil, and pea are symptomless carriers of FOC. Usage of resistant cultivar is the best option to manage this disease, but it is limited by the presence of eight races in FOC (Jimenez-Gasco et al. 2002). *Fusarium* wilt can be managed to some extent with a blend of benomyl 30% and thiram 30% at 1.5 g kg^{-1} seed (Haware et al. 1996). However, fungicides are not economical against soil-borne pathogens such as FOC. Further, it has also led to environmental degradation and pollution, imbalance in the microbial community in the rhizosphere soil, pathogen resistance, and increased risk to animal and human health (Li et al. 2012; On et al. 2015).

5.1.2 Collar Rot

It is caused by *Sclerotium rolfsii* Sacc. This pathogen attacks more than 100 crop species including chickpea, groundnut, vegetables, fruits, and ornamental crops (Aycock 1966). Collar rot generally occurs where the soil moisture is high, temperature above 30 °C, and up to 6 weeks after sowing. It is one of the most important diseases of chickpea that causes mortality of seedlings up to 95% under favorable conditions (Sharma and Ghosh 2017). This is one of the important reasons why collar rot is a major problem in areas where chickpea follows rice. *S. rolfsii* survives in the infected plant tissues and crop debris for years. Sclerotia usually attack collar region of the host plants. The disease symptoms of collar rot include dark-brown lesions on the root surface (at collar region), and yellowing and wilting of leaves leads to death of the plant (Nene et al. 1991). Elimination of *S. rolfsii* in the field is not possible due to its ability to produce large numbers of sclerotia in short time and persist in the soil for decades (Punja 1988). Synthetic chemicals such as thiram are used for collar rot management, but yield losses still persist (Singh and Gaur 2016). High levels of resistant cultivar against collar rot of chickpea were not available (Tarafdar et al. 2018). Hence, combination of cultivar with low level of genetic resistance and a biocontrol product or fungicide may be advised as an alternate strategy for the management of collar rot.

5.1.3 Dry Root Rot

It is caused by *Rhizoctonia bataticola* in chickpea. It is endemic in both tropical and temperate regions of the world and cause the disease in over 500 different crops. Dry root rot causes up to 100% yield losses under favorable conditions. *R. bataticola* normally causes the disease when the day temperature is high (above 30 °C) and under dry conditions. In chickpea, the dry root rot symptoms include brown discoloration of lower leaves and stems, black-colored taproots, rotten, and devoid of the lateral and fine roots. When the dried chickpea stem (of the collar region) is split open vertically, sclerotia (minute and dark colored) are seen along with sparse mycelium in the pith (Nene et al. 1991). Although many control measures are available to manage dry root rot in chickpea, soil-borne nature, persistence in soil, and a wide host range of *R. bataticola* make this disease difficult to control. High levels of resistant cultivar are not available. Many biological control products are being evaluated to control dry root rot disease in chickpea.

5.2 Soil-Borne Pathogens of Pigeonpea (*Cajanus cajan* L.)

Diseases are one of the major concerns for the production of pigeonpea as more than 60 pathogens such as bacteria, fungi, nematodes, mycoplasma, and viruses are reported to infect it (Reddy et al. 2012). The four most important soil-borne diseases of pigeonpea include *Fusarium* wilt, collar rot, and dry root rot, and *Phytophthora* blights are briefed here.

5.2.1 Fusarium Wilt

Wilt in pigeonpea is caused by *Fusarium udum* Butler. It is distributed in India, Bangladesh, Ghana, Kenya, Malawi, Myanmar, and Nepal. The annual losses due to *Fusarium* wilt have been estimated at US$ 5 million in Eastern Africa and US$

71 million in India (Reddy et al. 2012). *Fusarium* wilt in pigeonpea is seed- and soil-borne and can survive in plant debris or soil for more than a decade. Disease incidence of wilt is more severe on Vertisols (black soils) than Alfisols (red soils) and occurs when plants are 1–2 months old and when plants are flowering or podding. Symptoms of *Fusarium* wilt include patches of dead plants during flowering/podding of the crop, purple band extending upward from the base of the main stem (seen with green stems), and partial wilting of the plant. When the main stem or branches are split open, browning or blackening of the xylem is visible. Fungicides such as benomyl, bavistin, and thiram are extensively sprayed to manage wilt of pigeonpea but with less success (Meena et al. 2002). Since *F. udum* is a soil-borne, synthetic chemical control is not useful (Maisuria et al. 2008). Hence, much reliable and sustainable control measure needs to be explored.

5.2.2 Collar Rot

It is caused by a soil-borne pathogen called *Sclerotium rolfsii* Saccardo. It is mainly distributed in South Asian countries such as India, Sri Lanka, and Pakistan and other countries including Venezuela, Puerto Rico, the USA, and Trinidad and Tobago. Collar rot causes severe yield losses when undecomposed organic matter is left in the field. It can be a serious problem when the soil is having high moisture and temperature above 30 °C. Collar rot appears within a month of sowing under field conditions. The symptoms are rotting in the collar region and turn chlorotic when they die. Brown or white sclerotia are usually found in the collar region of the plant (Reddy et al. 2012). Collar rot disease can be reduced by removing the stubble of previous crop, selecting well drained fields and doing deep summer ploughing. Biocontrol agents are being explored for controlling this disease without much success.

5.2.3 Dry Root Rot

It is caused by *Macrophomina phaseolina* (Tassi) Goidanich and/or *Rhizoctonia bataticola* (Taub.) Butler. It is distributed in Sri Lanka, Nepal, Myanmar, India, Trinidad and Tobago, and Jamaica. This is a serious problem for short-duration pigeonpea sown in rainy season, particularly in the reproductive stage. Temperature above 30 °C and dry weather helps disease development. Dry root rot is found more on Vertisols than on Alfisols. Symptoms include sudden and premature dry up and rotten and shredded roots upon uprooting. Dark sclerotial bodies can be seen underneath the bark. Spindle-shaped lesions are seen on stems which coalesce and cause the entire plants to dry out (Reddy et al. 2012). Dry root rot disease can be reduced by removing the stubble of previous crop and planting on time. Biocontrol agents are being explored for controlling this disease.

5.3 *Phytophthora* Blight

The causal organism of *Phytophthora* blight is *Phytophthora drechsleri* Tucker f. sp. *cajani* in pigeonpea. *Phytophthora* blight is the third important disease of pigeonpea after wilt and sterility mosaic disease (Kannaiyan et al. 1984). It is distributed in India, Kenya, Dominican Republic, Puerto Rico, and Panama. The pathogen survives as oospores, chlamydospores, and dormant mycelia on soil and plant debris. It is predominant on Alfisols than Vertisols. Continuous drizzling, cloudy weather, and water-logged soils with temperature around 25 °C favor this disease infection and spread. Plants infected with *Phytophthora* blight show symptoms of brownish black and sunken lesions on stems and petioles and water-soaked lesions on leaves. Infected leaves lose turgidity and desiccate, while stems and branches break leaving the foliage above the lesion to dry up (Reddy et al. 2012). Low level of resistance in cultivated and wild germplasms is available against *Phytophthora* blight (Pande et al. 2011). Not many management options are available, for this disease.

5.4 Management of Soil-Borne Diseases of Chickpea and Pigeonpea

Management of soil-borne diseases are difficult, as not one control measure is completely effective. Traditionally, soil-borne pathogens of chickpea and pigeonpea were managed through synthetic chemicals which lead to the development of resistance to wide range of synthetic chemicals, environmental degradation, and contamination (On et al. 2015). Soil-borne pathogens were managed to some extent with a blend of benomyl 30% and thiram 30% at 1.5 g kg^{-1} seed (Haware et al. 1996). Advanced sowing date, solarization of soil, and use of disease-free seed are the cultural control measures usually used but with limited success. High levels of resistant cultivars against majority of soil-borne pathogens are also not available (Tarafdar et al. 2018). Hence, there is an urgent need for looking environmentally viable and human- and animal health-friendly approaches for managing soil-borne diseases. The use of microbial biocontrol products could be one of such alternative options for the management of these diseases (Jiménez-Fernández et al. 2015). Biocontrol of soil-borne diseases has been addressed using antagonistic bacteria and fungi such as *Bacillus* spp., *Pseudomonas* spp., *Pantoea* spp., and *Trichoderma* spp. These agents were reported effective not only to manage plant pathogens but also to help the plants to mobilize and acquire macro- and micronutrients (Postma et al. 2003; Perner et al. 2006; Maisuria et al. 2008; Kothasthane et al. 2017). Microbial biocontrol agents reported for soil-borne diseases of chickpea and pigeonpea are summarized in Table 5.1. However, this chapter is focused only on actinomycetes.

Table 5.1 Biocontrol agents against soil-borne pathogens of chickpea and pigeonpea

Crop	Disease	Causal organism	Biocontrol agents	References
(a) Biocontrol agents against soil-borne pathogens of chickpea				
Chickpea	*Fusarium* Wilt	*Fusarium solani* f. sp. *Pisi*	*Streptomyces* sp.	Mahboobeh et al. (2016)
		Fusarium oxysporum f. sp. *ciceri*	*Streptomyces* sp.	Amini et al. (2016)
			Streptomyces sp.	Gopalakrishnan et al. (2011)
			Streptomyces sp.	Anusha et al. (2018, submitted)
			Streptomyces sp.	Sreevidya and Gopalakrishnan (2013)
			Trichoderma viride, *Trichoderma harzianum*, *Trichoderma virens*, *Bacillus subtilis* *Aspergillus niger AN27*	Singh (2014)
			B. subtilis	Kumar (1999)
			Pseudomonas aeruginosa PNA 1	Anjaiah et al. (2003)
			Pseudomonas fluorescens	Vidhyasekaran and Muthamilan (1995), Saikia et al. (2009)
	Collar rot	*Sclerotium rolfsii*	*Streptomyces griseus*	Singh and Gaur (2016)
			Streptomyces sp.	Sreevidya and Gopalakrishnan (2013)
			T. viride, T. harzianum, P. fluorescens	Singh (2014)
	Dry root rot	*Rhizoctonia bataticola*	*Streptomyces* sp.	Anusha et al. (2018, submitted)
			T. harzianum, T. viride	Mishra et al. (2018)
		Macrophomina phaseolina	*T. viride* and *P. fluorescens*	Manjunatha et al. (2013)
	Wet root rot	*Rhizoctonia solani*	*T. harzianum* (PDBCTH 10) and *T. viride* (PDBCTV)	Prasad et al. (2002a)

(continued)

Table 5.1 (continued)

Crop	Disease	Causal organism	Biocontrol agents	References
(b) Biocontrol agents against soil-borne pathogens of pigeonpea				
Pigeonpea	*Fusarium* Wilt	*Fusarium udum*	*T. harzianum*	Prasad et al. (2002a, b)
			Alcaligenes xylosoxydans	Vaidya et al. (2003)
			T. harzianum, T. viride, P. fluorescens, B. subtilis	Goudar and Srikanth (2002)
			Pantoea dispersa	Maisuria et al. (2008)
			B. subtilis	Siddiqui and Mahmood (1995)
			P. fluorescens	Siddiqui et al. (1998)
			B. subtilis AF1	Manjula and Podile (2001)
			T. harzianum, T. hamatum, T. viride, T. koningii, B. subtilis	Mishra et al. (2018)
		Fusarium Oxysporum	*P. aeruginosa* PNA 1	Anjaiah et al. (2003)
	Phytophthora stem blight	*Phytophthora drechsleri* f. sp. *Cajani*	*T. harzianum, T. hamatum, Glomus mosseae, P. fluorescens, B. subtilis*	Mishra et al. (2018)

5.5 Actinomycetes and Their Role in Biocontrol and Plant Growth-Promoting Traits

Actinomycetes, a group of Gram-positive (Gram +ve) bacteria with a high G+C content, are found commonly in marine and fresh water, compost, and soil. They are known to decompose organic residues and produce secondary metabolites of commercial interest in agriculture and medical field. Actinomycetes are also reported to play a key role in the plant protection and plant growth promotion (PGP). The PGP actinomycetes directly influence crop growth by producing growth hormones (including auxin and gibberellins), fixing nitrogen in the roots of leguminous plants (through symbiotic nitrogen fixation), and solubilizing phosphorous, iron, zinc, and potassium, thereby increasing nutrient availability for the plants. They also promote crop growth indirectly by producing 1-aminocyclopropane-1-carboxylate (ACC) deaminase (a stress-relieving enzyme), antibiotics (such as pyocyanin, pyoluteorin, viscosinamide, 2,4-diacetylphloroglucinol, streptomycin, kanosamine, pyrrolnitrin, phenazine-1-carboxylic acid, and neomycin A), siderophores (for iron uptake and antagonistic traits against plant pathogens in the vicinity), hydrocyanic acid (HCN; inhibits electron transport system and disrupts energy supply to the plant pathogenic cells), hydrolytic enzymes (including chitinase, cellulase, β-1,3-glucanase, lipase and protease; lyse cell walls of plant pathogens), and inducing systemic resistance (by producing phenylalanine ammonia-lyase (PAL) and antioxidant enzymes such as catalase, peroxidase, lipoxygenase, polyphenol oxidase, superoxide dismutase, and ascorbate peroxidase) in plants (Gopalakrishnan et al. 2016).

5.6 Broad-Spectrum Actinomycetes Having Biocontrol and PGP Traits

Soil-borne pathogens such as *F. oxysporum*, *S. rolfsii*, and *R. bataticola* affect many agriculturally important crops including chickpea and pigeonpea that lead to significant yield losses. For instance, *S. rolfsii* and *R. bataticola* attack more than 100 and 500 crop species, respectively. It is very hard to breed variety with resistance to a wide range of plant pathogens; however, identifying a biocontrol agent with broad-spectrum of activities is relatively easy. Therefore, there is an urgent need to identify broad-spectrum biocontrol and PGP microorganisms for the control of multiple plant diseases in a single crop, and thereby the productivity (yield traits) can also be enhanced. This is very important as one microbial treatment solves more than one pathogen problem apart from promotion of plant growth. For instance, endophytic *Streptomyces* spp. were reported to colonize on the nodules of soybean with *Rhizobium* to increase nodulation, nitrogen fixation, and grain yield (Soe et al. 2010). Strains of *Streptomyces* spp. were demonstrated to provide broad-spectrum of antagonistic traits and protect mycorrhizal roots from mycorrhizal fungal competitors and parasites (Schrey et al. 2012). The usefulness of actinomycetes (such as *Streptomyces*) and their secondary metabolites for biocontrol of plant pathogens and PGP are widely reported in many crops (Jetiyanon and Kloepper 2002; Ryu et al. 2007; El-Tarabily 2008; Soe et al. 2010; Sadeghi et al. 2012; Alekhya and Gopalakrishnan 2016, 2017; Sathya et al. 2017; Vijayabharathi et al. 2018).

Actinomycetes are widely reported as inducers of host-plant resistance and plant immunization against many soil-borne plant pathogens including *Phytophthora*, *Fusarium*, *Rhizoctonia*, *Colletotrichum*, and *Pythium* (Raaijmakers et al. 2009). Endophytic actinomycetes such as *Streptomyces* were reported to manage diseases of potato scab and wheat under the field conditions (Liu et al. 1996; Coombs et al. 2004). Endophytic *Streptomyces* sp. EN 27 and *Micromonospora* sp. EN 43 were reported to induce resistance in *Arabidopsis thaliana* by upregulating genes involved in systemic acquired resistance (SAR; Conn et al. 2008). Studies on *Streptomyces* induced host-plant resistance was demonstrated on wide range of crops including vegetable crops such as potato (Arseneault et al. 2014) and *Arabidopsis* (Bernardo et al. 2013) and woody trees such as oak (Kurth et al. 2014) and *Eucalyptus* (Salla et al. 2016). Such actinomycetes are needed for soil-borne pathogens of grain legumes including chickpea and pigeonpea.

5.7 ICRISAT's Experience in Dealing with Broad-Spectrum Actinomycetes Having Biocontrol and PGP Traits

ICRISAT, Patancheru, Hyderabad, and Telangana, India, reported five strains of *Streptomyces* spp. (CAI-24, KAI-32, KAI-90, CAI-121, and CAI-127), isolated from compost of bitter guard, chrysanthemum, and garlic foliage and rice straw, having antagonistic potential against *Fusarium* wilt of chickpea under both glasshouse and field experiments (Gopalakrishnan et al. 2011). The details of the five strains including their NCBI accession numbers and sources of isolation were shown in Table 5.2.

The five FOC antagonistic *Streptomyces* spp. strains were also reported to produce HCN, siderophore, β-1,3-glucanase, lipase, indole acetic acid (IAA; except KAI-90), and chitinase (except CAI-121 and CAI-127), whereas two strains produced protease (CAI-24 and CAI-127) and cellulase (KAI-32 and KAI-90) (Gopalakrishnan et al. 2011, 2013). Three of the *Streptomyces* strains (CAI-24, CAI-127 and KAI-90) were also found to produce ACC deaminase (unpublished data). These enzymatic activities and secondary metabolites are well known for their role in biocontrol and have been demonstrated against many plant pathogens. Actinomycetes having these biocontrol and PGP traits can be used for managing soil-borne pathogens.

Two of the selected five *Streptomyces* spp. strains (KAI-32 and KAI-90) inhibited *Macrophomina phaseolina* (causes charcoal rot in sorghum), whereas three strains

Table 5.2 *Streptomyces* spp. strains used for the management of *Fusarium* wilt of chickpea

Strains	Scientific name	NCBI accession number	Source of isolation
CAI-24	*Streptomyces tsusimaensis*	JN400112	Bitter gourd foliage compost
CAI-121	*Streptomyces caviscabies*	JN400113	*Chrysanthemum* foliage compost
CAI-127	*Streptomyces setonii*	JN400114	Garlic foliage
KAI-32	*Streptomyces africanus*	JN400115	Rice-straw compost
KAI-90	*Streptomyces* sp.	JN400116	Rice-straw compost

(KAI-90, CAI-24, and KAI-32) inhibited *R. bataticola* (causes dry root rot in chickpea) in the dual culture assay (Gopalakrishnan et al. 2011). All five strains were also found to inhibit *Botrytis cinerea*, which causes *Botrytis* gray mold (BGM) disease in chickpea (unpublished data). Hence, it is concluded that the selected five FOC antagonistic strains displayed broad-spectrum biocontrol potential.

Under in vitro conditions, the selected five FOC antagonistic *Streptomyces* spp. strains were able to grow in temperatures between 20 and 40 °C, NaCl up to 6% and at pH values between 5 and 13. All were sensitive to captan, thiram, benlate, Radonil, and benomyl but highly tolerant to bavistin at field application levels (Gopalakrishnan et al. 2013). It is concluded that the selected five actinomycetes are capable of surviving in harsh environmental conditions and compatible with the fungicide bavistin and thus can be used in integrated pest and/or disease management (IPM/IDM) programs.

Under field conditions, on 2 consecutive years in chickpea, the selected five FOC antagonistic *Streptomyces* spp. strains significantly enhanced nodule number and weight (up to 70% and 82%, respectively), root weight (up to 7%), and shoot weight (up to 21%) over the control (uninoculated) plots (Table 5.3). At crop maturity, the five strains significantly enhanced stover and grain yields (up to 39% and 12%, respectively) over the control plots (Fig. 5.1). All other traits including pod number, pod weight, total dry matter, seed number, and seed weight were also found to enhance over the uninoculated control. The detailed results and discussion of this experiment can be seen in Gopalakrishnan et al. (2015). Similar results were obtained in pigeonpea under field conditions. In pigeonpea, the traits including nodule weight, root weight, shoot weight, stover yield, and grain yield were found significantly enhanced over the uninoculated control (Table 5.3; Fig. 5.1; Unpublished data). Soe et al. (2010) reported similar kind of results in soybean where endophytic *Streptomyces* spp. were shown to colonize on the nodules of soybean and have beneficial association with *Rhizobium* and increased nodules, nitrogen fixation, and grain yield.

5.8 Conclusion

Among the important diseases of chickpea and pigeonpea, soil-borne diseases including *Fusarium* wilt, collar rot, dry root rot, and *Phytophthora* blight are important and cause severe damages to the crop. Management of such soil-borne diseases are extremely difficult, as no one control measure is completely effective. Biocontrol of soil-borne plant pathogenic fungi using actinobacteria such as *Streptomyces* spp. was reported effective not only to manage soil-borne pathogens but

Table 5.3 Evaluation of the five *Streptomyces* spp. strains for their nodulation and shoot and root weights of chickpea and pigeonpea, at 30 days after sowing

Strains	Chickpea				Pigeonpea			
	Nodule number (plant^{-1})	Nodule weight (mg plant^{-1})	Root weight (g plant^{-1})	Shoot weight (g plant^{-1})	Nodule number (plant^{-1})	Nodule weight (mg plant^{-1})	Root weight (g plant^{-1})	Shoot weight (g plant^{-1})
CAI-24	13	58	1.75	1.60	6	133	2.00	2.03
CAI-121	25	53	1.76	1.90	5	200	1.90	2.03
CAI-127	17	24	1.46	1.43	5	167	2.23	2.38
KAI-32	17	48	1.84	1.74	5	167	2.20	2.11
KAI-90	31	75	1.96	1.51	5	133	2.24	2.33
Control	12	29	1.71	1.35	5	130	1.84	2.02
Mean	21	46	1.75	1.58	5	155	2.07	2.15
SE±	1.5***	4.4***	0.08***	0.070**	0.3**	14.0*	0.058**	0.061**
LSD (5%)	4.8	14.0	0.25	0.222	0.8	43.0	0.184	0.191
CV%	14	16	8	8	10	15	5	5

* statistically significant at 0.05, ** statistically significant at 0.01, *** statistically significant at 0.001

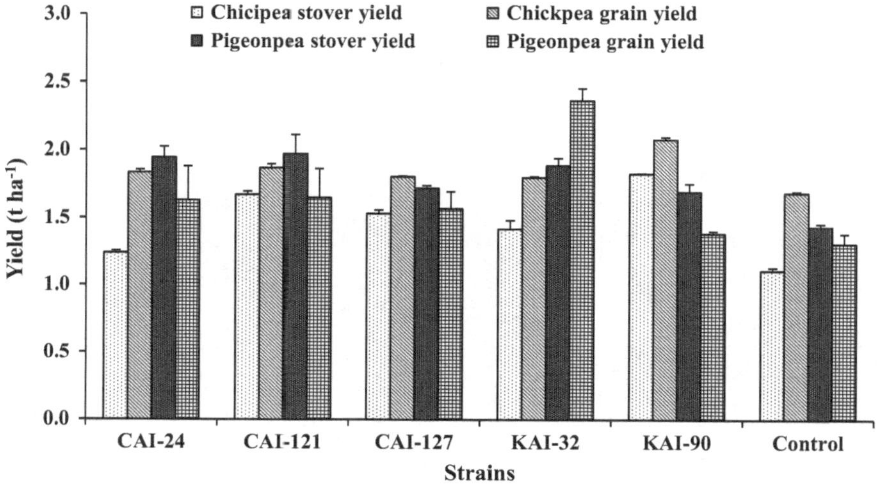

Fig. 5.1 Influence of the five *Streptomyces* spp. strains on stover and grain yields (t ha^{-1}) of chickpea and pigeonpea under field conditions

also promote crop growth and yield. ICRISAT reported five strains of *Streptomyces* spp. (CAI-24, KAI-32, KAI-90, CAI-121, and CAI-127), not only having antagonistic potential against *Fusarium* wilt of chickpea under both greenhouse and field experiments but also having PGP potentials on chickpea and pigeonpea. Such novel bacteria with broad-spectrum PGP and biocontrol traits needs to be exploited for sustainable agriculture.

References

Alekhya G, Gopalakrishnan S (2016) Exploiting plant growth-promoting *Amycolatopsis* sp. in chickpea and sorghum for improving growth and yield. J Food Legumes 29:225–231

Alekhya G, Gopalakrishnan S (2017) Biological control and plant growth-promotion traits of *Streptomyces* species under greenhouse and field conditions in chickpea. Agric Res 6:410–420

Amini J, Agapoor Z, Ashemgroph M (2016) Evaluation of *Streptomyces* spp. against *Fusarium oxysporum* f. sp. *ciceri* for the management of chickpea wilt. J Plant Protect Res 56:257–264

Anjaiah V, Cornelis P, Koedam N (2003) Effect of genotype and root colonization in biological control of *Fusarium* wilts in pigeonpea and chickpea by *Pseudomonas aeruginosa* PNA1. Can J Microbiol 49:85–91

Anusha BG, Gopalakrishnan S, Naik MK, Sharma M (2018) Evaluation of *Streptomyces* spp. and *Bacillus* spp. for biocontrol of *Fusarium* wilt in chickpea (*Cicer arietinum* L.). Submitted in J Arch Phytopathol Plant Protect

Arseneault T, Pieterse CM, Gérin-Ouellet M, Goyer C, Filion M (2014) Long-term induction of defense gene expression in potato by *Pseudomonas* sp. LBUM223 and *Streptomyces scabies*. Phytopathology 104:926–932

Aycock RA (1966) Stem rot and other diseases of caused by *Sclerotium rolfsii*. N C Agric Exp Stn Tech Bull 174:202

Bernardo KT, Franco CMM, Chataway T (2013) Proteomic analysis in *Arabidopsis thaliana* inoculated with a beneficial endophytic *Streptomyces* sp. in response to a challenge by the pathogen *Fusarium oxysporum*. In: Schneider C, Leifert C, Feldmann F (eds) Endophytes for plant protection: the state of the art. In: Proceedings of the 5th international symposium on plant protection and plant health in Europe, 26?29 May, 2013, Humboldt University Berlin, Berlin-Dahlem, pp 78–87

Conn VM, Walker AR, Franco CMM (2008) Endophytic actinobacteria induce defense pathways in *Arabidopsis thaliana*. Mol Plant-Microbe Interact 21:208–218

Coombs JT, Michelsen PP, Franco CMM (2004) Evaluation of endophytic actinobacteria as antagonists of *Gaeumannomyces graminis* var. *tritici* in wheat. Biol Control 29:359–366

Dubey SC, Suresh M, Singh B (2007) Evaluation of *Trichoderma* species against *Fusarium oxysporum* f sp. *ciceri* for integrated management of chickpea wilt. Biol Control 40:118–127

El-Tarabily KA (2008) Promotion of tomato (*Lycopersicon esculentum* Mill.) plant growth by rhizosphere competent 1-aminocyclopropane-1-carboxylic acid deaminase–producing *Streptomycete* actinomycetes. Plant Soil 308:161–174

Gopalakrishnan S, Strange RN (2005) Identity and toxicity of *Fusarium* species isolated from wilted chickpea. Phytopathol Mediterr 44:180–188

Gopalakrishnan S, Beale MH, Ward JL, Strange RN (2005) Chickpea wilt: identification and toxicity of 8-O-methyl-fusarubin from Fusarium acutatum. Phytochemistry 66:1536–1539

Gopalakrishnan S, Pande S, Sharma M, Humayun P, Kiran BK, Sandeep D, Vidya MS, Deepthi K, Rupela O (2011) Evaluation of actinomycete isolates obtained from herbal vermicompost for biological control of *Fusarium* wilt of chickpea. Crop Prot 30:1070–1078

Gopalakrishnan S, Srinivas V, Vidya MS, Rathore A (2013) Plant growth-promoting activities of *Streptomyces* spp. in sorghum and rice. Springer Plus 2:1–8

Gopalakrishnan S, Srinivas V, Alekhya G, Prakash B, Kudapa H, Varshney RK (2015) Evaluation of *Streptomyces* sp. obtained from herbal vermicompost for broad spectrum of plant growth-promoting activities in chickpea. Org Agric 5:123–133

Gopalakrishnan S, Sathya A, Vijayabharathi R (2016) A book entitled "plant growth-promoting actinobacteria: a new avenue for enhancing the productivity & soil fertility of grain legumes". Springer, Singapore. ISBN 978-981-10-0705-7

Goudar SB, Srikanth K (2002) Bioassay of antagonist against *Fusarium udum*- causal agent of pigeonpea wilt. Karnataka J Agric Sci 13:64–67

Haware MP, Nene YL, Natarajan M (1996) Survival of *Fusarium oxysporum* f. sp. *ciceri* in soil in the absence of chickpea. Phytopathol Mediterr 35:9–12

Jetiyanon K, Kloepper JW (2002) Mixtures of plant growth-promoting rhizobacteria for induction of systemic resistance against multiple plant diseases. Biol Control 24:285–291

Jiménez-Fernández D, Montes-Borrego M, Jiménez Díaz RM, Navas-Cortés JA, Landa BB (2015) In planta and soil quantification of *Fusarium oxysporum* f. sp. *ciceri* and evaluation of *Fusarium* wilt resistance in *Fusarium oxysporum* f. sp. *ciceri* in the soil through seed transmission. Plant Pathol 23:3–6

Jimenez-Gasco MM, Milgroom MG, Jimenez-Diaz RM (2002) Gene genealogies support *Fusarium oxysporum* f. sp. *ciceri* as a monophyletic group. Plant Pathol 51:72–77

Kannaiyan J, Nene YL, Reddy MV, Ryan JG, Raju TN (1984) Prevalence of pigeonpea diseases and associated crop losses in Asia, Africa and the Americas. Trop Pest Manag 30:62–71

Kothasthane AS, Agrawal T, Zaidi NW, Singh US (2017) Identification of siderophore producing and cynogenic fluorescent *Pseudomonas* and a simple confrontation assay to identify potential bio-control agent for collar rot of chickpea. 3 Biotech 7:137

Kumar BSD (1999) Fusarial wilt suppression and crop improvement through two rhizobacterial strains in chickpea growing in soils infested with *Fusarium oxysporum* f. sp. *ciceri*. Biol Fertil Soils 29:87–91

Kurth F, Mailänder S, Bönn M, Feldhahn L, Herrmann S, Große I, Buscot F, Schrey SD, Tarkka MT (2014) *Streptomyces*-induced resistance against oak powdery mildew involves host plant responses in defense, photosynthesis, and secondary metabolism pathways. Mol Plant-Microbe Interact 27:891–900

Landa BB, Navas-Cortes JA, Jimenez-Diaz RM (2004) Integrated management of *Fusarium* wilt of chickpea with sowing date, host resistance and biological control. Phytopathology 94:946–960

Li Q, Ning P, Zheng L, Huang J, Li G, Hsiang T (2012) Effects of volatile substances of *Streptomyces globisporus* JK-1 on control of *Botrytis cinerea* on tomato fruit. Biol Control 61:113–120

Liu D, Anderson NA, Kinke LL (1996) Selection and characterization of strains of *Streptomyces* suppressive to the potato scab pathogen. Can J Microbiol 42:487–502

Mahboobeh S, Meysam SN, Shahidi GHB (2016) Application of soil-borne actinomycetes for biological control against *Fusarium* wilt of chickpea (*Cicer arietinum*) caused by *Fusarium solani* f. sp. *pisi*. J Phytopathol 164:967–978

Maisuria VB, Gohel V, Ashvinkumar NM, Rakesh RP, Hari SC (2008) Biological control of *Fusarium* wilt of pigeonpea by *Pantoea dispersa*, a field assessment. Ann Microbiol 58:411–419

Manjula K, Podile AR (2001) Chitin supplemented formulations improve biocontrol and plant growth promoting efficiency of *Bacillus subtilis* AF1. Can J Microbiol 47:618–625

Manjunatha SV, Naik MK, Khan MFR, Goswami RS (2013) Evaluation of bio-control agents for management of dry root rot of chickpea caused by *Macrophomina phaseolina*. Crop Prot 45:147–150

Meena B, Radhajeyalakshmi R, Marimuthu T, Vidhyasekaran P, Velazhahan R (2002) Biological control of groundnut late leaf spot and rust by seed and foliar application of powder formulation of *Pseudomonas fluorescens*. Biocontrol Sci Technol 12:195–204

Mishra RJ, Bohra A, Kamaal N, Kumar K, Gandhi K, Sujayanand GK, Saabale PR, Satheesh Naik SJ, Sarma BK, Kumar D, Mishra M, Srivastava DK, Mishra NPS (2018) Utilization of biopesticides as sustainable solutions for management of pests in legume crops: achievements and prospects. Egypt J Biol Pest Control 28:3

Nene YL, Reddy MV, Haware MP, Ghanekar AM, Amin KS (1991) Field diagnosis of chickpea diseases and their control. In: Information bulletin, vol 28. International Crops Research Institute for the Semi-Arid Tropics, Patancheru

On A, Wong F, Ko Q, Tweddell RJ, Antoun H, Avis TJ (2015) Antifungal effects of compost tea microorganisms on tomato pathogens. Biol Control 80:63–69

Pande S, Sharma M, Mangla UN, Ghosh R, Sundaresan G (2011) *Phytophthora* blight of pigeonpea [*Cajanus cajan* (L.) Millsp.]: an updating review of biology, pathogenicity and disease development. Crop Prot 30:951–957

Perner H, Schwarz D, George E (2006) Effect of mycorrhizal inoculation and compost supply on growth and nutrient uptake of young leek plants grown on peat-based substrates. Hortic Sci 41:628–632

Postma J, Montanari M, Van den Boogert PHJF (2003) Microbial enrichment to enhance disease suppressive activity of compost. Eur J Soil Biol 39:157–163

Prasad RD, Rangeshwaran R, Anuroop CP, Rashni HJ (2002a) Biological control of wilt and root rot of chickpea under field conditions. Ann Plant Prot Sci 10:72–75

Prasad RD, Rangeshwaran R, Hegde SV, Anuroop CP (2002b) Effect of soil and seed application of *Trichoderma harzianum* on pigeonpea wilt caused by *Fusarium udum* under field conditions. Crop Prot 21:293–297

Punja ZK (1988) *Sclerotium* (Athelia) *rolfsii*, a pathogen of many plant speices. Adv Plant Pathol 6:523–534

Raaijmakers JM, Paulitz TC, Steinberg C, Alabouvette C, Moenne-Loccoz Y (2009) The rhizosphere: a playground and battlefield for soil borne pathogens and beneficial microorganisms. Plant Soil 321:341–361

Reddy MV, Raju TN, Sharma SB, Nene YL, McDonald D, Pande S, Sharma M (2012) Hand book of pigeonpea diseases (revised). Information bulletin number 42, Patancheru, AP 502324. International Crops Research Institute for the Semi-Arid Tropics, India

Ryu CM, Murphy CF, Reddy MS, Kloepper JW (2007) A two strain mixture of rhizobacteria elicits induction of systemic resistance against *Pseudomonas syringae* and cucumber mosaic virus coupled to promotion of plant growth on *Arabidopsis thaliana*. J Microbiol Biotechnol 17:280–286

Sadeghi A, Karimi E, Dahazi PA, Javid MG, Dalvand Y, Askari H (2012) Plant growth-promoting activity of an auxin and siderophore producing isolate of *Streptomyces* under saline soil condition. World J Microbiol Biotechnol 28:1503–1509

Saikia R, Varghese S, Singh BP, Arora DK (2009) Influence of mineral amendment on disease suppressive activity of *Pseudomonas fluorescens* to *Fusarium* wilt of chickpea. Microbiol Res 164:365–373

Salla TD, Astarita LV, Santarém ER (2016) Defense responses in plants of *Eucalyptus* elicited by *Streptomyces* and challenged with *Botrytis cinerea*. Planta 243:1055–1070

Sathya A, Vijayabharathi R, Gopalakrishnan S (2017) Plant growth-promoting actinobacteria: a new strategy for enhancing sustainable production and protection of grain legumes. 3 Biotech 7:102

Schrey SD, Erkenbrack E, Fruh E, Fengler S, Hommel K, Horlacher N, Schulz D, Ecke M, Kulik A, Fiedler H, Hampp R, Tarkka MT (2012) Production of fungal and bacterial growth modulating secondary metabolites is widespread among mycorrhiza-associated *Streptomyces*. BMC Microbiol 12:164

Sharma M, Ghosh B (2017) Heat and soil moisture stress differentially impact chickpea plant infection with fungal pathogens. In: Senthil Kumar M (ed) Plant tolerance to individual and concurrent stresses. Springer, New Delhi, pp 47–57

Siddiqui ZA, Mahmood I (1995) Biological control of *Heterodera cajani* and *Fusarium udum* by *Bacillus subtilis, Bradyrhizobium japonicum* and *Glomus fasciculatum* on pigeonpea. Fundam Appl Nematol 18:559–556

Siddiqui ZA, Mahmood I, Hayat S (1998) Biocontrol of *Heterodera cajani* and *Fusarium udum* on pigeonpea using *Glomus mosseae*, *Paecilomyces lilacinus* and *Pseudomonas fluorescens*. Thai J Agric Sci 31:310–321

Singh HB (2014) Management of plant pathogens with microorganisms. Proc Indian Natl Sci Acad 80:443–454

Singh SP, Gaur R (2016) Evaluation of antagonistic and plant growth-promoting activities of chitinolytic endophytic actinomycetes associated with medicinal plants against *Sclerotium rolfsii* in chickpea. J Appl Microbiol 121:506–518

Soe KM, Bhromsiri A, Karladee D (2010) Effects of selected endophytic actinomycetes (*Streptomyces* sp.) and *Bradyrhizobia* from Myanmar on growth, nodulation, nitrogen fixation and yield of different soybean varieties. CMU J Nat Sci 9:95–109

Sreevidya M, Gopalakrishnan S (2013) *Bacillus* and *Streptomyces* spp. as potential biocontrol agents to control soil-borne pathogens of chickpea and sorghum. Acta Phytopathol Sin 43:520

Tarafdar A, Rani TS, Chandran USS, Ghosh R, Chobe DR, Sharma M (2018) Exploring combined effect of abiotic (soil moisture) and biotic (*Sclerotium rolfsii* Sacc.) stress on collar rot development in chickpea. Front Plant Sci 9:1154

Vaidya RJ, Macmil SL, Vyas PR, Ghetiya LV, Thakor KJ, Chhatpar HS (2003) Biological control of *Fusarium* wilt of pigeonpea *Cajanus cajan* (L.) Millsp with chitinolytic *Alcaligenes xylosoxydans*. Indian J Exp Biol 41:1469–1472

Vidhyasekaran P, Muthamilan M (1995) Development of formulations of *Pseudomonas fluorescens* for control of chickpea wilt. Plant Dis 79:782–786

Vijayabharathi R, Gopalakrishnan S, Sathya A, Kumar MV, Srinivas V, Sharma M (2018) Tri-dimensional deciphering of endophytic *Streptomyces* sp. on chickpea for plant growth-promotion, helper effect with *Mesorhizobium ciceri* on nodulation and host-plant resistance induction against *Botrytis cinerea*. Microb Pathog 122:98–107

Chapter 6
Soil–Microbes–Plants: Interactions and Ecological Diversity

Prem Chandra and Enespa

Abstract In interactions between plants and soil, microorganisms have significant roles. Ecological stability is contributed by the biogeochemical cycling of elements. An emerging body of research is distinguishing the impacts that root-associated microbial communities can have on plant fitness and growth. Rocks and minerals are weathered by the activities of plants, which exude various types of hormones, with a crucial role in the supply of organic matter and formation of soils. Various types of plant species have distinctive biological characteristics that show constraint to precise soil types. Plant–microbe interactions in soil are contributing to a new, microbially based perspective on plant community and ecology. These microorganisms are soil dwellers, diverse, and their interactions with plants vary with respect to specificity, environmental heterogeneity, and fitness impact. The key influences on plant community structure and dynamics are effected by two microbial procedures: microbial intervention of niche diversity in resource use and response dynamics among the soil community and plants. The hypothesis of niche diversity is based on various interpretations that the nutrients of soil are found in different chemical forms: the plant requires accessing these enzymes and nutrients, and the microorganisms of the soil are a major source of these enzymes. Plant–microbe interactions are a significant establishing force for extensive spatial gradients in species abundance. The positive response (a homogenizing force) and negative response (a diversifying force) of virtual balance may contribute to detected latitudinal (and altitudinal) diversity patterns. The microbially based perception for the ecology of plants promises to contribute to our understanding of long-standing issues in ecology and to disclose new areas of future investigation.

P. Chandra · Enespa (✉)
Department of Environmental Microbiology, School for Environmental Sciences, Babasaheb Bhimrao Ambedkar (A Central) University, Lucknow, UP, India

A. Varma et al. (eds.), *Plant Microbe Interface*,
https://doi.org/10.1007/978-3-030-19831-2_6

6.1 Introduction

The soil is the most important factor for plants and their associated microorganisms, which have a crucial role in the modification and formation of soil (Marschner 2012). According to the origin of the parent material, climate, and vegetation, soil has various properties formed by the weathering of rocks and minerals (Jenny 1980). Directly or indirectly, plants derive soil carbon; mostly the plants are involved in the weathering process by microbial activities that depend on root-derived carbon, and physical and chemical processes are also involved secondarily (Six et al. 2004). Microbial diversity and soil functionality are the important regulating factors in the elements of biogeochemical cycling, that is, carbon, nitrogen, and phosphorus, and as such contribute to the stability and quality of the ecosystem (Van Der Heijden et al. 2008). Microbial ecology has developed rapidly in general and in soil–plant interactions in particular, led by molecular and isotope labeling technologies (Dawson et al. 2002). Concern for preserving ecosystem services and food provision are in urgent need of solutions in East Asian countries such as Korea, Japan, and China (Davis et al. 2011). The East Asian Federation of Ecological Societies (EAFES), which promote collaboration of ecological and environmental sciences, was established by these three countries in 2003 (Kim et al. 2018). Maintenance and establishment of microbial populations in rhizospheric soil have been completed successfully by the transfer of molecular and genetic information. The microbial soil communities also have an important major function in the protection of plants from abiotic stresses and phytopathogens (Johansson et al. 2004). In the environment, the microorganisms are rarely encountered as single-species populations, but observations have shown an enormous biodiversity and abundance variability in a small quantity of samples from various habitats (Sogin et al. 2006). It is suggested that the establishment of microbial populations in the rhizospheric environment (physical, chemical, and biological) is inherent in the interactions of microbes (Bais et al. 2006). The coevolution of various species has led to specialization and adaptation; consequently, the relationships in large variety can expedite sharing, whether symbiotic, mutualistic and antagonistic, endosymbiotic, parasitic, or pathogenic relationships (Toby et al. 2010). In microbial interactions, various secondary metabolites, also known as bioactive compounds that execute various important functions in the interactions of rhizospheric ecology, have been detected (Quiñones et al. 2005). A quorum-sensing mechanism resides in a stimuli–response system which is widely related in cellular concentration studies, and the microbes interact with each other. Production of the signaling molecules (auto-inducers) permits messaging of cells and reacting to the environment in a synchronized way (Waters and Bassler 2005). The microbially associated molecular patterns (MAMP) during interactions with the host cells are preserved in various taxa of microbes, permitting continuous proliferation during interactions with the animal and plant cells and adaptability to the interactions of microbes with various hosts (Table 6.1) (Braga et al. 2016).

The belowground microbial communities have a significant function in the productivity, microbial diversity, and composition of various plants (Van Der

Table 6.1 Microbial interaction studies

Organisms involved	Type of interaction	Compounds/mechanisms involved	Findings	References
Moniliophthora roreri and *Trichoderma harzianum*	Phytopathogen–endophyte	T39 butenolide, harzianolide, sorbicillinol	Compounds dependent on the phytopathogen presence and were spatially localized in the interaction zone.	Braga et al. (2016)
Trichoderma atroviride and *Arabidopsis* sp.	Endophyte–plant	Indole acetic acid-related indoles	Plant root colonization promotes growth and enhances systemic disease resistance in the plant by endophytes.	Salas-Marina et al. (2011)
Xylella fastidiosa and *Methylobacterium mesophilicum*	Phytopathogen–endophyte	Hydroxamate type	Genes related to energy production, stress, transport, and motility were upregulated in the phytopathogen, but genes related to growth were downregulated.	Lacava et al. (2004)
Burkholderia gladioli, *B. seminalis*, and orchid	Phytopathogen–endophyte–plant	Extracellular polysaccharides; altering hormone metabolism	By using extracellular polysaccharides and by altering hormone metabolism, the endophyte strain probably interacts with the plant, as was suggested by genomic analysis.	Araújo et al. (2016)
Bradyrhizobium diazoefficiens and *Aeschynomene afraspera*	Symbiont–plant	C35 hopanoids	C35 hopanoids are essential for symbiosis and are related to evasion of plant defense, utilization of host photosynthates, and nitrogen fixation.	Barrière et al. (2017)
Stachybotrys elegans and *Rhizoctonia solani*	Mycoparasite–host	Trichothecenes and atranones	Mycoparasite-induced alterations in *Rhizoctonia solani* metabolism, growth, and development by the production of mycotoxins. The biosynthesis of many antimicrobial compounds by *R. solani* was downregulated.	Chamoun et al. (2015)
Candida albicans and *Pseudomonas aeruginosa*	Microbial community	Quorum sensing	The *Pseudomonas aeruginosa* QS system may block the yeast-to-hypha transition or activate the hypha-to-yeast reversion of *Candida albicans*. Farnesol produced by *C. albicans* downregulates the QS system of *P. aeruginosa*.	Polke et al. (2017)

(continued)

Table 6.1 (continued)

Organisms involved	Type of interaction	Compounds/mechanisms involved	Findings	References
Vibrio fischeri and fishes or squids	Symbiont–fish	Quorum sensing	In symbiotic association with fishes and squids, the auto-inducer molecule reaches a threshold and luminescence genes are activated.	Fuqua et al. (1994)
Rhizobium leguminosarum and plants	Symbiont–plant	Quorum sensing	The quorum-sensing system in these bacteria is related to different functions: nodulation efficiency, growth inhibition, nitrogen fixation, and plasmid transfer.	Gonzalez and Marketon (2003)
Xanthomonas or *Xylella* and grapevines or citrus	Pathogen–host	Quorum sensing	Quorum-sensing signaling molecules control the expression of virulence factor as well a s biofilm formation.	Mansfield et al. (2012)
Pantoea stewartii and *Zea mays*	Pathogen–host	Quorum sensing	Quorum-sensing mutants of *Pantoea stewartii* were not able to disperse and migrate in the vasculature, consequently decreasing the disease.	Koutsoudis et al. (2006)
Pseudomonas syringae and tabacco and bean	Phytopathogen–plant	Quorum sensing	Quorum-sensing system allows this bacterium to control motility and exopolysaccharide synthesis essential on biofilm formation and leaves colonization.	Quiñones et al. (2005)
Streptomyces coelicolor and other *Actinomycetes* spp.	Microbial community	Prodiginines, ctinorhodins, coelichelins, acyl-desferrioxamines, and other compounds	The 227 compounds differentially produced in the interactions were unique.	Bentley et al. (2002)
Aspergillus nidulans and *Streptomyces rapamycinicus*	Microbial community	Aromatic polyketides	Activation of fungal secondary metabolite genes that were otherwise silent led physical interaction between the microorganisms. The actinomycete triggered alterations in fungal histone acetylation.	Bertrand et al. (2014)

Pseudomonas sp.	Microbial community	Pyoverdines (siderophore)	Pyoverdines act as signaling molecules, activating a cascade that results in the production of several virulence factors. It is essential to infection and biofilm formation.	Jimenez et al. (2012)
Burkholderia sp., *Rhizopus* sp., and rice	Symbiont phytopathogen plant	Rhizoxin, bongkrekic acid, enacyloxins	In the absence of the endosymbiont the fungus does not form spores. The phytotoxin rhizoxin is the causal agent of rice seedling blight produced by the endosymbiont; fungus induces the growth of the endosymbiont.	Depoorter et al. (2016)
Vibrio sp. and diverse marine bacteria strains	Microbial community	*N,N*-bis-(2,3-Dihydroxybenzoyl)-*O*-serylserine: exogenous siderophore	Siderophores and iron-regulated outer membrane proteins produced by marine bacteria and other species only in the presence of exogenous siderophores.	Kanoh and Kamino (2001)

Heijden et al. 2008). As such, experimental observations have confirmed that microbial diversity in the rhizospheric regions influences plant growth plants and the efficiency, nutrient accessibility, and functioning of an ecosystem (Delgado-Baquerizo et al. 2016). Furthermore, the significance of the soil–plant response is demonstrated by various observations, whereby the changes in the composition of community microbial response allow the simultaneity of plant and community arrangements. Although the effects of soil and plant–microbial interactions on ecological dynamics have been widely apparent, a brief observation indicates how the rhizospheric microbial communities stimulate the evolutionary process of plant communities (Lambers et al. 2009). A short generation time and a very high degree of genetic diversity are found in the microbes, with the capability to develop on ecologically significant timescales (Jessup et al. 2004). The microbial community structure is altered very rapidly because of these characteristics, which in turn may shape the way that the populations of plants react to innovative selective pressures in their environment (Whitham et al. 2006). Genetic differentiation in fully associated microbes derived from local adaptation in plants has been verified by various new observations (Richardson et al. 2009). For example, fungal endophytes have been recognized to colonize in stress conditions such as high temperatures, salinity, and drought, and heavy metal-resistant strains of mycorrhizae have been shown to expedite plant colonization of adulterated mine tailings (Calvo et al. 2014). The complex interactions with plant-associated microbes affect plant ecology. The functions of both plant-associated microbes and the hosts in the ecosystem have been identified, but the mechanism is not clarified (Hardoim et al. 2015). The immobile plants have developed a number of mechanisms that restrain the product of their interactions. The wide range of chemical compounds synthesized, secreted, and accumulated by the plant roots pass into the soil as root exudates (Chapin et al. 2002). These root exudates include various carbon-containing primary and secondary compounds, enzymes, water content, H^+ ions, and mucilaginous substances (Berg and Smalla 2009). These exudates of the plant roots structure and shape the bacterial community. In bulky soil, the density of microbes is 100 times less than in the rhizospheric soil (Baudoin et al. 2003). The composition of microbes alters and produces a response in the related plant routines.

Long-term effects of the soil microbes on their synchronicity with that plant species are described (Jentsch et al. 2007). There are two types: response-positive soil–plant microbial content emphasizes the spatial splitting of the microbial populations, whereas a negative response results in plant replacement, which requires recolonization of locally explicit roots (Lambers et al. 2008). The genome-wide association observed by experimental methods has facilitated our discoveries of the interactions of plant loci and associated communities. The diversity of the microbiome and its functions potentially affects the performance of host plants (Bodenhausen et al. 2014). The plant microbiome system performs important roles among plants and the community of microbes. An intricate assortment of volatile compounds is produced by the plant growth-promoting rhizobacteria (PGPR), which are distinctive among microbial species (Swamy et al. 2016). Plant growth is stimulated by the volatile compounds released by PGPR that suppress the disease-motivating induced systemic resistance (ISR) or alienate phytopathogens such as nematodes or insects. Biotic and abiotic stress factors affect the worldwide crop production and cause millions of dollars in losses (Wang et al. 2013). PGPR or plant-associated microbes were found to assist plants by producing various enzymes and hormones, improving the nutrient uptake, stimulating the root and shoot growth such as indole acetic acid, 1-aminocyclopropane-1-carboxylate (ACC) deaminases, and solubilizing phosphate, and enhancing the uptake of nutrients from environmental strains such as heavy metals, salts, nutrient deficiency, and drought (Gravel et al. 2007). The stress of biotic factors includes interactions with other organisms and infection by pathogens or damage by insect pests, and some plant growth-promoting bacteria have been used as bio-control agents against phytopathogens. This chapter explores symbiotic microbial communities and plant–host relationships, and these interactions require communication between various microorganisms that are involved in the rhizospheric system.

6.2 Fundamentals of Plant–Microbe Interactions

6.2.1 Functions and Diversity

The microorganisms that reside in the soil include a variety of phylogenetic groups and other major functional groups such as producer, consumer, and decomposer. Exceptionally, thousands of genomes are found in each gram of soil and form the genetic diversity (Bardgett et al. 2008). Interactions among soil microbes and plants span the range from mutualistic to pathogenic (Reynolds et al. 2003). The soil microbes are ultimately accountable for the bulk of terrestrial vegetation and annual nutrient demand as decomposers. In this order, the plant photo-synthetically fixed carbon is the major source for decomposition by microbes (Gougoulias et al. 2014). The microbes and plants form their relationship simultaneously, mutualistic and competitive, and also compete for soil nutrients (Van Der Heijden et al. 2008). The mycorrhizal fungi and PGPR increase the fitness of the host plant by providing mineral resources and safeguarding against other pests (Parnell et al. 2016; Prasad et al. 2015). Various other nonmycorrhizal fungi, rhizospheric bacteria, protozoa, and nematodes have also been revealed to protect plants from soil pathogens such as fungi, bacteria, actinobacteria, protozoa, nematodes, and viruses (Igiehon and Babalola 2018; Singh et al. 2019). These types of pathogens are responsible for various diseases such as damping-off, vascular wilt, and root rot diseases (Weller 1988).

6.2.2 The Interconnection Between Plants and Soil Microbes

The soil is the source of nutrients for plants: it is an intricate ecosystem accommodating various bacteria, fungi, protists, and animals. Plants demonstrate a diverse arrangement of interactions with these soil-dwelling microorganisms, which span the full range of environmental potentials such as competitive, exploitative, neutral, commensal, and mutualistic (Ratnadass et al. 2013). Various interactions were observed, which focused on improving the effects of pathogens such as herbivory and infection or tempering abiotic stress conditions through modern plant science (Shoresh et al. 2010), although the ecological interactions positively stimulate the growth of plants by long-standing interest in characterization. For example, in the second half of the nineteenth century the root symbiotic relationship was documented as mycorrhizal fungi and bacteria (Smith and Read 2010), and the bacterial cultures such as *Azotobacter chroococcum* or *Bacillus megaterium*, which improve the growth and crop yield when the seeds were coated by this culture (Burr et al. 1978). *Pseudomonas* sp. and *Azospirillum* sp. had been described as having plant growth-promoting effects, isolated in the 1980s. The diversity and abundance of the root microbiome is documented through metagenomics, which has shifted from individual microbial strains since the twentieth century (Hartmann et al. 2009). The rhizospheric niche is a hotspot of ecological richness observed from such types of sequencing, with plant roots hosting a massive array of microbial taxa (Van Der Heijden et al. 2008). In the current scenario, research has altered toward accumulating reasonably premeditated uninspired groupings that comprise the strains representing the overriding rhizospheric taxa, with the aim of reiterating the advantageous functions of microbes under controlled experimental conditions. To gain a systematic understanding in this research field, how soil microbes boost the plant growth and defense is a major goal, and to use these facts to inform the best strategy of microbial societies design to carry out specific functions (Fig. 6.1) (Johnson 2010).

6.2.3 Soil and Plant

For host health and improvement, the microbiota colonized in soil by microorganisms such as Archaea, Bacteria, Fungi, and viruses offer key functions (Xu 2006). The association of the microbiome with plants is measured as its second genome. It is also a contributing factor for the health of plants, growth, suitability, and productivity consequently (Björkman et al. 2011). The rhizosphere, endosphere, and phyllosphere have a specific microbial community with specific functions associated with the environment. The densities of the plant microbiome are greater than the number of plant cells and also contain better expressed genes than the host cells, as illustrated by these culture-independent methods (Hardoim et al. 2015). The next-generation sequencing technologies used for metagenomics analysis demonstrate that the current methods have cultured only 5% of bacteria, revealing that many microorganisms and their functions remain unknown (Walker et al. 2014). The microbial gratitude of plant exudates in the region of rhizosphere occurs by the collaboration of plant microbes (Sugiyama et al. 2014). The plants have the capability to recruit the microorganisms by their exudates (Compant et al. 2010).

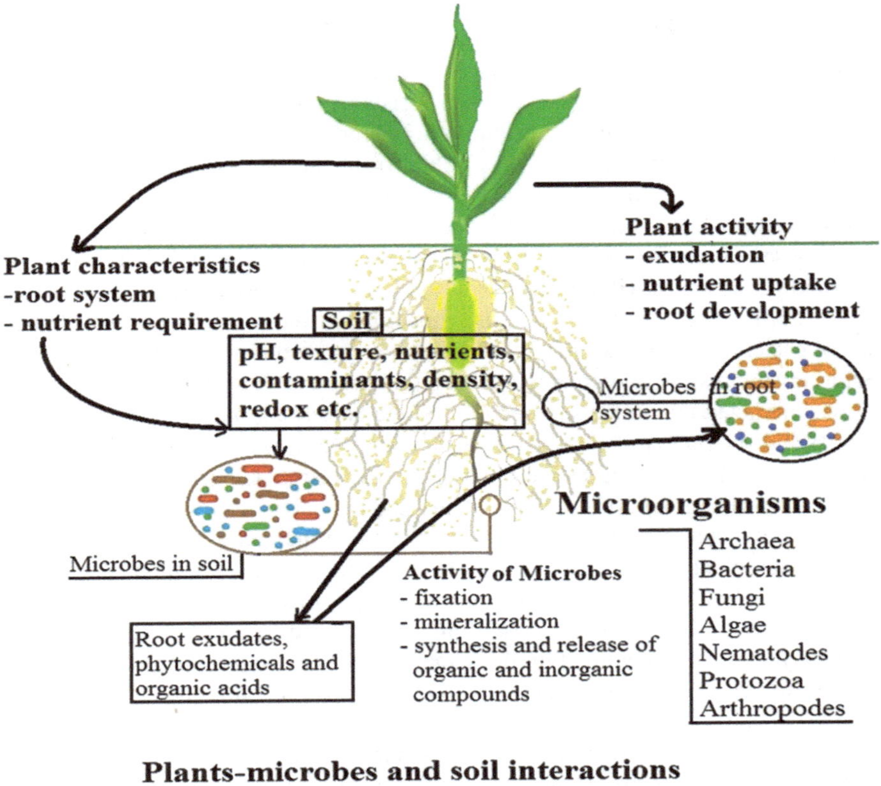

Fig. 6.1 Interactions among plants, microbes, and soil with microbial activity

The exudates are composed of amino acids, carbohydrates, and organic acids and vary according to the plant and its biotic or abiotic condition (Badri and Vivanco 2009). The specific microbial communities are selected from different plants when matching with rhizospheric settlement of two medicinal plants, *Matricaria chamomilla* (chamomile) and (*Solanum distichum*) nightshade, even with being harvested under the same conditions, and they access various analyzing 16*S* rRNA genes (structural) and analyzing the nitrogen fixing-*nifH* genes (functional) microbial community. Furthermore, the plant exudate of the same plant contrasts according to the developmental stages of plants, which selects the exact communities of microbes (Pérez-Montaño et al. 2014). The plant legume–Rhizobia secreted flavonoids and strigolactone as a signal molecule for arbuscular mycorrhizal fungi (AMF) showing specific interactions (Badri et al. 2009). A great diversity of the microbial community exists in the bulk soil, produced only by environmental factors and soil textures (Zornoza et al. 2015). The more specialized community of some species is found in the rhizospheric regions and the plant root exudates, and some species have the capability to enter the roots of plants and establish therein

(Chandra and Enespa 2019). Moreover, the microbial community varies among different organs after entering the plant, such as roots, top leaves, fruits, bottom leaves, flowers, and stems. Plants can be protected from pathogens by mutualistic microorganisms, either by inducing plant resistance or by antibiosis (De Coninck et al. 2015). High tolerance to pathogens is produced in plants by the induced systemic resistance (ISR) mechanisms. Whether the mechanisms of ISR suppressed the growth of the pathogen or the disease does not occur in the soil is still being investigated (Pineda et al. 2010). The disease of damping-off in several agricultural crops caused by the *Rhizoctonia solani* fungal pathogen is suppressed in soil by the microbiome observed (Berendsen et al. 2012). The oligonucleotide microarray (PhyloChip) using 16*S* rDNA has the capability to recognize more than 33,000 taxa of bacteria and Archaea in the seedlings of sugar beet grown in the rhizospheric region in suppressive and conductive soils (Hoitink and Fahy 1986). These observations predicted that in the suppressive soil the bacterial groups would be present. The Proteobacteria, particularly the Pseudomonadaceae, were all more abundant in suppressive soil than in conducive soil as noted by various authors, focusing thereby in this bacterial group (Choudhary et al. 2009). Using the random transposon mutagenesis techniques they were able to recognize the genes accountable for the biosynthesis of nine amino acid-chlorinated lipopeptides in *Pseudomonas* sp., an antifungal manufactured by *Pseudomonas* sp. that controls the pathogen (Ramaswamy et al. 2007). Other antifungals produced by rhizosphere-associated *Streptomycetes* were identified by the same PhyloChip diversity analysis (Chandra and Enespa 2017). These isolates were capable of producing chemically diverse volatile organic compounds (VOCs) with an antifungal effect as well as the plant growth-promoting properties. The various bacterial groups can perform similar roles in the same environment (Mendes et al. 2013). Strains of *Methylobacterium* that protect the plants against pathogen attack and affected communities of endophytes after inoculation were reported by various investigators (Ardanov et al. 2012). Consequently, investigators started inoculating plants with a pool of microorganisms with complementary traits using these concepts, such as various control mechanisms, with the aim of defining the various methods that were used for the inoculation of microorganisms (Lynch et al. 2004). The core microbiome of a healthy host is defined by the first methods, or the function of microbiomes can be understood by sequencing methods. In this manner, the modulation of microbial communities is beginning to be studied as "microbiome engineering." Plant breeding programs select a beneficial interaction between the plant lines and the rhizospheric microbiome or by redirecting the rhizospheric microbiome or stimulating beneficial microorganisms (Pieterse et al. 2012). The ecological processes can be altered by microbiome engineering such as the variation in diverse communities and the changing structure of microbe interaction networks and the extinction of microbial species in the microbiome, transfer of genes horizontally, and transmutations which can restructure the genomes of microorganisms by modifying the processes of evolution (Mueller and Sachs 2015). Briefly described, the sum of plant solutions for the environment and to the present microbiome such as endophytes and pathogens are the phenotype of the plant; this microbiome also replies to the environment, and these interact with each other (Gaiero et al. 2013). The gut and plant rhizospheric microbiomes demonstrate similarity with each other, as has been observed. Both are open systems, with oxygen gradient, H_2O, and pH resulting in huge numbers and microorganism diversity because of the altered conditions of survival (Herbst et al. 2016). The compositions of plant rhizosphere and gut microbiome differ between each other, but the acquisitions of nutrients, modulation of the immune system, and protection against infections have several similarities (Hacquard et al. 2015). The similarities between host-associated microbiome ecology include various conditions of abiotic shape and the microbial community structure; the microbiome of host coevolution (Braga et al. 2016); the microbiome of the core region can be vertically transmitted; the microbiome structure varies during the life cycle; the microbiomes associated with the host are self-possessed prokaryotic and eukaryotic microorganisms; functional diversity is the key in a microbiome; and the diversity of microbes is destroyed by human interventions (Lloyd-Price et al. 2016).

6.2.4 Secondary Metabolism

Microorganisms produce a large variety of compounds, secondary metabolites that are not necessary for the growth, improvement, and reproduction of the manufacturing organisms (Bourgaud et al. 2001). However, these metabolites are known as bioactive compounds and work in defence mechanisms, competition, and signaling, as well as in ecological interactions (Kliebenstein 2004). Microorganisms react by exchange of metabolic activity to establish a microbial interaction network, which leads to difficult regulatory replies and connecting to the biosynthesis of secondary metabolites (Zhao et al. 2005). The metabolites involved for interactions can be parasitic, antagonistic, or competitive, and their functions are being studied especially just now as a result of the beginning of new methods, for example, the technology of imaging mass spectrometry (IMS) and metabolomics (Gemperline et al. 2017). The competitive and cooperative microbial interactions are related to siderophores and can also have other functions such as signaling and antibiotic activity (West et al. 2007). In the interaction of bacteria, tolerance and discussing and improving the adaptation of bacteria in various environments, hopanoids are important (Sikkema et al. 1995). The compounds synchronized in fungi differentially in collaboration are often bioactive secondary metabolites, for example, diketopiperazines, trichothecenes, atranones, and polyketides. We present examples of such studies in these sections on secondary metabolites that elaborate various types of interactions in microbes.

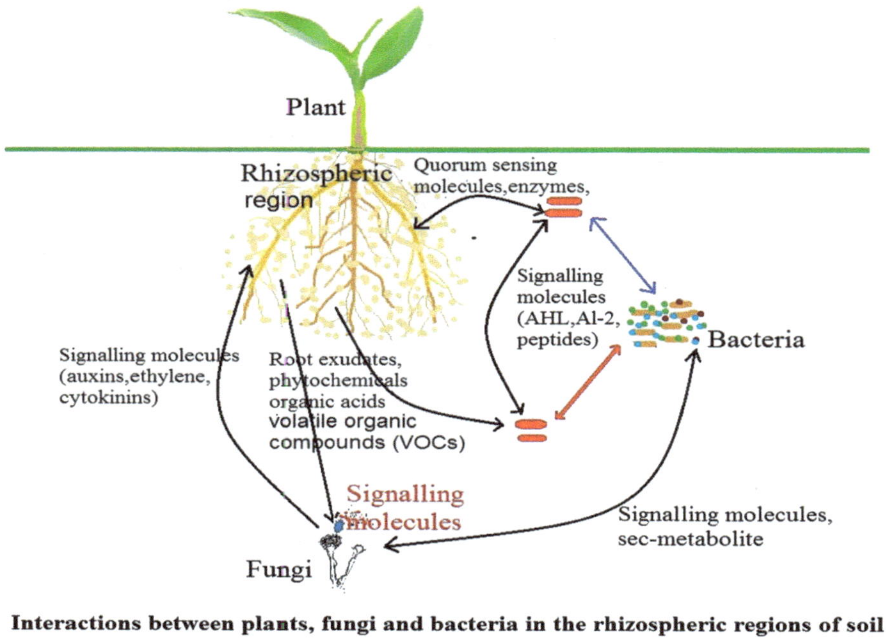

Fig. 6.2 Microbes and plant interactions using signaling molecules between rhizosphere and soil

6.2.5 Plant Community Ecology and Soil Symbiotic Interactions

Global climate changes can affect the ecology of plants in a terrestrial ecosystem in terms of both above- and belowground diversity (Bobbink et al. 1998). The dynamic changes in soil microbial ecology drive the plant communities of terrestrial regions that may result in adaptations in the functions of the ecosystem and stabilize the mechanisms essential for the maintenance of species diversity and synchronicity (Shade et al. 2012). These mechanisms of plant–plant interactions are contributed in the entire plant ecosystem by the microbes (Bever et al. 2010). Habitually, challenging the plant species have been supposed to have resilient intraspecific interactions negative for the high overlap in resource tradition, although the cohabitation of opposing plant species in local diversity explains the mechanism of success in finding the community (Hausch et al. 2018). Feedback might be one reason for neglecting a soil microbial community. In the rhizospheric region, the microbial communities communicate with each other and with the variety of mechanisms of the plant root, with bacterial AHLs (*N*-acylhomoserine lactones) and AI-2 (autoinducer-2) (Bogino et al. 2013). The composition of microbial communities is influenced directly by this and in some circumstances leads to the growth of improved plant health when the roots of plants establish beneficial interactions with root microbes (Fig. 6.2) (Berendsen et al. 2012).

To produce the responses in eukaryotes such as plants and fungi, bacteria employ signaling molecules. The plant growth-promoting rhizobacteria (PGPR) produce volatile organic compounds and initiate the induced systemic resistance (ISR) in plants, promoting growth in *Arabidopsis thaliana* (Choudhary and Johri 2009); thus, the expression of defence genes is stimulated that can be effective against microbes such as fungi, bacteria, oomycetes, and viruses (Pieterse et al. 2009). Although regulating the activities of bacteria in a density-dependent fashion, the molecules of quorum sensing (QS) also stimulate a plant-beneficial response range in the host plants (Lareen et al. 2016). The plant "priming" is included by them in which disclosure to signaling molecules of quorum primes the plant to return more strongly and quickly to biotic challenges (Mauch-Mani et al. 2017).

The exposure to AHL produced by *Serratia liquefaciens* MG1 and *Pseudomonas putida* IsoF increased the systemic resistance of tomato plants in contrast to the fungal foliar pathogen *Alternaria alternata* by inducing the ethylene and salicylic acid-dependent defence genes (Lareen et al. 2016). In *Arabidopsis* the AHL *N*-3-oxo-tetradecanoyl-L-homoserine lactone also supports pathogen defense of the accumulation of phenolic compounds by enhanced deposition of callose, cell wall lignification, and stomata closure in response to *Pseudomonas syringae* infection. Significantly, the increases in salicylic acid and oxylipin levels are associated with AHL activities. Plant hormone activity stimulation is used further by several microbes to accept the processes elaborated in the early stages of legume–rhizobia interactions and root mycorrhization that indicate the inter-domain communication for their successful establishment. A potential of microbial species or derivatives of metabolites modified the improvement of plants and fungi as observed in this study (Bloemberg and Lugtenberg 2001).

6.2.6 Soil Communities Affected by Bacteria and Fungi Interaction

The site enriched as plant nutrients in rhizospheric regions is a highly reasonable environment for the microorganisms (Hodge 2004). Secondary metabolites such as antibiotics, toxins, lytic enzymes, and siderophores are produced by the microbes to outcompete competitors to establish inside roots at the rhizosphere and occupy the similar niches (Lareen et al. 2016). The large gene clusters that are involved in detoxification, secretion of antibiotics, and siderophores are possessed by some rhizospheric microbes (Compant et al. 2010). Some rhizosphere microbes possess these large gene clusters involved in detoxification, production/release of antibiotics, and siderophores, such as *Bacillus amyloliquefaciens* and *Pseudomonas fluorescens* (Tewari and Arora 2013; Chandra and Singh 2016). Various common antibiotic compounds are secreted by the microbes such as 2,4-diacetylphloroglucinol (DAPG), hydrogen cyanide, oomycin A, and phenazine. In the soil these antibiotics (phenazines) inhibit the growth of pathogens such as *Fusarium oxysporum* produced by *Pseudomonas chlororaphis* (Lee et al. 2013; Raaijmakers et al. 2002; Enespa and Chandra 2017). Furthermore, various microbes released the lower concentrations of antibiotic compounds and have led to the recommendation that the primary functions of these molecules is in statement rather than inhibition or exclusion of opponents (Taylor et al. 2007). This range of functions in the soil suggests antimicrobial compounds as key in establishing microbial communities (Torsvik and Øvreås 2002). In the rhizospheric regions these antimicrobial compounds are the key to establishing microbial communities subjected to a wide range of functions in the soil (Badri and Vivanco 2009). The microorganisms produce the secondary metabolites in addition to antibiotic compounds to modify the signaling of the plant and its metabolism (Berendsen et al. 2012) to obtain nutrients. The plant can modify the arrangement of root exudates and encourage the release of more positive exudates by this reprogramming, in the rhizospheric region, which leads to a selective improvement of respective microbes in the rhizosphere (Shtark et al. 2010). This action recommends that in the regions of the rhizosphere the secondary metabolites and the antimicrobial compounds establish the microbial communities, which assist in the competitive niche exclusion (Hibbing et al. 2010). As a prerequisite the competitiveness for the formation and the dominancy of these communities required a harmonized message among the microbes as well as the sensitivity and the transformation of ecological signals (Cray et al. 2013).

6.2.7 Soil Resources and Microbial Interactions

Plant species synchronicities govern the soil resources by resource partitioning and sharing (Schoener 1974). The efficiency of nutrient uptake is increased by the root symbionts and allows the host to maintain in a low-nutrient environment, thus directly subsidizing to the competitive exclusion of other plants (Langley and Hungate 2003). Different forms of nitrogen or phosphorus in the soil can modify availability by the rhizosphere microbes and affect the plant–plant interactions through the intervention of source separating (Palacios et al. 2014). Common mycorrhizal networks (CMNs) formed by symbiotic fungi can also transfer micronutrients. In nature, different plant species commonly share the broadly specific mycorrhizal fungi (Brooker et al. 2015). CMNs with labeled carbon, nitrogen, and phosphorus transfer the resources directly from one plant to another. Soil microbial mediation resource distributing and involvement are driven by the community dynamics of the plant (Selosse et al. 2006).

6.2.8 *Host Response to Microbes and Soil Community Feedback*

In plant populations, the coexistence of plant species is affected indirectly via the feedback of dynamic density and composition of the rhizospheric microbes, such as the competition or inhibition of symbionts (Bever 2003). Three hypotheses have been proposed by ecologists to explain the mechanism that produces the low diversity of plant communities (Wright 2002). The novel symbionts reside in those regions that were invaded by invasive plants, as suggested by the empty niche hypothesis (Hierro et al. 2005). These symbionts are well organized for obtaining the resources and favor accompanying hostile plants rather than other plants (Johnson 2010). The invasive plants and their symbionts obstruct the capability of the native community of symbionts to obtain the positions and reduce the acts of native plants indirectly, suggested by the hypothesis of degraded mutualistism (Downer 2014). Growth improvement and survival of exotic seedlings near the natural recognized symbionts is demonstrated by the positive feedback (Bever et al. 2010). The interactions of symbiotic microbes have a high relationship with the monodominancy of plant coexistence and the ecology of invasion (Kulmatiski et al. 2008).

6.3 Role of Root Exudates Shaping the Rhizospheric Microbial Community

The microbes of soil are attracted to the plant root exudates, the volatile organic carbon, and rhizodeposition chemotactically, and then thrive in the carbon-rich environment (Somers et al. 2004; Shrivastava et al. 2014). The exudates of plant root differ among the species of plants, so in various plant species the differences in the rhizospheric microbiome are expected. Strong confirmation for species-specific microbiomes of plant have been provided by more recent observations (Fiehn et al. 2008). The root exudates of plants can be shaped via the microbial community. Various types of sugars, amino acids, organic acids, nucleotides, flavonoids, antimicrobial compounds, and enzymes are characterized in root exudates (Haas and Défago 2005).

6.4 Types of Root Exudates

6.4.1 *Amino and Organic Acids*

The growth of soil-borne pathogens affected the compositions of root exudates from different cultivars (Berendsen et al. 2012). Analysis of root exudate and its evaluation for the responses of soil-borne pathogens such as *Fusarium oxysporum* and *Fusarium solani* to the susceptible peanut cultivar Ganhua-5 (GH) and the mid-resistant cultivar Quanhua-7 (QH) were selected (Li et al. 2013). Ingredients such as total amino acids, sugars, and alanine in the root exudates of the mid-resistant cultivar were significantly less than in the susceptible cultivars, but substances such as total phenolic acids, *p*-hydroxybenzoic acid, benzoic acid, and *p*-coumaric acid were significantly higher than in the susceptible cultivars. In the composition of root exudates, these differences of susceptible and resistant cultivars might be presumed to adjust the mechanism of resistance in the rhizospheric regions of peanut (Lattanzio et al. 2006). Soil-borne pathogens such as *Fusarium oxysporum* and *F. solani* considerably enhanced the germination of spores and the growth of mycelia from both the susceptible and mid-resistant cultivars by treatment with root exudates compared with a control (Balendres et al. 2016). The effects of other factors must be measured if the root exudates do not inhibit the growth of pathogens directly (Jones et al. 2004). The colonization of organic acids controlled and improved the formation of the biofilm of the root microbiome observed previously (Lugtenberg and Kamilova 2009). The bacterium *Bacillus amyloliquefaciens* facilitated the colonization of banana root exudates, which released organic acids as demonstrated by Yuan et al. (2015). The organic acids from the root exudates of banana were significant in attracting and beginning the colonization of PGPR in plant roots (Haas and Défago 2005). Biofilm formation was induced significantly by fumaric acid, although the greatest chemotactic response was evoked by malic acid (Enespa and Dwivedi 2014). The residues of various amino acids, such as histidine, proline, valine, alanine, and glycine, and the carbohydrates such as glucose, arabinose, mannose, galactose, and glucuronic acid, are secreted primarily by the rice plant, and a higher chemotactic response is induced by the endophytic bacteria *Corynebacterium flavescens* and *Bacillus pumilus* (Kong et al. 2004).

6.4.2 Sugars

The plant pathogens effected the infection by the secretion of a quantity of sugar. The Glc-derived carbon efflux restricted the *Arabidopsis* vacuolar sugar transporter SWEET2 from roots and inhibited infection by *Pythium*. The secretion of SWEET 2 modulated sugar in restraining the loss of carbon to rhizosphere was proposed by some scientists. The decreased substrate availability in the rhizosphere subsidized the resistance to *Pythium* (Nega 2014).

6.4.3 Other Antimicrobial Compounds

Root exudates also participated in belowground plant defense. The "phytoanticipins" and "phytoalexins" are formed by antimicrobial action of low molecular weight (Baetz and Martinoia 2014). A biotic stress such as pathogen infection of the plant root that was previously produced released a defensive compound, phytoanticipins. The manufacture of diterpene rhizathalene-A deficiency in the roots of *Arabidopsis* was found to be more vulnerable to insect herbivory, as observed in a new study. Thus, the root is considered as a part of a constitutive direct defense system by rhizathalene-A (Enespa and Dwivedi 2014). Phytoalexins are not detected in healthy plants: these are known as inducible defensive compounds. The soil-borne pathogen *Fusarium graminearum* is inhibited by five phenylpropanoid root-derived aromatic root exudates, demonstrated as antifungal activity (Lanoue et al. 2010). Generally, the phenolic defensive compounds and terpenoid secreted by the root have very strong antibacterial and antifungal activity. Terpenoids form the largest class of plant defensive compounds both above and below ground (Bais et al. 2004). Volatile organic compounds (VOCs) are secreted from the roots as plant protective compounds, and the rhizospheric regions secrete terpenoids, which are nonvolatile. Also, the root exudates secrete a group of plant defensive phenolics known as phenylpropanoids (Massalha et al. 2017). Phenylpropanoids are cinnamic acid rapidly accumulated and secreted by barley plant resist fungal attack infection such as *Fusarium graminearum*. The exudates of roots are phenolic compounds that have antimicrobial activity and also attract soil-borne microorganisms, which also affects the soil microbial community natively (Barsainya et al. 2016). The amino acid canavanine is also a chemical compound that stimulates a specific group of microbes which inhibit the growth of other soil microbes. The specific rhizosphere microbial community shapes the plants via root exudates (Venturi and Keel 2016).

6.5 Effects of Environmental Factors on Root Exudates

The root exudates of various compositions are produced with various genotypes by plants. The exudates of roots are also affected by abiotic and biotic factors (Wardle et al. 2011). Nutrient availability, organic matter content, structure, pH, and texture are the physicochemical properties of soil affecting the microbial requirement and availability of root exudates by root plants (Schmidt et al. 2011). The secondary metabolism of soil microbes can also affect the exudates by some biotic factors (Badri and Vivanco 2009).

6.5.1 Temperature

Since the beginning of global warming and climate change, cold waves and extreme heat consequentially have affected the harvesting of various crops (Wassmann et al. 2009). To explicate the effects of temperature on root exudates, it was also studied that the strawberry plant grew at 5–10 °C and were linked at 20–30 °C growing-stage plants (Vigo et al. 2000). In the plants growing at low soil temperature, more amino acids were found in exudates, significantly affecting the pathogenicity of *Rhizoctonia fragariae* (Harrier and Watson 2004). The content of organic acid increased with the elevation of temperature in Japanese cucumber grown hydroponically in a growth chamber at high and low temperatures, and a few other compounds identified significantly inhibited the growth of root and germination as demonstrated in this plant (Enespa and Dwivedi 2014).

6.5.2 Soil Moisture Content

Cereal harvesting globally is reduced by flooding and drought conditions: the moisture content of soil affects the release of root exudates, confirmed in various reports (Römheld and Kirkby 2010). Wilting of plants increased temporarily with the discharge of amino acids from the plant roots, which might be related to the incidence of pathogens in the rhizosphere. Plants that grow normally in moist and dried sand include peas, soybeans, wheat, barley, and tomatoes; the sand is remoistened for the liberation of amino acids (Curl and Truelove 2012). In temporarily dried sand, the total quantity of amino nitrogen was many times higher than in the normal moist sand (Certini 2005).

6.5.3 Nutrition and Soil pH

The status of soil pH and the availability of nutrients such as C, N, and P have been found to affect the initiation of plant root exudates and the formation of specified chemical niches in the soil region, along with the abundance of phytopathogens and beneficiary microbes (Kõljalg et al. 2013). The composition of arbuscular mycorrhizal (AM) fungi and bacteria was significantly affected and correlated with changes in pH, phosphate, and soil carbon content in maize plants in a long-term fertilization trial (Toljander et al. 2008). The fungal community of arbuscular mycorrhiza (AM) in 425 individuals and 28 other plant species modified the structure of the AM fungal community by soil pH. In Fe-deficient soil, the phenolic of root secretion was induced and rehabilitated the microbial community in the rhizosphere (Lindsay and Norvell 1978).

6.5.4 Other Soil Microorganisms

The growth of plants and their exudates was improved and developed by the microbial activity of soil. The permeability of root cells and root metabolism was affected by the plant exudations, which were also affected by microbes (Bais et al. 2006). The root exudates and other released compounds are absorbed by microorganisms. Secondary metabolites produced by the soil microbes affected the metabolism and the signaling of plants and are considered as a "plant secondary genome," which is delivered to the plant hosts with derived microbial compounds. Antibiotics such as penicillin and polymyxin and the microbes increased the exudation of organic materials, rehabilitated cell permeability, and amplified the outflow (Sparbier et al. 2012). The microbes of soil induced the exudation of phenolic compounds for improving the Fe absorption of plant in low-Fe availability soil (Marschner 2012).

6.6 Ecological Diversity and Its Interactions

Four major groups of organisms have the greater impact on the performance of plants and the ecological processes that form very close associations with the plant roots. In specific groups of organisms, a plant is exposed to more than one of these groups at any time, and the interactions between them can change the outcome for the plant.

6.6.1 Nitrogen Fixation Symbiotically

In terrestrial ecosystems, N_2 is the most limiting nutrient for plant growth. Although molecular nitrogen is found abundantly in the atmosphere, eukaryotes do not have the capability to fix the atmospheric nitrogen into ammonia (Chandra et al. 2019; Codispoti et al. 2001). In actual fact, this capability is limited to various species of bacteria and Archaea with various life strategies (Bäckhed et al. 2005). Several of these reside in free form in soil and water, such as *Azotobacter* and *Clostridium*, in the intercellular spaces of plants, the phyllosphere, and rhizosphere regions occupied by other microbes such as *Azospirillum*, *Azoarcus*, and *Gluconacetobacter*; and still others are highly specific symbionts such as *Frankia*, connected to *Alnus*, *Myrica*, *Ceanothus*, *Elaeagnus*, and *Casuarina*; and legume symbionts mutually known as rhizobia (Beattie 2007). In the terrestrial ecosystem, the diazotrophs are the main providers for biological nitrogen fixation symbiotically. In the field of agriculture, legume plants form symbiotic relationships, so it is mainly focused on exploration (Lambers et al. 2009). The wide distribution of legume plants in temperate, tropical, and arid regions is significant in natural ecosystems. The order Rhizobiales is mostly known for legume symbionts, but some species known as nodulate legumes are in the order Burkholderiales (Peix et al. 2015). The genera of *Rhizobium*, *Sinorhizobium*, *Mesorhizobium*, *Bradyrhizobium*, *Azorhizobium*, and *Allorhizobium* have the capability to nodulate the legumes and fix nitrogen. *Ensifer*, *Blastobacter*, *Burkholderia*, and *Ralstonia*, similar to other genera, contain both legume symbionts and nonsymbiotic species (Sawada et al. 2003; Rasolomampianina et al. 2005). The relationship among the bacterial symbionts and legumes depends on fine molecular communication. The legume roots produce flavonoids and respond to nodulating bacteria by producing *N*-acetyl-D-glucosamine, which promotes the physiological changes in the roots of the host to nodulation, known as Nod factors (Chandra and Enespa 2017). The structure of Nod factors has dissimilarities, dependent on the various strains and species to determine the host specificity, although in nature symbiotic promiscuity is common and could be beneficial for inhabiting new soils (Shamseldin 2013). The highly immoral legumes are successful invasive species as observed by some scientists. For the institution and growth of many pioneer species of legumes, this is a crucial association. From these relationships the growth of other plant species in nitrogenous enrichment soil is consequently expedited (Sharpley et al. 1994). So, it promotes the succession of plants. In consequence, when the level of nitrogen increased, it displaced the other species and promoted spatial heterogeneity. Thus, it is shown that the functioning of terrestrial ecosystems depends on the key symbiosis relationship (Laliberté et al. 2013).

6.6.2 *Fungal Mycorrhiza*

A plant root and a specialized fungus show a symbiotic, nonpathogenic, and permanent association known as mycorrhiza in both natural and cultivated environments. This is an ancient symbiotic association found in plants and evolved by primitive plants during the colonization of land (Rodrıguez-Echeverrıa et al. 2007). The carbohydrates exchanged by plants for mineral nutrients such as phosphorus, nitrogen, potassium, calcium, and zinc are repossessed by the mycelia of fungi from enormous soil volumes in the symbiotic relationship (Korhonen et al. 2015). The mycorrhizal fungi are involved in other various processes such as plant protection against abiotic stress (Rapparini and Peñuelas 2014). The essential nutrients are available to the plants by the breakdoown of complex organic molecules, which synthesized the plant growth hormones such as auxins, citokinins, and gibberellins (Pieterse et al. 2009). On the basis of morphology and physiology, the mycorrhizal association is classified into three main groups. (1) The fungal mycelia surround the root and penetrate the intercellular spaces, known as ectomycorrhizae. (2) The mycelium does not coat the root; however, there is contact among the fungi and the root through constructions inside the root cells that are specialized for nutrient exchange and storage, known as endomycorrhizae (Taylor and Peterson 2005). (3) Both ecto- and endomycorrhizae are intermediate and include the ectendo-, arbutoid, monotropoid, and orchid mycorrhizae. The mycorrhizal associations most widespread by far are the ectomycorrhizae and the arbuscular mycorrhizae (Rodrıguez-Echeverrıa et al. 2007). Seed plants belonging to the families Betulaceae, Fagaceae, Pinaceae, Rosaceae, Myrtaceae, Mimosaceae, and Salicaceae show ectomycorrhizal association. Although there are more ectomycorrhizal plant species than endomycorrhizal plants, the dominant species of boreal, temperate, and many subtropical forests involve ecologically significant associations (Brundrett 2002). In this symbiosis the fungi involved are almost exclusively basidiomycetes and ascomycetes. Both hypogeous and epigeous genera, such as *Amanita*, *Boletus*, *Leccinum*, *Suillus*, *Hebeloma*, *Gomphidius*, *Paxillus*, *Clitopilus*, *Lactarius*, *Russula*, *Laccaria*, *Thelephora*, *Rhizopogon*, *Pisolithus*, and *Scleroderma*, are common genera of basidiomycetous fungi (Read et al. 2000).

The arbuscular mycorrhizae are ubiquitous, occurring over a broad ecological range. Almost all the natural and cultivated plant species have arbuscular mycorrhizae and are ubiquitous (Öpik et al. 2010). Also, all the angiospermic families form the associations of endomycorrhizae, with some exceptions. Some gymnospermic plant species of *Taxus* and *Sequoia* show the infection. The bryophytes and pteridophytes are also infected by these fungi, phylogenetically. The phylum Glomeromycota shows obligate symbionts that talso form these associations (arbuscular mycorrhizal fungi, AMF) (Redecker et al. 2013). In this association little specificity has traditionally been recognized, but recent observations have revealed more genetic and functional diversity than was estimated formerly (Romero et al. 2012). Ecosystem productivity and plant diversity are increased by the presence of AMF. The high functional diversity of AMF and the specificity are explained by this and the consequence of the interaction with various plant species (Bever et al. 2001). A wider range of plant species benefited by a rich AMF community is more competent at exploiting soil resources (Pérez-Jaramillo et al. 2016). For the positive correlation between AMF and plant diversity, an alternative explanation comes from the observation that plant growth has a detrimental effect by AMF. At the site, the plant diversity increased by a richer fungal community because no plant has a greater advantage with all AMF (McCann 2000). Plant diversity is increased in the absence of mycorrhizal fungi in some circumstances. The highly mycotrophic species dominate plant communities in this case or by ectomycorrhizal species. Removal of the mycorrhizal fungi leads to a decrease of the dominant species and the subsequent modest issue of the subsidiary species (Van Der Heijden and Horton 2009). An underground net links several plants established by the external mycelium of mycorrhizal fungi. Nutrients loss reduces this fungal network also by the sequestering of nitrogen, phosphorus, and carbon within their biomass (Smith and Smith 2012). The external mycelium moves nutrients according to fungal needs, but there is also a transfer of nutrients between plants through the hyphal network (Smith et al. 2010). In ectomycorrhizae the carbon transfers between plants, but these also occur through arbuscular mycorrhizae (Jeffries et al. 2003). The nutrient flow between plants contrasts with the colonization of mycorrhizal content of soil nutrients and the physiological status of plants (Marschner 2011). So, in the greenhouse studies the obtained results have been very adjustable. The external hyphae of mycorrhizal fungi transfer of nutrient rates were high, having important ecological consequences such as establishment of the growth of new seedlings of mycorrhizal plants that augments the transfer of nutrients (Smith et al. 2010).

6.6.3 Pathogenic Fungi

Fungus inhabitants in the soil can have lethal outcomes on the growth of plants, and historically focus has been on agricultural systems for obvious economic reasons. Ecologists have been on track to explore the diversity of microorganisms and the role of fungal pathogens in the natural ecosystem in the past two decades (Pringle et al. 2009). The ascomycetes group found in soil are fungal pathogens that attack plants. In agricultural systems, various genera of pathogenic ascomycetes have been identified and isolated from natural systems (Boer et al. 2005). Sand dunes studies in coastal areas have focused on the deterioration of innovator plant species. In the Netherlands, *Verticillium* and *Fusarium* species were isolated from declining stands of the dune grass *Ammophila arenaria*, and species of *Fusarium*, *Cladosporium*, *Phoma*, and *Sporothrix* were involved in the degeneration of *Leymus arenarius* in Iceland. Dieback of the endemic Hawaiian tree koa (*Acacia koa*), a keystone species in the upper-elevation forests, caused by the systemic wilt pathogen *Fusarium oxysporum* f. sp. *koae*, is another example (Gilbert 2002). Killing of big trees and opening gaps in the forest by other root rot fungi is part of the dynamic of temperate forests. The basidiomycete *Phellinus weirii* that specifically damages *Pseudotsuga menziesii* in North American temperate forests is a well-studied example (Hansen and Goheen 2000). *Pythium* and *Phytophthora*, two genera of Oomycota, are also fungal-related species. *Pythium* is accountable for seedlings mortality in tropical and temperate forests of natural and agricultural systems. A high mortality of new seedlings was caused in the vicinity of parent trees by the pathogenic *Pythium* spp., which is correlated to parent trees in the rhizospheric regions. The life stage of the plants not only depends on the impact of pathogenic fungi and oomycetes, but also depends on the overall history of pathogens, virulence, and also on specificity (Packer and Clay 2003). The seed decay, seedling diseases, foliage diseases, systemic infections, cankers, wilts, and diebacks, root and butt rots, and floral diseases are classified by the fungal pathogens of non-crop plants. The plant populations affected by the fungal pathogens are assumed to be subsidized by the genetic and species diversity of plants and succession in natural systems (Simard et al. 2015).

6.6.4 Nematodes

The most abundant metazoans are known as nematodes. On the basis of ecological observations, they are usually classified by their feeding habits: they can be bacterial feeders, fungal feeders, omnivores, or plant feeders (Bongers and Ferris 1999). We focus on the plant feeders in this section: to feed on plant roots, one group of nematodes have specialized mouth structures (stylets). Nematodes feeding on the roots of plants are a highly specialized obligate parasite that evolved through close interactions with plants, which explains their high impact on the plant populations that they attack (Rodrıguez-Echeverrıa et al. 2007). The plant-parasitic nematodes belowground are subdivided into four groups: sedentary endoparasites, sedentary semi-endoparasites, migratory endoparasites, and ectoparasites (De Deyn et al. 2004). Sedentary endoparasites, such as *Meloidogyne* spp. and *Heterodera* spp., are completely surrounded and protected by their host's root tissue for most of their life cycle, (Nicol et al. 2011). To cultivate permanent and highly specialized feeding sites, they interact with the plant root within the root tissues that act as nutrient sinks (Vovlas et al. 2005). *Rotylenchulus* spp. are known as sedentary semi-endoparasites exposed partially in the root tissue for part of their life cycles while the juveniles and young females feed ectoparasitically in the rhizosphere (Khan 2015). In *Pratylenchus* spp., the migratory endoparasites hatch and develop to maturity inside the root tissues of the plant host: causing extensive damage, they migrate within the roots and do not establish a permanent feeding site (Jones et al. 2013). Additive and synergistic interactions are developed by plant-feeding nematodes with pathogenic fungi and bacteria, and some, such as *Xiphinema* and *Longidorus*, are vectors of plant viruses. The changes caused by nematodes promote the nutrient influx of soil, and the microbial biomass of soil increases and constructs the damaged root growth of neighboring plants (Jones et al. 2016).

6.7 Conclusions

The various studies discussed in this chapter indicate that the structure of microbial communities of the soil certainly does have an effect on the suppression of phytopathogens. Niches can be created by microbes for selected plant species and niche space limited for others, depending on the properties of pathogens, symbionts (mycorrhizal fungi), and accumulated nutrients such as N-fixation and weathering of rock phosphate microbially, with the successful establishments of C- and N-fixing microbes such as Cyanobacteria and the concurrent weathering of parent material by organic acids. In the same way, we expect that high variability in the distribution of microbes that are known symbionts with species of plants, such as mycorrhizal fungi, will strongly influence which plants can invade particular sites. Conclusively, plant communities move into sparsely vegetated regions and alter the functioning of these ecosystems. The microbial community is limited by carbon in plant-free areas with numerous groups such as zoosporic fungi particularly linked with increased amine-containing compounds and nitrification. The barren soil is colonized by the plants, and by competitive exclusion of algae we anticipate the plants to decrease the plentitude of zoosporic fungi. The available nitrate is mostly taken up by plants; therefore, the plants should decrease the export of nitrate from the watershed. Thus, the rhizospheric regions of plants provide better interactions for crop productivity.

References

Araújo WL, Creason AL, Mano ET, Camargo-Neves AA, Minami SN, Chang JH, Loper JE (2016) Genome sequencing and transposon mutagenesis of *Burkholderia seminalis* TC3.4.2 R3 identify genes contributing to suppression of orchid necrosis caused by *B. gladioli*. Mol Plant Microbe Interact 29:435–446

Ardanov P, Sessitsch A, Häggman H, Kozyrovska N, Pirttilä AM (2012) *Methylobacterium*-induced endophyte community changes correspond with protection of plants against pathogen attack. PLoS One 7:1–8

Bäckhed F, Ley RE, Sonnenburg JL, Peterson DA, Gordon JI (2005) Host-bacterial mutualism in the human intestine. Science 307:1915–1920

Badri DV, Vivanco JM (2009) Regulation and function of root exudates. Plant Cell Environ 32:666–681

Badri DV, Weir TL, van der Lelie D, Vivanco JM (2009) Rhizosphere chemical dialogues: plant–microbe interactions. Curr Biotechnol 20:642–650

Baetz U, Martinoia E (2014) Root exudates: the hidden part of plant defense. Trends Sci 19 (2):90–98

Bais HP, Park SW, Weir TL, Callaway RM, Vivanco JM (2004) How plants communicate using the underground information superhighway. Trends Sci 9:26–32

Bais HP, Weir TL, Perry LG, Gilroy S, Vivanco JM (2006) The role of root exudates in rhizosphere interactions with plants and other organisms. Annu Rev Plant Biol 57:233–266

Balendres MA, Nichols DS, Tegg RS, Wilson CR (2016) Metabolomes of potato root exudates: compounds that stimulate resting spore germination of the soil-borne pathogen *Spongospora subterranea*. J Agric Food Chem 64:7466–7474

Bardgett RD, Freeman C, Ostle NJ (2008) Microbial contributions to climate change through carbon cycle feedbacks. ISME J 2:805–814

Barrière Q, Guefrachi I, Gully D, Lamouche F, Pierre O, Fardoux J, Mergaert P (2017) Integrated roles of BclA and DD-carboxypeptidase 1 in *Bradyrhizobium* differentiation within NCR-producing and NCR-lacking root nodules. Sci Rep 7:1–13

Barsainya M, Chandra P, Singh DP (2016) Investigation of Cr (VI) uptake in saline condition using psychrophilic and mesophilic *Penicillium* sp. Int J Curr Microbiol App Sci 5(1):274–288

Baudoin E, Benizri E, Guckert A (2003) Impact of artificial root exudates on the bacterial community structure in bulk soil and maize rhizosphere. Soil Biol Biochem 35:1183–1192

Beattie GA (2007) Plant-associated bacteria: survey, molecular phylogeny, genomics and recent advances. In: Plant-associated bacteria. Springer, Dordrecht, pp 1–56

Bentley SD, Chater KF, Cerdeño-Tárraga AM, Challis GL, Thomson NR, James KD, Bateman A (2002) Complete genome sequence of the model actinomycete *Streptomyces coelicolor* A3 (2). Nature 417:141–147

Berendsen RL, Pieterse CM, Bakker PA (2012) The rhizosphere microbiome and plant health. Trends Sci 17:478–486

Berg G, Smalla K (2009) Plant species and soil type cooperatively shape the structure and function of microbial communities in the rhizosphere. FEMS Microbiol Ecol 68:1–13

Bertrand S, Bohni N, Schnee S, Schumpp O, Gindro K, Wolfender JL (2014) Metabolite induction via microorganism co-culture: a potential way to enhance chemical diversity for drug discovery. Biotechnol Adv 32:1180–1204

Bever JD (2003) Soil community feedback and the coexistence of competitors: conceptual frameworks and empirical tests. New Phytol 157:465–473

Bever JD, Schultz PA, Pringle A, Morton JB (2001) Arbuscular mycorrhizal fungi: more diverse than meets the eye, and the ecological tale of why. BioScience 51:923–931

Bever JD, Dickie IA, Facelli E, Facelli JM, Klironomos J, Moora M, Zobel M (2010) Rooting theories of plant community ecology in microbial interactions. Trends Ecol Evol 25:468–478

Björkman M, Klingen I, Birch AN, Bones AM, Bruce TJ, Johansen TJ, Stewart D (2011) Phytochemicals of Brassicaceae in plant protection and human health: influences of climate, environment and agronomic practice. Phytochemistry 72:538–556
Bloemberg GV, Lugtenberg BJ (2001) Molecular basis of plant growth promotion and biocontrol by rhizobacteria. Curr Biol 4:343–350
Bobbink R, Hornung M, Roelofs JG (1998) The effects of air-borne nitrogen pollutants on species diversity in natural and semi-natural European vegetation. J Ecol 86:717–738
Bodenhausen N, Bortfeld-Miller M, Ackermann M, Vorholt JA (2014) A synthetic community approach reveals plant genotypes affecting the phyllosphere microbiota. PLoS Genet 10:1–12
Boer WD, Folman LB, Summerbell RC, Boddy L (2005) Living in a fungal world: impact of fungi on soil bacterial niche development. FEMS Microbiol Rev 29:795–811
Bogino P, Abod A, Nievas F, Giordano W (2013) Water-limiting conditions alter the structure and biofilm-forming ability of bacterial multispecies communities in the alfalfa rhizosphere. PLoS One 8:1–17
Bongers T, Ferris H (1999) Nematode community structure as a bioindicator in environmental monitoring. Trends Ecol Evol 14:224–228
Bourgaud F, Gravot A, Milesi S, Gontier E (2001) Production of plant secondary metabolites: a historical perspective. Plant Sci 161:839–851
Braga RM, Dourado MN, Araújo WL (2016) Microbial interactions: ecology in a molecular perspective. Braz J Microbiol 47(suppl 1):86–98
Brooker RW, Bennett AE, Cong WF, Daniell TJ, George TS, Hallett PD, Li L (2015) Improving intercropping: a synthesis of research in agronomy. New Phytol 206:107–117
Brundrett MC (2002) Coevolution of roots and mycorrhizas of land plants. New Phytol 154:275–304
Burr TJ, Schroth MN, Suslow T (1978) Increased potato yields by treatment of seed pieces with specific strains of *Pseudomonas fluorescens* and *P. putida*. Phytopathology 68:1377–1383
Calvo P, Nelson L, Kloepper JW (2014) Agricultural uses of plant biostimulants. Plant Soil 383:3–41
Certini G (2005) Effects of fire on properties of forest soils: a review. Oecologia (Berl) 143:1–10
Chamoun R, Aliferis KA, Jabaji S (2015) Identification of signatory secondary metabolites during mycoparasitism of *Rhizoctonia solani* by *Stachybotrys elegans*. Front Microbiol 6:1–11
Chandra P, Enespa (2017) Microbial volatiles as chemical weapons against pathogenic fungi. In: Volatiles and food security. Springer, Singapore, pp 227–254
Chandra P, Enespa (2019) Fungal Community for Novel Secondary Metabolites. In: Recent advancement in white biotechnology through fungi. Springer, Cham, pp 249–283
Chandra P, Singh E (2016) Applications and mechanisms of plant growth-stimulating rhizobacteria. In: Plant–microbe interaction: an approach to sustainable agriculture. Springer, Singapore, pp 37–62
Chandra P, Enespa, Mukesh K (2019) Contribution of microbes in the renovation of wetlands. Springer Nature, Singapore, pp 105–026
Chapin FS, Matson PA, Mooney HA (2002) Terrestrial decomposition. Springer, New York, pp 151–175
Choudhary DK, Johri BN (2009) Interactions of *Bacillus* spp. and plants–with special reference to induced systemic resistance (ISR). Microbiol Res 164:493–513
Choudhary DK, Prakash A, Wray V, Johri BN (2009) Insights of the fluorescent pseudomonads in plant growth regulation. Curr Sci 97:170–179
Codispoti LA, Brandes JA, Christensen JP, Devol AH, Naqvi SWA, Paerl HW, Yoshinari T (2001) The oceanic fixed nitrogen and nitrous oxide budgets: moving targets as we enter the anthropocene. Sci Mar (Barc) 65:85–105
Compant S, Clément C, Sessitsch A (2010) Plant growth-promoting bacteria in the rhizo- and endosphere of plants: their role, colonization, mechanisms involved and prospects for utilization. Soil Biol Biochem 42:669–678

Cray JA, Bell AN, Bhaganna P, Mswaka AY, Timson DJ, Hallsworth JE (2013) The biology of habitat dominance: can microbes behave as weeds. Microb Biotechnol 6:453–492

Curl EA, Truelove B (2012) The rhizosphere. Advanced Series in Agricultural Sciences, vol 15. Springer Science, Berlin

Davis MA, Chew MK, Hobbs RJ, Lugo AE, Ewel JJ, Vermeij GJ, Thompson K (2011) Don't judge species on their origins. Nature 474:153–154

Dawson TE, Mambelli S, Plamboeck AH, Templer PH, Tu KP (2002) Stable isotopes in plant ecology. Annu Rev Ecol Syst 33:507–559

De Coninck B, Timmermans P, Vos C, Cammue BP, Kazan K (2015) What lies beneath: belowground defense strategies in plants. Trends Sci 20:91–101

De Deyn GB, Raaijmakers CE, Van Ruijven J, Berendse F, Van Der Putten WH (2004) Plant species identity and diversity effects on different trophic levels of nematodes in the soil food web. Oikos 106:576–586

Delgado-Baquerizo M, Maestre FT, Reich PB, Jeffries TC, Gaitan JJ, Encinar D, Singh BK (2016) Microbial diversity drives multifunctionality in terrestrial ecosystems. Nat Commun 7:1–8

Depoorter E, Bull MJ, Peeters C, Coenye T, Vandamme P, Mahenthiralingam E (2016) *Burkholderia*: an update on taxonomy and biotechnological potential as antibiotic producers. Appl Microbiol Biotechnol 100:5215–5229

Downer CC (2014) The horse and burro as positively contributing returned natives in North America. Am Life Sci 2:5–23

Enespa, Chandra P (2017) Microbial volatiles as chemical weapons against pathogenic fungi. In: Choudhary D, Sharma A, Agarwal P, Varma A, Tuteja N (eds) Volatiles and food security. Springer, Berlin

Enespa, Dwivedi SK (2014) Effectiveness of some antagonistic fungi and botanicals against *Fusarium solani* and *Fusarium oxysporum* f. sp. *lycopersici* infecting Brinjal and tomato plants. Asian J Plant Patho 8(1):18–25

Fiehn O, Wohlgemuth G, Scholz M, Kind T, Lee DY, Lu Y, Nikolau B (2008) Quality control for plant metabolomics: reporting MSI-compliant studies. Plant J 53:691–704

Fuqua WC, Winans SC, Greenberg EP (1994) Quorum sensing in bacteria: the LuxR-LuxI family of cell density-responsive transcriptional regulators. J Bacteriol 176:269

Gaiero JR, McCall CA, Thompson KA, Day NJ, Best AS, Dunfield KE (2013) Inside the root microbiome: bacterial root endophytes and plant growth promotion. Am Bot 100:1738–1750

Gemperline E, Horn HA, DeLaney K, Currie CR, Li L (2017) Imaging with mass spectrometry of bacteria on the exoskeleton of fungus-growing ants. ACS Chem Biol 12:1980–1985

Gilbert GS (2002) Evolutionary ecology of plant diseases in natural ecosystems. Annu Rev Phytopathol 40:13–43

Gonzalez JE, Marketon MM (2003) Quorum sensing in nitrogen-fixing rhizobia. Microbiol Mol Biol Rev 67:574–592

Gougoulias C, Clark JM, Shaw LJ (2014) The role of soil microbes in the global carbon cycle: tracking the below-ground microbial processing of plant-derived carbon for manipulating carbon dynamics in agricultural systems. J Sci Food Agric 94:2362–2371

Gravel V, Antoun H, Tweddell RJ (2007) Growth stimulation and fruit yield improvement of greenhouse tomato plants by inoculation with *Pseudomonas putida* or *Trichoderma atroviride*: possible role of indole acetic acid (IAA). Soil Biol Biochem 39:1968–1977

Haas D, Défago G (2005) Biological control of soil-borne pathogens by fluorescent pseudomonads. Nat Rev Microbiol 3:307

Hacquard S, Garrido-Oter R, González A, Spaepen S, Ackermann G, Lebeis S, Schulze-Lefert P (2015) Microbiota and host nutrition across plant and animal kingdoms. Cell Host Microbe 17:603–616

Hansen EM, Goheen EM (2000) *Phellinus weirii* and other native root pathogens as determinants of forest structure and process in western North America. Annu Rev Phytopathol 38:515–539

Hardoim PR, Van Overbeek LS, Berg G, Pirttilä AM, Compant S, Campisano A, Sessitsch A (2015) The hidden world within plants: ecological and evolutionary considerations for defining functioning of microbial endophytes. Microbiol Mol Biol Rev 79:293–320
Harrier LA, Watson CA (2004) The potential role of arbuscular mycorrhizal (AM) fungi in the bioprotection of plants against soil-borne pathogens in organic and/or other sustainable farming systems. Pest Manag Sci 60:149–157
Hartmann A, Schmid M, Van Tuinen D, Berg G (2009) Plant-driven selection of microbes. Plant Soil 321:235–257
Hausch S, Vamosi SM, Fox JW (2018) Effects of intraspecific phenotypic variation on species coexistence. Ecology 99:1453–1462
Herbst FA, Lünsmann V, Kjeldal H, Jehmlich N, Tholey A, von Bergen M, Nielsen PH (2016) Enhancing metaproteomics: the value of models and defined environmental microbial systems. Proteomics 16:783–798
Hibbing ME, Fuqua C, Parsek MR, Peterson SB (2010) Bacterial competition: surviving and thriving in the microbial jungle. Nat Rev Microbiol 8:15–25
Hierro JL, Maron JL, Callaway RM (2005) A biogeographical approach to plant invasions: the importance of studying exotics in their introduced and native range. J Ecol 93:5–15
Hodge A (2004) The plastic plant: root responses to heterogeneous supplies of nutrients. New Phytol 162:9–24
Hoitink HA, Fahy PC (1986) Basis for the control of soilborne plant pathogens with composts. Annu Rev Phytopathol 24:93–114
Igiehon NO, Babalola OO (2018) Below-ground-above-ground plant-microbial interactions: focusing on soybean, rhizobacteria and mycorrhizal fungi. Microbiol J 12:261–279
Jeffries P, Gianinazzi S, Perotto S, Turnau K, Barea JM (2003) The contribution of arbuscular mycorrhizal fungi in sustainable maintenance of plant health and soil fertility. Biol Soil 37:1–16
Jenny H (1980) The soil resource: origin and behavior. Ecological studies, vol 37. Springer, New York
Jentsch A, Kreyling J, Beierkuhnlein C (2007) A new generation of climate-change experiments: events, not trends. Front Ecol Environ 5:365–374
Jessup CM, Kassen R, Forde SE, Kerr B, Buckling A, Rainey PB, Bohannan BJ (2004) Big questions, small worlds: microbial model systems in ecology. Trends Ecol Evol 19:189–197
Jimenez PN, Koch G, Thompson JA, Xavier KB, Cool RH, Quax WJ (2012) The multiple signaling systems regulating virulence in *Pseudomonas aeruginosa*. Microbiol Mol Biol Rev 76:46–65
Johansson JF, Paul LR, Finlay RD (2004) Microbial interactions in the mycorrhizosphere and their significance for sustainable agriculture. FEMS Microbiol Ecol 48:1–13
Johnson NC (2010) Resource stoichiometry elucidates the structure and function of arbuscular mycorrhizas across scales. New Phytol 185:631–647
Jones DL, Hodge A, Kuzyakov Y (2004) Plant and mycorrhizal regulation of rhizodeposition. New Phytol 163:459–480
Jones JT, Haegeman A, Danchin EG, Gaur HS, Helder J, Jones MG, Perry RN (2013) Top 10 plant-parasitic nematodes in molecular plant pathology. Mol Plant Pathol 14:946–961
Jones MG, Iqbal S, Fosu-Nyarko J (2016) Belowground defence strategies against migratory nematodes. In: Belowground defence strategies in plants. Springer, Cham, pp 253–278
Kanoh K, Kamino K (2001) Effect of exogenous siderophores on iron uptake activity of marine bacteria under iron-limited conditions. Appl Environ Microl 67:1710–1717
Khan MR (2015) Nematode diseases of crops in India. In: Recent advances in the diagnosis and management of plant diseases. Springer India, Cham, pp 183–224
Kim ES, Trisurat Y, Muraoka H, Shibata H, Amoroso V, Boldgiv B, Ohte N (2018) The International Long-Term Ecological Research-East Asia-Pacific Regional Network (ILTER-EAP): history, development, and perspectives. Ecol Res 33:19–34
Kliebenstein DJ (2004) Secondary metabolites and plant/environment interactions: a view through *Arabidopsis thaliana* tinged glasses. Plant Cell Environ 27:675–684

Kõljalg U, Nilsson RH, Abarenkov K, Tedersoo L, Taylor AF, Bahram M, Douglas B (2013) Towards a unified paradigm for sequence-based identification of fungi. Mol Ecol 22:5271–5277
Kong C, Liang W, Xu X, Hu F, Wang P, Jiang Y (2004) Release and activity of allelochemicals from allelopathic rice seedlings. J Agric Food Chem 52:2861–2865
Korhonen A, Lehto T, Repo T (2015) Frost hardiness of mycorrhizal and non-mycorrhizal Scots pine under two fertilization treatments. Mycorrhiza 25(5):377–386
Koutsoudis MD, Tsaltas D, Minogue TD, von Bodman SB (2006) Quorum-sensing regulation governs bacterial adhesion, biofilm development, and host colonization in *Pantoea stewartii* subspecies *stewartii*. Proc Natl Acad Sci U S A 103:5983–5988
Kulmatiski A, Beard KH, Stevens JR, Cobbold SM (2008) Plant–soil feedbacks: a meta-analytical review. Ecol Lett 11:980–992
Lacava PT, Araújo WL, Marcon J, Maccheroni W Jr, Azevedo JL (2004) Interaction between endophytic bacteria from citrus plants and the phytopathogenic bacteria *Xylella fastidiosa*, causal agent of citrus-variegated chlorosis. Lett Appl Microbiol 39:55–59
Laliberté E, Grace JB, Huston MA, Lambers H, Teste FP, Turner BL, Wardle DA (2013) How does pedogenesis drive plant diversity. Trends Ecol Evol 28:331–340
Lambers H, Chapin FS, Pons TL (2008) Interactions among plants. In: Plant physiological ecology. Springer, New York, pp 505–531
Lambers H, Mougel C, Jaillard B, Hinsinger P (2009) Plant–microbe–soil interactions in the rhizosphere: an evolutionary perspective. Plant Soil 321:83–115
Langley JA, Hungate BA (2003) Mycorrhizal controls on belowground litter quality. Ecology 84:2302–2312
Lanoue A, Burlat V, Henkes GJ, Koch I, Schurr U, Röse US (2010) De novo biosynthesis of defense root exudates in response to *Fusarium* attack in barley. New Phytol 185:577–588
Lareen A, Burton F, Schäfer P (2016) Plant root-microbe communication in shaping root microbiomes. Plant Mol Biol 90:575–587
Lattanzio V, Lattanzio VM, Cardinali A (2006) Role of phenolics in the resistance mechanisms of plants against fungal pathogens and insects. Phytol Adv Res 661:23–67
Lee KJ, Oh B, Seralathan KK (2013) Advances in plant growth promoting rhizobacteria for biological control of plant diseases. In: Bacteria in agrobiology: disease management. Springer, Berlin, pp 1–13
Li XG, Zhang TL, Wang XX, Hua K, Zhao L, Han ZM (2013) The composition of root exudates from two different resistant peanut cultivars and their effects on the growth of soil-borne pathogen. Int J Biol Sci 9:164–173
Lindsay WL, Norvell WA (1978) Development of a DTPA soil test for zinc, iron, manganese, and copper. Soil Sci Am J 42:421–428
Lloyd-Price J, Abu-Ali G, Huttenhower C (2016) The healthy human microbiome. Gen Med 8:1–11
Lugtenberg B, Kamilova F (2009) Plant-growth-promoting rhizobacteria. Annu Rev Microbiol 63:541–556
Lynch JM, Benedetti A, Insam H, Nuti MP, Smalla K, Torsvik V, Nannipieri P (2004) Microbial diversity in soil: ecological theories, the contribution of molecular techniques and the impact of transgenic plants and transgenic microorganisms. Biol Fertil Soil 40:363–385
Mansfield J, Genin S, Magori S, Citovsky V, Sriariyanum M, Ronald P, Toth IAN (2012) Top 10 plant pathogenic bacteria in molecular plant pathology. Mol Plant Pathol 13:614–629
Marschner H (2011) Marschner's mineral nutrition of higher plants. Academic press, London
Marschner P (2012) Rhizosphere biology. In: Marschner's mineral nutrition of higher plants, 3rd edn. Springer, New York, pp 369–388
Massalha H, Korenblum E, Tholl D, Aharoni A (2017) Small molecules below-ground: the role of specialized metabolites in the rhizosphere. Plant J 90:788–807
Mauch-Mani B, Baccelli I, Luna E, Flors V (2017) Defense priming: an adaptive part of induced resistance. Annu Rev Plant Bio 68:485–512
McCann KS (2000) The diversity stability debate. Nature 405:228–233

Mendes R, Garbeva P, Raaijmakers JM (2013) The rhizosphere microbiome: significance of plant beneficial, plant pathogenic, and human pathogenic microorganisms. FEMS Microbiol Rev 37:634–663

Mueller UG, Sachs JL (2015) Engineering microbiomes to improve plant and animal health. Trends Microbiol 23:606–617

Nega A (2014) Review on concepts in biological control of plant pathogens. J Biol Agric Health 4:33–54

Nicol JM, Turner SJ, Coyne DL, Den Nijs L, Hockland S, Maafi ZT (2011) Current nematode threats to world agriculture. In: Genomics and molecular genetics of plant-nematode interactions. Springer, Dordrecht, pp 21–43

Öpik M, Vanatoa A, Vanatoa E, Moora M, Davison J, Kalwij JM, Zobel M (2010) The online database MaarjAM reveals global and ecosystemic distribution patterns in arbuscular mycorrhizal fungi (Glomeromycota). New Phytol 188:223–241

Packer A, Clay K (2003) Soil pathogens and *Prunus serotina* seedling and sapling growth near conspecific trees. Ecology 84:108–119

Palacios OA, Bashan Y, de-Bashan LE (2014) Proven and potential involvement of vitamins in interactions of plants with plant growth-promoting bacteria: an overview. Biol Fertil Soils 50:415–432

Parnell JJ, Berka R, Young HA, Sturino JM, Kang Y, Barnhart DM, DiLeo MV (2016) From the lab to the farm: an industrial perspective of plant beneficial microorganisms. Front Sci 7:1–12

Peix A, Ramírez-Bahena MH, Velázquez E, Bedmar EJ (2015) Bacterial associations with legumes. Crit Rev Sci 34:17–42

Pérez-Jaramillo JE, Mendes R, Raaijmakers JM (2016) Impact of plant domestication on rhizosphere microbiome assembly and functions. Plant Mol Biol 90:635–644

Pérez-Montaño F, Alías-Villegas C, Bellogín RA, Del Cerro P, Espuny MR, Jiménez-Guerrero I, Cubo T (2014) Plant growth promotion in cereal and leguminous agricultural important plants: from microorganism capacities to crop production. Microbiol Res 169:325–336

Pieterse CM, Leon-Reyes A, Van der Ent S, Van Wees SC (2009) Networking by small-molecule hormones in plant immunity. Nat Chem Biol 5:308–316

Pieterse CM, Van der Does D, Zamioudis C, Leon-Reyes A, Van Wees SC (2012) Hormonal modulation of plant immunity. Annu Rev Cell Biol 28:489–521

Pineda A, Zheng SJ, Van Loon JJ, Pieterse CM, Dicke M (2010) Helping plants to deal with insects: the role of beneficial soil-borne microbes. Trends Sci 15:507–514

Polke M, Sprenger M, Scherlach K, Albán-Proaño MC, Martin R, Hertweck C, Jacobsen ID (2017) A functional link between hyphal maintenance and quorum sensing in *Candida albicans*. Mol Microbiol 103:595–617

Prasad R, Kumar M, Varma A (2015) Role of PGPR in soil fertility and plant health. In: Egamberdieva D, Shrivastava S, Varma A (eds) Plant growth-promoting rhizobacteria (PGPR) and medicinal plants. Springer International, Switzerland, pp 247–260

Pringle A, Bever JD, Gardes M, Parrent JL, Rillig MC, Klironomos JN (2009) Mycorrhizal symbioses and plant invasions. Annu Rev Ecol Evol Syst 40:699–715

Quiñones B, Dulla G, Lindow SE (2005) Quorum sensing regulates exopolysaccharide production, motility, and virulence in *Pseudomonas syringae*. Mol Interact 18:682–693

Raaijmakers JM, Vlami M, De Souza JT (2002) Antibiotic production by bacterial biocontrol agents. Antonie van Leeuwenhoek 81:537–547

Ramaswamy AV, Sorrels CM, Gerwick WH (2007) Cloning and biochemical characterization of the hectochlorin biosynthetic gene cluster from the marine cyanobacterium *Lyngbya majuscula*. J Nat Prod 70:1977–1986

Rapparini F, Peñuelas J (2014) Mycorrhizal fungi to alleviate drought stress on plant growth. In: Use of microbes for the alleviation of soil stresses, vol 1. Springer, New York, pp 21–42

Rasolomampianina R, Bailly X, Fetiarison R, Rabevohitra R, Béna G, Ramaroson L, Avarre JC (2005) Nitrogen-fixing nodules from rose wood legume trees (*Dalbergia* spp.) endemic to

Madagascar host seven different genera belonging to α-and β-Proteobacteria. Mol Ecol 14:4135–4146

Ratnadass A, Blanchart É, Lecomte P (2013) Ecological interactions within the biodiversity of cultivated systems. In: Cultivating biodiversity to transform agriculture. Springer, Dordrecht, pp 141–179

Read DJ, Duckett JG, Francis R, Ligrone R, Russell A (2000) Symbiotic fungal associations in 'lower' land plants. Philos Trans R Soc Lond B Biol Sci 355:815–831

Redecker D, Schüßler A, Stockinger H, Stürmer SL, Morton JB, Walker C (2013) An evidence-based consensus for the classification of arbuscular mycorrhizal fungi (Glomeromycota). Mycology 23:515–531

Reynolds HL, Packer A, Bever JD, Clay K (2003) Grassroots ecology: plant–microbe-soil interactions as drivers of plant community structure and dynamics. Ecology 84:2281–2291

Richardson AE, Barea JM, McNeill AM, Prigent-Combaret C (2009) Acquisition of phosphorus and nitrogen in the rhizosphere and plant growth promotion by microorganisms. Plant Soil 32:305–339

Rodrıguez-Echeverrıa S, Costa SR, Freitas H (2007) Biodiversity and interactions in the rhizosphere: effects on ecosystem functioning. In: Functional plant ecology. 7488. CRC Press, Boca Raton, FL, pp 581–595

Romero IG, Ruvinsky I, Gilad Y (2012) Comparative studies of gene expression and the evolution of gene regulation. Nat Rev Genet 13(7):505–516

Römheld V, Kirkby EA (2010) Research on potassium in agriculture: needs and prospects. Plant Soil 335:155–180

Salas-Marina MA, Silva-Flores MA, Uresti-Rivera EE, Castro-Longoria E, Herrera-Estrella A, Casas-Flores S (2011) Colonization of *Arabidopsis* roots by *Trichoderma atroviride* promotes growth and enhances systemic disease resistance through jasmonic acid/ethylene and salicylic acid pathways. Eur J Plant Pathol 131:15–26

Sawada H, Kuykendall LD, Young JM (2003) Changing concepts in the systematics of bacterial nitrogen-fixing legume symbionts. J Appl Microbiol 49:155–179

Schmidt MW, Torn MS, Abiven S, Dittmar T, Guggenberger G, Janssens IA, Nannipieri P (2011) Persistence of soil organic matter as an ecosystem property. Nature 478:49–56

Schoener TW (1974) Resource partitioning in ecological communities. Science 185:27–39

Selosse MA, Richard F, He X, Simard SW (2006) Mycorrhizal networks: des liaisons dangereuses? Trends Ecol Evol 21:621–628

Shade A, Peter H, Allison SD, Baho D, Berga M, Bürgmann H, Matulich KL (2012) Fundamentals of microbial community resistance and resilience. Front Microbiol 3:1–19

Shamseldin A (2013) The role of different genes involved in symbiotic nitrogen fixation. Review. J Biotechnol Biochem 8:84–94

Sharpley AN, Chapra SC, Wedepohl R, Sims JT, Daniel TC, Reddy KR (1994) Managing agricultural phosphorus for protection of surface waters: issues and options. J Environ Qual 23:437–451

Shoresh M, Harman GE, Mastouri F (2010) Induced systemic resistance and plant responses to fungal biocontrol agents. Annu Rev Phytopathol 48:21–43

Shrivastava S, Prasad R, Varma A (2014) Anatomy of root from eyes of a microbiologist. In: Morte A, Varma A (eds) Root engineering. Springer, Berlin, pp 3–22

Shtark OY, Borisov AY, Zhukov VA, Provorov NA, Tikhonovich IA (2010) Intimate associations of beneficial soil microbes with host plants. In: Soil microbiology and sustainable crop production. Springer, Dordrecht, pp 119–196

Sikkema J, de Bont JA, Poolman B (1995) Mechanisms of membrane toxicity of hydrocarbons. Microbiol Rev 59:201–222

Simard S, Asay A, Beiler K, Bingham M, Deslippe J, He X, Teste F (2015) Resource transfer between plants through ectomycorrhizal fungal networks. In: Horton TR (ed) Mycorrhizal networks. Ecological studies, vol 224. Analysis and synthesis. Springer, Dordrecht, pp 133–176

Singh D, Raina TK, Kumar A, Singh J, Prasad R (2019) Plant microbiome: a reservoir of novel genes and metabolites. Plant Gene. https://doi.org/10.1016/j.plgene.2019.100177

Six J, Bossuyt H, Degryze S, Denef K (2004) A history of research on the link between (micro) aggregates, soil biota, and soil organic matter dynamics. Soil Res 79:7–31

Smith SE, Read DJ (2010) Mycorrhizal symbiosis. Academic Press, New York

Smith SE, Smith FA (2012) Fresh perspectives on the roles of arbuscular mycorrhizal fungi in plant nutrition and growth. Mycologia 104:1–13

Smith SE, Facelli E, Pope S, Smith FA (2010) Plant performance in stressful environments: interpreting new and established knowledge of the roles of arbuscular mycorrhizas. Plant Soil 326:3–20

Sogin ML, Morrison HG, Huber JA, Welch DM, Huse SM, Neal PR, Herndl GJ (2006) Microbial diversity in the deep sea and the underexplored "rare biosphere.". Proc Natl Acad Sci U S A 103:12115–12120

Somers E, Vanderleyden J, Srinivasan M (2004) Rhizosphere bacterial signalling: a love parade beneath our feet. Crit Rev Microbiol 30:205–240

Sparbier K, Schubert S, Weller U, Boogen C, Kostrzewa M (2012) Matrix-assisted laser desorption ionization–time of flight mass spectrometry-based functional assay for rapid detection of resistance against β-lactam antibiotics. J Clin Microbiol 50:927–937

Sugiyama A, Ueda Y, Zushi T, Takase H, Yazaki K (2014) Changes in the bacterial community of soybean rhizospheres during growth in the field. PLoS One 9:1–9

Swamy MK, Akhtar MS, Sinniah UR (2016) Root exudates and their molecular interactions with rhizospheric microbes. In: Hakeem KR, Akhtar MS (eds) Plant, soil and microbes: vol 2. Mechanisms and molecular interactions. Springer, Cham, pp 59–77

Taylor JH, Peterson CA (2005) Ectomycorrhizal impacts on nutrient uptake pathways in woody roots. New For 30:203–214

Taylor MW, Radax R, Steger D, Wagner M (2007) Sponge-associated microorganisms: evolution, ecology, and biotechnological potential. Microbiol Mol Biol Rev 71:295–347

Tewari S, Arora NK (2013) Transactions among microorganisms and plant in the composite rhizosphere habitat. In: Plant microbe symbiosis: fundamentals and advances. Springer, New Delhi, pp 1–50

Toby Kiers E, Palmer TM, Ives AR, Bruno JF, Bronstein JL (2010) Mutualisms in a changing world: an evolutionary perspective. Ecol Lett 13:1459–1474

Toljander JF, Santos-González JC, Tehler A, Finlay RD (2008) Community analysis of arbuscular mycorrhizal fungi and bacteria in the maize mycorrhizosphere in a long-term fertilization trial. FEMS Microbiol Ecol 65:323–338

Torsvik V, Øvreås L (2002) Microbial diversity and function in soil: from genes to ecosystems. Currt Opin Microbiol 5:240–245

Van Der Heijden MG, Horton TR (2009) Socialism in soil. The importance of mycorrhizal fungal networks for facilitation in natural ecosystems. J Ecol 97:1139–1150

Van Der Heijden MG, Bardgett RD, Van Straalen NM (2008) The unseen majority: soil microbes as drivers of plant diversity and productivity in terrestrial ecosystems. Ecol Lett 11:296–310

Venturi V, Keel C (2016) Signaling in the rhizosphere. Trends Sci 21:187–198

Vigo C, Norman JR, Hooker JE (2000) Biocontrol of the pathogen *Phytophthora parasitica* by arbuscular mycorrhizal fungi is a consequence of effects on infection loci. Plant Pathol 49:509–514

Vovlas N, Rapoport HF, Jiménez Díaz RM, Castillo P (2005) Differences in feeding sites induced by root-knot nematodes, *Meloidogyne* sp., in chickpea. Phytopathology 95:368–375

Walker AW, Duncan SH, Louis P, Flint HJ (2014) Phylogeny, culturing, and metagenomics of the human gut microbiota. Trends Microbiol 22:267–274

Wang M, Zheng Q, Shen Q, Guo S (2013) The critical role of potassium in plant stress response. Int J Mol Sci 14(4):7370–7390

Wardle DA, Karban R, Callaway RM (2011) The ecosystem and evolutionary contexts of allelopathy. Trends Ecol Evol 26:655–662

Wassmann R, Jagadish SVK, Sumfleth K, Pathak H, Howell G, Ismail A, Heuer S (2009) Regional vulnerability of climate change impacts on Asian rice production and scope for adaptation. Adv Agron 102:91–133

Waters CM, Bassler BL (2005) Quorum sensing: cell-to-cell communication in bacteria. Annu Rev Cell Dev Biol 21:319–346

Weller DM (1988) Biological control of soilborne plant pathogens in the rhizosphere with bacteria. Annu Rev Phytopathol 26:379–407

West SA, Diggle SP, Buckling A, Gardner A, Griffin AS (2007) The social lives of microbes. Annu Rev Ecol Evol Syst 38:53–77

Whitham TG, Bailey JK, Schweitzer JA, Shuster SM, Bangert RK, LeRoy CJ, Wooley SC (2006) A framework for community and ecosystem genetics: from genes to ecosystems. Nat Rev Genet 7:510–523

Wright JS (2002) Plant diversity in tropical forests: a review of mechanisms of species coexistence. Oecologia (Berl) 130:1–14

Xu J (2006) Invited review: microbial ecology in the age of genomics and metagenomics: concepts, tools, and recent advances. Mol Ecol 15:1713–1731

Yuan J, Zhang N, Huang Q, Raza W, Li R, Vivanco JM, Shen Q (2015) Organic acids from root exudates of banana help root colonization of PGPR strain *Bacillus amyloliquefaciens* NJN-6. Sci Rep 5:1–8

Zhao J, Davis LC, Verpoorte R (2005) Elicitor signal transduction leading to production of plant secondary metabolites. Biotechnol Adv 23:283–333

Zornoza R, Acosta JA, Bastida F, Domínguez SG, Toledo DM, Faz A (2015) Identification of sensitive indicators to assess the interrelationship between soil quality, management practices and human health. Soil 1:173–185

Chapter 7
Pathogen and Management of Fungal Wilt of Banana Through Biocontrol Agents

Narendra Kumar and S. M. Paul Khurana

Abstract Banana is an important and affordable source of food for millions of people of the developing tropical countries. It comes in the category of export fruits in the world. Banana wilt is a limiting factor for growth of the banana industry. It is one of the most destructive diseases of banana. The main causal organism of wilt of banana is *Fusarium oxysporum* f. sp. *cubense* (Foc). Right from the discovery of *Fusarium* wilt in banana, many control measures like fumigation of soil and crop rotation along with organic amendments have been attempted. But the problem could not be resolved fully except by planting resistant cultivars. Use of resistant varieties also is not effective and can't be implemented easily because of lack of consumer preference. Due to these problems, use of antagonistic agents is being adopted largely. They have the potential to protect and promote plant growth by colonizing and multiplying both in the rhizosphere and plant system. They are useful as an effective eco-friendly alternative for field management of the banana wilt.

To control *Fusarium* wilt, biocontrol method is now gaining popularity. It is eco-friendly in nature. Biocontrol has potential and various mechanisms for plant protection. *Trichoderma* sp. acts as a major interactive agent in root, soil, and foliar environments through releasing an array of compounds producing localized or systemic resistance in plants. *Pseudomonas* sp. also has varying mechanisms towards the control of phytopathogens through release of a wide range of antagonistic metabolites. This chapter records updated information about the pathogen and how *Fusarium* can be managed through the biocontrol agents.

7.1 Introduction

Banana (*Musa* spp.) flourishes well under tropical and moisture-rich farmlands and propagates from the underground rhizome. The whole plant is a false stem and consists of broad leaves with their long petioles arranged in disc-like fashion. During

N. Kumar · S. M. Paul Khurana (✉)
Amity Institute of Biotechnology, Amity University Haryana, Manesar, Gurgaon, India
e-mail: smpkhurana@ggn.amity.edu

A. Varma et al. (eds.), *Plant Microbe Interface*,
https://doi.org/10.1007/978-3-030-19831-2_7

2011 banana worth $44 billion, having production of 145 million metric tons, was produced in over 130 countries. *Fusarium* wilt (also known as Panama disease) is the most severe and destructive disease of the crop (Ploetz 2015). *Fusarium* wilt of banana is caused by the soil-borne fungus *Fusarium oxysporum* f. sp. *cubense* (Foc) (Stover 1962). It is a major constraint for banana production throughout the world. A strain of the fungus that affects Cavendish and other dessert bananas in the tropics, called Foc tropical race 4 (TR4; VCG 01213/16), has been confined to five Asian countries (Indonesia, Malaysia, Philippines, mainland China and Taiwan) for more than three decades. Cavendish banana production is expanding in Asia to Laos, Myanmar and Vietnam where local varieties still dominate the market. This is due to an increase in Cavendish banana consumption and a decline in areas of production caused by Foc TR4 in China (Hung and Hung 2018). The fungus enters the plant through the roots and spreads its colony in the xylem vessels which blocks the flow of water and nutrients. The pathogen had evolved along with its host in the region of Indo-Malaya. Then it spread to other banana-growing places through infected planting material (Mostert et al. 2017).

The pathogen that originated in Asia and evolved with the host, the wild banana plant *Musa acuminate* Foc, consists of 8 lineages and 3 races along with 24 VCGs (vegetative compatibility groups). Maximum damage occurs due to Foc tropical race 4 (TR4) found only in Asia (Li et al. 2013).

It causes serious problems in parts of Africa and in Central and South America (Viljoen 2002), Indonesia, Burma, Sri Lanka, Thailand and the Philippines causing in enormous losses every year (Stover and Simmonds 1987; Ploetz 1994). In the last 10 years, banana plantations in China have decreased dramatically due to *Fusarium* wilt in Hainan, Guangdong and Fujian provinces. The fungus produces external symptoms like gradual wilting, progressive yellowing of banana leaves and resultantly collapse at the petiole (Yin et al. 2011). Distinguishing symptoms for the disease are discoloration of vascular tissues (fruit stalk, pseudostem, roots, corm) varying from dark brown to light yellow appearing first on the outer or oldest leaf sheath and then extending to pseudostem (Ploetz 2006), and finally the disease causes death of banana plants. The Foc isolates have been grouped into four physiological races based on pathogenicity to host cultivars in fields (Fourie et al. 2009). *F. oxysporum* f. sp. *cubense* races 1, 2, 3 and 4 attack many cultivars (Table 7.1), and all cultivars show susceptibility to Foc1 and Foc2 (Persley 1987; Ploetz 1990).

Once Foc R1 threatened banana production most severely. Which ruined the 'Gros Michel' industry of banana in the Central America and Caribbean in the mid-1900s (Ploetz 1994). This persisted till the development of Cavendish banana

Table 7.1 Infection of *F. oxysporum* f. sp. *cubense* on different cultivars

SN	Race	Infected cultivar
1.	Race 1 (Foc R1)	'Gros Michel', Lady Finger (AAB) and Silk (AAB) varieties
2.	Foc R2	Bluggoe (ABB)
3.	Foc R3	*Heliconia* spp.
4.	Foc R4	Cultivar Cavendish

cultivars which had no susceptibility for Foc R1 and Foc R2 in tropical regions (Fernández-Falcón et al. 2003). However, it was not long-lasting. Cavendish cultivars soon became highly susceptible to Foc R4, a new race (Ploetz and Pegg 2000) causing worldwide losses. It brings the spiteful change that the banana industry is once again threatened by *Fusarium* wilt. *F. oxysporum* is able to infect more than 100 plant species which can be divided into more than 120 host-specific forms called *formae speciales* (Minerdi et al. 2008). Foc is one of the most destructive pathogen f. sp. of *F. oxysporum* (Ploetz 1990) easily soil-borne and strongly saprophytic. This survives in soil up to several decades through spores (specifically as chlamydospores) that reinfect the susceptible varieties of banana (Stover 1962). This creates problem in disease management. The results of the wilt management studies have been disappointing, i.e. effective and efficient control strategies are still not available to meet the world demand.

7.2 Symptoms

The *Fusarium* wilt symptoms usually become prominent at the time of flowering. The fungus infects the plant roots through forming colony in the vascular tissues in the rhizome and pseudostem. Wilt symptoms are induced after 5–6 months of planting when symptoms get expressed both internally and externally (Figs. 7.1 and 7.2) (Wardlaw 1961; Stover 1962). Usually wilt-infected plants bear no bunches or have very few small fruits that ripen irregularly, have pithy flesh and are acidic. Purplish brown discoloration of the vascular bundles, which can be seen in cross sections of the corm and pseudostem (Fig. 7.2), is the typical internal symptom. In the corm, the discoloration appears as a collection of tiny dots. When root portions

Fig. 7.1 Collapsed leaves of banana plant due to fungal infection

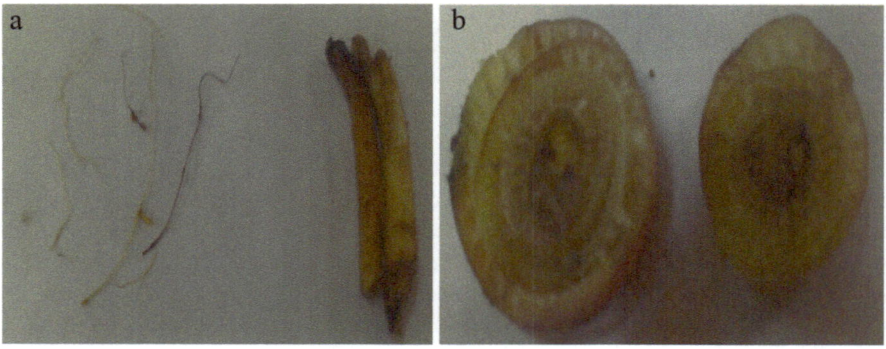

Fig. 7.2 Discolouration in wilted banana roots and pseudostems

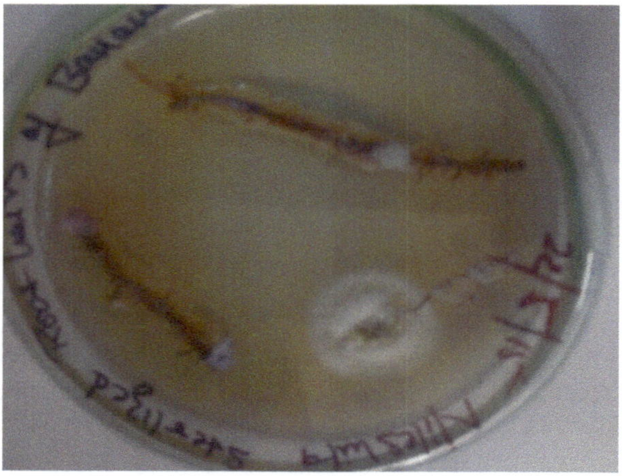

Fig. 7.3 Colonization of *Fusarium oxysporum* on banana roots and root hairs

are cultured on PDA medium, fungal colonies develop (Fig. 7.3). The mycelia of *Fusarium oxysporum* are delicate white and may be sparse to abundant. The fungus develops three types of spores: microconidia, macroconidia and chlamydospores. Microconidia arise laterally on simple phialides and are abundant. They are ellipsoid to oval and straight to curved 4–11 × 2.1–3.4 μm and nonseptate. Macroconidia are also sparse to abundant, borne on branched conidiophores or on the surface of sporodochia. They are thin-walled, three- to five-septate, subulate-fusoid and pointed at both ends with pedicellate base. Three septate are the most common conidia measuring 26–45 × 3–4 μm while five-septate conidia measure 32–57 × 3–5 μm. Chlamydospores may be both smooth and rough walled, produced abundantly and formed either terminally or on intercalary basis (Fig. 7.4) (Kumar and Saxena 2015).

The severity of the disease depends on host susceptibility, fungal virulence and environmental conditions such as rainfall and temperature. In highly susceptible

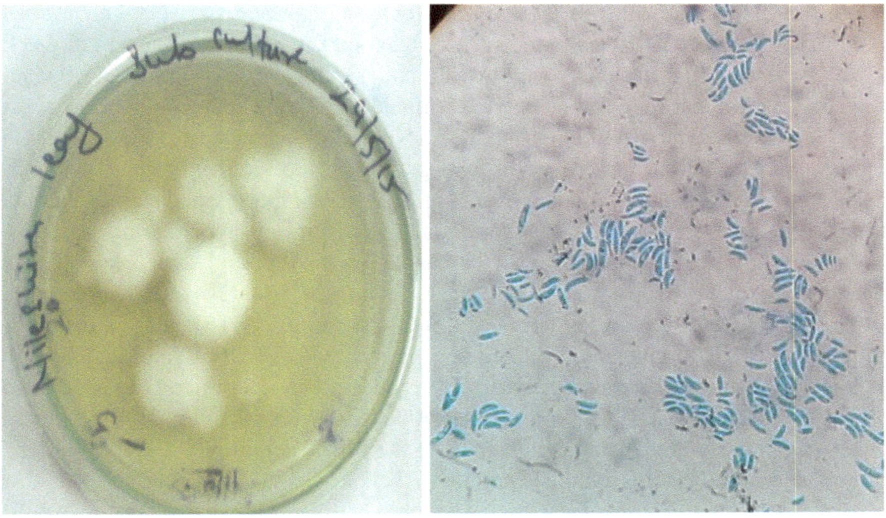

Fig. 7.4 Pure colonies of *Fusarium oxysporum* on PDA medium and the conidia

cultivars, under the conditions of water stress or waterlogging, the entire foliage may become yellow, the growth may cease, emergence of bunch may stop and finally the whole plant may collapse. The pathogen spreads to suckers, producing internal symptoms within 1 or 2 months of emergence of suckers (Moore et al. 1995).

Fusaric acid (FA) is a phytotoxin produced by *F. oxysporum* f. sp. *cubense*, and different members of the *F. oxysporum* species complex cause leaf chlorosis (Dong et al. 2012, 2014). Chloroplast damage reduces photochemical efficiency of plant photosynthesis and net CO_2 assimilation. It has been confirmed through artificially inoculated plants of Gros Michel FA. The pathogen however occurs in symptomatic leaves, but chlorosis is induced upon injecting FA into the leaf lamina. There was a reduction in transpiration in diseased leaves due to stomatal closure with reduction in hydraulic conductivity. This was detected in diseased stems associated with development of the above symptoms (Dong et al. 2012, 2014). Biochemical and structural alterations occur in affected plants showing senescence (Dong et al. 2014).

7.3 Diversity of the Disease

For the first time, Smith (1910) isolated *Fusarium oxysporum* f. sp. *cubense*. It has abundant ellipsoid oval microconidia on short lateral phialidic conidiophores. In due course of time, clusters of typical fusoid 3–5 macroconidia are produced. In culture, the fungus produces a reddish pigment and upon ageing develops globose chlamydospores. These characteristics remarkably separate this from other similar-looking species *F. solani* and *F. moniliforme* also having abundant microspores

(Booth 1971). So far, four races of *F. oxysporum* f. sp. *cubense* (Foc) have been reported (Moore et al. 1995) based on pathogenicity to different banana cultivars: race 1, which occurs throughout the world, attacks cultivars like Pome (AAB) and Silk (AAB) groups; race 2 is widely distributed throughout banana-growing areas which is pathogenic to Monthan, Bluggoe and also cooking bananas; race 3 occurs in Honduras, Costa Rica and Australia and pathogenic to *Heliconia* spp.; and race 4 occurs in maximum banana-growing countries like Malaysia, Australia, South Africa, Canary Islands, Taiwan, Brazil, etc. but not in India, and it attacks Cavendish group of banana (AAA) and also the race 1- and race 2-susceptible banana varieties (Pushpavathi et al. 2015). A sum of 594 *F. oxysporum* isolates from ten (10) Asian countries were identified and grouped on vegetative compatibility group (VCG) analysis. The isolates were divided in DNA lineages through PCR-RFLP analysis. For representing 3 Foc races, they identified 6 lineages and 14 VCGs in this study. VCG complex 0124/5 was the most common in the Indian subcontinent, Cambodia and Vietnam, but the VCG complex 01213/16 showed dominance in the rest of Asia (Mostert et al. 2017).

7.4 Survival and Disease Cycle of the Wilt Pathogen

The pathogen *F. oxysporum* f. sp. *cubense* (Foc) is a facultative parasite capable of saprophytic growth and classified as a root-inhabiting fungus with populations unevenly distributed which decline speedily in absentia of the host (Gowen 1995). However, Moore et al. (1995) reported that the fungus can survive in the field for up to 30 years as chlamydospores in the infested plant debris or in the roots of alternative hosts. It also survives in the roots of several species of common grasses and weeds such as *Paspalum*, *Panicum*, *Ixophorus* and *Commelina* which are the nonsymptomatic hosts (Gowen 1995). Ramakrishnan and Damodaran (1956) reported that liming of soil reduced the survival period of the pathogen to 2 months. The Indian strain of the pathogen could survive under water stagnation for a month (Rawal 2000). The texture and organic matter content of the soil greatly influence the survival of the pathogen. The pathogen colony tends to be higher and survive for long in light texture soils but not in heavy alkaline soils. Certain crop residues may enhance antagonistic microflora which reduce pathogen survival (Sequiera 1992). However, Thangavelu et al. (2001) observed the incidence of the wilt disease from loose soil to heavy clay soil with the pH ranging from 4.80 to 8.45 and EC of 0.12 to 1.10 dsm^{-1}. In the suppressive soils having high microbial populations, the pathogen development gets inhibited. Such soil is common in Australia, Central America, South Africa and Canary Islands (Moore et al. 1995). Figure 7.5 depicts naturally occurring cycle of banana wilt pathogen.

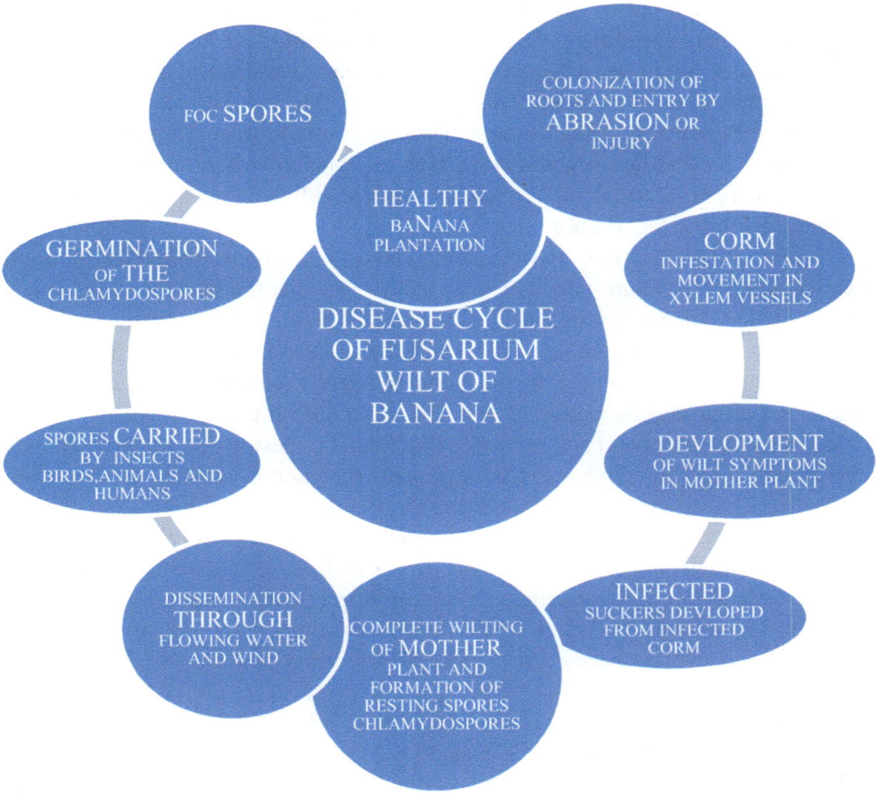

Fig. 7.5 Disease cycle of *Fusarium* wilt of banana

7.5 Management of the *Fusarium* Wilt

The exclusion of pathogen is essential for having pathogen-free areas. *F. oxysporum* f. sp. *cubense* can never be eradicated completely from a region when it gets infested. So quarantine regulations are essentially needed. The regional awareness and contingency programmes should be kept in mind in all threatened regions (Ploetz 2015).

To prevent Foc infection, many control methods were developed, viz. crop rotation, soil fumigation, fungicidal application, flooding fallow lands (Wardlaw 1961) and organic amendments. But none proved effective. The only effective method is cultivation of wilt-resistant banana plantations. Strict quarantine procedures regulate spread into areas where this pathogen does not occur yet possible only with great vigilance (Dita et al. 2010). Through involvement of quarantine practices, checking of infection in plant materials and along with farm implements cleaning can reduce disease spread (Nel 2004). Destruction of infected rhizomes and suckers can retard the flow of chlamydospore distribution in soil. The burning and sterilization of the diseased plant parts such as pseudostem and rhizome along with leaves in soils help in reducing the pathogen populations (Thakker et al. 2013). One of the alternates is to use pathogen-free tissue-cultured banana plants because this may be a substitute of suckers. This will reduce the flow of infection. Another investigation revealed an enhancement of susceptibility in micropropagated Cavendish cultivars for Foc race 4 with comparison to plants grown through conventional planting materials (Smith et al. 1998). This study highlights the value of agronomic practices in solving the disease.

The accumulation of host-specific pathogens in the soil can be reduced through intercropping and rotations which in turn together can alter the microbiological niche of the soil. The banana plantations can be changed to crops like sugarcane, paddy, cereals and cassava (Buddenhagen 2009). Besides, Huang et al. (2012) investigated that crop rotation of Chinese leek-banana efficiently controls *Fusarium* wilt in banana, reducing wilt incidence by 88–97%. The crop rotation practices however are less effective for potent soil microbes like Foc. The destruction of asymptomatic alternate hosts and effective weed management schedule in infested fields prevent and easily check the spread of *Fusarium* wilt (Hennessy et al. 2005).

Chemical methods have proved effective in field control of the *Fusarium* wilt of banana. The fungicides of benzimidazole group such as carbendazim, benomyl and thiabendazole showed effectiveness in suppressing Foc under in vitro and even in greenhouse conditions (Nel et al. 2007). Chemical compounds, viz. propiconazole, prochloraz and cyproconazole, also reduced *Fusarium* banana wilt incidence by about 80% (Nel 2004).

Fusarium population is maximum in the rhizosphere (83.4%) but least in the rhizoplane (46.3%). Its occurrence was 71.6% in the infected stems while only 51.6% in collars. The pathogenicity tests confirmed it to be dominantly responsible for wilt in banana. Extracts of *Ranunculus sceleratus* showed the highest inhibition of mycelial growth (97.3%) (Kumar 2016).

7.5.1 Biological Control

In the developing countries, banana comes in the category of staple food and consumed daily for breakfast or lunch and even exported widely to the developed countries. Since *Fusarium* wilt is a big problem for banana, so control strategies like soil fumigation, fungicides (Lakshmanan et al. 1987), crop rotation (Su et al. 1986), flooding fallow lands (Stover 1962), organic amendments (Stover and Simmonds 1987) and plant extracts (Kumar 2016) were tried, but the problem could be resolved only by planting the resistant cultivars (Moore et al. 1999). Resistant varieties however are not acceptable everywhere due to lack of consumers' preference (Viljoen 2002).

Use of antagonistic microbes is a potential alternative and has become increasingly popular (Weller et al. 2002). Therefore the search is on for novel mechanisms of plant protection and for other bioagents (Pushpavathi et al. 2016). Biological control of many soil-borne pathogens like *Fusarium oxysporum* is well studied (Thangavelu et al. 2004). There are reports demonstrating the successful use of different species of *Trichoderma*, *Pseudomonas*, *Streptomyces* and nonpathogenic *Fusarium* (npFo) of both rhizospheric and endophytic nature against *Fusarium* wilt under both glasshouse and field conditions (Rajappan et al. 2002; Getha et al. 2005). Results of the glasshouse evaluations of Nel et al. (2006) revealed that two of the nonpathogenic *F. oxysporum* isolates, CAV 255 and CAV 241, reduced *Fusarium* wilt incidence by 87.4 and 75.0%, respectively.

Pushpavathi et al. (2015) reported that sucker treatment before planting with biocontrol agents either *Trichoderma viride* or *Pseudomonas fluorescens* and soil drenching with the same biocontrol agent (twice at 30 and 180 DAP as booster application) effectively reduced the disease incidence and intensity thereby significantly increasing the yields.

7.5.1.1 *Trichoderma* spp.

It is a free living and common fungus in soil and root ecosystems. This interacts well in root, soil and foliar environments and releases a variety of compounds causing localized or systemic resistance in plants. It has long been known as biological agent used in the disease management having the ability for increasing root development, growth, productivity, resistance to abiotic stresses, nutrient uptake and use. This may be efficiently applied as spores which are tolerant to odd conditions for formulation and field application in comparison to their mycelia and chlamydospores (Amsellem et al. 1999). The mycelial mass produces antagonistic metabolites (Yedidia et al. 2000). The investigations have revealed that *Trichoderma* species effectively suppress *Fusarium* wilt (Thangavelu et al. 2004). Thangavelu (2002) reported application of *T. harzianum* Th-10 preparation @ 10 g/plant having 4×10^{31} cfu/g in basal and top dressing in 2, 4 and 6 months after planting showed maximum reduction of wilt incidence (51.16%).This was succeeded by *Bacillus subtilis* and *Pseudomonas fluorescens* (41.17%) applications as talc-based preparations both in glasshouse and fields. *T. harzianum* Th-10 talc-based preparation and 'carbendazim' (0.1%) showed only 40.1% and 18.1% reduction of the wilt problem. In the *Fusarium* wilt-nematode complex, soil application of biocontrol agents significantly checked the wilt severity and the root lesions and root-knot index also and besides caused 50–82% of reduction in population of nematode, viz. *Pratylenchus coffeae* and *Meloidogyne incognita*. The maximum reduction was because of *T. harzianum* use (Thangavelu 2002). Raguchander et al. (1997) found *T. viride* and *P. fluorescens* to be equally effective in reducing the wilt problem. The potted abaca plants when inoculated with *T. viride* and yeast together also showed 81.76% and 82.52% decrease of wilt disease incidence (Bastasa and Baliad 2005).

Soil application of chaffy grain formulation of *T. viride* NRCB1 greatly checked the external (up to 78%) and internal (up to 80%) symptoms of *Fusarium* wilt in tissue-cultured as well as sucker-derived banana plants also increasing the plant

Table 7.2 Bioagents and mechanisms of action the control of banana *Fusarium* wilt

S. no.	Bioagent	Mode of action	References
1.	*Streptomyces violaceusniger*	On incubation in agar plates showed in vitro antibiosis, microscopic studies evidenced lysis of hyphal ends in the inhibited fungal colonies It also evidenced in vitro antagonistic actions against *F. oxysporum* f. sp. *cubense* in the form of swelling and distortion, also showing excessive branching in hyphae finally inhibition of spore germination	Getha and Vikineswary (2002)
2.	*P. fluorescens*	Banana roots precolonization with *Pseudomonas fluorescens* could reduce *Fusarium oxysporum* f. sp. *cubense* colonization by 72%. Unusual deposits at sites to prevent fungal entry with a number of defence activity against spread of pathogen	Sukhada et al. (2004)
3.	*Streptomyces violaceusniger*	Through antibiotics produced	Getha et al. (2005)
4.	*Pseudomonas fluorescens*	Antibiotics (2,4-diacetylphloroglucinol) production inhibiting the growth and spore germination of *F. oxysporum* f. sp. *cubense*	Saravanan and Muthusamy (2006)
5.	*Serratia* sp.	Shows promising growth-promoting properties	Ting et al. (2008)
6.	*Proteobacteria*	Increase in PO and SOD wilt for defence	Jie et al. (2009)
7.	*Trichoderma viride*	Induction of PO and PAL activities and increase of defence-related enzymes and antibiosis	Thangavelu and Mustaffa (2010)
8.	*P. fluorescens*	Induction of defence enzymes such as PO and PAL	Akila et al. (2011)
9.	*Bacillus subtilis*	Induction of defence enzymes PO and PAL	Akila et al. (2011)
10.	*Pseudomonas* spp.	Production of volatile compounds such as methanethiol, 2-pentane 3-methyl and 3-undecene	Ting et al. (2011)
11.	*Herbaspirillum* spp.	Production of volatile compounds 2-pentane 3-methyl, methanethiol and 3-undecene	Ting et al. (2011)
12.	*Trichoderma viride*	Through production of antibiotics	Pushpavathi et al. (2015)
13.	*P. fluorescens*	Through production of metabolites and antibiotics	Pushpavathi et al. (2015)

growth parameters significantly both under pot culture and field conditions (Thangavelu and Mustaffa 2010).

The mechanisms reducing the *Fusarium* wilt severity due to *Trichoderma* spp. may be mycoparasitism, spatial and nutrient competition, antibiosis because of enzymes and secondary metabolites and improvement of plant defence system (Table 7.2). The mycoparasitism shows coiling, disorganization of cell contents and penetration of the host (Papavizas 1985). *Trichoderma* sp. parasitizes the pathogen hyphae and produces proteolytic enzymes like 1,3-glucanolytic, chitinase, etc. resulting in lysis. They produce metabolites, extracellular enzymes, volatiles and antibiotics (e.g. gliotoxin, viridian) which are fungistatic (Weindling 1941) causing antibiosis. Further *Trichoderma* spp. can also compete by forming siderophores (Srinivasan et al. 1992). Thangavelu and Mustaffa (2010) highlighted the application of *T. viride* in the form of rice chaffy grain formulation upon infection produces defence-related enzymes [peroxidase and phenylalanine ammonia lyase (PAL)]. The phenolic content was also significantly higher (>50%) in comparison. Increased activities of lytic enzymes and also phenols in the *T. viride*-treated plants impart resistance to Foc by making physical barrier stronger or chemically impervious to the pathogen (Thangavelu and Mustaffa 2010).

7.5.1.2 *Pseudomonas* spp.

Pseudomonas spp. belong to the category of useful biocontrol agents in agriculture because they can use many exudates/compounds serving as nutrient (Lugtenberg et al. 1999). They show high rate of growth and have diverse mechanisms against plant pathogens by producing a broad range of antagonistic metabolites (Lugtenberg et al. 1999). They are easy to culture in vitro. They can be easily inoculated in the rhizosphere (Rhodes and Powell 1994) inducing systemic resistance (Pieterse et al. 2001). *P. fluorescens* also suppresses *Fusarium* wilt disease in banana. Fluorescent pseudomonads such as *P. fluorescens* (Sakthivel and Gnanamanickam 1987), *P. putida* (de Freitas and Germida 1991), *P. chlororaphis* (Chin-A-Woeng et al. 1998) and *P. aeruginosa* (Anjaiah et al. 2003) have been found to inhibit pathogens and promote better growth and yield of many crops. Sivamani and Gnanamanickam (1988) observed that *P. fluorescens*-treated seedlings of *Musa balbisiana* expressed less severe wilting and internal discoloration because of Foc infection. The bacterized seedlings have enhanced root system enhancing plant height. *P. fluorescens* strain pf10 from banana rhizosphere is potent in detoxifying fusaric acid of Foc race 1 reducing wilt incidence up to 50% (Thangavelu et al. 2001). Dipping suckers in the *P. fluorescens* suspension and biocontrol agent effectively checked *Fusarium* wilt of banana (Raguchander et al. 1997). Rajappan et al. (2002) also found that the talc-based preparation was effective in the field against Foc. *P. fluorescens* strain WCS 417 suppressed other *Fusarium* wilts by 87.4% in Cavendish bananas under glasshouse (Nel et al. 2006). Basal application of neem cake @ 0.5 kg/plant + sucker through dipping in spore suspension of *P. fluorescens* for 15 min + soil application of *P. fluorescens* @10 g/plant at 3, 5 and 7 months after planting also checked the wilt (Saravanan et al. 2003).

Fishal et al. (2010) evaluated capability of *Pseudomonas* sp. (UPMP3) and *Burkholderia* sp. (UPMB3) obtained from healthy oil palm roots in induction of resistance fighting against *F. oxysporum* f. sp. *cubense* race 4 on susceptible Berangan banana under glasshouse. Preinoculation of banana through *Pseudomonas* sp. UPMP3 had 51% less in wilt incidence while in combined use of either UPMP3 + UPMB3 or only UPMB3 led to 39% and 38% decrease, respectively.

Two isolates of *Herbaspirillum* spp. and *Pseudomonas* spp. produced volatile compounds having potential to inhibit growth of Foc race 4 (Ting et al. 2011). The

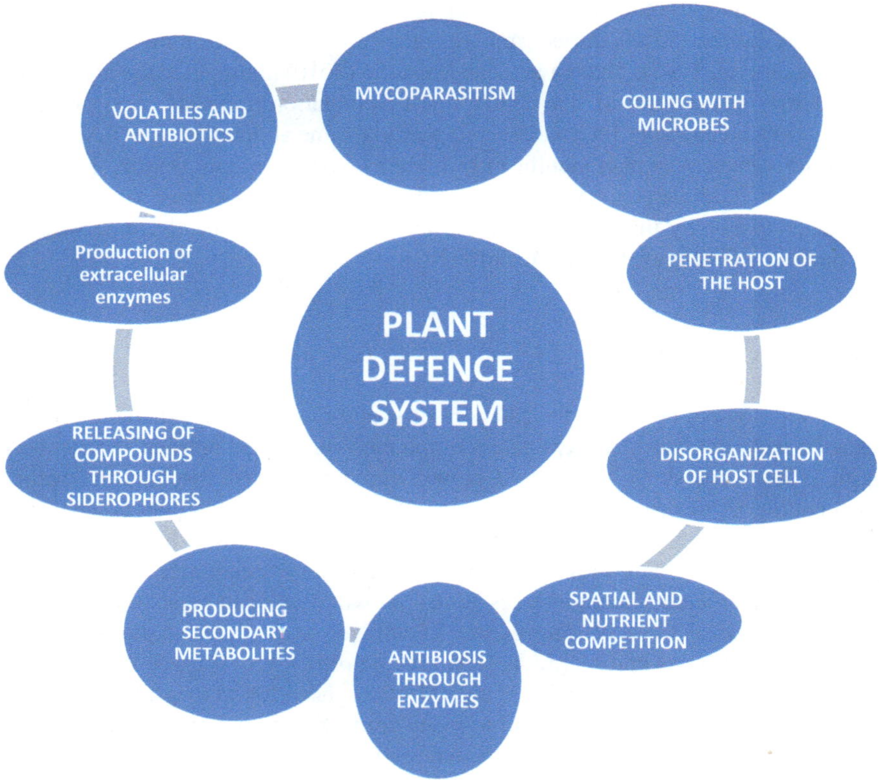

Fig. 7.6 Mechanism of action of bioagents for invoking plant defence

identified compounds were methanethiol, 3-undecene and 2-pentene 3-methyl. *Herbaspirillum* spp. isolate had 20.3% check of growth of Foc race 4 as its volatiles have all the three compounds, while *Pseudomonas* isolate AVA02 showed only 1.4% inhibition because it contained only 3-undecene and methanethiol. All three compounds, such as 2-pentane 3-methyl, in high dose have value for the antifungal activity against Foc. *Pseudomonas aeruginosa* strain FP10 proved to be the most potent in antibiosis and the plant growth-promoting activity (Ayyadurai et al. 2006).

Talc-based preparation of *P. fluorescens* when applied in soil @ 15 g/plant in banana significantly checked wilt disease (Saravanan and Muthusamy 2006). The power in *P. fluorescens* for suppressing *Fusarium* depends on its potential to produce antibiotic 2,4-diacetylphloroglucinol (DAPG). DAPG obtained from *P. fluorescens* when applied to soil significantly inhibited spore germination and growth of Foc.

The biocontrol agents cause, viz. coiling, penetration in the mycelia, mycoparasitism, disorganization in contents of host cell, spatial, competition for nutrients and antibiosis. This all is through production of secondary metabolites, released compounds by siderophores, volatiles, extracellular enzymes and antibiotics and thus induces plant defence system (Fig. 7.6).

7.6 Strategies to Enhance the Biocontrol Potential

There has to be a systematic approach for use of bioagents for the control of banana *Fusarium* wilt. For the best results, the following points are considered: (1) types of bioagents and their various attributes and (2) the problems in initial colonization of antagonists and related variations after initial colonization.

The first step to find effective biocontrol agents is to check their potential. The foremost thing is the known type of antagonists and properties that result in their production, maintain efficacy and transportation. Cost of BCAs needs to be low with yield viable and very effective propagules in high concentration and long-term storage under dry condition (Jackson 1997). *Bacillus* species strains are ideal candidates for viable BCAs (Farhana et al. 2011; Govindasamy et al. 2011; Tan et al. 2013) and advantageous as they survive in adverse environments by producing endospores (Schallmey et al. 2004). A large number of strains of *Bacillus* spp. are widely used as BCAs for soil-borne pathogens (Gurr et al. 2005), including *Rhizoctonia* (Yu et al. 2002) and *Fusarium* (Sun et al. 2011) which have proved to be highly effective after long storage and transportation (Schallmey et al. 2004). Nonpathogenic *F. oxysporum* (Nel et al. 2006) and *Trichoderma* spp. (Thangavelu et al. 2004) have also been demonstrated but preferred less due to their difficulty in storage and transportation.

7.7 Hurdles in Colonization of the Antagonists

The second factor which needs to be taken into consideration is the efficacy or potential of the antagonistic microbes for initially colonizing the rhizosphere and production of substances inhibiting the pathogens. There are natural barriers which hamper colonization of antagonistic microbes indicating problems encountered upon soil application of BCAs. This includes predation and phagocytosis from soil protozoa (Ronn et al. 2002) and suppression of microbes of the niche (Bolwerk et al. 2003) or in plant roots (Chao et al. 1986), fighting with local microbes in different ecosystems available for nutrients. They drastically reduce the population of most antagonistic microbes in the first 2 to 3 days after application of BCAs (Christoffersen et al. 1995). The BCAs must maintain a certain level to result into acceptable levels of pathogen control (Wang et al. 2011). To promote promising efficacy of bioagent measures need to be taken to help BCA tide over such initial adverse phase, the recommendation therefore has to be by repeatedly applying BCAs.

7.8 Conclusions

The causal organism of banana wilt is soil-borne *Fusarium oxysporum* f. sp. *cubense* (Foc) and a strain of the fungus that affects Cavendish and other dessert bananas in the tropics. It is called Foc tropical race 4. This affects the roots of banana plants by colonizing the vascular system of the rhizome and pseudostem producing wilt. For its management a lot of biocontrol agents have been tried but it cannot be controlled completely. The bioagents are under evaluation both under lab and greenhouse conditions. The field experiments have been conducted only in a few cases and however need large-scale confirmation. Most effective biological control methods are (1) the Foc pathogen present in a particular area or country may be tried under both in vitro and in vivo conditions, (2) select biocontrol agents in having multiple actions and functions, (3) test the compatibility of bioagents and their tolerance to the chemicals, (4) suitable easy method for mass production and delivery system need to be standardized for producing maximum number of durable propagules, (5) bioagents need to be incorporated in the right quantity in the right place (in the soil near the rhizosphere) at the appropriate time and physiological state, (6) use of bioagents through other organic amendments and (7) incorporation of biocontrol agents with easily adaptable agronomic practices to effectively manage the disease.

Acknowledgements We are grateful to the authorities of Amity University Haryana for the facilities and encouragement.

Conflict of Interest Statement We declare that we have no conflict of interest.

References

Akila R, Rajendran S, Harish K, Saveetha Raguchander T, Samiyappan R (2011) Combined application of botanical formulations and biocontrol agents for the management of *Fusarium oxysporum* f. sp. cubense (Foc) causing *Fusarium* wilt in banana. Biol Control 57:175–183

Amsellem Z, Zidack NK, Quimby PCJ, Gressel J (1999) Long-term dry preservation of viable mycelia of two mycoherbicidal organisms. Crop Prot 18:643–649

Anjaiah V, Cornelis P, Koedam N (2003) Effect of genotype and root colonization in biological control of *Fusarium* wilts in pigeonpea and chickpea by *Pseudomonas aeruginosa* PNA1. Can J Microbiol 49:85–91

Ayyadurai N, Naik P, Rao R, Sunish R, Samrat SK, Manohar M, Sakthivel N (2006) Isolation and characterization of a novel banana rhizosphere bacterium as fungal antagonist and microbial adjuvant in micropropagation of banana. J Appl Microbiol 100:926–937

Bastasa GN, Baliad AA (2005) Biological control of *Fusarium* wilt of abaca (*Fusarium oxysporum*) with *Trichoderma* and yeast. Philippine J Crop Sci 30:29–37

Bolwerk A, Lagopodi AL, Wijfjes AH, Lamers GE, Chin-A-Woeng TF, Lugtenberg BJ, Bloemberg GV (2003) Interactions in the tomato rhizosphere of two Pseudomonas biocontrol strains with the phyto pathogenic fungus *Fusarium oxysporum* f. sp. radicis-lycopersici. Mol Plant-Microbe Interact 16:983–893

Booth C (1971) The genus Fusarium. Commonwealth Mycological Institute, Kew, Surrey, p 237

Buddenhagen IW (2009) Understanding strain diversity in *Fusarium oxysporum* f. sp. cubense and history of introduction of 'tropical race 4' to better manage banana production. ISHS Acta Hortic 828:193–204

Chao WL, Nelson EB, Harman GE, Hoch HC (1986) Colonization of the rhizosphere by biological control agents applied to seeds. Phytopathology 76:60–65

Chin-A-Woeng TFC, Bloemberg GV, Vander Bij AJ, Vander Drift KMGM, Schripsema J, Kroon B, Scheffer RJ, Keel C (1998) Biocontrol by phenazine-1-carboxamide-producing *Pseudomonas chlororaphis* PCL1391 of tomato root rot caused by *Fusarium oxysporum* f. sp. radicis lycopersici. Mol Plant-Microbe Interact 11:1069–1077

Christoffersen K, Ahl T, Nybroe O (1995) Grazing of nonindigenous bacteria by nano-sized protozoa in a natural coastal system. Microb Ecol 30:67–78

de Freitas JR, Germida JJ (1991) *Pseudomonas cepacia* and *Pseudomonas putida* as winter wheat inoculants for biocontrol of *Rhizoctonia solani*. Can J Microbiol 37:780–784

Dita MA, Waalwijk C, Buddenhagen IW, Souza MT, Kema GHJ (2010) A molecular diagnostic for tropical race 4 of the banana Fusarium wilt pathogen. Plant Pathol 59:348–357

Dong X, Ling N, Wang N, Shen Q, Guo S (2012) Fusaric acid is a crucial factor in the disturbance of leaf water imbalance in *Fusarium*-infected banana plants. Plant Physiol Biochem 60:171–179

Dong X, Xiong Y, Ling N, Shen Q, Guo S (2014) Fusaric acid accelerates the senescence of leaf in banana when infected by *Fusarium*. World J Microbiol Biotechnol 30:1399–1408

Farhana MSN, Bivi MR, Khairulmazmi A (2011) Effect of carbon sources on bacterial production of metabolites against *Fusarium oxysporum* and *Colletotrichum gloeosporioides*. Int J Agric Biol 13:1–8

Fernández-Falcón M, Borges AA, Borges-Pérez A (2003) Induced resistance to *Fusarium* wilt of banana by exogenous applications of indoleacetic acid. Phytoprotection 84:149–153

Fishal EMM, Meon S, Yun WM (2010) Induction of tolerance to Fusarium wilt and defense-related mechanisms in the plantlets of susceptible Berangan banana pre-inoculated with *Pseudomonas* sp. (UPMP3) and *Burkholderia* sp. (UPMB3). Agric Sci China 9:1140–1149

Fourie G, Steenkamp ET, Gordon TR, Viljoen A (2009) Evolutionary relationships among the *Fusarium oxysporum* f. sp. cubense vegetative compatibility groups. Appl Environ Microbiol 75:4770–4781

Getha K, Vikineswary S (2002) Antagonistic effects of *Streptomyces violaceusniger* strain G10 on *Fusarium oxysporum* f. sp. cubense race 4: indirect evidence for the role of antibiosis in the antagonistic process. J Ind Microbiol Biotechnol 28:303–310

Getha K, Vikineswary S, Wong WH, Seki T, Ward A, Goodfellow M (2005) Evaluation of *Streptomyces* sp. strain G10 for suppression of *Fusarium* wilt and rhizosphere colonization in pot grown banana plantlets. J Ind Microbiol Biotechnol 32:24–32

Govindasamy V, Senthilkumar M, Magheshwaran V, Kumar U, Bose P, Sharma V, Annapurna K (2011) *Bacillus* and *Paenibacillus* spp.: potential PGPR for sustainable agriculture. Microbiol Monogr 18:333–364

Gowen S (1995) Bananas and plantains. Chapman and Hall, London, p 612

Gurr GM, Wratten SD, Kehrli P, Scarratt S (2005) Cultural manipulations to enhance biological control in Australia and New Zealand: progress and prospects. In: Mark SH (ed) Proceedings of the second international symposium on biological control of arthropods, Davos, Switzerland, pp 154–166

Hennessy C, Walduck G, Daly A, Padovan A (2005) Weed hosts of *Fusarium oxysporum* f. sp. cubense tropical race 4 in northern Australia. Australas Plant Pathol 34:115–117

Huang YH, Wang RC, Li CH, Zuo CW, Wei YR, Zhang L, Yi GJ (2012) Control of *Fusarium* wilt in banana with Chinese leek. Eur J Plant Pathol 134:87–95

Hung TN, Hung NQ (2018) First report of Fusarium wilt on Cavendish bananas, caused by *Fusarium oxysporum* f. sp. cubense tropical race 4 (VCG 01213/16), in Vietnam. Plant Dis 102:448

Jackson MA (1997) Optimizing nutritional conditions for the liquid culture production of effective fungal biological control agents. J Ind Microbiol Biotechnol 19:180–187

Jie L, Zifeng W, Lixiang C, Hongming T, Patrik I, Zide J, Shining Z (2009) Artificial inoculation of banana tissue culture plantlets with indigenous endophytes originally derived from native banana plants. Biol Control 51:427–434

Kumar N (2016) Control of *Fusarium oxysporum* causing wilt in Banana through plant extracts. Ann Plant Protect Sci 24:104–108

Kumar N, Saxena S (2015) *Fusarium* wilt of Banana at Panchgaon. Int J Curr Microbiol App Sci 4:724–730

Lakshmanan P, Selvaraj P, Mohan S (1987) Efficiency of different methods for the control of Panama disease. Trop Pest Manage 33:373–376

Li CY, Mostert G, Zuo CW, Beukes I, Yang QS (2013) Diversity and distribution of the banana wilt pathogen *Fusarium oxysporum* f. sp. cubense in China. Fungal Genomics Biol 3:111. https://doi.org/10.4172/2165-8056.1000111

Lugtenberg BJJ, Kravchenko LV, Simons M (1999) Tomato seed and root exudates sugars: composition, utilization by *Pseudomonas* bio-control strains and role in rhizosphere colonization. Environ Microbiol 1:439–446

Minerdi D, Moretti M, Gilardi G, Barberio C, Gullino ML, Garibaldi A (2008) Bacterial ectosymbionts and virulence silencing in a *Fusarium oxysporum* strain. Environ Microbiol 10:1725–1741

Moore NY, Bentley S, Pegg KG, Jones DR (1995) *Fusarium* wilt of banana. Musa disease fact sheet. INBAP, Montpellier, France, 5:4

Moore NY, Pegg KG, Bentley S, Smith LJ (1999) *Fusarium* wilt of banana: global problems and perspectives. In: Molina AB, Masdek NHN, Liew KW (eds) Banana Fusarium wilt management: towards sustainable cultivation. Proceedings of the international workshop on banana Fusarium wilt disease. INIBAP, Kuala Lumpur, pp 11–30

Mostert D, Molina AB, Daniells J, Fourie G, Hermanto C, Chao CP (2017) The distribution and host range of the banana *Fusarium* wilt fungus, *Fusarium oxysporum* f. sp. cubense, in Asia. PLoS One 12(7):e0181630. https://doi.org/10.1371/journal.pone.0181630

Nel B (2004) Management of *Fusarium* wilt of banana by means of biological and chemical control and induced resistance [MSc thesis]. University of Pretoria, Pretoria

Nel B, Steinberg C, Labuschagne N, Viljoen A (2006) The potential of nonpathogenic *Fusarium oxysporum* and other biological control organisms for suppressing *Fusarium* wilt of banana. Plant Pathol 55:217–223

Nel B, Steinberg C, Labuschagne N, Viljoen A (2007) Evaluation of fungicides and sterilants for potential application in the management of *Fusarium* wilt of banana. Crop Prot 26:697–705

Papavizas GC (1985) *Trichoderma* and *Gliocladium*: biology, ecology and potential for biocontrol. Annu Rev Phytopathol 23:23–54

Persley GJ (1987) Banana and plantain breeding strategies. In: Persley GJ, De Langhe EA, (eds) Proceedings of an international workshop held at cairns, Australia, 13–17 Oct 1986

Pieterse CMJ, van Pelt JA, van Wees SCM, Ton J, Leon-Kloosterziel KM, Keurentjes JJB, Verhagen BWM, van Knoester M, Bakkerand LC, Loon V (2001) Rhizobacteria-mediated induced systemic resistance: triggering, signalling and expression. Eur J Plant Pathol 107:51–61

Ploetz RC (1990) Population biology of *Fusarium oxysporum* f. sp. cubense. In: Ploetz RC (ed) Fusarium wilt of banana. APS Press, St Paul, pp 63–67

Ploetz RC (1994) Panama disease: return of the first banana menace. Int J Pest Manage 40 (4):326–336

Ploetz RC (2006) *Fusarium* Wilt of banana is caused by several pathogens referred to as *Fusarium oxysporum* f. sp. Cubense. Phytopathology 96:653–656

Ploetz RC (2015) *Fusarium* wilt of banana. Phytopathology 105:1512–1521

Ploetz RC, Pegg KG (2000) Fungal disease of the root, corm, and pseudostem: *Fusarium* wilt. In: Jones DR (ed) Diseases of banana, abaca and enset. CAB International, London, pp 143–157

Pushpavathi Y, Dash SN, Mishra MK, Triveni V (2015) Management of *Fusarium* wilt of banana under coastal Odisha conditions. Int J Farm Sci 5:241–247

Pushpavathi Y, Dash SN, Madhavi N, Deepika D (2016) Biological control of *Fusarium* wilt disease in banana with emphasis on *Trichoderma* spp. and *Pseudomonas* spp. Plant Archives 16:51–59

Raguchander T, Jayashree K, Samiyappan R (1997) Management of *Fusarium* wilt of banana using antagonistic microorganisms. J Biol Control 11:101–105

Rajappan K, Vidhyasekaran P, Sethuraman K, Baskaran TL (2002) Development of powder and capsule formulations of *Pseudomonas fluorescens* strain Pf-1 for the control of banana wilt. Zeitschrift für Pflanzenkrankheiten und Pflanzenschutz 109:80–87

Ramakrishnan TS, Damodaran S (1956) Observation on the wilt disease of banana. Proc Indian Acad Sci 43:213–222

Rawal RD (2000) Fungal diseases of banana. Current scenario in India. In: Singh HP, Chadha KL (eds) Banana improvement, production and utilization. National Research Centre for Banana (NRCB), Trichy, p 23

Rhodes DJ, Powell KA (1994) Biological seed treatments – the development process. BCPC Monogr 57:303–310

Ronn R, McCaig AE, Griffiths BS, Prosser JI (2002) Impact of protozoan grazing on bacterial community structure in soil microcosms. Appl Environ Microbiol 68:6094–6105

Sakthivel N, Gnanamanickam SS (1987) Evaluation of *Pseudomonas fluorescens* for suppression of sheath rot disease and for the enhancement of grain yields in rice (*Oryza sativa* L.). Appl Environ Microbiol 53:2056–2059

Saravanan T, Muthusamy M (2006) Influence of *Fusarium oxysporum* f. sp. cubense (e.f. smith) Snyder and Hansen on 2,4-diacetylphloroglucinol production by *Pseudomonas fluorescens* migula in banana rhizosphere. J Plant Protect Res 46:241–254

Saravanan T, Muthusamy M, Marimuthu T (2003) Development of integrated approach to manage the Fusarial wilt of banana. Crop Prot 22:1117–1123

Schallmey M, Singh A, Ward OP (2004) Developments in the use of *Bacillus* species for industrial production. Can J Microbiol 50:1–17

Sequiera L (1992) Influence of organic amendments on survival of *Fusarium oxysporum* f. sp. cubense in the soil. Phytopathology 52:976–982

Sivamani E, Gnanamanickam SS (1988) Biological control of *Fusarium oxysporum* f.sp. cubense in banana by inoculation with *Pseudomonas fluorescens*. Plant Soil 107:3–9

Smith MK, Whiley AW, Searle C, Langdon PW, Schaffer B, Pegg KG (1998) Micropropagated bananas are more susceptible to *Fusarium* wilt than plants grown from conventional material. Aust J Agric Res 49:1133–1139

Srinivasan U, Staines HJ, Bruce A (1992) Influence of media type on antagonistic modes of *Trichoderma* spp. against wood decay basidiomycetes. Mater Org 27:301–321

Stover RH (1962) Fusarial wilt (panama disease) of bananas and other musa species. Commonwealth Mycological Institute, Kew

Stover RH, Simmonds NW (1987) Bananas, 3rd edn. Longmans, London, pp 29–30

Su HJ, Hwang SC, Ko WH (1986) Fusarium wilt of Cavendish bananas in Taiwan. Plant Dis 70:814–818

Sukhada M, Manamohan M, Rawal RD, Chakraborty S, Sreekantappa H, Manjula R, Lakshmikantha HC (2004) Interaction of *Fusarium oxysporum* f.sp. cubense with *Pseudomonas fluorescens* precolonized to banana roots. World J Microbiol Biotechnol 20:651–655

Sun B, Peng M, Wang YG, Zhao J, Xia QY (2011) Isolation and characterization of antagonistic bacteria against fusarium wilt and induction of defense related enzymes in banana. Afr J Microbiol Res 5:509–515

Tan S, Dong Y, Liao H, Huang J, Song S, Xu Y, Shen Q (2013) Antagonistic bacterium *Bacillus amyloliquefaciens* induces resistance and controls the bacterial wilt of tomato. Pest Manag Sci 69(11):1245–1252. https://doi.org/10.1002/ps.3491

Thakker JN, Patel S, Dhandhukia PC (2013) Induction of defense-related enzymes in banana plants: effect of live and dead pathogenic strain of *Fusarium oxysporum* f. sp. cubense. ISRN Biotechnol 2013:601303

Thangavelu R (2002) Characterization of *Fusarium oxysporum* schlecht. f. sp. cubense (E.F. Smith) Snyd. & Hans. and molecular approaches for the management of Fusarium wilt of banana. Ph.D. thesis. Tamil Nadu Agricultural University, Coimbatore, Tamil Nadu, India, 245 p

Thangavelu R, Mustaffa MM (2010) A potential isolate of *Trichoderma viride* NRCB1 and its mass production for the effective management of *Fusarium* wilt disease in banana. Tree Forestry Sci Biotechnol 4:76–84

Thangavelu R, Palaniswami A, Ramakrishnan G, Sabitha D, Muthukrishnan S, Velazhahan R (2001) Involvement of Fusaric acid detoxification by *Pseudomonas fluorescens* strain Pf10 in the biological control of *Fusarium* wilt of banana caused by *Fusarium oxysporum* f.sp. cubense. J Plant Dis Protect 108:433–445

Thangavelu R, Palaniswami A, Velazhahan R (2004) Mass production of *Trichoderma harzianum* for managing *Fusarium* wilt of banana. Agric Ecosyst Environ 103:259–263

Ting ASY, Meon S, Kadir J, Son S, Radu and Singh G. (2008) Endophytic microorganisms as potential growth promoters of banana. BioControl 53:541–553

Ting ASY, Mah SW, Tee CS (2011) Detection of potential volatile inhibitory compounds produced by endobacteria with biocontrol properties towards *Fusarium oxysporum* f. sp. cubense race 4. World J Microbiol Biotechnol 22:229–235

Viljoen A (2002) The status of *Fusarium* wilt (Panama disease) of banana in South Africa. S Afr J Sci 98:341–344

Wang T, Duan YJ, Liu BT, Guo G, Zhou DB, Tan X, Zhang XY, Huang MJ (2011) The colonization of two strains of antagonistic bacteria of *Fusarium oxysporum* in Banana. Gen Appl Biol 30:342–350 (in Chinese)

Wardlaw CW (1961) Banana diseases, including plantains and abaca, vol 648. Longmans, Green and Co. Ltd, London

Weindling R (1941) Experimental consideration of the mold toxin of Gliocladium and Trichoderma. Phytopathology 31:991–1003

Weller DM, Raaijmakers JM, McSpadden Gardener BB, Thomashow LS (2002) Microbial populations responsible for specific soil suppressiveness to plant pathogens. Annu Rev Phytopathol 40:309–348

Yedidia I, Benhamou N, Kapulnik Y, Chet I (2000) Induction and accumulation of PR proteins activity during early stages of root colonization by the mycoparasite *Trichoderma harzianum* strain T-203. Plant Physiol Biochem 38:863–873

Yin XM, Jin ZQ, Xu BY, Ma WH, Fu YG, Wang JB (2011) Characterization of early events in banana root infected with the GFP-tagged *Fusarium oxysporum* f. sp. cubense. Acta Hortic 897:371–376

Yu GY, Sinclair JB, Hartman GL et al (2002) Production of iturinA by *Bacillus amyloliquefaciens* suppressing *Rhizoctonia solani*. Soil Biol Biochem 34:955–963

Chapter 8
Potential of Plant-Microbe Interactions in Management of Pesticide-Riddled Soil

Narendra Kumar, Sarika Chaturvedi, and S. M. Paul Khurana

Abstract Pesticides have been widely used after the Second World War in management of weeds, diseases and pests of plants. Most of these have persistent nature and cause serious environmental concerns. They can be managed only through the biological agents for remediation of agricultural soils. The crop fields are normally over polluted through pesticides. Biodegradation of pesticides has been an ecofriendly, cost-effective, highly efficient approach in comparison to the physical and chemical means. The chemical means are not only expensive but also not ecofriendly. Biodegradation is sensitive to temperature and pH conversions. The researches hit that bioremediation has more potential than physicochemical approaches. Rhizosphere bacteria and fungi degrade organic pollutants known as bioremediation/rhizodegradation. If selected vegetation is used, there may be enhancement of pollutant decomposers in terms of numbers and action potential in the rhizosphere, which can result in speedy rhizodegradation of toxic pesticides. This is directly related with human health. It needs a very careful understanding of the mechanisms of pollutant degradation in the rhizosphere environment. Recent investigations revealed that plant-related microorganisms in the rhizosphere produce pesticide-decomposing enzymes. This mineralizes toxic pesticides. This rhizoremediation may be a promising technology in removal of pesticides in polluted soil. The chapter deals mainly with microbial interaction of rhizosphere.

8.1 Introduction

Nowadays, pesticides are widely used in preventing and controlling the crop diseases and pests, but at the same time, pesticide residues have brought serious harm to human health and the environment. It is an important subject to study microbial degradation of pesticides in soil environment for environmental restoration (Huang et al. 2018). Environmental pollution is now a global concern. The modern innovations have resulted in wastes having very complex inorganic and organic chemicals.

N. Kumar · S. Chaturvedi · S. M. Paul Khurana (✉)
Amity Institute of Biotechnology, Amity University Haryana, Manesar, Gurgaon, India

A. Varma et al. (eds.), *Plant Microbe Interface*,
https://doi.org/10.1007/978-3-030-19831-2_8

Pesticide wastes generated by the pesticide industry also have enhanced hazardous environmental contaminants. More than one kind of pesticide may be used in the management of different types of pests, which may include insecticides for insect control, fungicides against fungi, rodenticide for rodents and defoliant for leaf weeds. These are classified on the basis of their chemical constitution, for example, organochlorides, organophosphates, pyrethroids, etc. These are artificially synthesized which are non-biodegradable and can enhance environmental pollution. It has been reported that about three million people suffer per annum due to pesticides (WHO) (Veiga et al. 2006). The decomposition of persistent pesticides has been very essential in the decontamination of soil and waterbodies (Hodaifa et al. 2009).

Rate of pesticide degradation is influenced by several factors such as structure of pollutants, pH of soil, concentration of hydrogen peroxide and even concentration of iron. The photocatalytic degradation, biodegradation, ozonation and photo-Fenton reactions are normally applied in pesticide removal studies (Wyss et al. 2006). The effects of soil moisture, temperature, aeration, pH and organic matter content in pesticide biodegradation have been well investigated (Johnson and Ware 1992; Rasmussen and Olsen 2004; Bending et al. 2006; Charnay et al. 2005). Microbes are very useful in bioremediation of pesticides. Through applications of strains of microbes (*Aspergillus*), it is possible to remove endosulfan pesticides from the environment (Bhalerao and Puranik 2007). Often it has been seen that there is a natural balance among microbial evolution in respect to bioremediation (Hodgson et al. 1993). The organic matter gets decomposed through microbial action. But the degradation is slow in case of synthetic pesticides (due to structural variations and less compatibility with metabolic pathways of applied microbes) (Surekha et al. 2008).

The degradation through bio-organisms has been widely used in treatment of xenobiotics in soil. It is very economical and also ecofriendly (Enrica 1994; Vidali 2001; Bhupathiraju et al. 2002). Approaches like land filling, recycling and incineration have their own limitations, and many toxic intermediates may be formed through these processes (Sayler et al. 1990). This may project an upper hand of bioremediation over physicochemical methods (Muhammad et al. 2016).

Rhizoremediation is helpful in facing the problem of remediation for management of hazardous chemicals. The search for microbial consortium has been found useful in solving pesticides of soil and water environment (Fulekar 2005).

In rhizoremediation there is a use of plant roots and related microbes in management of environmental pollutants/toxins in soil. Rhizoremediation involves the benefits of plant roots and related natural microbes for enhanced decomposition of toxic pollutants of rhizosphere. It has been recorded that 1 g of soil bears 100 million bacteria of 5000–7000 different species. It bears more than 10,000 colonies of fungi (Melling 1993).

Rhizomicrobia may speed up remediation processes through humification (Salt et al. 1998). It catalyses polymerization with release of oxidoreductase enzymes, e.g. peroxidase. Bacterial populations in nature can degrade pollutants in an effective way with respect to single species/strain (Kuiper et al. 2004). The development of colony in different niches of plant roots through different strains has been studied (Kuiper et al. 2004; Dekkers et al. 2000). But a very few studies report the directed introduction of a microbial strain on consortium in xenobiotic degradation (bioaugmented rhizoremediation) which may form colony in roots (Kuiper et al. 2004; Korade and Fulekar 2009).

In the present chapter, different mechanisms involved in biological degradation of pesticides have been projected in light of various factors which affect the process of bioremediation. Detoxification of pesticides through indigenous soil microbes or enzymes has also been reviewed. The objective of the present chapter is to summarize information of the mode of rhizoremediation of pesticides in the rhizosphere region. This also records various aspects of plant-associated microbes and their relevant remediation effects.

8.2 Pesticides and Their Effect

A pesticide is used to destroy, prevent, repel and mitigate any pest, viz. insects, mites, nematodes, weeds, rats, etc. Pesticides, viz. herbicides, insecticides, fungicides and various other substances, are useful in controlling pest populations (EPA 2015). Nowadays millions of tons of pesticides are applied annually all over the globe worth billions of dollars. The expenditure on pesticides is increasing day by day. Herbicides are useful in management of pest populations (EPA 2015). One of the primary concerns is to minimize harmful effects caused by the target organisms including viruses, bacteria, fungi and insects (Liu et al. 2001).

The total registered pesticides in India are 234 (Faridi 2014). The states of India are consuming pesticides in the years 2005–2006 to 2009–2010 which has been recorded by the Directorate of Plant Protection, Quarantine and Storage, Government of India, shown in Table 8.1.

The periodic use of pesticides makes the situation particularly perturbing. This leads to an accumulation of pesticides with residues in the environment with multifaceted toxicity (Bouziani 2007). This shows a direct relationship in contamination of pesticides with residual detection (Calderbank 1989). There are 24 pesticides registered and used in India (Fig. 8.1) which comes in the US EPA list as potential carcinogens, viz. atrazine (C), acephate (C), alachlor (B2), bifenthrin (C), benomyl (C), chlorothalonil (B2), captan (B2), dichlorvos (C), cypermethrin (C),

Table 8.1 Consumption of pesticides in India (Faridi 2014)

Sl. no.	State	Total pesticides consumed (mt)	Sl. no.	State	Total pesticides consumed (mt)
1	Uttar Pradesh	39,948	5	Rajasthan	15,239
2	Punjab	29,235	6	Gujarat	13,430
3	Haryana	21,908	7	Tamil Nadu	12,851
4	Maharashtra	16,480		All India	210,600

Fig. 8.1 Total carcinogenic pesticides used in India are 24

dicofol (C), diclofop-methyl (C), oxadiazon (C), mancozeb (B2), methomyl (C), oxyfluorfen (C), metolachlor (C), propoxur (B2), permethrin (C), phosphamidon (C), propiconazole (C), trifluralin (C), thiodicarb (C), thiophanate-methyl (C) and triadimefon (C) (Cox 1992).

They have high risk of contamination in ecosystem and can affect nontarget vegetation resulting in contamination of air, soil and nontarget plants. It may result in hormonal disruption, diminished intelligence and reproductive abnormalities (Gupta 2004). They also result in contamination of different levels in the environment (Nawab et al. 2003).

8.3 Biodegradation and Bioremediation of Pesticide

The degradation of xenobiotic chemicals/pesticide through a microorganism takes place naturally which is essential for their survival. These microbes work in their natural environment, but some changes are still needed to degrade a pesticide in a faster rate. This potential of microbes may be used in the form of technology in removing the contaminants from an actual place. So details of biochemistry, genetics and physiology of the useful microbe can help in enhancing the microbial process of achieving faster bioremediation. The genes which encode enzymes have been studied and identified for many pesticides which provide a new input forgetting capability in microbes for degrading pesticides. This may develop a better strain for achieving the desired speed of bioremediation (Singh 2008).

8.4 Rhizoremediation

Rhizoremediation is an application of rhizomicrobial populations of plant rhizosphere for bioremediation (Kuiper et al. 2004). It consists of rhizodegradation and microbial stimulation. This describes the use of both microbes and plants for useful interaction.

8.4.1 Factors Affecting Pesticide Bioremediation

The physicochemical parameters of soil, viz. organic matter, nutrients, temperature, pH, moisture, redox conditions, amount and nature of clay, have a direct effect on the success of bioremediation. Soil water is responsible for moisture availability to microorganisms. Due to this redox reactions occur in soil that lead to various biochemical processes. Schroll et al. (2006) investigated the potential of soil moisture in aerobic microbial mineralization of some pesticides (glyphosate and benzolin-ethyl) on various soils. They found a linear correlation $p < 0.0001$ during increasing soil moisture in a range of −20 and −0.015 MPa and reported enhanced relative pesticide degradation.

8.4.2 Temperature

Temperature and pH are the main factors which affect biodegradation of pesticides in soil. The biochemical reactions depend on temperature for microbial activities having a direct impact on cell physiology changing proteins and permeability of cell membrane (Alberty 2006). The bacteria normally degrade chlorpyrifos (CP) and fenamiphos at temperatures 15–35 °C, but its degradation power was reduced sharply at 5 or 50 °C (Singh et al. 2006). Siddique et al. (2002) found similar results during the study of biodegradation of HCH isomers of soil slurry. For α- and γ-HCH isomers, an incubation temperature of 30 °C was optimum for degradation.

8.4.3 pH

Microorganisms secrete specific enzymes which are useful in biodegradation of a compound. The activity of enzymes depends on pH. For bacteria the optimum pH is 6.5 to 7.5 which is their intracellular pH. *Pandoraea* sp. obtained through enrichment culture method (Okeke et al. 2002) has the capacity to degrade HCH isomers at pH 4 to 9 (Siddique et al. 2002). Singh et al. (2006) found the same observations of biodegradation of organophosphate pesticides of soil. At lower pH degradation speed was slower in soils in comparison with alkaline and neutral soils. The soil pH directly affects the biochemical processes. It influences adsorption/desorption of soil pesticides matrix and finally biodegradation. The adsorption of weakly acidic pesticides occurs at lower soil pH.

Boivin et al. (2005) studied the adsorption and desorption processes in 5 pesticides in 13 contrasting field soils of very weak base to weakly acidic chemicals. They reported a significant relation between soil pH and bentazone adsorption.

There was no significant adsorption in weakly acidic pesticide 2,4-D; this was because of greater repulsion between electronegative charges of soil and full ionization at lower pH.

8.5 Soil Organic Matter

Degradation of herbicides in soils amended with rice straw, compost and NPK chemical fertilizers was studied under upland, oxidative-flooded (aerobic-flooded) and reductive-flooded (anaerobic-flooded) conditions through benthiocarb and MCPA. Rice straw, compost and NPK amendments accelerated degradation in herbicides under upland and oxidative-flooded conditions. But in reductive-flooded conditions, herbicide degradation was remarkably slow. Benthiocarb on degradation resulted in the formation of 4-chlorobenzoic acid, desethyl benthiocarb, benthiocarb sulfoxide and 4-chlorobenzyl methyl sulfone. Rice straw amendments enhanced the amount of benthiocarb sulfoxide. The amount of desethyl benthiocarb was reduced through rice straw and compost in upland conditions. But 4-chloro-2-methylphenol which was a major degradation product of MCPA resulted in large amounts upon rice straw amendments in oxidative-flooded and NPK amendment under upland conditions (Stephen and Shozo 1980).

This is the crop residue which is a source of organic matter and provides nutrients. Pesticides in soil may behave differently. Boivin et al. (2005) studied correlation of pesticides, viz. isoproturon, trifluralin and atrazine, in relation to organic matter content of soil. Singh et al. (2006) studied fenamiphos and chlorpyrifos for its biodegradation but could not observe any potential of soil organic matter in biodegradation of pesticide. In a study Fenlon et al. (2007) found that diazinon appreciably mineralized in two types of the organic soils when assessed organically. This

Table 8.2 Major consumed pesticides in India in the years 2005–2006 to 2009–2010 (Faridi 2014)

Sl. no.	Pesticide used	Consumed (in metric tonnes)	Sl. no.	Pesticide used	Consumed (in metric tonnes)
1	Sulphur (fungicide)	16,424	9	Chlorpyrifos (insecticide)	07163
2	Endosulfan (insecticide)	15,537	10	Malathion (insecticide)	07103
3	Mancozeb (fungicide)	11,067	11	Carbendazim (fungicide)	06767
4	Phorate (insecticide)	10,763	12	Butachlor (herbicide)	06750
5	Methyl parathion (insecticide)	08408	13	Quinalphos (insecticide)	06329
6	Monocrotophos (insecticide)	08209	14	Copper oxychloride	06055
7	Cypermethrin (insecticide)	07309	15	Dichlorvos (insecticide)	05833
8	Isoproturon (herbicide)	07163			

conventionally managed soils in comparison to cypermethrin (CMP) which degraded in all observed soils (Table 8.2).

8.6 Pesticide Degradation Mechanism of Rhizosphere

Plants release chemicals which may speed up xenobiotic decomposition. So it may be useful in managing contamination of soils. The rhizosphere covers soil near to plant roots and action mechanism of living roots. It creates a complex environment which may support metabolically active microorganisms. Rhizosphere is the zone of soil around the root. The population of microbes are affected through root system. This forms a dynamic root-soil interaction (Kuiper et al. 2004; Barea et al. 2005; Shrivastava et al. 2014). Involvement of plant species in facilitating microbial degradation of pesticides in the rhizosphere is recorded in Table 8.3.

The rhizosphere can be separated in three different components:

1. Rhizosphere (soil): this has soil which is effected by activities of roots and releases various substrates which affects microbial activity.
2. Rhizoplane: the surface of the root which strongly adheres soil particles.
3. Root tissue: deals with endophytic microorganisms which have the potential to form a colony (Barea et al. 2005).

The difference between physical, chemical and biological parameters of soil associated with the root in comparison to bulk soil is responsible for alterations of microbial diversity. This increases various metabolic activities of microorganisms in

Table 8.3 Involvement of plant species in facilitating rhizospheric microbial decomposition

Pesticide involved	Rhizosphere of plant	Report	Investigators
2,4-D	Sugarcane	High density of 2,4-D degrading microbes in the rhizosphere of sugarcane	Sandman and Loos (1984)
Benthiocarb	Rice	Eightfold enhancement of heterotrophic bacteria in the rhizosphere	Sato (1989)
Atrazine	Corn	Increases the formation of atrazine degradation	Seibert et al. (1981)
Atrazine, metolachlor and trifluralin	Kochia	Increases mineralization in comparison to non-rhizosphere soils	Anderson et al. (1994)
Mefenoxam	*Zinnia angustifolia*	*Pseudomonas fluorescens* and *Chryseobacterium indologenes*	Pai et al. (2001)
Chlorpyrifos	Rye grass	Enhanced degradation in rhizosphere soils	Korade and Fulekar (2009)
Chlorpyrifos Fenvalerate Cypermethrin	*Pennisetum pedicellatum*	Selective upgradation of degraders in rhizosphere soil	Dubey and Fulekar (2011a, b)

the rhizosphere area. This alteration of microenvironment is known as the rhizosphere effect (Barea et al. 2005; Kuiper et al. 2004).

Densities of rhizospheric bacteria can be two to four orders of magnitude greater than populations in the surrounding bulk soils. This displays a greater range of metabolic capabilities which includes their ability to degrade a number of xenobiotics (Pilon-Smits 2005). An accelerated speed of decomposition of organic pollutants has been observed in vegetated soils in comparison to nonvegetated soils. This has been well studied for many compounds. But the exact action on how some plants speed up biodegradation is not well known. The differences in tolerance may be due to toxic compounds of soils which may have some relation with plants. For chlorophenols and chlorinated solvent phytoremediation, *Morus alba* L. and *Populus deltoides* trees have been tried successfully (Stomp et al. 1994).

Phytoremediation is an innovative as well as a cost-effective technology. It is based on the use of plants cleaning a wide range of inorganic and organic wastes (Licht and Isebrands 2005). Plants can accumulate xenobiotic chemicals at their aboveground portions. This may be harvested for removal. Plants may be useful in remediation, viz. aerating soil, reducing the leaching of contaminants, phytodegradation/transformation, evapotranspiration, phytovolatilization and rhizoremediation (Amos and Younger 2003; Chang et al. 2005). A study revealed that phytoremediation alone is not a viable process in removing organic pollutants of hydrophobic nature without the use of microbes (Chaudhry et al. 2005).

8.7 Investigations on Rhizospheric Pesticide Bioremediation

Pai et al. (2001) investigated the future of mefenoxam fungicide in rhizosphere system. They tried *Zinnia angustifolia* where microbes in the rhizosphere were inoculated to establish for 3 weeks before addition of fungicides 20 μg per g mix. Degradation product was analysed through high-performance liquid chromatography (HPLC) or capillary electrophoresis. Seventy-eight percent of the fungicide was decomposed after 21 days. The degradation product was *N*-(2,6-dimethylphenyl)-*N*-(methoxyacetyl)-DL-alanine. There was an increase in bacterial populations in the rhizosphere in a 30-day period in comparison to the parent compound. Pure cultures of *Pseudomonas fluorescens* and *Chryseobacterium indologenes* were isolated from the rhizosphere. This decomposed fungicide 10 μg/ml fully to the free acid in time of 54 h.

Drakeford et al. (2003) reported decomposition of isoxaben{*N*-[3-(1-ethyl-1-methylpropyl)-5-isoxazolyl]-2,6-dimethoxybenzamide} in sterile, bulk and rhizosphere soil having a potting mix of 80% bark and 20% sand. The rhizosphere used was *Panicum virgatum* (switch grass). These were kept for establishing up to 14 days when isoxaben 10 μg/g of potting mix was added. This got degraded to 0.5 μg/g within 60 days. The decomposition products were 3-nitrophthalic acid in the rhizosphere and 4-methoxyphenol in the sterile regime.

Yu et al. (2003) studied degradative features of butachlor in wheat rhizosphere, non-rhizosphere and inoculated rhizosphere soils. They reported that the decomposition of butachlor may be enhanced in wheat rhizosphere and mainly in the bacterial community inoculated rhizosphere which has the capability to degrade butachlor. It was recorded that rhizosphere soil inoculated through microbes degrades target herbicides in a useful pathway for achieving more rapid decomposition of herbicides of soil.

Sun et al. (2004) observed decomposition of an oxime carbamate insecticide—aldicarb—in non-sterile and even sterile plant-grown soils. They reported that aldicarb disappeared in soil very quickly in plants' presence. They reported that t1/2 of the pesticide were 1.6, 1.4 and 1.7 days in the soil on mix cultivation with mung bean, corn and cowpea, respectively. The comparison of plant-promoted degradation and plant uptake recorded an increased aldicarb removal in plant-grown soil. Shaw and Richards (2004) studied mineralization of [U-^{14}C]2,4-dichlorophenoxyacetic acid (2,4-D) in rhizosphere soil. He found a rhizosphere effect in [^{14}C]2,4-D mineralization (50 μg g^{-1}) when compared to nonplanted soil. That showed $P < 0.05$ reductions. There were enhancements in mineralization rate in 25- and 60-day using *Trifolium pratense* in soil. Liao and Xie (2008) studied characteristics of simazine degradation in inoculated non-rhizosphere soil, uninoculated rhizosphere soil and inoculated rhizosphere soil. They observed higher degradation rates of SIM in rhizosphere soils, especially in inoculated rhizosphere soil. The decomposition characteristics of simazine showed to be closely related to microbial process. Korade and Fulekar (2009) reported the use of ryegrass rhizosphere in bioremediation of chlorpyrifos. This was conducted through pot culture in greenhouse. The pot-cultured soil amended with 10 mg/kg of chlorpyrifos was observed and degraded in 7 days. This bioremediation of chlorpyrifos in soil is mainly due to the microbes which are in close connection with roots of ryegrass rhizosphere. The degrader identified by 16 s rDNA analysis was *Pseudomonas nitroreducens* PS-2 through BLAST technique.

The mixture of *Staphylococcus cohnii* subspecies *urealyticus* in the presence of *Withania somnifera*, a tolerant plant cultivated in lindane-dressed soil, was studied for rhizoremediation potential (Abhilash et al. 2011). *Withania* was cultivated in garden soil when amended with 20 mg kg^{-1} of lindane. This was then inoculated with microbial culture of 100 ml which was having 8.1×10^6 CFU. The potential of microbial inoculation on lindane uptake, plant growth, dehydrogenase activity, microbial biomass carbon, lindane dissipation percentage and residual lindane concentration was studied. Microbial inoculation significantly increased growth and lindane uptake potential in test plant $p < 0.05$. The study result showed that the integrated use of rhizospheric microbial inoculation and plant species increased the spread of lindane and had the potential for in situ remediation of polluted soils. Dubey and Fulekar (2011a, b) performed an experiment to evaluate the application of grass species in rhizospheric bioremediation of pesticides. They found that chlorpyrifos was more toxic than cypermethrin and fenvalerate at higher concentrations of 75 and 100 mg/kg, respectively, for survival, germination and subsequent growth of *Pennisetum pedicellatum* and *Cenchrus setigerus*. The microbes were found higher in mycorrhizosphere soil when co-cropped with *Pennisetum pedicellatum* and *Cenchrus setigerus*. They also recorded increased chlorpyrifos degradation in the rhizosphere of *Pennisetum*.

8.8 Degradation Through Bacteria

Production of carbon dioxide and water occurs by the oxidative degradation of pesticides. The pesticide's degrading ability of bacteria is recorded in Table 8.4. The bacteria involved in the degradation process get energy by degradation of products. The efficiency of degradation is based on optimum atmospheric conditions such as soil pH, moisture and temperature. The modifications of different bacterial specimens via genetic mutations also enhanced effectiveness of applied microbes (Qiao et al. 2003; Li-Qing et al. 2008). If removal of pesticides occurs through microbes, it shows positive effects on agricultural soil fertility. Chlorpyrifos has a massive effect on contaminating soil and waterbodies. Microbial degradation is very useful for the detoxification of such (chloroorganic) pesticides. These are specific enzymes and genes which are very specific in cleavage of functional groups in pesticide. The optimization of environmental conditions and an effective microbial community in the contaminated site is very essential for the degradation of pesticides (Chishti et al. 2013).

Table 8.4 Pesticide-degrading bacteria

Pesticides	Degrading bacteria	References
Organochlorine insecticide	*Escherichia coli*	Singh et al. (2003)
Endosulfan	Different genera have different potential to degrade them from which *Micrococcus* and *Pseudomonas* were highly active compared to others	Li et al. (2004)
Lindane, methyl parathion, carbofuran	Bacterial consortium like *Bacillus* sp. and *Chryseobacterium joostei*	Foster et al. (2004)
Ethion (OPs)	*Pseudomonas* and *Azospirillum*	Zhang et al. (2007)
Allethrin	*Acidomonas* sp.	Paingankar et al. (2005)
Triazophos, chlorpyrifos and phoxim	Photosynthetic bacteria	Zhang et al. (2005)
Methyl parathion	*Shewanella* and *Vibrio* bacteria	Liu et al. (2006a, b)
OPs, chlorinated pesticides, herbicides and fungicides	*Rhodobacter sphaeroides*	Harada et al. (2006)
Endosulfan	*Pseudomonas* bacterium	Prabakaran and Peterson (2006)
Hexachlorocyclohexane	*Sphingobium japonicum*	Liu et al. (2007)
DDT and endosulfan	*Stenotrophomonas maltophilia*	Barragán-Huerta et al. (2007)
Sumithion	Indigenous bacteria	Savadogo et al. (2007)
Fenamiphos	Cyanobacteria and blue green algae	Cáceres et al. (2008)
Esbiothrin	*Acinetobacter* sp.	Ha et al. (2009)
Malathion	*Bacillus thuringiensis*	Zeinat et al. (2008)
Organophosphate pesticides	*Bacillus*, *Actinobacteria* and *L-Proteobacteria* strains	Sabdono and Radjasa (2008)
Me-parathion	*Psychrotrophic* bacterium	Krishna and Philip (2009)
Para-nitrophenol	Rhodococcus bacteria	Zhang et al. (2009)
Malathion	*Acinetobacter johnsonii*	Xie et al. (2009)
Organophosphorus insecticides	Lactic acid bacteria	Kye et al. (2009)
Pyrethroid	*Enterobacter aerogenes*	Liao and Xie (2009)
Pyridine	*Paracoccus* sp.	Qiao and Wang (2010)
Bifenthrin, fenvalerate and fenpropathrin	Actinomycete strain HP-S-01	Chen et al. (2011)
Permethrin and cypermethrin	*Pseudomonas putida* and *Pseudomonas mendocina*	Mendoza et al. (2011)
Imidacloprid and metribuzin	*Burkholderia cepacia* strain CH-9	Madhuban et al. (2011)

(continued)

Table 8.4 (continued)

Pesticides	Degrading bacteria	References
Cypermethrin	Photosynthetic bacterium (GJ-22)	Yin et al. (2012)
Methomyl and carbofuran	*Flavobacterium*, *Alcaligenes* and *Pseudomonas*	Mbogo et al. (2012), Kumar et al. (2011)
Chlorpyrifos (CP)	*Streptomycetes* strains	Briceño et al. (2012)
Bifenthrin (BF)	*Acinetobacter calcoaceticus*	Tingting et al. (2012)
Paraoxon	*Pseudomonas plecoglossicida*	Farivar et al. (2017)
Monocrotophos	*Bacillus subtilis* (BAGN005), *B. licheniformis* (BKGN007) and *Pseudomonas stutzeri* (BVGN010)	Buvaneswari et al. (2017)
Chlorpyrifos	*E. coli* (EC1) and *Pseudomonas fluorescens* (PF1)	Altaf (2018)

Researchers have reported the presence of naturally occurring chlorpyrifos-resistant bacteria in the environments of Kashmir Valley. Chlorpyrifos is an organophosphorus insecticide widely used in the region. Scientists have identified and isolated two different types of bacteria *E. coli* (EC1) and *Pseudomonas fluorescens* (PF1) living in waterbodies and soil, respectively, which are highly efficient in biodegrading chlorpyrifos into simpler and non-toxic chemicals. These microorganisms use chlorpyrifos as their source of energy, growth and other metabolic activities. *E. coli* occurs in the Dal Lake and Anchar in Srinagar, and *Pseudomonas fluorescens* was found in soil samples of Ganderbal district adjoining Srinagar (Altaf 2018).

For the management of organophosphate pesticides (OPs), genetically engineered *Pseudomonas plecoglossicida* potential was studied. Genetically engineered *P. plecoglossicida* was prepared by transferring PCR product of opd gene of *Flavobacterium* sp. ATCC 27551 in the chromosome of *P. plecoglossicida*. This has a potential to hydrolyse paraoxon into p-nitrophenol and di-ethylphosphate into paraoxon supplemented in complete supplement mixture (CSM) medium. The isolate uses paraoxon as the source of carbon. The bacteria degrade organophosphate pesticides and utilize the degraded nutrient products for their nutrients. The observed potential in genetically engineered *P. plecoglossicida* in biodegradation of xenobiotics revealed that this will be useful in bioremediation of contaminated industrial and agricultural sites (Farivar et al. 2017).

Excessive levels of organophosphorus pesticides degrade soil quality, reduction of crop yield and inferior agricultural products, consequently posing significant hazards to ecosystem and health of humans and animals. Buvaneswari et al. (2017) found bacterial strains of *Bacillus subtilis* (BAGN005), *B. licheniformis* (BKGN007) and *Pseudomonas stutzeri* (BVGN010) to be very effective in degrading monocrotophos.

Tian and Chen (2012) mentioned that microbial strain isolation and screening are very useful in carbendazim degradation in mineral culture media. *Rhizobium meliloti* coating on *Medicago sativa* seeds and pesticide-degrading bacteria are effective in repairing soil polluted through organic phosphorus pesticides (Zhao et al. 2012). *Sphingobium japonicum* has a potential in degradation of chlorinated pesticide—hexachlorocyclohexane. This strain (*Sphingobium japonicum* LZ-2) degrades 20 mg/L of lindane in 10 h (Liu et al. 2007). An aerobic bacterium (*Burkholderia cepacia* strain CH-9) degrades imidacloprid and metribuzin. Sixty-nine percent of degradation of imidacloprid and 86% degradation of metribuzin can be obtained in 20 days having initial concentration of 50 mg/L in mineral salt medium (Madhuban et al. 2011). Bifenthrin (BF) is a synthetic pesticide. It is degraded by pyrethroid bacteria (*Acinetobacter calcoaceticus*). The degradation rate could be achieved up to 56.4% with initial concentration of 100 mg/L with pH range of 6.0–8.0 and 5% inoculation (Tingting et al. 2012).

Streptomycetes strains have enormous applications for degradation of chlorpyrifos (CP) pesticide. The degradation potential of these strains can be evaluated by performing a study in agar medium plates. The pH alterations can affect the efficiency of degradation process (Briceño et al. 2012). Tert-Bu mercaptan (TMB) undergoes biodegradation in water under aerobic conditions. First-order kinetics are involved in biodegradation process. There is a slight increase in the rate of reaction by addition of TMB and a slight decrease with addition of phenol (Karthikeyan and Hutchinson 2012).

Bacterial strains which are capable of degrading methomyl and carbofuran can be studied by high-performance liquid chromatography (HPLC) in biodegradation analysis. Acetonitrile and water were used as mobile phases. The closeness of carbofuran-degrading strains to the genera *Flavobacterium* and *Alcaligenes* and that of methomyl-degrading strains to genera *Pseudomonas* and *Alcaligenes* were observable by using 16S rDNA sequence analysis (Mbogo et al. 2012; Kumar et al. 2011). Photosynthetic bacterium (GJ-22) is capable of degrading cypermethrin (CMP). That CMP degradation by GJ-22 is very productive at 25–35 °C and at pH of 7.0. By performing gas chromatography/mass spectrometry (GC-MS), the metabolic products are detected. Degradation of CMP proceeds through oxidative or/and through hydrolytic pathways by GJ-22 yielding five metabolites (Yin et al. 2012). Removal of organochlorine pesticides from soil is performed by microbial applications under optimum environmental conditions. Better results may be obtained by addition of potassium humate for increasing concentration of microorganisms (Soromotin et al. 2012). *Pseudomonas putida* and *Pseudomonas mendocina* strains have a great capacity of biodegrading permethrin and cypermethrin pesticides. Bioremediation up to 90% can be achieved with the help of these bacterial strains within the period of 15 days (Mendoza et al. 2011).

Actinobacteria sp. TW and *Sphingomonas* sp. TY strains are novel and very useful for the disposal of tobacco waste in the temperature range of 25–37 °C and pH range of 7.0–8.0 (Wang et al. 2011). The actinomycete strain HP-S-01 is isolated from activated sludge for its application to degrade deltamethrin. The degradation results in 3-phenoxybenzaldehyde as a major hydrolysis product. This strain is highly efficient in degrading bifenthrin, fenvalerate and fenpropathrin. This process undergoes first-order kinetics and provides an effective tool for bioremediation of environmental contamination from pesticides (Chen et al. 2011). Diazinon-degrading bacteria utilize it as a source of carbon and phosphorus under different culture conditions. The addition of carbon sources, as glucose or succinate, causes a decrease in degradation rate (Abo-Amer 2011). Biodegradation of profenofos is conducted by bacterial strains isolated by enrichment technique. About 90% concentration of profenofos can be degraded in 90 h (Malghani et al. 2009). *Paracoccus* sp. strain was applied for the biodegradation studies of pyridine. It was observed that, at the concentration of pyridine <0.9 mg/L, the rate of degradation is higher, while at the concentration >0.9 mg/L, the rate is lower (Qiao and Wang 2010). A bacterial consortium which degrades tetrachlorvinphos is isolated from agricultural soil. It is composed of six pure strains. The study reveals that these strains have a potential to degrade organophosphate pesticides (Ortiz-Hernandez and Sanchez-Salinas 2010).

Lactic acid bacteria have the potential to degrade organophosphorus insecticides through fermentation. This uses organophosphate as a source of carbon and phosphorus (Kye et al. 2009). *Enterobacter aerogenes* can degrade bifenthrin and cypermethrin (Liao and Xie 2009). *Acinetobacter johnsonii* (MA-19) strain can degrade organophosphate pesticides by enrichment culture method. Four additional compounds were added to enhance efficiency out of which Na succinate was very effective. An increase in its concentration the rate of degradation of malathion was increased (Xie et al. 2009). *Rhodococcus* bacteria can degrade para-nitrophenol (Zhang et al. 2009). Similarly, organophosphate pesticide degradation can be done by *Bacillus*, *Actinobacteria* and *L-Proteobacteria* (Sabdono and Radjasa 2008). Bacterium *Bacillus thuringiensis* is effective in degrading malathion in minimum salt media. Through addition of glucose and yeast, the growth of bacteria increases up to 10^5-fold which degrades more than 99% malathion within 30 days. Residues were studied by HPLC and GC-MS (Zeinat et al. 2008). Esbiothrin can be degraded by immobilized *Acinetobacter* on magnetic polyurethane (Ha et al. 2009). By using immobilized bacteria on Ca-alginate gel beads, organophosphate insecticide degradation was studied, along with hydrolysed products (Zeinat et al. 2008).

Cyanobacteria can convert fenamiphos into the number of its stable non-toxic components (Cáceres et al. 2008). Indigenous bacteria degrade sumithion OPs through anaerobic decomposition. They decompose them into CH_4, N_2, CO_2 and H_2S (Savadogo et al. 2007). Beans of green coffee can be used for the support and growth of *Stenotrophomonas maltophilia* which degrades DDT and endosulfan. A medium amended with glucose is used as a supplement (Barragán-Huerta et al. 2007). *Pseudomonas* bacterium can degrade endosulfan. Whenever it bioaccumulates in fishes—*Cyprinus carpio*, it uses endosulfan as a carbon source (Prabakaran and Peterson 2006). Endosulfan is metabolized into endosulfan sulphate by bacterial action (Shivaramaiah and Kennedy 2006). Microscopic organisms (three bacterial strains) potentially degrade mefenacet and many other amide pesticides such as propanil and metolachlor by hydrolysis (Harada et al. 2006).

Different types of pesticides (OPs, chlorinated pesticides, herbicides and fungicides) are effectively degraded by the fermentation process carried out by *Rhodobacter sphaeroides* (Harada et al. 2006). *Vibrio* and *Shewanella* bacteria can effectively degrade methyl parathion. Its biodegradation mechanism is entirely different from photocatalytic process (Liu et al. 2006a, b). Photosynthetic bacteria have the capacity to degrade multiple types of pesticides (chlorpyrifos, phoxim and triazophos) (Zhang et al. 2005). Allethrin is a pyrethroid insecticide, and its degradation is achieved by *Acidomonas* sp. (Paingankar et al. 2005). Eight bacterial strains potentially degrade PCNP pesticide. Better results were obtained when all these strains were collectively used (Ning et al. 2005). Two bacteria cad1 and cad2 degrade cadusafos in mineral salts medium with nitrogen (MSMN). They have the potential to degrade ethoprophos nematicide completely (Karpouzas et al. 2005).

Immobilized bacteria have the capacity to degrade multiple pesticides such as herbicides, fungicides and carbamates under different environmental conditions (Pattanasupong et al. 2004). Ethion (OPs) is anaerobically degraded through *Azospirillum* and *Pseudomonas* (Zhang et al. 2007). Bacterial consortium like *Bacillus* sp. and *Chryseobacterium joostei* can degrade lindane, methyl parathion and carbofuran in individual and mixed pesticide-enriched cultures by using biokinetic parameters (Foster et al. 2004). *Psychrotrophic* bacterium can degrade Me-parathion. This biodegradation is sensitive to pH and temperature variations (Krishna and Philip 2009). Six genera are able to degrade organochloride pesticides, that is, endosulfan. Different genera have different potential to degrade them from which *Micrococcus* and *Pseudomonas* were highly active compared to others (Li et al. 2004). Immobilized *Escherichia coli* could degrade organochlorine insecticide that contains ester bond (Singh et al. 2003).

The same bacterium is highly efficient in degrading a number of pesticides including BHC, DDT, endosulfan, HCH isomers and 2,4-D (Xue-Dong et al. 2003; Qiao et al. 2003; Gupta 2005; Santacruz et al. 2005). DLL-1 bacterial strain biologically degrades pesticide that is present in soil and plant system. GPRB (growth-promoting rhizobacteria) strains are effective in degrading fungicide and herbicide compared to *Azotobacter* and *Bacillus*. The purpose was to determine the capacity of different bacteria to effectively degrade fungicides and herbicides (Atia et al. 2002).

8.9 Degradation Through Fungi

A lot of fungi are present in their natural habitats which may be isolated and screened for biodegradation of toxic pesticides. This will act as an effective tool. *Fusarium verticillioides* have the potential to use lindane as a source of carbon and energy in aerobic conditions. This can be obtained from *Agave tequilana* leaves through enrichment techniques. The environmental factors and yeast extract improved the power of the biodegradation process (Pinto et al. 2012). There is a great potential in fungal strains, viz. *Fusarium oxysporum*, *Lentinula edodes*, *Penicillium brevicompactum* and *Lecanicillium saksenae*, in biodegradation of the pesticides like terbuthylazine, pendimethalin and difenoconazole in batch liquid cultures. These fungal strains are active microbes in pesticide degradation (Hai et al. 2012). Nonacclaimatized mixed culture of white-rot fungus and bacteria has applications for biodegradation of atrazine, aldicarb and alachlor in the liquid phase. With incubation period of 14 days, the mixed culture achieved 47%, 98% and 62% removal. Removal of these pesticides may follow phenomena of biosorption and biodegradation (Nyakundi et al. 2011).

Methomyl and diazinon (pesticides) are biodegraded with the help of rot fungi. The rate of degradation is higher if using different mixtures of fungal isolates (Sagar and Singh 2011). Different fungal strains are observed for their degradation ability of DDD pesticide (Ortega et al. 2011). Endosulfan-decomposing aerobic fungal strains are useful in soil contaminated with organochlorine pesticides. These strains (*Mortierella* sp. strains W8 and Cm1–45) produced 50–70% degradation at 25 °C in 28 days. There is diol formation first in endosulfan, and then conversion in endosulfan lactone takes place during degradation (Kataoka et al. 2010). The mixed fungal strains have the possibility of degrading mixed insecticides such as chlorpyrifos and DDT. It has been observed that if using low mixed insecticide concentration, the efficiency of decomposition was found to be high. The efficiency is observed in 26.94% and 24.94% degradation of DDT and chlorpyrifos, respectively (Kulshrestha and Kumari 2010). *Sphingomonas yanoikuyae* strain under harsh conditions can decompose carbamate and pyrethrin (OPs) with high efficiency in enrichment culture method (Ouyang et al. 2008). Salt-resistant actinomycete may degrade carbofuran. *S. alanosinicus* is most efficient and gave rise to 95% decomposition. It utilizes carbofuran for carbon source and is useful in saline soils for its power (Chougale and Deshmukh 2007).

It has been observed that more than 30 microbes have the potential to decompose pesticide. The *Gliocladium* genus has a maximum power to degrade carbofuran (Slaoui et al. 2007). Fungus utilizes chlorpyrifos for carbon and energy resulting in rapid decomposition. Another basidiomycete fungus decomposes chlorpyrifos very powerfully (Yu et al. 2006). A fungus, *C. elegans*, degrades DEET, an insecticide, into different less toxic metabolites analysed by HPLC-MS (Seo et al. 2005). Phytopathogenic fungi easily degrade herbicides. This fungus grows easily on organophosphonate herbicides and decomposes them (Lipok et al. 2003). *Trichoderma harzianum* and *T. viride* have a high potential in degrading pirimicarb. Degradation power increases when activated charcoal is added (Romeh 2001).

8.10 Degradation by Enzymes

Plants as well as microbes are storehouse of many potential enzymes. Their production occurs in different metabolic pathways which is useful in bioremediation of pesticides. Optimum conditions favour removal of toxic intermediates. The engineered bacteria were used to produce esterase gene which specifically act on substrate and degrade more than 65% of methyl parathion within 3 h (Li et al. 2004). Carbofuran present in contaminated soil can be treated with *Paracoccus* sp. This enzymatically degrades carbofuran into its metabolites which were analysed by HPLC. This bacterium uses carbofuran as the sole source of carbon (Peng et al. 2008). Genetically modified *Escherichia coli* can enzymatically degrade methyl parathion (Zhang et al. 2008). *Micrococcus* sp. has the ability to degrade OP pesticide like cypermethrin by enzymatic action (Tallur et al. 2008). Lindane can be degraded by *Conidiobolus* fungus through enzyme action. GC-ECD and GC/MS confirm that there is no metabolite; this proved that lindane is completely degraded by this fungus (Nagpal et al. 2008). In a study of atrazine (AT) and alachlor (AL), their degradation by treating them with extracellular enzyme extracted from fungi was determined (Chirnside et al. 2007). FDS-1 strain of *Burkholderia* sp. can degrade nitrophenyl enzymatically at 30 °C and pH of 7.0 taken as optimized conditions (Lan et al. 2006). Strains of genetically modified bacteria contain enzymes that potentially degrade pesticides OPs, carbamates and pyrethroids (Liu et al. 2006a, b).

Different enzymes can specifically degrade different pesticides (OPs) in wheat kernels (Yoshii et al. 2006). Thirty fungal strains were used to find out degradation rate of diuron and pyrithiobac-sodium. The study revealed that the highest degradation was done through ligninolytic enzymes (Gondim-Tomaz et al. 2005). *Enterobacter* enzymatically degrades chlorpyrifos and uses them as carbon and phosphorus source (Singh et al. 2004). Some Gram-negative bacteria have the ability to degrade dimethoate as a sole source of carbon. Bacteria hydrolyse insecticide through enzymes phosphatases and esterases (Kadam et al. 2003). More than 15 fungal strains can degrade different OPs up to 96% by enzyme-catalysed pathways (Jauregui et al. 2003). The amino acid sequence of phosphotriesterase mutant is useful in application in organophosphorus pesticide degradation (Xiang-Ming and Ping-Ping 2012).

8.11 Conclusions

The management of soil contamination because of pesticide usage in polluted areas is the necessity of modern era. Now it is not possible to manage this burning problem through conventional means. We may say that physicochemical means are not efficient. These are even costly and even not ecofriendly for different ecosystems. For removal of hazardous chemicals from the environment, one should use biological agents such as bacteria, fungi and their enzymes. This will be ecofriendly and will bear low cost. This will degrade pesticides into lesser toxic components. It needs more study to find out mechanisms and their enzyme secretion during degradation process. What is the exact role of rhizomicrobial population and plants in facilitating microbial degradation for in situ bioremediation needs more attention. There is a need of more study of fundamental microbial ecology of the rhizosphere. It even needs documentation of microbial decomposition of agricultural pesticides in the root zone. Further to know the factors which influence plant microbe and toxicant association in soil will permit more realization of in situ bioremediation.

Acknowledgements Authors are thankful to the Amity University Haryana authorities for the facilities and constant encouragement.

Conflict of Interest Statement We declare that we have no conflict of interest.

References

Abhilash PC, Srivastava S, Srivastava P, Singh B, Jafri A, Singh S (2011) Influence of rhizospheric microbial inoculation and tolerant plant species on the rhizoremediation of lindane. Environ Exp Bot 74:127–130

Abo-Amer AE (2011) Biodegradation of diazinon by Serratia marcescens DI101 and its use in bioremediation of contaminated environment. J Microbiol Biotechnol 21:71–80

Alberty RA (2006) Biochemical reactions at specified temperature and various pHs. In Biochemical thermodynamics. Book Series: Methods of Biochemical Analysis. https://onlinelibrary.wiley.com

Altaf R (2018) Kashmir's Dal lake harbours bacteria that can degrade pesticides. Science (TWC India). https://weather.com/en-IN/india/science/news

Amos PW, Younger PL (2003) Substrate characterisation for a subsurface reactive barrier to treat colliery spoil leachate. Water Res 37:108–120

Anderson JA, Kruger ER, Coats JR (1994) Enhanced degradation of a mixture of three herbicides in the rhizosphere of a herbicide tolerant plant. Chemosphere 28:1551–1557

Atia M, Awad NM, Turky AZ (2002) Associative action of growth promoting Rhizobacteria and phytoremediationon the biodegradation of certain pesticides in soil. Bull Natl Res Centre Egypt 27:469–480

Barea JM, Pozo MJ, Azcón R, Azcón-Aguilar C (2005) Microbial co-operation in the rhizosphere. J Exp Bot 56:1761–1778

Barragán-Huerta BE, Costa-Pérez C, Peralta-Cruz J, Barrera-Cortés J, Esparza-García F, Rodríguez-Vázquez R (2007) Biodegradation of organochlorine pesticides by bacteria grown in microniches of the porous structure of green bean coffee. Int Biodeterior Biodegrad 59:239–244

Bending GD, Lincoln SD, Edmondson RN (2006) Spatial variation in the degradation rate of the pesticides isoproturon, azoxystrobin and diflufenican bioremediation and phytoremediation of pesticides in soil and its relationship with chemical and microbial properties. Environ Pollut 139:279–287

Bhalerao TS, Puranik PR (2007) Biodegradation of organochlorine pesticide, endosulfan, by a fungal soil isolate, Aspergillus niger. Int Biodeterior Biodegrad 59:315–321

Bhupathiraju VK, Krauter P, Holman H-YN, Conrad ME, Daley PF, Templeton AS, Hunt JR, Hernandez M, Alvarez-Cohen L (2002) Assessment of in-situ bioremediation at a refinery waste-contaminated site and an aviation gasoline contaminated site. Biodegradation 13:79–90

Boivin A, Cherrier R, Schiavon M (2005) A comparison of five pesticides adsorption and desorption processes in thirteen contrasting field soils. Chemosphere 61:668–676

Bouziani M (2007) L'usage immodéré des pesticides. De graves conséquences sanitaires. Le Guide de la Médecine et de la Santé, Santémaghreb, 2007

Briceño G, Fuentes MS, Palma G, Jorquera MA, Amoroso MJ, Diez MC (2012) Chlorpyrifos biodegradation and 3,5,6-trichloro-2-pyridinol production by actinobacteria isolated from soil. Int Biodeterior Biodegrad 73:1–7

Buvaneswari G, Thenmozhi R, Nagasathya A, Thajuddin N (2017) Screening of efficient monocrotophos degrading bacterial isolates from paddy field soil of Sivagangai District, Tamil Nadu, India. J Environ Sci Technol 10:13–24
Cáceres TP, Megharaj M, Naidu R (2008) Biodegradation of the pesticide fenamiphos by ten different species of green algae and cyanobacteria. Curr Microbiol 57:643–646
Calderbank A (1989) The occurrence and significance of bound pesticide residues in soil. Rev Environ Contam Toxicol 108:71–103
Chang SW, Lee SJ, Je CH (2005) Phytoremediation of atrazine by poplar trees: Toxicity, uptake, and transformation. J Environ Sci Health 40:801–811
Charnay M, Tuis S, Coquet Y, Barriuso E (2005) Spatial variability in 14C-herbicide degradation in surface and subsurface soils. Pest Manag Sci 61:845–855
Chaudhry Q, Blom-Zandstra M, Gupta S, Joner EJ (2005) Utilising the synergy between plants and rhizosphere microorganisms to enhance breakdown of organic pollutants in the environment. Environ Sci Pollut Res 12:34–48
Chen S, Lai K, Li Y, Hu M, Zhang Y, Zeng Y (2011) Biodegradation of deltamethrin and its hydrolysis product 3-phenoxybenzaldehyde by a newly isolated *Streptomyces aureus* strain HP-S-01. Appl Microbiol Biotechnol 90:1471–1483
Chirnside AEM, Ritter WF, Radosevich M (2007) Bioremediation strategies for pesticide-contaminated Soil: IV. Biodegradation using a selected microbial consortium following pretreatment with fungal Enzymes. In: Proceedings of the 9th international in situ and on-site bioremediation symposium, Baltimore, MD, pp 1233–1240
Chishti Z, Hussain S, Arshad KR, Khalid A, Arshad M (2013) Microbial degradation of chlorpyrifos in liquid media and soil. J Environ Manag 114:372–380
Chougale VV, Deshmukh AM (2007) Biodegradation of carbofuran pesticide by saline soil actinomycetes. Asian J Microbiol Biotechnol Environ Sci 9:1057–1061
Cox C (1992) A new list of carcinogenic pesticides used in food. Northwest Coalition for Alternatives to Pesticides for the Indian context 12(4):28
Dekkers LC, Mulders IHM, Phoelich CC, Chin-A-Woeng TFC, Wijfjes AHM, Lugtenberg BJJ (2000) The sss colonization gene of the tomato-*Fusarium oxysporum* f. radicis-lycopersici biocontrol strain *Pseudomonas fluorescens* WCS365 can improve root colonization of other wild-type *Pseudomonas* spp. bacteria. Mol Plant-Microbe Interact 13:1177–1183
Drakeford CE, Dwight NC, Riley MB (2003) Fate of isoxaben in containerized rhizosphere system. Chemosphere 50:1243–1247
Dubey KK, Fulekar MH (2011a) Effect of pesticides on the seed germination of *Cenchrus setigerus* and *Pennisetum pedicellatum* as monocropping and co-cropping system: implications for rhizospheric bioremediation. Romanian Biotechnol Lett 16:5909–5919
Dubey KK, Fulekar MH (2011b) Chlorpyrifos bioremediation in *Pennisetum* rhizosphere by a novel potential degrader *Stenotrophomonas maltophilia* MHF ENV20. World J Microbiol Biotechnol 28:1715–1725. https://doi.org/10.1007/s11274-011-0982-1
Enrica G (1994) The role of microorganisms in environmental decontamination. In: Renzoni A (ed) Contaminants in the environment: a multidisciplinary assessment of risks to man and other organisms, vol 25. Lewis Publishers, Boca Raton, FL, pp 235–246
EPA. Pesticides industry, sales and usage. April 2015. http://www.epa.gov
Faridi R (2014). Pesticide use in India. http://indiaforsafefood.in/farminginindia
Farivar TN, Peymani A, Najafipour R, Aslani MM, Ali S (2017) Biodegradation of paraoxan as an organophosphate pesticide with *Pseudomonas plecoglocissida* transfected by opd Gene. Biotech Health Sci 4:e45055
Fenlon KA, Jones KC, Semple KT (2007) Development of microbial degradation of cypermethrin and diazinon in organically and conventionally managed soils. J Environ Monit 9:510–515
Foster LJR, Kwan BH, Vancov T (2004) Microbial degradation of the organophosphate pesticide. FEMS Microbiol Lett 240:49–53
Fulekar MH (2005) Bioremediation technologies for environment. Indian J Environ Protect 25:358–364

Gondim-Tomaz MA, Franco TT, Durrant LR (2005) Biodegradation of diuron and pyruthiobac-sodium by white-rot and soil fungi. Contamin Soils Sediments Water 9:21–32

Gupta PK (2004) Pesticide exposure—Indian scene. Toxicology 198:83–90

Gupta US (2005) Biodegradation of chlorinated hydrocarbon insecticide by Pseudomonas species. Himalayan J Environ Zool 19:1–10

Ha J, Engler CR, Wild JR (2009) Biodegradation of coumaphos, chlorferon, and diethylthiophosphate using bacteria immobilized in Ca-alginate gel beads. Bioresour Technol 100:138–1142

Hai FI, Modin O, Yamamoto K, Fukushi K, Nakajima F, Nghiem LD (2012) Pesticide removal by a mixed culture of bacteria and white-rot fungi. J Taiwan Inst Chem Eng 43:459–462

Harada N, Takagi K, Harazono A, Fujii K, Iwasaki A (2006) Isolation and characterization of microorganisms capable of hydrolysing the herbicide mefenacet. Soil Biol Biochem 38:173–179

Hodaifa G, Nieto LM, Casanova MS (2009) Elimination of pesticide residues from virgin olive oil by ultraviolet light: preliminary results. J Hazard Mater 168:555–559

Hodgson E, Roe RM, Goh DKS, Rock GC, Rose RL (1993) Insect cytochrome P-450: metabolism and resistance to insecticides. Biochem Soc Trans 21:1060–1065

Huang Y, Xiao L, Li F, Xiao M, Lin D, Long X, Wu Z (2018) Microbial degradation of pesticide residues and an emphasis on the degradation of Cypermethrin and 3-phenoxy benzoic acid: a review. Molecules 23:2313. https://doi.org/10.3390/molecules23092313

Jauregui J, Valderrama B, Albores A, Vazquez-Duhalt R (2003) Microsomal transformation of organophosphorus pesticides by white rot fungi. Biodegradation 14:397–406

Johnson J, Ware WG (1992) Pesticide litigation manual edition, Clark Boardman Callaghan Environmental Law Series, New York, NY

Kadam TA, Jadhav VD, Gyananath G (2003) Microbial degradation of dimethoate by gram negative soil isolates from cotton field. Pollut Res 22:443–445

Karpouzas DG, Fotopoulou A, Menkissoglu-Spiroudi U, Singh BK (2005) Non-specific biodegradation of the organophosphorus pesticides, cadusafos and ethoprophos, by two bacterial isolates. FEMS Microbiol Ecol 53:369–378

Karthikeyan R, Hutchinson SL (2012) Biodegradation of tertiary butyl mercaptan in water. J Bioremed Biodegr 3(6):156

Kataoka R, Takagi K, Sakakibara F (2010) A new endosulfan-degrading fungus, Mortierella species, isolated from a soil contaminated with organochlorine pesticides. J Pestic Sci 35:326–332

Korade DL, Fulekar MH (2009) Rhizoremediation of Chlorpyrifos in mycorrhizospheric soil using rye grass. J Hazard Mater 172:1344–1350

Krishna KR, Philip L (2009) Biodegradation of mixed pesticides by mixed pesticide enriched cultures. J Environ Sci Health B Pesticides Food Contam Agric Wastes 44:18–30

Kuiper I, Lagendijk EL, Bloemberg GV, Lugtenberg BJJ (2004) Rhizoremediation: a beneficial plant microbe interaction. Mol Plant-Microbe Interact 17:6–15

Kulshrestha G, Kumari A (2010) Simultaneous degradation of mixed insecticides by mixed fungal culture isolated from sewage sludge. J Agric Food Chem 58:852–856

Kumar S, Anthonisamy A, Arunkumar S (2011) Biodegradation of methyl parathion and endosulfan using: *Pseudomonas aeruginosa* and *Trichoderma viridae*. J Environ Sci Eng 53:115–122

Kye CM, Reukaradhya M, Islam K et al (2009) Biodegradation of chlorpyrifos by lactic acid bacteria during kimchi fermentation. J Agric Food Chem 57:1882–1889

Lan WS, Gu JD, Zhang JL (2006) Coexpression of two detoxifying pesticide-degrading enzymes in a genetically engineered bacterium. Int Biodeterior Biodegrad 58:70–76

Li H, Liang W, Wu X, Liu Y (2004) Research on biodegradation of organophosphorus insecticide by a novel psychrotrophic bacterium SA-8. Zhongshan Daxue Xuebao, Ziran Kexueban 43:131–132

Liao M, Xie X (2008) Effects of combination of plant and microorganism on degradation of simazine in soil. J Environ Sci 20:195–198

Licht LA, Isebrands JG (2005) Linking phytoremediated pollutant removal to biomass economic opportunities. Biomass Bioenergy 28:203–218

Lio M, Xie X (2009) Application of enterobacteraerogenes in degrading pyrethroid pesticide residue, and preparation with enterobacteraerogenes. Faming Zhuanli Shenqing, CN102021135 A, 2009

Lipok J, Dombrovska I, Wieczorek P (2003) The ability of fungi isolated from stored carrot seeds to degrade organophosphonate herbicides. In: Pesticide in air, plant, soil and water system, proceedings of the 12th symposium pesticide chemistry, Piacenza, Italy, June 4–6, La Goliardica Pavese, pp 575–580

Li-Qing BZ, Hong-Xia J, Yan-Chun Y (2008) Degradation of organophosphorus pesticide by engineered bacteria and immobilized cell. Huanjing Kexue Yu Jishu 31:45–48

Liu YH, Chung YC, Xiong Y (2001) Purification and characterization of a dimethoate-degrading enzyme of *Aspergillus niger* ZHY256, isolated from sewage. Appl Environ Microbiol 67:3746–3749

Liu J, Wang L, Zheng L, Wang X, Lee FSC (2006a) Analysis of bacteria degradation products of methyl parathion by liquid chromatography/electrospray time-of-flight mass spectrometry and gas chromatography/mass spectrometry. J Chromatogr 37:180–187

Liu Z, Hong Q, Xu JH, Jun W, Li SP (2006b) Construction of a genetically engineered microorganism for degrading organophosphate and carbamate pesticides. Int Biodeterior Biodegrad 58:65–69

Liu Z, Yang C, Qiao C-L (2007) Biodegradation of p-nitrophenol and 4-chlorophenol by *Stenotrophomonas* sp. FEMS Microbiol Lett 277:150–156

Madhuban G, Debashis D, Das SK (2011) Biodegradation of imidacloprid and metribuzin by *Burkholderia cepacia* strain CH9. Pestic Res J 23:36–40

Malghani S, Chatterjee N, Yu HX, Luo Z (2009) Isolation and identification of profenofos degrading bacteria. Braz J Microbiol 40:893–900

Mbogo OK, Magoma G, Ngamau K et al (2012) Characterization of methomyl and carbofuran degrading-bacteria from soils of horticultural farms in Rift Valley and Central Kenya. Afr J Environ Sci Technol 6:104–114

Melling FB (1993) Soil microbial ecology: applications in agricultural and environmental management. Marcel Dekker, New York

Mendoza JC, Perea Y, Salvador JA (2011) Bacterial biodegradation of permetrina and cipermetrina pesticides in a culture assemblage. Avances en Ciencias e Ingenieria 2:45–55

Muhammad KJ, Mehrban A, Muhammad T (2016) Potential of biological agents in decontamination of agricultural soil. Scientifica, 1598325, 9 pages,. doi:https://doi.org/10.1155/2016/1598325

Nagpal V, Srinivasan MC, Paknikar KM (2008) Biodegradation of γ-hexachlorocyclohexane (Lindane) by a non-white rot fungus conidiobolus isolated from litter. Indian J Microbiol 48:134–141

Nawab A, Aleem A, Malik A (2003) Determination of organochlorine pesticides in agricultural soil with special reference to γ-HCH degradation by Pseudomonas strains. Bioresour Technol 88:41–46

Ning L, Jing-ying C, Li L (2005) Isolation and selection of strains used to degrade organic PCNB pesticides and their application effects. Jilin Nongye Daxue Xuebao 27:205–208

Nyakundi WO, Magoma G, Ochora J, Nyende AB (2011) Biodegradation of diazinon and methomyl pesticides by white rot fungi from selected horticultural farms in rift valley. J Appl Technol Environ Sanitation 1:107–124

Okeke BC, Siddique T, Arbestain MC, Frankenberger WT (2002) Biodegradation of γ-hexachlorocyclohexane and α-hexachlorocyclohexane in water and soil slurry by *Pandoraea* sp. J Agric Food Chem 50:2548–2555

Ortega SN, Nitschke M, Mouad AM (2011) Isolation of Brazilian marine fungi capable of growing on DDD pesticide. Biodegradation 22:43–50

Ortiz-Hernandez M, Sanchez-Salinas E (2010) Biodegradation of the organophosphate pesticide tetrachlorvinphos by bacteria isolated from agricultural soil in Mexico. Revista Internacional de Contaminacion Ambiental 26:27–38

Ouyang ZC, Wang YH, Li XN (2008) Test of pesticide degradability by *Sphingomonas yanoikuyae* XJ strain. Huanan Nongye Daxue Xuebao 29:47–49

Pai SG, Riley MB, Camper ND (2001) Microbial degradation of mefenoxam in rhizosphere of *zinnia anguistifolia*. Chemosphere 44(4):577–582

Paingankar M, Jain M, Deobagkar D (2005) Biodegradation of allethrin, a pyrethroid insecticide, by an *Acidomonas* sp. Biotechnol Lett 27:1909–1913

Pattanasupong A, Nagase H, Inoue M (2004) Biodegradation of persistent pesticides by immobilized microbial consortia for use in paddy fields. Biotechnol Sust Util Biol Resour Tropics 17:421–425

Peng X, Zhang JS, Li YY, Li W, Xu GM, Yan YC (2008) Biodegradation of insecticide carbofuran by *Paracoccus* sp. YM3. J Environ Sci Health 43:588–594

Pilon-Smits E (2005) Phytoremediation. Annu Rev Plant Biol 56:15–39

Pinto AP, Serrano C, Pires T (2012) Degradation of terbuthylazine, difenoconazole and pendimethalin pesticides by selected fungi cultures. Sci Total Environ 43:402–410

Prabakaran V, Peterson A (2006) Effect of *Pseudomonas* on biodegradation of pesticide in the fish *Cyprinus carpio*. J Ecotoxicol Environ Monit 16:459–464

Qiao L, Wang J (2010) Biodegradation characterization of a pyridine-degrading strain. Qinghua Daxue Xuebao 50:869–872

Qiao CL, Yan YC, Shang HY, Zhou XT, Zhang Y (2003) Biodegradation of pesticides by immobilized recombinant *Escherichia coli*. Bull Environ Contam Toxicol 71:370–374

Rasmussen G, Olsen RA (2004) Sorption and biological removal of creosote-contaminants from groundwater in soil/sand vegetated with orchard grass (*Dactylis glomerata*). Adv Environ Res 8:313–327

Romeh A (2001) Biodegradation of carbosulfan, pirimicarb and diniconazole pesticides by *Trichoderma* spp. J Environ Res 3:162–172

Sabdono A, Radjasa OK (2008) Phylogenetic diversity of organophosphorous pesticide-degrading coral bacteria from Mid-West Coast of Indonesia. Biotechnology 7:694–701

Sagar V, Singh DP (2011) Biodegradation of lindane pesticide by non white-rots soil fungus *Fusarium* sp. World J Microbiol Biotechnol 27:1747–1754

Salt DE, Smith RD, Raskin I (1998) Phytoremediation. Annu Rev Plant Physiol 49:643–668

Sandman E, Loos MA (1984) Enumeration of 2, 4-D degrading microorganisms in soils and crop plant rhizospheres using indicator media: high populations associated with sugarcane (*Saccharum officinarum*). Chemosphere 13:1073–1084

Santacruz G, Bandala ER, Torres LG (2005) Chlorinated pesticides (2,4-D and DDT) biodegradation at high concentrations using immobilized *Pseudomonas fluorescens*. J Environ Sci Health B Pesticides Food Contam Agric Wastes 40:571–583

Sato K (1989) In: Vancura V, Kunc F (eds) Interrelationships between microorganisms and plants in soil. Elsevier, New York, pp 335–342

Savadogo PW, Savadogo A, Ouattara AS, Sedogo MP, Traoré AS (2007) Anaerobic biodegradation of sumithion an organophosphorus insecticide used in Burkina Faso agriculture by acclimatized indigenous bacteria. Pak J Biol Sci 10:1896–1905

Sayler GS, Hooper SW, Layton AC, King JMH (1990) Catabolic plasmids of environmental and ecological significance. Microb Ecol 19:1–20

Schroll R, Becher HH, Dorfler U, Gayler S, Grundmann S, Hartmann HP, Ruoss J (2006) Quantifying the effect of soil moisture on the aerobic microbial mineralization of selected pesticides in different soils. Environ Sci Technol 40:3305–3312

Seibert K, Fuehr F, Cheng HH (1981) In theory and practical use of soil-applied herbicides symposium. Eur Weed Resour Soc, Paris, pp 137–146

Seo J, Lee YG, Kim SD, Cha CJ, Ahn JH, Hur HG (2005) Biodegradation of the insecticide N, N-diethyl-m-toluamide by fungi: identification and toxicity of metabolites. Arch Environ Contam Toxicol 48:323–328

Shaw LJ, Richard GB (2004) Enhanced mineralization of [U-14C] 2,4-dichlorophenoxyacetic acid in soil from the Rhizosphere of *Trifolium pratense*. Appl Environ Microbiol 70:4766–4774

Shivaramaiah HM, Kennedy IR (2006) Biodegradation of endosulfan by a soil bacterium. J Environ Sci Health B 41:895–905

Shrivastava S, Prasad R, Varma A (2014) Anatomy of root from eyes of a microbiologist. In: Morte A, Varma A (eds) Root Engineering. Springer, Berlin, pp 3–22

Siddique T, Okeke BC, Arshad M, Frankenberger WT Jr (2002) Temperature and pH effects on biodegradation of hexachlorocyclohexane isomers in water and soil slurry. J Agric Food Chem 50:5070–5076

Singh DK (2008) Biodegradation and bioremediation of pesticide in soil: concept, method and recent developments. Indian J Microbiol 48:35–40

Singh A, Ranjit JA, Dhasarathan P (2003) Isolation and characterization of pesticide metabolizing bacteria from groundnut soil system. Asian J Microbiol Biotech Environ Sci 5:323–332

Singh BK, Walker A, Morgan JAW, Wright DJ (2004) Biodegradation of chlorpyrifos by Enterobacter strain B-14 and its use in bioremediation of contaminated soils. Appl Environ Microbiol 70:4855–4863

Singh BK, Walker A, Wright DJ (2006) Bioremedial potential of fenamiphos and chlorpyrifos degrading isolates: influence of different environmental conditions. Soil Biol Biochem 38:2682–2693

Slaoui M, Ouhssine M, Berny E, Elyachioui M (2007) Biodegradation of the carbofuran by a fungus isolated from treated soil. Afr J Biotechnol 6:419–423

Soromotin A, Ogurtsova L, Morozova T, Zhdanova EB (2012) Method of microbiological degradation of organochlorine pesticides. Russian Patent 2448786

Stephen DY, Shozo K (1980) Effect of organic matter and chemical fertilizers on the degradation of benthiocarb and mcpa herbicides in the soil. Soil Sci Plant Nutr 26:541–549

Stomp AM, Han KH, Wilbert S, Gordon MP, Cunningham SD (1994) Genetic strategies for enhancing phytoremediation. Ann N Y Acad Sci 721:481–492

Sun H, Xu J, Yang S, Liu G, Dai S (2004) Plant uptake of aldicarb from contaminated soil and its enhanced degradation in the rhizosphere. Chemosphere 54:569–574

Surekha RM, Lakshmi PKL, Suvarnalatha D (2008) Isolation and characterization of a chlorpyrifos degrading bacterium from agricultural soil and its growth response. Afr J Microbiol 2:26–31

Tallur PN, Megadi VB, Ninnekar HZ (2008) Biodegradation of Cypermethrin by *Micrococcus* sp. strain CPN 1. Biodegradation 19:77–82

Tian LS, Chen F (2012) Biological characteristics and degradation performance of a degrading strain. Yangzhou Daxue Xuebao 33:86–90

Tingting L, Kunming D, Miao L et al (2012) Isolation, identification and biodegradation characteristics of a bacterial strain able to degrade bifenthrin. Nongye Huanjing KexueXuebao 31:1147–1152

Veiga MM, Silva DM, Veiga LBE, de Castro Faria MV (2006) Pesticide pollution in water systems in a small rural community in Southeast Brazil. Cad Saude Publica 22:2391–2399

Vidali M (2001) Bioremediation: an overview. Pure Appl Chem 73:1163–1172

Wang M, Yang G, Wang X, Yao Y, Min H, Lu Z (2011) Nicotine degradation by two novel bacterial isolates of *Acinetobacter* sp. TW and *Sphingomonas* sp. TY and their responses in the presence of neonicotinoid insecticides. World J Microbiol Biotechnol 27:1633–1640

Wyss A, Boucher J, Montero A, Marison I (2006) Micro-encapsulated organic phase for enhanced bioremediation of hydrophobic organic pollutants. Enzym Microb Technol 40:25–31

Xiang-Ming L, Ping-Ping C (2012) Progress in degradation of organo-chlorinated pesticides by microorganism. Huanjing Kexue Yu Jishu 35:89–93

Xie S, Liu J, Li L, Qiao C (2009) Biodegradation of malathion by Acinetobacter johnsonii MA19 and optimization of cometabolism substrates. J Environ Sci 21:76–82

Xue-Dong W, Xiao-Ming O, Hui-Li W (2003) Optimized cultivation of highly-efficient bacterial strains and their biodegradation ability towards imazapyr. Nongye Huanjing Kexue Xuebao 22:102–105

Yin L, Li X, Liu Y, Zhang D, Zhang S, Luo X (2012) Biodegradation of cypermethrin by rhodopseudomonas palustris GJ-22 isolated from activated sludge. Fresenius Environ Bull A 21:397–405

Yoshii K, Tonogai Y, Ueno H, Nakamuro K (2006) Malathion residue in wheat kernels is degraded by thion organophosphorus pesticide-specific carboxylesterase. J Health Sci 52:221–227

Yu YL, Chen YX, Luo YM, Pan XD, He YF, Wong MH (2003) Rapid degradation of butachlor in wheat rhizosphere soil. Chemosphere 50:771–774

Yu YL, Fang H, Wang X, Wu XM, Shan M, Yu JQ (2006) Characterization of a fungal strain capable of degrading chlorpyrifos and its use in detoxification of the insecticide on vegetables. Biodegradation 17:487–494

Zeinat K, Nashwa AH, Ibrahim M (2008) Biodegradation and detoxification of malathion by of *Bacillus thuringiensis* MOS-5. Aust J Basic Appl Sci 2:724–732

Zhang D, Xinqiu T, Xiangwen L (2005) Isolation of photosynthetic bacteria HP-1 with degradation of organic-phosphorus insecticides and studies on its biodegradation ability and capacity of increasing growth. Shengming Kexue Yanjiu 9:247–253

Zhang X, Wu W, Zhang Y et al (2007) Screening of efficient hydrocarbon-degrading strains and study on influence factors of degradation of refinery oily sludge. Ind Eng Chem Res 46:8910–8917

Zhang H, Yang C, Li C, Li C, Zhao Q, Qiao C (2008) Functional assembly of a microbial consortium with autofluorescent and mineralizing activity for the biodegradation of organophosphates. J Agric Food Chem 56:7897–7902

Zhang J, Sun Z, Li Y, Peng X, Li W, Yan Y (2009) Biodegradation of p-nitrophenol by *Rhodococcus* sp. CN6 with high cell surface hydrophobicity. J Hazard Mater 163:723–728

Zhao P, Wen-jing Q, Wang Y (2012) Isolation and identification of two strains capable of degrading of organophosphorus pesticides and their biodegradation abilities. Lanzhou LigongDaxue Xuebao 36:86–90

Chapter 9
Algae and Cyanobacteria as Biocontrol Agents of Fungal Plant Pathogens

Hillary Righini and Roberta Roberti

Abstract Since long time, algae are used in agriculture as soil amendment for their beneficial effects on plant health and productivity. In fact, algae contain several molecules such as plant growth hormones (cytokinins, auxins, abscisic and gibberellic acid), polysaccharides, betaines and micronutrients. The research on algae, their compounds and their effects on plants have started in the middle 1950s and brought to the formulations of liquid products containing extracts with compounds readily available for plants. The algae extracts, besides having effects on plant growth, have demonstrated to improve plant resistance to both abiotic and biotic stresses. Among biotic stresses, algae showed antifungal activity against different pathogens especially of horticultural plants. From the middle of last century, plant management has always been dependent from the market demand that required growing quantity of 'perfect' fruits and vegetables over the year. In this scenario, the chemical industry of fertilizers and pesticides developed new products that have been used for years. In particular, pesticides have represented the base of the management of fungal plant pathogens. During the last decades, the use of both pesticides and chemical fertilizers has represented a serious risk for human health and brought disorder of ecosystem equilibrium. Consequently, algae for their biostimulant and antifungal effects may be considered useful tools to reduce the input of chemicals in integrated pest management strategies. In line with these strategies, the European Regulation EC 1107/2009, concerning the placing of plant protection products on the market and repealing Council Directives 79/117/EEC and 91/414/EEC, recommends that priority should be given to non-chemical and natural alternatives wherever possible.

H. Righini · R. Roberti (✉)
Department of Agricultural and Food Sciences, Alma Mater Studiorum, University of Bologna, Bologna, Italy
e-mail: roberta.roberti@unibo.it

A. Varma et al. (eds.), *Plant Microbe Interface*,
https://doi.org/10.1007/978-3-030-19831-2_9

9.1 Introduction

As stated by the European Commission, the traditional strategy for preventing, destroying or controlling a harmful organism ('pest') or disease or protecting plants or plant products during production, storage and transport is mainly based on the use of synthetic 'pesticides'. This term includes herbicides, fungicides, insecticides, acaricides, nematicides, molluscicides, rodenticides, growth regulators, repellents, rodenticides and biocides (EU Pesticide Database). Upon the recent EC regulation No. 1107/2009, concerning the placing of plant protection products on the market and repealing the Council Directives 91/414/EEC, the use of many synthetic pesticides is currently outlawed or limited for their harmful effects on nontarget organisms and environment and for health hazards. Therefore, the management of fungal plant pathogens has become problematic and greatly stimulated the research of alternative solutions to chemicals. Moreover, under the same EC regulation, no synthetic soil fumigants are allowed, making the control of soilborne pathogens extremely difficult. In this regard, the research on plant disease control is focusing on alternative means to chemical ones, considering also consumer demands of healthy foods obtained with cultivation system with low ambient impact. Since long time, a sustainable approach to crop protection is represented by alternatives to synthetic pesticides such as microorganisms and compounds derived from plants (Gwinn 2018). Among alternatives, algae and cyanobacteria can be promising bioactive agents, which have recently received considerable attention by scientists.

This chapter presents a brief overview of the recent scientific literature on the antifungal activity of algae and cyanobacteria against several plant pathogens that affect root system, stem, leaf and fruit, their potential as biocontrol agents and their role in activating systemic plant defences, which results in the increase of defence-related enzymes and their gene expression.

9.1.1 *Algae and Cyanobacteria Against Plant Pathogens*

Algae and cyanobacteria have long been used in human consumption, as a source of natural compounds of pharmaceutical and cosmetic interest. Moreover, they are used for the production of biofuels and substrates for microbiological culture media. In agriculture, algae and cyanobacteria extracts are commonly used for their beneficial effects, because they stimulate plant vigour and productivity, but their effect against fungal pathogens is scarcely known. Fungal pathogens can infect plants in all cultivation systems and can be responsible of fruit decay and severe losses in postharvest. *Sclerotinia sclerotiorum* and several species of *Rhizoctonia*, *Phytophthora*, *Pythium*, *Fusarium* and *Verticillium* are among the most important polyphagous soilborne fungi (Husaini and Neri 2016; Pastrana et al. 2016). They attack the root system, hampering the nutrient and water absorption from soil, causing yellowing, wilting, damping-off, root rot and collar rot. Among foliar pathogens, fungi belonging to *Erysiphales* order are agents of powdery mildew that cause important economic losses (Jarvis et al. 2002) and require several chemical applications (Romero et al. 2007). Harvested fruits, such as tomato, cucumber and strawberry, are perishable fresh food. *Botrytis cinerea*, *Colletotrichum* spp., agent of grey mould and anthracnose, respectively, and several species of *Penicillium*, *Rhizopus* and *Mucor* are common fungal pathogens responsible of postharvest decay (Husaini and Neri 2016). These pathogens are usually controlled with the application of fungicides during the crop growing cycle, from flowering until harvest time.

On the base of the above-cited EU restriction on synthetic pesticides, application of algae and cyanobacteria extracts can be considered an alternative to chemical treatment in crop protection. Indeed, crop protection in Europe has become problematic due to the recent restrictions in pesticide authorization (Regulation 2009/1107/EC) and sustainable use (Directive 2009/128/EC and subsequent amendments) and due to the lack of effective alternatives, in particular for horticultural plants in open field and in soil or soilless cultivation under greenhouse conditions. The existing findings show the potential of algae and cyanobacteria as biocontrol agents of fungal plant pathogens through application of their extracts or substances derived from the extracts.

9.2 Algae

Algae are a large diversified group of aquatic photosynthetic organisms, which include unicellular organisms such as green algae, for example, *Chlorella*, and marine multicellular algae (seaweeds), such as *Sargassum*, a brown alga that can reach 1–3 m in length. Algae are classified in several phyla (Table 9.1) among which Chlorophyta (green algae), Rhodophyta (red algae) and Ochrophyta are the major ones. Ochrophyta phylum includes the class Phaeophyceae that is commonly named brown algae (Guiry 2012). The season of harvest and geographic location influence algal composition (Black 1950; Painter 1983; Westermeier et al. 2012) and the content of compounds such as polysaccharides (Rioux et al. 2007; Schiener et al. 2015). Other algal compounds are essential nutrients; trace of metals such as Cu, Co, Zn, Mn, Mo, etc. (Cabrita et al. 2016); and plant hormones like abscisic acid, auxins and cytokinins (Rayorath et al. 2008; Craigie 2011). Moreover, bioactive compounds with antimicrobial, antiviral, anticancer, antioxidant and antifungal properties occur in algae. They have been studied for several applications such as in production of pigments, bioactive substances, pharmaceuticals and cosmetics (Sharma and Sharma 2017). Due to algae composition, their extracts have long been used in agriculture to increase soil fertility and crop productivity (Khan et al. 2009; Craigie 2011; Arioli et al. 2015). Beneficial effects after application of algae extracts are reported on several crops such as apple, strawberry, tomato, wheat and winter rapeseed (Crouch and van Staden 1992; Basak 2008; Kumar and Sahoo 2011; Alam et al. 2013; Jannin et al. 2013). Moreover, several authors (Craigie 2011;

Table 9.1 Classification of algae cited in this chapter

Phylum/Class	Alga species
Chlorophyta	*Caulerpa sertularioides*
	Chlorella
	Ulva lactuca
	Zygnema czurdae
	Zygnema stellinum
	Zygnema tenue
Phaeophyceae	*Ascophyllum nodosum*
	Cystoseira myriophylloides
	Ecklonia sp.
	Ecklonia kurome
	Durvillaea potatorum
	Fucus spiralis
	Laminaria digitata
	Leathesia nana
	Padina gymnospora
	Pelvetia canaliculata
	Sargassum
	Sargassum filipendula
	Sargassum liebmannii
	Stypopodium zonale
	Undaria pinnatifida
Rhodophyta	*Corallina* sp.
	Eucheuma denticulatum
	Gelidium pusillum
	Gracilaria edulis
	Halopithys sp.
	Kappaphycus alvarezii
	Porphyra umbilicalis
	Rhodomela confervoides

Calvo et al. 2014; Ibrahim et al. 2014; Latique et al. 2014) report that extracts from algae increased crop tolerance to a wide range of abiotic and biotic stresses and prolonged the postharvest shelf life of fruits. Extracts from algae have also great potential as plant disease protection products. They can act both directly against several fungal plant pathogens and indirectly by inducing plant resistance through the synthesis of various defence-related enzymes that can help the plants to hinder fungal invasion (Jayaraman et al. 2011; Hernández-Herrera et al. 2014; Roberti et al. 2016; Esserti et al. 2017).

9.2.1 Antifungal Activity

There are numerous articles on showing the direct effect of extracts from algae against pathogens. The antifungal activity of extracts from brown algae is widely reported (Righini et al. 2018). Aqueous and cyclohexanic extracts from *Sargassum* sp. inhibited *Aspergillus* spp. mycelial growth by 37.0% and 54.5%, respectively (Mabrouk et al. 1985; Khallil et al. 2015). The cyclohexanic extract also reduced the growth of *Fusarium oxysporum* and *Penicillium* spp. (Khallil et al. 2015). Methanolic extract from *Sargassum latifolium* and *Padina gymnospora* inhibited *Fusarium solani* and *Rhizoctonia solani* colony growth up to 84.0% (Ibraheem et al. 2017). The occurrence of phenols and terpenes in extracts from *Sargassum muticum*, *Ascophyllum nodosum*, *Fucus spiralis*, *Stypopodium zonale* and *Pelvetia canaliculata* was related to the inhibition of *Colletotrichum lagenarium* growth (Fernandes Peres et al. 2012). Complete inhibition of *Botrytis cinerea* mycelial growth and spore germination was obtained with extracts from *Laminaria digitata* and *Undaria pinnatifida* (De Corato et al. 2017). The same authors compared the antifungal activity of different extract fractions obtained by three solvents, hexane, water and ethanol, having different affinity towards fatty acids, water-soluble polysaccharides and phenolic compounds, respectively. Only the hexane-soluble extracts from algae showed to inhibit *B. cinerea* mycelial growth and spore germination especially, mainly the extract from the brown algae *L. digitata* and *U. pinnatifida* and the red alga *Porphyra umbilicalis*. Generally, brown algae extracts showed higher antifungal activity (100% for both mycelial growth and spore germination) with respect to the extracts from red algae (average of 67.0% for mycelial growth and 73.3% for spore germination). This finding is in accordance with the higher antioxidant activity of brown algae than that of green and red algae (Kelman et al. 2012). Although the lowest antioxidant activity of red algae, a methanolic extract from *Gracilaria edulis* reduced *Macrophomina phaseolina* mycelial growth by 36.7% (Ambika and Sujatha 2015), and water extracts from *Corallina* sp. and *Halopithys* reduced *Podosphaera xanthii* infection by 87.8% and 96.8%, respectively, on detached zucchini cotyledons (Roberti et al. 2016; Figs. 9.1 and 9.2). An antifungal substance, namely, bromophenol *bis* (2,3-dibromo-4,5-dihydroxybenzyl) ether (BDDE), was extracted from *Rhodomela confervoides* (red alga) and from *Leathesia nana* (brown algae) and showed to inhibit *B. cinerea* mycelial growth, spore germination and germ tube elongation (Liu et al. 2014). BDDE was also able to inhibit *Colletotrichum gloeosporioides* mycelial growth.

Among green microalgae, *Chlorella vulgaris* extract obtained by enzymatic digestion highly reduced the mycelial growth and sporulation of *B. cinerea* (El-ghanam et al. 2015). A methanolic extract also inhibited *Aspergillus niger* mycelial growth (Ghasemi et al. 2007). The antifungal activity of *C. vulgaris* extract was related to antioxidant compounds, such as phenolics and flavonoids that are very abundant in the alga (Ahmed 2016). The composition of extracts from other green microalgae, such as *Zygenma czurdae*, *Z. stellinum* and *Z. tenue*, showed different content in saturated and unsaturated fatty acids, sterols and terpenes. These

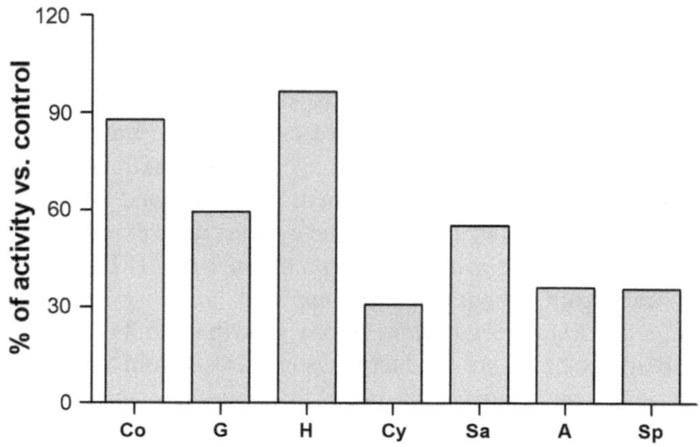

Fig. 9.1 Effective extracts from algae and cyanobacteria against disease severity caused by *Podosphaera xanthii* on zucchini (adapted from Roberti et al. 2016)

compounds were correlated to the antifungal activity of the extracts on mycelial growth of *Curvularia lunata*, *Fusarium sporotrichioides*, *M. phaseolina*, *R. solani* and *Sclerotium rolfsii* (Ghazala and Shameel 2005). A direct antifungal activity was also showed by extracts from the green macroalga *Ulva lactuca*, which highly inhibited *A. niger*, *Penicillium digitatum* and *R. solani* mycelial growth (Abbassy et al. 2014). The highest antifungal activity was showed by the fraction containing phthalic acid, aromatic compounds and fatty acids.

9.2.2 Disease Control

The disease suppressive activity of extracts from algae has been demonstrated on several crops after their application through soil irrigation or spray treatment onto the leaves or fruits. The disease control by brown algae extracts is widely reported (Righini et al. 2018). Soil treatment with extract from *E. maxima* significantly controlled *Verticillium* wilt disease of pepper (Rekanović et al. 2010), and a spray treatment on zucchini cotyledons with *Ecklonia* sp. was able to slightly reduced powdery mildew caused by *P. xanthii*, hampering fungal sporulation (Roberti et al. 2016). The antifungal activity showed by *Ecklonia* extracts can be correlated to the antioxidant activities of secondary bioactive compounds (e.g. phenols) as demonstrated against both human and plant pathogens. Algae phenols, such as phlorotannins from *Ecklonia kurome*, were active against food-borne pathogenic bacteria, especially versus strains of methicillin-resistant *Staphylococcus aureus* (Nagayama et al. 2002). Phenols from *Ecklonia cava* also displayed high antioxidant and anticancer activities against murine colon cancer cell line CT-26 (Athukorala et al. 2006). Again as soil treatment, powder of *P. gymnospora*, *S. latifolium* and *Hydroclathrus clathratus* significantly decreased by 83.6% and 27.0%, respectively,

Fig. 9.2 Effect of water extracts from the algae *Spirulina platensis* (S) and *Corallina* sp. (C) and from the cyanobacterium *Anabaena* sp. (A) on the control of powdery mildew caused by *Podosphaera xanthii* on zucchini cotyledons (from Roberti et al. 2016). IC, infected control (from Roberti et al. 2016)

the root rotting caused by *F. solani* on eggplant, in pot experiment (Ibraheem et al. 2017). In a study, De Corato et al. (2017) examined the effect of application of extracts from the brown algae, *L. digitata* and *U. pinnatifida*, and red algae, *Eucheuma denticulatum*, *Gelidium pusillum* and *P. umbilicalis*, on strawberry fruits against the grey mould disease caused by *B. cinerea*. Both preventive and curative treatments with extracts from *L. digitata*, *P. umbilicalis* and *U. pinnatifida* highly suppressed the disease severity by 60.0% (curative activity) and by 65.0% (preventive activity), while extracts from *E. denticulatum* and *G. pusillum* slightly reduced grey mould. Again on strawberry, a mixture of the brown algae *A. nodosum* + *Laminaria* sp. applied as spray treatments decreased grey mould incidence by 48.0% (Boček et al. 2012).

Species of brown alga *Laminaria* are source of laminarin, a polysaccharide that is exploited for the production of a plant protection product, which is already commercialized in many EU countries (EU Pesticide Database) for its capacity to induce plant resistance. Indeed, studies focused on the activity of laminarin from *Laminaria* sp. against foliar and fruit pathogens. On grapevine, the application of laminarin reduced powdery mildew caused by *Erysiphe necator* by an average of 58.2% in a 2-year experiment (Pugliese et al. 2018). Under field conditions, treatments of several strawberry cultivars with laminarin controlled the infection caused by *B. cinerea*, *Podosphaera aphanis* (powdery mildew) and *M. fragariae* (leaf spot) (Meszka and Bielenin 2011). Repeated preharvest applications of laminarin on strawberry reduced decay incidence and severity caused by *B. cinerea* and *Rhizopus* sp. (Feliziani et al. 2015). Krawiec et al. (2016) obtained effective control of raspberry grey mould and hypothesized that by replacing one or two chemical fungicide application with a product based on laminarin, it is possible to reduce the phenomenon of *B. cinerea* resistance to fungicides. Another bioactive compound of algae is bromophenol BDDE, which has been extracted from the brown alga *L. nana* and from the red alga *R. confervoides*. Fresh strawberry fruits treated by soaking with BDDE showed low decay incidence by *B. cinerea*. BDDE was able to interact with the fungus DNA by binding in the minor groove as well as by intercalation (Liu et al. 2014). With regard to green algae, *C. vulgaris* applied on strawberry controlled the disease severity rot caused by *B. cinerea* (El-ghanam et al. 2015).

As mentioned above, algal extracts work against fungal pathogens not only by direct antifungal activity but also through the induction of plant resistance. Cell wall and storage polysaccharides from brown, red and green algae can activate defence responses in plants increasing protection against pathogens (Vera et al. 2011). This characteristic has been investigated by several authors (Table 9.2). Among brown algae, extract from *A. nodosum* has been mostly studied on several plant species. On carrot, Jayaraj et al. (2008) showed that the extract applied by spray treatment reduced infections caused by *B. cinerea* and *Alternaria radicina* and increased the transcript levels of pathogenesis-related proteins (PRs) which are molecules associated to plant-induced resistance. On cucumber, *A. nodosum* extract sprayed on plants or applied by root drench reduced disease incidence caused by *Alternaria cucumerinum*, *B. cinerea* and *F. oxysporum* (Jayaraman et al. 2011) and by *Phytophthora melonis*, in greenhouse environment (Abkhoo and Sabbagh 2016). The extract from *A. nodosum* enhanced the synthesis of PRs including those with enzymatic activity such as chitinase, β-1,3-glucanase, lipoxygenase, peroxidase, phenylalanine ammonia lyase and polyphenol oxidase and increased content of phenolic compounds in treated cucumber plants (Jayaraman et al. 2011; Abkhoo and Sabbagh 2016). In particular, the increase of lipoxygenase activity is involved in the oxidation of membrane lipids that generate signals leading to the production of several antifungal compounds, cell wall lignification and systemic acquired resistance (Rustérucci et al. 1996; Guo et al. 1998). The application of a mixture of *A. nodosum* + *Durvillaea potatorum* extracts on soil controlled *Plasmodiophora brassicae* infection on broccoli, reducing the number of plasmodia formed in the root hairs (Wite et al. 2015). These authors supposed that the disease control was due to the stimulation of resistance mechanisms in the plant because of the presence of

Table 9.2 Effective applications of alga extracts against fungal pathogens correlated to the increase of plant defence responses

Alga	Plant	Increase of plant defence responses	Pathogen	References
Ascophyllum nodosum	Cucumber	Enzymes: chitinase, β-1,3-glucanase, peroxidase, polyphenol oxidase, phenylalanine ammonia lyase and lipoxygenase Genes: chitinase, lipoxygenase, glucanase, peroxidase and PAL	*Alternaria cucumerinum*, *Didymella applanata*, *Fusarium oxysporum*, *Botrytis cinerea*	Jayaraman et al. (2011)
A. nodosum	Cucumber	Enzymes: β-1,3-glucanase, peroxidase, polyphenol oxidase Genes: lipoxygenase, PAL, galactinol synthase	*Phytophthora melonis*	Abkhoo and Sabbagh (2016)
A. nodosum	Carrot	Enzymes: peroxidase, polyphenol oxidase, PAL, chitinase, β-1,3-glucanase Genes: PR-1, chitinase, lipid transfer protein, PAL, chalcone synthase, non-expressing pathogenesis-related protein, PR-5	*Alternaria radicina*, *B. cinerea*	Jayaraj et al. (2008)
Cystoseira myriophylloides, *Laminaria digitata*, *Fucus spiralis*	Tomato	Enzymes: polyphenol oxidase, peroxidase	*Verticillium dahliae*	Esserti et al.
Ulva lactuca, *Sargassum filipendula*, *Gelidium serrulatum*	Tomato	Genes: PR-1, PIN II	*Alternaria solani*	Ramkissoon et al. (2017)
Ulva lactuca, *Caulerpa sertularioides*, *Padina gymnospora*, *Sargassum liebmannii*	Tomato	Genes: wound-induced proteinase inhibitor I, wound-induced proteinase inhibitor II, wound-inducible carboxypeptidase, PAL; cysteine proteinase changes were observed for PI–PII	*A. solani*	Hernández-Herrera et al. (2014)

(continued)

Table 9.2 (continued)

Alga	Plant	Increase of plant defence responses	Pathogen	References
Kappaphycus alvarezii	Tomato	Hormones: ABA, IAA, SA and zeatin Genes: PR-1b1, PR-3 and PR-5	*Macrophomina phaseolina*	Agarwal et al. (2016)

PR pathogenesis-related protein, *PIN II* marker gene of the jasmonate-mediated defence pathway, *PAL* phenylalanine ammonia lyase, *ABA* abscisic acid, *IAA* indole-3-acetic acid, *SA* salicylic acid

polysaccharides in the extract. Indeed, several compounds are contained in *A. nodosum* extracts, namely, polysaccharides and oligosaccharides, which act as elicitor molecules in plants by inducing the expression of various PRs and proteinase inhibitors (Cluzet et al. 2004; Walters et al. 2005).

An extract from *Sargassum filipendula* reduced disease symptoms caused by *Alternaria solani* and *Xanthomonas campestris* pv. *vesicatoria* on tomato plants (Ramkissoon et al. 2017). Again, in tomato, protection against *A. solani* was observed after application of a polysaccharide-enriched extract from the green algae, *Ulva lactuca* (43.0%) and *Caulerpa sertularioides* (39.0%), and the brown algae, *P. gymnospora* (42.0%) and *Sargassum liebmannii* (31.0%) (Hernández-Herrera et al. 2014). In particular, *U. lactuca* extracts increased the expression of defence-related genes, such as phenylalanine ammonia lyase, that could justify the highest reduction of necrotic lesions caused by *A. solani*. Again, in tomato, extracts from *Cystoseira myriophylloides*, *L. digitata* and *F. spiralis* controlled *Verticillium dahliae* wilt and enhanced defence enzyme activities such as polyphenol oxidases and peroxidases (Esserti et al. 2017). In tomato seedlings infected by *M. phaseolina*, the application of the red alga *Kappaphycus alvarezii* increased transcription of PR genes such as PR-1b1, PR-3 and PR-4 and the levels of phytohormones, namely, abscisic acid, indole-3-acetic acid, salicylic acid and the cytokinin zeatin (Agarwal et al. 2016). Accordingly, a high concentration of cytokinin triggers accumulation of salicylic acid and defence gene expression (Argueso et al. 2012).

9.3 Cyanobacteria

Cyanobacteria, otherwise known as *Cyanophyta* or blue-green algae, are among the most various and ubiquitous prokaryotes (Walter et al. 2017) included in Bacteria superkingdom (Oren and Garrity 2014). Their origin dates back to 3 billion years ago, during the transition from anoxygenic to oxygenic conditions through photosynthesis (Schirrmeister et al. 2011). These conditions influenced their evolution by leading to their development in unicellular and multicellular, photosynthetic and non-photosynthetic free-living, symbiotic, toxic and predatory organisms (Schirrmeister et al. 2011; Di Rienzi et al. 2013; Soo et al. 2014, 2015).

Cyanobacteria occupy many niches such as terrestrial, planktonic and benthic habitats (Walter et al. 2017). They are often called 'blue-green algae' due to the blue pigment phycocyanin that together with chlorophyll *a* and other pigments is used to capture light for photosynthesis. They are divided in eight orders: *Gloeobacterales*, *Synechococcales*, *Spirulinales*, *Chroococcales*, *Pleurocapsales*, *Oscillatoriales*, *Chroococcidiopsidales* and *Nostocales* (Komárek et al. 2014). *Chroococcales*, *Oscillatoriales* and *Nostocales* produce the majority of compounds with toxic activity. In particular, *Anabaena* sp. and *Nostoc* sp. among the *Nostocales* synthetize many secondary metabolites with anticancer, antimalarial, antimicrobial and antifungal activity (Burja et al. 2001).

In the last few years, considering the harmful effect on the environment of agrochemical products such as fertilizers and pesticides, the research is focusing on alternatives to chemicals, and cyanobacteria can be a promising candidate. The research on application in agriculture of cyanobacteria highlights their potential as source of both biofertilizers and bioactive compounds for plant disease control. The capacity of cyanobacteria to carry out nitrogen fixation has long been used for rice fertilization (Yanni 1991; Jha et al. 1999; Jha and Prasad 2006; Sinha et al. 2002). In addition, as a natural fertilizer for rice, the symbiotic association between *Anabaena azollae* and the water fern *Azolla*, both capable of fixing atmospheric nitrogen, has long been used in China and in other countries in the Far East (Shi and Hall 1988; Kannaiyan 2000). Along with their capacity of fixing atmospheric N, they improved plant growth and crop yield since they are sources of organic matter to soil (Zaccaro et al. 1999; Maqubela et al. 2009), thus ameliorating soil structure (De Caire et al. 2000; De Cano et al. 2002; Pandey et al. 2005; Maqubela et al. 2009). Beneficial effects on crop yield were attributed to various substances released in the soil such as amino acids, vitamins, polypeptides and antibacterial and antifungal substances, especially exopolysaccharides (Zaccaro et al. 1999; Singh et al. 2005; Maqubela et al. 2009). Cyanobacteria, in particular *Anabaena* sp. and *Scytonema hofmanni*, are also able to synthesize phytohormones, such as indole-3-acetic acid and gibberellin-like substances (Prasanna et al. 2008, 2009; Rodríguez et al. 2006). Bioactive compounds synthetized by cyanobacteria may play a role in the control of fungal plant pathogens by acting directly against pathogens and indirectly by activating genes responsible of plant systemic resistance.

9.3.1 Antifungal Activity

Cyanobacteria have been recognized as one of the most promising sources of natural bioactive compounds with antimicrobial (Burja et al. 2001), anti-protozoal (Simmons et al. 2008), anticancer (Russo and Cesario 2012), antiviral, antibacterial and antiproliferative activities (Dixit and Suseela 2013) (Fig. 9.3). Among their various activities, several authors showed efficacy against fungal colony growth of several plant pathogens (Table 9.3). Most of the studies have regarded several species of *Anabaena*, *Nostoc* and *Microcystis*. A crude ethanolic extract from

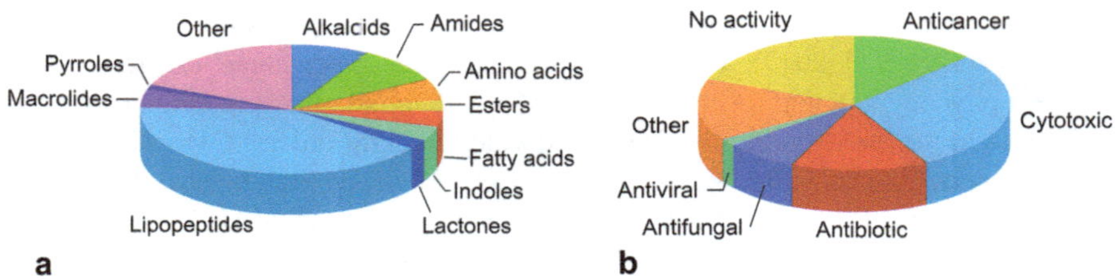

Fig. 9.3 Compounds isolated from cyanobacteria (**a**) and their main biological activity (**b**) (adapted from Burja et al. 2001)

Anabaena laxa showed inhibitory effect against different fungi, namely, *Aspergillus oryzae*, *Candida albicans*, *Penicillium notatum*, *Saccharomyces cerevisiae* and *Trichophyton mentagrophytes* (Frankmölle et al. 1992a, b). From a screening conducted by Prasanna et al. (2008), 23 strains of *Anabaena* reduced *Fusarium moniliforme* colony growth, while 17 strains inhibited that of *A. solani*. *Nostoc muscorum* is known to be effective against *R. solani* (de Caire et al. 1990), and several strains of *Anabaena* and *Calothrix* exerted activity against species of *Pythium*, *Fusarium* and *Rhizoctonia* (Moon and Martin 1981; Prasanna et al. 2008; Radhakrishnan et al. 2009; Manjunath et al. 2010). Against *R. solani*, also *N. muscorum* was effective and inhibited the colony growth more than *N. entophytum* (Osman et al. 2011). Among all the compounds synthetized by cyanobacteria, chitosanase homologues, endoglucanase and benzoic acid were detected, and their presence was correlated to the activity against fungi (Gupta et al. 2010, 2011; Natarajan et al. 2012; Prasanna et al. 2010). Chitosanase enzymes selectively decompose chitosan and chitin by hydrolysis of the β-1,4-glycosidic bonds that link N-acetyl glucosamine residues of chitin. This degradation is the mechanism of chitosanase antifungal activity. In addition to hydrolytic enzymes, cyanobacteria are able to produce phenolic compounds that were active against *C. albicans* mycelial growth (de Cano et al. 1990). Moreover, the terpenoid noscomin, extracted from *N. commune*, showed activity against the bacteria *Bacillus cereus*, *Staphylococcus epidermidis* and *Escherichia coli* (Jaki et al. 2000). Several studies have regarded the antifungal effect of *Microcystis aeruginosa* extracts. A methanolic extract from *M. aeruginosa* showed high antifungal activity against seven human pathogens, five plant pathogens (Table 9.3) and eight saprophytes (Khalid et al. 2010). Analysis of this extract revealed the presence of fatty acids, and, among them, those unsaturated were present in a larger proportion than saturated ones. Another strain of *M. aeruginosa* demonstrated to inhibit the growth of *A. flavus*, *A. niger*, *Fusarium verticillioides* and *F. proliferatum* (Marrez and Sultan 2016). The main substances which were responsible of the antifungal activity were identified as butylated hydroxytoluene, hexadecanoic acid and methyl ester.

Table 9.3 Antifungal activity of extracts from cyanobacteria on colony growth of plant pathogens

Cyanobacteria	Pathogen	References
Anabaena sp.	*Alternaria alternata, Botrytis cinerea, Colletotrichum gloeosporioides, Fusarium oxysporum*	Kim (2006)
Anabaena sp.	*Macrophomina phaseolina, Fusarium moniliforme, Alternaria solani, Pythium aphanidermatum, Fusarium solani*	Prasanna et al. (2008)
A. laxa	*F. moniliforme, F. oxysporum* f. sp. *lycopersici*	Prasanna et al. (2013)
Anabaena variabilis	*F. moniliforme, F. oxysporum* f. sp. *lycopersici*	Prasanna et al. (2013)
A. variabilis	*Aspergillus niger, A. solani*	Tiwari and Kaur (2014)
Calothrix sp.	*A. alternata, B. cinerea, C. gloeosporioides, F. oxysporum, Phytophthora capsici, Pythium ultimum*	Kim (2006)
Fischerella sp.	*A. niger, Aspergillus fumigatus*	Ghasemi et al. (2003)
Microcystis aeruginosa	*F. oxysporum, M. phaseolina, P. aphanidermatum, Pythium oedochilum, Rhizoctonia solani*	Khalid et al. (2010)
Microcystis aeruginosa	*Aspergillus flavus, Fusarium verticillioides, Fusarium proliferatum*	Marrez and Sultan (2016)
Nodularia sp.	*A. alternata, B. cinerea, C. gloeosporioides, F. oxysporum*	Kim (2006)
Nostoc sp.	*A. alternata, B. cinerea, C. gloeosporioides, F. oxysporum, P. capsici, P. ultimum, Rhizopus stolonifer*	Kim (2006)
Nostoc calcicola	*F. solani, Penicillium chrysogenum*	Yadav et al. (2016)
Nostoc commune	*F. oxysporum* f. sp. *lycopersici*	Kim and Kim (2008)
Nostoc entophytum	*R. solani*	Osman et al. (2011)
Nostoc linckia	*F. oxysporum* f. sp. *lycopersici*	Alwathnani and Perveen (2012)
N. muscorum	*R. solani*	Osman et al. (2011)
Oscillatoria sp.	*A. alternata, B. cinerea, C. gloeosporioides, F. oxysporum, P. ultimum, P. capsici*	Kim (2006)
Phormidium autumnale	*F. oxysporum* f. sp. *lycopersici*	Alwathnani and Perveen (2012)
Spirulina platensis	*A. niger, A. solani*	Tiwari and Kaur (2014)
Synechococcus elongates	*A. niger, A. solani*	Tiwari and Kaur (2014)
Stigonema sp.	*A. fumigatus, A. niger*	Ghasemi et al. (2003)

9.3.2 Disease Control

The activity of cyanobacteria against plant pathogens has been quite studied through their application on both soil and leaf. Soil application of *Nostoc muscorum* and *N. entophytum* significantly increased soybean seedling survival in soil infected with *R. solani* and enhanced plant length, root and shoot dry weight of infected seedlings (Osman et al. 2011). Alwathnani and Perveen (2012) reported that soil application of *Nostoc linckia* reduced wilt disease caused by *F. oxysporum* f. sp. *lycopersici* on tomato plants. Again, on tomato, the treatment with *N. commune* of seeds, subsequently infected with *F. oxysporum*, reduced the average number of diseased seedlings (Kim and Kim 2008). Cyanobacteria extracts showed efficacy also against foliar pathogens as shown by Roberti et al. (2015) following application of *Anabaena* sp. on zucchini cotyledons against *P. xanthii*, causal agent of powdery mildew. The treatment with *Anabaena* sp. extract increased defence enzyme activities such as endochitinase, β-*N*-acetylhexosaminidase, chitin 1,4-β-chitobiosidase, β-1,3-glucanase and peroxidases. Prasanna et al. (2015) demonstrated that inoculating a biofilm composed by *Anabaena* sp., the activity of defence enzymes such as peroxidases, phenylalanine ammonia lyase and polyphenol oxidases increased in maize roots and shoots within the levels of accumulation of Zn in flag leaf. Moreover, cyanobacteria were able to enhance glomalin-related soil proteins and polysaccharides in soil.

Polysaccharides are important compounds in triggering plant defence responses. Like micro- and macroalgae, cyanobacteria are abundant source of polysaccharides, but reports on the use of their polysaccharides against plant pathogens are very limited (Singh 2014; Righini et al. 2019). Polysaccharides form a mucilaginous external layer around the cyanobacteria cell that serves as a cover layer for cell protection (Pereira et al. 2009). Xu et al. (2013) reported that a polysaccharide obtained from *Phormidium tenue* elicited the growth of *Caragana korshinskii* and increased seed germination (Xu et al. 2013). This polysaccharide caused also a decrease in oxidative damage by producing the antioxidant superoxide dismutase, which is responsible for neutralizing free radicals and reducing oxidative stress.

9.4 Conclusion

The reduction of the application of fertilizers and pesticides in agricultural crops of commercial interest is a current topic for environment-friendly agriculture. Significant progress has been made in the development of biofertilizer products based on both algae and cyanobacteria. On the contrary, the use of algae and cyanobacteria in disease control is a relative new concept, as well as the study of commercial products for application in crops. The interesting antifungal capacities of algae and cyanobacteria extracts need further investigation to identify each singular component active against the several plant pathogens and to elucidate their specific mechanism of action.

References

Abbassy MA, Marzouk MA, Rabea EI, Abd-Elnabi AD (2014) Insecticidal and fungicidal activity of *Ulva lactuca* Linnaeus (Chlorophyta) extracts and their fractions. Annu Res Rev Biol 4:2252–2262

Abkhoo J, Sabbagh SK (2016) Control of *Phytophthora melonis* damping-off, induction of defense responses, and gene expression of cucumber treated with commercial extract from *Ascophyllum nodosum*. J Appl Phycol 28:1333–1342

Agarwal P, Patel K, Das AK, Ghosh A, Agarwal PK (2016) Insights into the role of seaweed *Kappaphycus alvarezii* sap towards phytohormone signalling and regulating defence responsive genes in *Lycopersicon esculentum*. J Appl Phycol 28:2529–2537

Ahmed EA (2016) Antimicrobial activity of microalgal extracts isolated from Baharia Oasis, Egypt. Glob Adv Res J Microbiol 5:033–041

Alam MZ, Braun G, Norrie J, Hodges DM (2013) Effect of *Ascophyllum* extract application on plant growth, fruit yield and soil microbial communities of strawberry. Can J Plant Sci 93:23–36

Alwathnani HA, Perveen K (2012) Biological control of *Fusarium* wilt of tomato by antagonist fungi and cyanobacteria. Afr J Biotechnol 11:1100–1105

Ambika S, Sujatha K (2015) Antifungal activity of brown, red and green alga seaweed extracts against *Macrophomina phaseolina* (Tassi) Goid., in pideonpea var. CO (Rg) 7. Int J Agric Sci 11:210–216

Argueso CT, Ferreira FJ, Epple P, To JP, Hutchison CE, Schaller GE, Dangl JL, Kieber JJ (2012) Two-component elements mediate interactions between cytokinin and salicylic acid in plant immunity. PLoS Genet 8:e1002448

Arioli T, Mattner SW, Winberg PC (2015) Applications of seaweed extracts in Australian agriculture: past, present and future. J Appl Phycol 27:2007–2015

Athukorala Y, Kim K-N, Jeon Y-J (2006) Antiproliferative and antioxidant properties of an enzymatic hydrolysate from brown alga, *Ecklonia cava*. Food Chem Toxicol 44:1065–1074

Basak A (2008) Effect of preharvest treatment with seaweed products, Kelpak® and Goëmar BM 86®, on fruit quality in apple. Int J Fruit Sci 8:1–14

Black W (1950) The seasonal variation in weight and chemical composition of the common British Laminariaceae. J Mar Biol Assoc UK 29:45–72

Boček S, Salaš P, Sasková H, Mokričková J (2012) Effect of Alginure® (seaweed extract), Myco-Sin®VIN (sulfuric clay) and Polyversum® (*Pythium oligandrum* Drechs.) on yield and disease control in organic strawberries. Acta Univ Agric Et Silvic Mendel Brun 40:19–28

Burja AM, Banaigs EB, Abou-Mansour A, Burgess JG, Wright PC (2001) Marine cyanobacteria – a prolific source of natural products. Tetrahedron 57:9347–9377

Cabrita ARJ, Maia MRG, Oliveira HM, Sousa-Pinto I, Almeida AA, Pinto E, Fonseca AJM (2016) Tracing seaweeds as mineral sources for farm-animals. J Appl Phycol 28:3135–3150

Calvo P, Nelson L, Kloepper JW (2014) Agricultural uses of plant biostimulants. Plant Soil 383:3–41

Cluzet S, Torregrosa C, Jacquet C, Lafitte C, Fournier J, Mercier L, Salamagne S, Briand X, Esquerre-Tugaye MT, Dumas B (2004) Gene expressing profiling and protection of *Medicago truncatula* against a fungal infection in response to an elicitor from green alga *Ulva* spp. Plant Cell Environ 27:917–928

Craigie JS (2011) Seaweed extract stimuli in plant science and agriculture. J Appl Phycol 23:371–393

Crouch IJ, van Staden J (1992) Effect of seaweed concentrate on the establishment and yield of greenhouse tomato plants. J Appl Phycol 4:291–296

De Caire GZ, de Cano MS, de Mule MCZ, de Halperin DR (1990) Antimycotic products from the Cyanobacterium *Nostoc muscorum* against *Rhizoctonia solani*. Phyton-Buenos-Aires 51:1–4

De Caire GZ, De Cano MS, Palma RM, De Mule MCZ (2000) Changes in soil enzyme activities following additions of cyanobacterial biomass and exopolysaccharide. Soil Biol Biochem 32:1985–1987

De Cano MMS, de Mule MCZ, de Caire GZ, de Halperin DR (1990) Inhibition of *Candida albicans* and *Staphylococcus aureus* by phenolic compounds from the terrestrial Cyanobacterium *Nostoc muscorum*. J Appl Phycol 2:79–81

De Cano MMS, De Caire GZ, De Mulé MCZ, Palma RM (2002) Effect of *Tolypothrix tenuis* and *Microchaete tenera* on biochemical soil properties and maize growth. J Plant Nutr 25:2421–2431

De Corato U, Salimbeni R, De Pretis A, Avella N, Patruno G (2017) Antifungal activity of crude extracts from brown and red seaweeds by a supercritical carbon dioxide technique against fruit postharvest fungal diseases. Postharvest Biol Technol 131:16–30

Di Rienzi SC, Sharon I, Wrighton KC, Koren O, Hug LA, Thomas BC et al (2013) The human gut and groundwater harbor non-photosynthetic bacteria belonging to a new candidate phylum sibling to Cyanobacteria. elife 2:e01102. https://doi.org/10.7554/eLife.01102

Dixit RB, Suseela MR (2013) Cyanobacteria: potential candidates for drug discovery. Antonie Van Leeuwenhoek 103:947–961

Eu Pesticide Database. http://ec.europa.eu/food/plant/pesticides/eu-pesticides-database/public/?event=activesubstance.selection&language=EN

El-ghanam AA, Farfour SA, Ragab SS (2015) Bio-suppression of strawberry fruit rot disease caused by Botrytis cinerea. J Plant Pathol Microbiol S3:005

Encarnação T, Pais AACC, Campos MG, Burrows HD (2015) Cyanobacteria and microalgae: a renewable source of bioactive compounds and other chemicals. Sci Prog 98:145–168

Esserti S, Smaili A, Rifai LA, Koussa T, Makroum K, Belfaiza M, Kabil EM, Faize L, Burgos L, Alburquerque N, Faize M (2017) Protective effect of three brown seaweed extracts against fungal and bacterial diseases of tomato. J Appl Phycol 29:1081–1093

Feliziani E, Landi L, Romanazzi G (2015) Preharvest treatments with chitosan and other alternatives to conventional fungicides to control postharvest decay of strawberry. Carbohyd Polym 132:111–117

Fernandes Peres JC, de Carvalho LR, Gonçalez E, Saggion Berian LO, D'arc Felicio J (2012) Evaluation of antifungal activity of seaweed extracts. Ciênc Agrotec Lavras 36:294–299

Frankmölle WP, Larsen LK, Caplan FR, Patterson GML, Knübel G, Levine IA, Moore RE (1992a) Antifungal cyclic peptides from the terrestrial blue-green alga *Anabaena laxa*. I. Isolation and biological properties. J Antibiot 45:1451–1457

Frankmölle WP, Patterson GML, Knübel G, Moore RE (1992b) Antifungal cyclic peptides from the terrestrial blue-green alga *Anabaena laxa*. II. Structures of laxaphycins. J Antibiot 45:1458–1466

Ghasemi Y, Tabatabaei Yazdi M, Shokravi S, Soltani N, Zarrini G (2003) Antifungal and antibacterial activity of paddy-fields cyanobacteria from the north of Iran. J Sci Islamic Republic of Iran 14:203–209

Ghasemi Y, Moradian A, Mohagheghzadeh A, Shokravi S, Morowvat MH (2007) Antifungal and antibacterial activity of the microalgae collected from paddy fields of Iran: characterization of antimicrobial activity of *Chroococcus disperses*. J Biol Sci 7:904–910

Ghazala B, Shameel M (2005) Phytochemistry and bioactivity of some freshwater green algae from Pakistan. Pharm Biol 43:358–369

Guiry MD (2012) How many species of algae are there? J Phycol 48:1057–1063

Guo ZJ, Lamb C, Dixon RA (1998) Potentiation of the oxidative burst and isoflavonoid phytoalexin accumulation by serine protease inhibitors. Plant Physiol 118:1487–1494

Gupta V, Prasanna R, Natarajan C, Srivastava AK, Sharma J (2010) Identification, characterization and regulation of a novel antifungal chitosanase gene (cho) in *Anabaena* sp. Appl Environ Microbiol 76:2769–2777

Gupta V, Natarajan C, Kumar K, Prasanna R (2011) Identification and characterization of endoglucanases for fungicidal activity in *Anabaena laxa*. J Appl Phycol 23:73–81

Gwinn KD (2018) Bioactive natural products in plant disease control. In: Atta-ur-Rahman (Ed) Studies in natural products chemistry. Elsevier 56:229–246

Hernández-Herrera RM, Virgen-Calleros G, Ruiz-López M, Zañudo-Hernández J, Délano-Frier JP, Sánchez-Hernández C (2014) Extracts from green and brown seaweeds protect tomato (*Solanum lycopersicum*) against the necrotrophic fungus *Alternaria solani*. J Appl Phycol 26:1607–1614

Husaini AM, Neri D (2016) Strawberry growth, development and diseases. CABI, Wallingford

Ibraheem IBM, Hamed SM, Abd elrhman AA, Farag FM, Abdel-Raouf N (2017) Antimicrobial activities of some brown macroalgae against some soil borne plant pathogens and in vivo management of *Solanum melongena* root diseases. Aust J Basic Appl Sci 11:157–168

Ibrahim WM, Refaat MA, Khaulood AH, Makram AS (2014) Role of *Ulva lactuca* extract in alleviation of salinity stress on wheat seedlings. Sci World J 2014:1–11

Jaki B, Orjala J, Heilmann J, Linden A, Vogler B, Sticher O (2000) Novel extracellular diterpenoids with biological activity from the cyanobacterium *Nostoc commune*. J Nat Prod 63:339–343

Jannin L, Arkoun M, Etienne P, Laîné P, Goux D, Garnica M, Fuentes M, Francisco SS, Baigorri R, Cruz F, Houdusse F, Garcia-Mina JM, Yvin JC, Ourry A (2013) Brassica napus growth is promoted by Ascophyllum nodosum (L.) Le Jol. seaweed extract: microarray analysis and physiological characterization of N, C, and S metabolisms. J Plant Growth Regul 32:31–52

Jarvis WR, Gubler WD, Grove GG (2002) Epidemiology of powdery mildews in agricultural pathosystems. In: Bélanger RR, Bushnell WR, Dik AJ, Carver TLW (eds) The powdery mildews: a comprehensive treatise. APS Press, St. Paul, pp 169–199

Jayaraj J, Wan A, Rahman M, Punja ZK (2008) Seaweed extract reduces foliar fungal diseases on carrot. Crop Prot 27:1360–1366

Jayaraman J, Norrie J, Punja ZK (2011) Commercial extract from the brown seaweed *Ascophyllum nodosum* reduces fungal diseases in greenhouse cucumber. J Appl Phycol 23:353–361

Jha MN, Prasad AN (2006) Efficacy of new inexpensive cyanobacterial biofertilizer including its shelf life. World J Microbiol Biotechnol 22:73–79

Jha MN, Sharma SG, Prasad AN (1999) Effect of cyanobacterial fertilization schedules on paddy yield and soil micronutrient status. Ann Agric Res 20:512–514

Kannaiyan S (2000) Biological fertilizers for sustainable rice (*Oryza sativa*) production. Adv Agric Res India 8:67–107

Kelman D, Kromkowski Posner E, McDermid KJ, Tabandera NK, Wright PR, Wright AD (2012) Antioxidant activity of Hawaiian marine algae. Mar Drugs 10:403–416

Khalid M, Shameel M, Ahmed V, Shahzad S, Lghani S (2010) Studies on the bioactivity and Phycochemistry of *Microcystis aeruginosa* (Cyanophyta) from Sindh. Pak J Bot 42 (4):2635–2646

Khallil AM, Daghman IM, Fady AA (2015) Antifungal potential in crude extracts of five selected brown seaweeds collected from the western Libya coast. J Micro Creat 1:103–107

Khan W, Rayirath UP, Subramanian S, Jithesh MN, Rayorath P, Hodges DM, Critchley AT, Craigie JS, Norrie J, Prithiviraj B (2009) Seaweed extracts as biostimulants of plant growth and development. J Plant Growth Regul 28:386–399

Kim J-D (2006) Screening of cyanobacteria (blue-green algae) from rice paddy soil for antifungal activity against plant pathogenic fungi. Mycobiology 34:138–142

Kim J, Kim J-D (2008) Inhibitory effect of algal extracts on mycelial growth of the tomato-wilt pathogen, *Fusarium oxysporum* f. sp. lycopersici. Mycobiology 36:242–248

Komárek J, Kaštovský J, Mareš J, Johansen JR (2014) Taxonomic classification of cyanoprokaryotes (cyanobacterial genera) 2014, using a polyphasic approach. Preslia 86:295–335

Krawiec M, Szot I, Krawiec P (2016) The use of laminarin as an effective tool for anti-resistance management in chemical control of grey mould in raspberry. Acta Hortic 1133:469–472

Kumar G, Sahoo D (2011) Effect of seaweed liquid extract on growth and yield of *Triticum aestivum* var. Pusa Gold. J Appl Phycol 23:251–255

Latique S, Elouaer MA, Chernane H, Hannachi C, Elkaou M (2014) Effect of seaweed liquid extract of Sargassum vulgare on growth of durum wheat seedlings (*Triticum durum* L.) under salt stress. Int J Innov Appl Stud 7:1430–1435

Liu M, Wang G, Xiao L, Xu X, Liu X, Xu P, Lin X (2014) Bis (2,3-dibromo-4,5-dihydroxybenzyl) ether, a marine algae derived bromophenol, inhibits the growth of *Botrytis cinerea* and interacts with DNA molecules. Mar Drugs 12:3838–3851

Mabrouk SS, El-Shayeb NMA, El-Refai AH, Sallam LAR, Hamdy AA (1985) Inhibitory activities of some marine algae on aflatoxin accumulation. Appl Microbiol Biotechnol 22:152–155

Manjunath M, Prasanna R, Nain L, Dureja P, Singh R, Kumar A, Jaggi S, Kaushik BD (2010) Biocontrol potential of cyanobacterial metabolites against damping off disease caused by *Pythium aphanidermatum* in solanaceous vegetables. Arch Phytopathol PFL 43:666–677

Maqubela MP, Mnkeni PNS, Malam Issa O, Pardo MT, D'Acqui LP (2009) *Nostoc* cyanobacterial inoculation in South African agricultural soils enhances soil structure, fertility and maize growth. Plant Soil 315:79–92

Marrez DA, Sultan YY (2016) Antifungal activity of the cyanobacterium *Microcystis aeruginosa* against mycotoxigenic fungi. J Appl Pharm Sci 6:191–198

Meszka B, Bielenin A (2011) Activity of laminarin in control of strawberry diseases. Phytopathologia 62:15–23

Moon RE, Martin DF (1981) Assay of diverse biological activities of materials elaborated by marine blue green algae, *Gomphosphaeria aponina*. Microbios Lett 18:103–110

Nagayama K, Iwamura Y, Shibata T, Hirayama I, Nakamura T (2002) Bactericidal activity of phlorotannins from the brown alga *Ecklonia kurome*. J Antimicrob Chemother 50:889–893

Natarajan C, Prasanna R, Gupta V, Dureja P, Lata M (2012) Dissecting the fungicidal activity of *Calothrix elenkinii* using chemical analyses and microscopy. Appl Biochem Microbiol 48:53–57

Oren A, Garrity GM (2014) Proposal to change general consideration 5 and principle 2 of the international code of nomenclature of prokaryotes. Int J Syst Evol Microbiol 64:309–310

Osman MEH, El-Sheekh MM, Metwally AM, AEA I, Ismail M (2011) Antagonistic activity of some fungi and cyanobacteria species against Rhizoctonia solani. Int J Plant Pathol 2 (3):101–114

Painter TJ (1983) Algal polysaccharides. In: Aspinall GO (ed) The polysaccharides. Academic, New York, pp 195–285

Pandey KD, Shukla PN, Giri DD, Kashyap AK (2005) Cyanobacteria in alkaline soil and the effect of cyanobacteria inoculation with pyrite amendments on their reclamation. Biol Fertil Soils 41:451–457

Pastrana AM, Basallote-Ureba MJ, Aguado A, Akdi K, Capote N (2016) Biological control of strawberry soil-borne pathogens *Macrophomina phaseolina* and *Fusarium solani*, using *Trichoderma asperellum* and *Bacillus* spp. Phytopathol Mediterr 55:109–120

Pereira I, Ortega R, Barrientos L, Moya M, Reyes G, Kramm V (2009) Development of a biofertilizer based on filamentous nitrogen-fixing cyanobacteria for rice crops in Chile. J Appl Phycol 21:135–144

Prasanna R, Nain L, Tripathi R, Gupta V, Chaudhary V, Middha S et al (2008) Evaluation of fungicidal activity of extracellular filtrates of cyanobacteria – possible role of hydrolytic enzymes. J Basic Microbiol 48:186–194

Prasanna R, Nain L, Ancha R, Shrikrishna J, Joshi M, Kaushik BD (2009) Rhizosphere dynamics of inoculated cyanobacteria and their growth-promoting role in rice crop. Egypt J Biol 11:26–36

Prasanna R, Sood A, Jaiswal P, Nayak S, Gupta V, Chaudhary V, Joshi M, Natarajan C (2010) Rediscovering cyanobacteria as valuable sources of bioactive compounds. Appl Biochem Microbiol 46:133–147

Prasanna R, Chaudhary V, Gupta V, Babu S, Kumar A, Nain L (2013) Cyanobacteria mediated plant growth promotion and bioprotection against *Fusarium* wilt in tomato. Eur J Plant Pathol 136:337–353

Prasanna R, Bidyarani N, Babu S, Hossain F, Shivay YS, Nain L (2015) Cyanobacterial inoculation elicits plant defense response and enhanced Zn mobilization in maize hybrids. Cogent Food Agric 1:998507. https://doi.org/10.1080/23311932.2014.998507

Pugliese M, Monchiero M, Gullino ML et al (2018) Application of laminarin and calcium oxide for the control of grape powdery mildew on *Vitis vinifera* cv. Moscato. J Plant Dis Prot 125:477. https://doi.org/10.1007/s41348-018-0162-8

Radhakrishnan B, Prasanna R, Jaiswal P, Nayak S, Dureja P (2009) Modulation of biocidal activity of *Calothrix* sp. and *Anabaena* sp. by environmental factors. Biologia 64:881–889

Ramkissoon A, Ramsubhag A, Jayaraman J (2017) Phytoelicitor activity of three Caribbean seaweed species on suppression of pathogenic infections in tomato plants. J Appl Phycol 29:3235–3244

Rayorath P, Jithesh MN, Farid A, Khan W, Palanisamy R, Hankins SD, Critchley AT, Prithiviraj B (2008) Rapid bioassays to evaluate the plant growth promoting activity of *Ascophyllum nodosum* (L.) Le Jol. using a model plant, *Arabidopsis thaliana* (L.) Heynh. J Appl Phycol 20:423–429

Rekanović E, Potočnik I, Milijašević-Marčić S, Stepanović M, Todorović B, Mihajlović M (2010) Efficacy of seaweed concentrate from *Ecklonia maxima* (Osbeck) and conventional fungicides in the control of *Verticillium* wilt of pepper. Pestic Phytomed 25:319–324

Righini H, Roberti R, Baraldi E (2018) Use of algae in strawberry management. J Appl Phycol 30:3551–3564. https://doi.org/10.1007/s10811-018-1478-2

Righini H, Baraldi E, García Fernández Y, Martel Quintana A, Roberti R (2019) Different antifungal activity of *Anabaena* sp., *Ecklonia* sp., and *Jania* sp. against *Botrytis cinerea*. Mar Drugs 17:299. https://doi.org/10.3390/md17050299

Rioux LE, Turgeon SL, Beaulieu M (2007) Characterization of polysaccharides extracted from brown seaweeds. Carbohydr Polym 69:530–537

Roberti R, Galletti S, Burzi PL, Righini H, Cetrullo S, Perez C (2015) Induction of defence responses in zucchini (*Cucurbita pepo*) by *Anabaena* sp. water extract. Biol Control 82:61–68

Roberti R, Righini H, Pérez Reyes C (2016) Activity of seaweed and cyanobacteria water extracts against *Podosphaera xanthii* on zucchini. Ital J Mycol 45:66–77

Rodríguez AA, Stella AM, Storni MM, Zulpa G, Zaccaro MC (2006) Effects of cyanobacterial extracellular products and gibberellic acid on salinity tolerance in *Oryza sativa* L. Saline Syst 2:7

Romero D, de Vicente A, Zeriouh H, Cazorla FM, Fernández-Ortuño D, Torés JA, Pérez-García A (2007) Evaluation of biological control agents for managing cucurbit powdery mildew on greenhouse-grown melon. Plant Pathol 56:976–986

Russo P, Cesario A (2012) New anticancer drugs from marine cyanobacteria. Curr Drug Targets 13 (8):1048–1053

Rustérucci C, Stallaert V, Milat ML, Pugin A, Ricci P, Blein JP (1996) Relationship between active oxygen species, lipid peroxidation, necrosis, and phytoalexin production induced by elicitins in *Nicotiana*. Plant Physiol 111:885–891

Schiener P, Black KD, Stanley MS, Green DH (2015) The seasonal variation in the chemical composition of the kelp *species Laminaria digitata, Laminaria hyperborea, Saccharina latissima* and *Alaria esculenta*. J Appl Phycol 27:363–373

Schirrmeister BE, Antonelli A, Bagheri HC (2011) The origin of multicellularity in cyanobacteria. Evol Biol 11:45. https://doi.org/10.1186/1471-2148-11-45

Sharma P, Sharma N (2017) Industrial and biotechnological applications of algae: a review. J Adv Plant Biol 1:1–25

Shi D, Hall DO (1988) The *Azolla-Anabaena* Association: historical perspective, symbiosis and energy metabolism. Bot Rev 54:353–386

Simmons TL, Engene N, Ureña LD, Romero LI, Ortega-Barría E, Gerwick L, Gerwick WH (2008) Viridamides A and B, lipodepsipeptides with antiprotozoal activity from the marine cyanobacterium *Oscillatoria nigro-viridis*. J Nat Prod 71:1544–1550

Singh S (2014) A review on possible elicitor molecules of cyanobacteria: their role in improving plant growth and providing tolerance against biotic or abiotic stress. J Appl Microbiol 117:1221–1244

Singh S, Bhushan NK, Banerjee UC (2005) Bioactive compounds from cyanobacteria and microalgae: an overview critical reviews in biotechnology. CRC Crit Rev Biotechnol 25:73–95

Sinha SK, Verma DC, Dwivedi CP (2002) Role of green manure (*Sesbania rostrata*) and biofertilizers (Blue-green algae and *Azotobacter*) in rice-wheat cropping system in state of Uttar Pradesh, India. Physiol Mol Biol Plants 8:105–110

Soo RM, Skennerton CT, Sekiguchi Y, Imelfort M, Paech SJ, Dennis PG et al (2014) An expanded genomic representation of the phylum Cyanobacteria. Genome Biol Evol 6:1031–1045. https://doi.org/10.1093/gbe/evu073

Soo RM, Woodcroft BJ, Parks DH, Tyson GW, Hugenholtz P (2015) Back from the dead; the curious tale of the predatory cyanobacterium *Vampirovibrio chlorellavorus*. Peer J 3:e968. https://doi.org/10.7717/peerj.968

Tiwari A, Kaur A (2014) Allelopathic impact of cyanobacteria on pathogenic fungi. Int J Pure Appl Biosci 2:63–70

Vera J, Castro J, Gonzalez A, Moenne A (2011) Review-Seaweed polysaccharides and derived oligosaccharides stimulate defense responses and protection against pathogens in plants. Mar Drugs 9:2514–2525

Walter JM, Coutinho FH, Dutilh BE, Swings J, Thompson FL, Thompson CC (2017) Ecogenomics and taxonomy of cyanobacteria phylum. Front Microbiol 8:2132. https://doi.org/10.3389/fmicb.2017.02132

Walters D, Walsh D, Newton A, Lyon G (2005) Induced resistance for plant disease control: maximizing the efficacy of resistance elicitors. Phytopathology 95:1368–1373

Westermeier R, Murúa P, Patiño DJ, Muñoz L, Ruiz A, Müller DG (2012) Variations of chemical composition and energy content in natural and genetically defined cultivars of *Macrocystis* from Chile. J Appl Phycol 24:1191–1201

Wite D, Mattner SW, Porter IJ, Arioli T (2015) The suppressive effect of a commercial extract from *Durvillaea potatorum* and *Ascophyllum nodosum* on infection of broccoli by *Plasmodiophora brassicae*. J Appl Phycol 27:2157–2161

Xu Y, Rossi F, Colica G, Deng S, De Philippis R, Chen L (2013) Use of cyanobacterial polysaccharides to promote shrub performances in desert soils: a potential approach for the restoration of desertified areas. Biol Fertil Soils 49:143–152

Yadav S, Agrawal M, Raipuria N, Agrawal MK (2016) Antimicrobial Activity of *Nostoc calcicola* (Cyanobacteria) isolated from central India against human pathogens. Asian J Pharm (Suppl) 10:S554

Yanni YG (1991) Efficiency of rice fertilization schedules including cyanobacteria under soil application of phosphate and molybdate. World J Microbiol Biotechnol 7:415–418

Zaccaro MC, De Caire GZ, De Cano MS, Palma RM, Colombo K (1999) Effect of cyanobacterial inoculation and fertilizers on rice seedlings and postharvest soil structure. Commun Soil Sci Plant Anal 30:97–107

Chapter 10
Non-target Effects of *Trichoderma* on Plants and Soil Microbial Communities

Monika Jangir, Satyawati Sharma, and Shilpi Sharma

Abstract Biocontrol agents are currently considered as promising alternative to chemical fungicides because of the latter's negative impacts on consumer health, plant health, and the environment. In the current biopesticide world, *Trichoderma* spp. has been globally accepted to prevent the invasion of pathogens, viz., *Fusarium oxysporum*, *Verticillium dahliae*, *Pythium aphanidermatum*, *Rhizoctonia solani*, etc. The antagonistic activity of *Trichoderma* spp. is attributed to several mechanisms, viz., mycoparasitism, antibiosis, induction of host systemic resistance, and production of hydrolytic enzymes. They not only have plant growth-promoting properties but also exert transient or long-term impact on the resident soil microbiome and may pose risk to beneficial non-target soil communities. Some compounds released by them in higher amount increase the sensitivity of the plant, and may pose negative impact on their growth. Additionally, *Trichoderma* spp. affects microbial community functions. The current chapter summarizes *Trichoderma*-pathogen-plant interaction, and the impact of *Trichoderma* spp. on plant growth, soil enzyme activities, and soil microbiome.

10.1 Introduction

Plant diseases caused by various phytopathogens often considerably deteriorate quantity and quality of agricultural products. Also, the production of toxins by these microbial pathogens during postharvest or field infestation can have severe effects on health of livestock as well as humans (Brimner and Boland 2003). For green revolution, the use of chemical or synthetic pesticides was seen as a substantial measure to control the invasion of plant pathogens. Although these methods for

M. Jangir · S. Sharma
Centre for Rural Development and Technology, Indian Institute of Technology Delhi, New Delhi, India

S. Sharma (✉)
Department of Biochemical Engineering and Biotechnology, Indian Institute of Technology Delhi, New Delhi, India
e-mail: shilpi@dbeb.iitd.ac.in

A. Varma et al. (eds.), *Plant Microbe Interface*,
https://doi.org/10.1007/978-3-030-19831-2_10

managing disease in plants are effective, they are not free from adverse impacts on the environment and human health (Naseby et al. 2000). Furthermore, these chemicals are non-biodegradable and expensive, and repeated use of chemical fungicides is leading toward the development of new resistant pathogenic population (Goldman et al. 1994; Naseby et al. 2000). This has prompted the exploration of some new methods and technologies for disease management to bring down or eradicate the use of chemical pesticides. One such technology is the integration of biocontrol agents that have minimal impact on environment, and potential to control diseases (Chet and Inbar 1994).

Bioinoculants play a major role in controlling plant diseases, with several properties of acting as phytostimulators, biofertilizers, etc. Though these biocontrol agents are considered to be safer as compared to chemical pesticides, their possible impact on environment is related to the application of high densities of viable cells in host plant rhizosphere (Trabelsi and Mhamdi 2013). Also, a large number of inoculants may pose threats to non-target resident soil microbiome (Brimner and Boland 2003), most likely by altering the abundance of other microorganisms or soil enzyme activities.

Trichoderma spp., known for their antagonistic efficacy, are avirulent plant symbionts and opportunistic in nature. Many species of *Trichoderma* have been broadly studied for their potential to reduce disease severity in plant disease by showing inhibition of plant pathogens (Ros et al. 2017). They promote plant growth (Harman 2006) and are tolerant to biotic and abiotic stresses (Hermosa et al. 2012). Nonetheless, direct interactions of antagonistic species of *Trichoderma* or other biocontrol agents are not only restricted to targeting particular phytopathogens but also affect other saprophytic microbes of resident soil community. While biology and potential benefits of *Trichoderma* spp. along with their other aspects of plant growth enhancement have been discussed in several reviews, their non-target impacts have been largely ignored. In recent years, growing interest toward the use of reliable and sustainable biological method has led to the risk assessment of biocontrol agents in terms of their effects on environment and non-target microbiome (Brimner and Boland 2003; Szczepaniak et al. 2015). The present chapter attempts to summarize the non-target effects of *Trichoderma* on plant health, soil enzyme activities, and fungal and bacterial communities.

10.2 *Trichoderma*-Plant-Pathogen Interactions

Trichoderma is one of the most popular commercial biological pesticides. Worldwide, it shares 60% of the total registered biocontrol-based pesticides (Verma et al. 2007). It is a soil-borne microbe and has the capability of penetration and colonization of roots of host plant (Howell 2006). It builds an opportunist/facultative symbiosis with plant, utilizes nutrients from the plant, boosts plant's immunity against phytopathogens, induces host systemic resistance, and enhances plant growth (Mukherjee et al. 2012) (Fig. 10.1). *Trichoderma* evokes synchronized

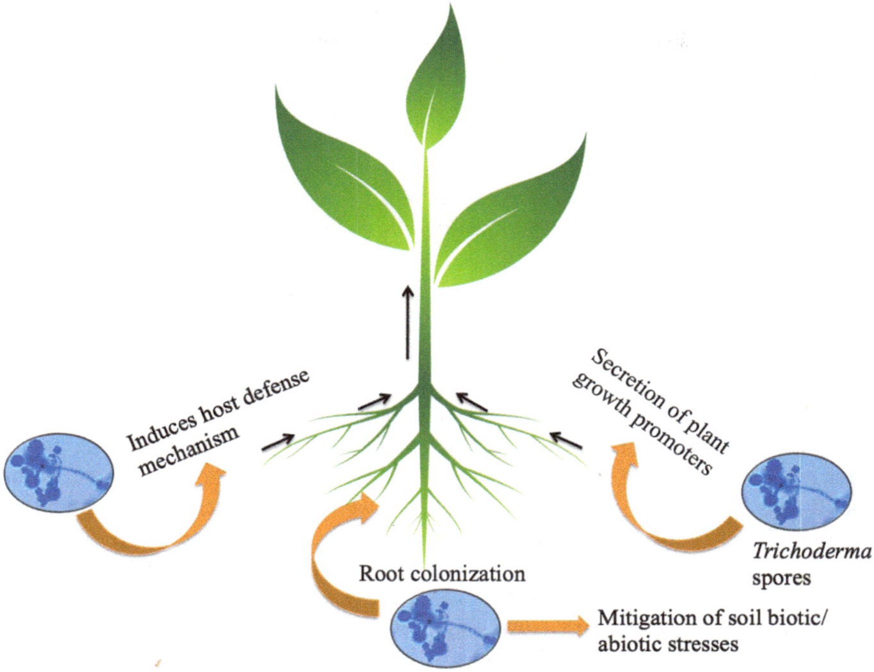

Fig. 10.1 Schematic representation of *Trichoderma*-plant interactions

metabolomic, proteomic and transcriptomic responses in plant system (Shoresh and Harman 2008; Lorito et al. 2010; Morán-Diez et al. 2012).

Among all its characteristics, mycoparasitism is one of the key factors for its antagonistic activity (Morán-Diez et al. 2012). Typically, mycoparasitic interactions between pathogen and *Trichoderma* involve sensing of the pathogen, attraction, attachment, hyphal coiling around pathogen hyphae, and production of lytic enzymes along with secondary metabolites for the lysis. The interactions between *Trichoderma* and plant pathogens are generally beneficial for plant disease management. However, *Trichoderma* exhibits some non-target impacts on soil environment also.

10.3 Non-target Effects on Plant

10.3.1 Effect on Plant Growth Parameters

It is a well-known fact that *Trichoderma* helps in plant growth promotion and enhances soil microbial population, but it also exhibits specific non-target effects. It was over two decades back when Cook et al. (1996) identified that toxicity and competitive displacement of non-target soil microbes are two chief unintended impacts correlated with the application of biocontrol agents. When a native microbe gets replaced or expelled by biocontrol agent for space and nutrients, this is called competitive displacement. Pathogenicity and toxicity may occur when alkaloids or antibiotics produced by these biocontrol agents start acting against beneficial or nontarget microbes (Brimner and Boland 2003).

Naseby et al. (2000) assessed the potential of some strains of *Trichoderma* for disease control in pea plant against *Pythium ultimum* (IMI 308273). Among all the strains, *Trichoderma* strain To10 showed minor inhibitory effect on plant growth. The production of fungistatic volatile organic compounds, viz., pentenyl and pentylpyrones by *Trichoderma harzianum* (Lumsden et al. 1990), had a negative impact on plant growth at high doses (Naseby et al. 2000). Two native ecosystems of New Zealand were evaluated for non-target effects of bioinoculant *T. atroviride* on plant health of trees (McLean et al. 2014). The application of biocontrol agent significantly modified the levels of photosynthetic pigments in case of some native forest seedlings. Nevertheless, variation in the levels of carotenoids and total chlorophyll was unusual. Generally, in stressed conditions, plants tend to decrease chlorophyll and increase carotenoid content in leaves (Peñuelas and Filella 1998; Sampson et al. 2003). McLean et al. (2014) reported that the addition of biocontrol agent in the rhizospheric soil of *Coprosma robusta* and *Asplenium gracillimum* decreased content of both carotenoids and chlorophyll. However, carotenoid content was increased and decreased with *Plagianthus regius* and *Dacrycarpus dacrydioides*, respectively. Also, there was no significant difference in percent seedling emergence after inoculation of biocontrol agent for each representative species of tussock grassland (*Chionochloa rigida*, *C. rubra*, *Festuca novae-zelandiae*, and *Poa cita*), and podocarp forest (*Pittosporum eugenioides*, *P. regius*, *C. robusta*, *D. dacrydioides*, *Melicytus ramiflorus*, and *A. gracillimum*) except *C. australis* (McLean et al. 2014).

In a study conducted by Hashem et al. (2014), abiotic stress of salt reduced dry weight and length of shoot and root of *Ochradenus baccatus*. The inoculation of *T. hamatum* mitigated the detrimental effect of salt stress and increased dry weight of shoot and root. Seven strains of *Trichoderma* were tested for their ability to enhance plant growth. *T.* sp. "*atroviride* B" LU660 and *T. atroviride* IMI 206040 enhanced fresh weight of *Arabidopsis* by approx. 72% (Nieto-Jacobo et al. 2017). In contrast *T. asperellum* LU1370 exerted significant inhibition of growth of *Arabidopsis* with 74% reduction in biomass over non-inoculated soil, despite a positive effect observed in plate assay. Also, *T. novaeharzianum* LU1328, *T.* sp. "*atroviride* B" LU668, *Trichoderma* sp. LU668, and *T. trixiae* had no significant effect on growth of *Arabidopsis*. *Trichoderma* spp. is known to produce antibiotics of a diverse range. *T. longibrachiatum* SMF2 produced a peptaibol, trichokonin VI (TK VI), which displayed a dose-dependent effect on growth promotion of *Arabidopsis* seedlings (Shi et al. 2016). Supply of spores in soil at low concentrations ($\leq 5 \times 10^6$ cm^{-3}) enhanced the growth, whereas high dose ($\geq 1 \times 10^7$ cm^{-3}) of SMF2 inhibited the growth of *Arabidopsis* seedlings. It further reduced the number of meristematic cells, compressed cortex cells of the roots, and also inhibited the growth of primary roots.

A mutant Δ*Tpx1*, strain that did not produce TK VI, had no negative impact on seedling growth. The response of *Trichoderma* on plant growth is dependent on its ability of survival and development in the rhizospheric soil of host plant (Kleifeld and Chet 1992), and dose of inoculum.

10.3.2 Effect of **Trichoderma** *spp. on Soil Microbial Communities*

Rhizosphere of a plant is an unrevealed world with complex soil community structure and functional dynamics, which presents it as a real challenge in soil ecology (Trabelsi and Mhamdi 2013; Shrivastava et al. 2014). Various culture-independent and culture-dependent techniques are used in assessing the change in soil microbial communities. However, the analysis using culture-dependent methods is constricted to restricted samples giving a biased image, whereas culture-independent methods generally do not allow unambiguous identification of taxonomic groups. Some high-throughput sequencing techniques can be used that are more explanatory, but have limited affordability on economical basis (Trabelsi and Mhamdi 2013). Few popular techniques used to evaluate the changes in microbial communities have been listed in Table 10.1.

10.3.2.1 Effect on Bacterial Population

Trichoderma, as a biocontrol agent, has fundamental functions in stimulating the plant beneficial microbiome and inhibiting the invading pathogen through different mechanisms (Saravanakumar et al. 2017). The effect of *Trichoderma* as a bioinoculant cannot be generalized because negative, positive, as well as no effects have been reported in earlier studies. Gupta et al. (2014) assessed the non-target effects of *T. harzianum* MTCC 801 alone and in combination with strains of *Bacillus megaterium* and *Pseudomonas fluorescens* on major bacterial groups (actinomycetes and β-proteobacteria) in rhizosphere of *Cajanus cajan*. They found that *T. harzianum* alone, and in combination with *B. megaterium*, enhanced the population of actinomycetes significantly with time showing maximum abundance at maturity stage of the plant. Furthermore, they assessed the rhizospheric bacterial community profiles using ARISA and reported that the non-target effects of the inoculant (*T. harzianum* alone and *T. harzianum* + *B. megaterium*) were quite evident from a cluster of peaks observed in rhizospheric soil from harvest stage, which was absent in control at the same time point.

Blaya et al. (2013) showed changes in bacterial community of compost after inoculating it with *T. harzianum*. On the basis of DGGE profiles of 16S rRNA as marker, they found that treatment of *T. harzianum* enhanced bacterial community, with six phyla and a group of unclassified bacteria in comparison to three phyla and a

Table 10.1 Selected techniques employed to evaluate the changes in soil microbial communities

Techniques	Description	References
Next-generation sequencing (NGS)	Performs sequencing of entire genome or millions of small fragments of DNA in parallel	Zhu et al. (2018), Ros et al. (2017), Soliman et al. (2017)
Automated ribosomal spacer analysis (ARISA)	Uses a fluorescence-tagged oligonucleotide primer for PCR amplification and for subsequent electrophoresis in an automated system	Ondreičková et al. (2018), Wood et al. (2016), Gupta et al. (2014)
q-PCR (quantitative PCR)	Uses the linearity of DNA amplification to determine relative or absolute quantities of a known gene in a sample	Epelde et al. (2018), Wang et al. (2018), Kleyer et al. (2017), Zhang et al. (2017), Gupta et al. (2014)
Denaturing gradient gel electrophoresis (DGGE)	Separates DNA fragments of identical length but different sequence on the basis of their mobilities under increasingly denaturing conditions	Pacwa-Płociniczak et al. (2018), Wang et al. (2018), McLean et al. (2014)
Diversity and evenness indices, viz., Shannon and Simpson indices	Accounts for diversity, evenness, and richness of microbial communities	Epelde et al. (2018), Pascual et al. (2018), Louis et al. (2016), Blaya et al. (2013)
Community-level physiological profiles (CLPP)	Compares different communities on the basis of sole carbon source utilization patterns gathered using BIOLOG microplates	Amaresan et al. (2018), Epelde et al. (2018), Garcia et al. (2018), Li et al. (2013), Lladó and Baldrian (2017), Frac et al. (2012)
Terminal restriction fragment length polymorphism (T-RFLP)	Uses size of terminal restriction fragment for differentiation of microbial populations	Wu et al. (2015), Araújo et al. (2016), Cordier and Alabouvette (2009)

group of unclassified bacteria of control. In case of compost treated with *T. harzianum*, largest phylum Proteobacteria was 48.7% as compared to 41.2% in control. Cordier and Alabouvette (2009) also assessed the impact of *T. atroviride* on the native resident bacterial community using T-RFLP analysis of 16S rRNA genes. They reported that the inoculation of *T. atroviride*I-1237 significantly increased the bacterial populations only after 3 days of postinoculation. Pang et al. (2017) evaluated the response of resident bacterial community to *Trichoderma*-enriched organic fertilizer. They found that *Trichoderma* treated organic fertilizer showed higher number of operational taxonomic units (OTUs). Moreover, it had the highest Shannon diversity index for bacterial community.

In a study conducted by Naseby et al. (2000), it was found that all the tested strains of *Trichoderma* reduced the total soil bacterial population in the absence of *P. ultimum*; but *Trichoderma* strain TH1 reduced the population of soil bacteria in the presence of *P. ultimum*. However, the presence and absence of *P. ultimum* showed no significant difference in the population of fluorescent *Pseudomonas*, whereas the inoculation of *Trichoderma* strain To10 in the presence of *P. ultimum* significantly increased the population of *Pseudomonas. P. ultimum* had pathogenic effect on plant by triggering nutrient leakage from the roots. The increased nutrient supplies to soil rhizosphere increased the population of soil bacteria including fluorescent *Pseudomonas.* Only strain TH1 was capable of lowering this effect of *P. ultimum* and hence reduced the bacterial population in its presence. It indicates that in the presence of *Trichoderma* strain, TH1 *P. ultimum* caused less damage to the roots of the host plant. However, the reduction of soil bacterial population in the absence of *P. ultimum* may be due to nutrient deficiency caused by inoculum of *Trichoderma* and rhizospheric soil fungal community that utilized a major portion of the soil niche (Naseby et al. 2000). In a study on PAH biodegradation, *Trichoderma viride* was inoculated together with bacterial consortium of 195 species (Szczepaniak et al. 2015). After the process of 12 months of biodegradation, only 73 bacterial species were identified from the consortium. *T. viride* proved to exert antagonistic effect on the bacterial consortium. However, the reason for this inhibitory effect was unexplained.

T. harzianum MTCC 5179 was inoculated in the rhizosphere of black pepper (*Piper nigrum* L.) for evaluating its impact on the population, and functional dynamics of soil microbiome (Umadevi et al. 2018). The α-diversity of the metagenome was 455,862 and 489,569 species for control and treatment (*T. harzianum* inoculated), respectively. Upon analyzing the relative abundance, ten most abundant bacterial species in treatment were found to be *Candidatus koribacter versatilis, Acidobacteriaceae bacterium* KBS 96, *Ktedonobacter racemifer, Pedosphaera parvula, Candidatus solibacterusitatus, Sphingomonas* sp., *Chthonomonas calidirosea, Pyrinomonas methylalipathogens, Gemmatimonadetes* bacterium, and uncultured bacteria (*C. koribacterversatilis* and *Acidobacteriaceae bacterium*), whereas in control sample uncultured bacteria were more. The abundance of these bacteria indicated that the bioinoculant imparts rhizospheric competency to bacteria to colonize the roots of the host plant. Functional-level dynamics suggested that the rhizosphere in treatment sample (with inoculation of *T. harzianum*) had more abundant reads for disease, virulence and defense, metabolism and ion acquisition, and chemotaxis and motility as compared to control sample. Gasoni et al. (2008) studied the effect of *T. harzianum* on functional diversity of soil microbial community after inoculating in rhizospheric soil of tobacco plant. They observed that control soil showed high bioactivity with higher Yokoyama CLPP diversity index (DI, 4.41) as compared to treated soil (DI, 3.68). However, *T. harzianum* showed significant changes in the metabolic profiles. Also, the application of *T. harzianum* resulted in less number of metabolized compounds, whereas absorbance was much higher for a specific set of compounds in control. This indicated that the inoculation of *T. harzianum* contributed to growth stimulation of a specific soil bacterial population that altered the microbial community of the host rhizosphere (Gasoni et al. 2008).

10.3.2.2 Effect on Fungal Community

Significant increase in the fungal population of rhizospheric soil was found upon inoculation of *Trichoderma* TH1, T4, and T12 in the absence of *P. ultimum* (Naseby et al. 2000). This increased fungal population utilizing higher amount of nutrients explained well the observed reduction of bacterial soil population in the absence of pathogen. *Trichoderma koningii* Oudemans reduced the germination of resting spores of arbuscular mycorrhizal fungus (AMF) *Glomus mosseae* (McAllister et al. 1994; Brimner and Boland 2003). Also, the volatiles produced by *T. koningii* reduced germination of *G. mosseae*. However, these volatiles had no effect on fungal mycelial growth of *G. mosseae* depicting that these compounds affected only the spores of resting stage. Inoculation of *Trichoderma* on mycorrhizal maize plants decreased population of *Azospirillum* in soil (Vázquez et al. 2000). However, this was observed only in natural AMF-mycorrhizal plants as compared to non-mycorrhizal control. Umadevi et al. (2018) analyzed the relative abundance of fungal species in the rhizospheric soil of black pepper (*Piper nigrum* L.). Upon treatment with *T. harzianum* MTCC 5179, most abundant fungi were observed to be *Fusarium oxysporum*, *Rhizophagus irregularis*, *Oidiodendron maius*, *Talaromyces stipitatus*, *Pseudogymnoascus pannorum*, *Pestalotiopsis fici*, *T. harzianum*, and *Mortierella verticillata*. *P. fici*, *T. stipitatus*, and *F. oxysporum* were abundant in treatment, whereas *R. irregularis* and *P. pannorum* (a human pathogen) and *O. maius* were higher in control sample. It depicted that the application of *T. harzianum* reduced effect of human pathogen in the amended soil, as compared to control (Umadevi et al. 2018).

Cordier and Alabouvette (2009) assessed the impact of *T. atroviride* I-1237 on native fungal soil community. They found that the inoculation of *T. atroviride* I-1237 resulted in a significant increase in the density of fungal community after 3 days of inoculation. However, T-RFLP analysis revealed that structure of fungal community evolved in a similar manner in both control and inoculated soil. Li et al. (2013) evaluated the impact of *T. longibrachiatum* T2 on functional diversity of rhizospheric soil microbiome with AWCD variations. They found that within 48 h of inoculation, the capacity of microbial community to utilize carbon source was highest in rhizospheric soil of treatment followed by non-rhizospheric soil of treatment and control. Moreover, the functional diversity index (Shannon index) was also highest in case of rhizospheric soil of treatment that depicted an enhancement in the richness of the soil microbiome. Pang et al. (2017) found higher number of OTUs in fungal community with *Trichoderma*-treated organic fertilizer. In addition, they found high diversity index for fungal community in *Trichoderma*-treated samples, as compared to chemical fertilizer and organic fertilizer.

Trichoderma-fortified compost was applied to pepper seedlings infected by *Phytophthora nicotianae* (Ros et al. 2017). Two successful techniques, viz., quantitative PCR and NGS, were used to evaluate the effect of *Trichoderma*-fortified compost on rhizospheric fungal and bacterial populations. The bacterial population showed no effect upon application of *Trichoderma* strain, whereas it was altered in response to the pathogen. However, fungal population was independent of pathogen and inoculation of *T. harzianum*, but substrate impacted the same. In another study, it was observed that the addition of *T. atroviride* promoted arbuscular mycorrhizal (AM) spore density in the rhizospheric soil of host plant *Dacrycarpus dacrydioides* in comparison to control sample (McLean et al. 2014). DGGE analysis suggested that the biocontrol agent had no effect on the genetic diversity of AM soil population. This indicated that *T. atroviride* possibly contributed to proliferation of existing AM population, but not in enhancing diversity.

10.3.3 Effect on Soil Enzyme Activities

In addition to quantitative and qualitative alteration exhibited in population of soil microbial communities, another factor for evaluating the changes in the soil system is soil enzymes (Vázquez et al. 2000). Soil enzyme activities have been used as indicators to evaluate the impact of bioinoculants and can be used for gaining a better understanding of natural agitations caused to the ecological system (Naseby and Lynch 1998). The soil inoculated with *P. ultimum* showed high activity of soil enzymes as compared to control that indicated a significant increase in leakage of carbon and other nutrients from the roots due to the damage caused by pathogen (Naseby et al. 2000). *Trichoderma* strains TH1, T4, T12, and N47 substantially decreased NAGase, β-glucosidase, and chitobiosidase activities that are related to disease control (Naseby et al. 2000). Reduced activity of alkaline phosphatase was found with *Trichoderma* strains T4, T12, and TH1 that was relative to *P. ultimum* control. Activity of acid phosphatase and cellobiosidase could not be affected by any of the strains. The varying levels of reduction in soil enzyme activities signified the lowering of pathogenic effect. The application of *Trichoderma* in the rhizosphere of plants affected the soil enzyme activity. Inoculation of *Trichoderma* in AMF-colonized plants showed reduction (89%) in phosphatase activity, whereas evident increase (188%) was observed on inoculation in *G. deserticola*-colonized plants (Vázquez et al. 2000). *Trichoderma* also increased (121%) the phosphatase activity in the rhizospheric soil of *G. mosseae*-colonized plants. Significant increase in chitinase activity was found on inoculating *Trichoderma* in soil of natural AMF-colonized plants and non-mycorrhizal plant (121% and 151%). However, it significantly reduced (47%) the activity of trehalase enzyme.

10.4 Conclusion and Future Prospects

Typically, inoculation of *Trichoderma* spp. can be an environment friendly approach for disease management as they impart less lethal effects on plant, soil, and human health as compared to synthetic pesticides. However, they are not entirely free of risks to non-target species. Volatiles, toxins, and antibiotics produced by *Trichoderma* spp.

might affect not only pathogenic species but also beneficial microbes. As per some reports, the intensity of such effects depends on the time of inoculation of biocontrol agents and concentration of the toxins secreted. There is competition for nutrients that might also be responsible for alteration in the population of soil microbiome. Soil enzyme activities are considered indicators for abiotic or biotic stresses, where the presence of pathogen increases their levels. Normally, addition of biocontrol agents have reported to decrease the biotic stress of pathogen by lowering enzyme activities. But, it has some non-target impacts due to which *Trichoderma* spp. has also been observed to increase enzyme activities. It is challenging to monitor the effect of a bioinoculant on non-target soil microbial communities in rhizosphere and to understand the functioning of a biological system. Still, there is a need for future research to assess the ecological effects correlated with the application of biocontrol agents in the soil, mechanisms responsible for their non-target impacts, and development of methods for determination of these impacts.

References

Amaresan N, Kumar K, Venkadesaperuma G, Srivathsa NC (2018) Microbial community level physiological profiles of active mud volcano soils in Andaman and Nicobar Islands. Nat Acad Sci Lett 41:1–4

Araújo ASF, De Souza DG, De Almeida Lopes AC (2016) T-RFLP analysis of soil bacterial structure from Cerrado within the Sete Cidades National Park, Brazil. Neotrop Biodiversity 2:163–170

Blaya J, López-Mondéjar R, Lloret E, Pascual JA, Ros M (2013) Changes induced by *Trichoderma harzianum* in suppressive compost controlling *Fusarium wilt*. Pestic Biochem Physiol 107:112–119

Brimner TA, Boland GJ (2003) A review of the non-target effects of fungi used to biologically control plant diseases. Agric Ecosyst Environ 100:3–16

Chet I, Inbar J (1994) Biological control of fungal pathogens. Appl Biochem Biotechnol 48:37–43

Cook RJ, Bruckart WL, Coulson JR, Goettel MS, Humber RA, Lumsden RD, Maddox JV, McManus ML, Moore L, Meyer SF, Quimby PC Jr, Stack JP, Vaughn JL (1996) Safety of microorganisms intended for pest and plant disease control: a framework for scientific evaluation. Biol Control 7:335–351

Cordier C, Alabouvette C (2009) Effects of the introduction of a biocontrol strain of *Trichoderma atroviride* on non-target soil microorganisms. Eur J Soil Biol 7:267–274

Epelde L, Jauregi L, Urra J, Ibarretxe L, Romo J, Goikoetxea I, Garbisu C (2018) Characterization of composted organic amendments for agricultural use. Front Sustain Food Syst 2:Article 44

Frac M, Oszust K, Lipiec J (2012) Community level physiological profiles (CLPP), characterization and microbial activity of soil amended with dairy sewage sludge. Sensors 12:3253–3268

Garcia DE, Lopez BR, de-Bashan LE, Hirsch AM, Maymon M, Bashan Y (2018) Functional metabolic diversity of the bacterial community in undisturbed resource island soils in the southern Sonoran Desert. Land Degrad Dev 29:1467–1477

Gasoni L, Khan N, Yokoyama K, Chiessa GH, Kobayashi K (2008) Impact of *Trichoderma harzianum* biocontrol agent on functional diversity of soil microbial community in tobacco monoculture in Argentina. World J Agric Sci 4:527–532

Goldman GH, Hayes C, Harman GE (1994) Molecular and cellular biology of biocontrol by *Trichoderma* spp. Trends Biotechnol 12:478–482

Gupta R, Mathimaran N, Wiemken A, Boller T, Bisaria VS, Sharma S (2014) Non-target effects of bioinoculants on rhizospheric microbial communities of *Cajanus cajan*. Appl Soil Ecol 76:26–33

Harman GE (2006) Overview of mechanisms and uses of *Trichoderma* spp. Phytopathology 96:190–194

Hashem A, Abd_Allah EF, Alqarawi AA, Al Huqail AA, Egamberdieva D (2014) Alleviation of abiotic salt stress in *Ochradenus baccatus* (Del.) by *Trichoderma hamatum* (Bonord.) Bainier. J Plant Interact 9:857–868

Hermosa R, Viterbo A, Chet I, Monte E (2012) Plant-beneficial effects of *Trichoderma* and of its genes. Microbiology 158:17–25

Howell CR (2006) Understanding the mechanisms employed by *Trichoderma virens* to effect biological control of cotton diseases. Phytopathology 96:178–180

Kleifeld O, Chet I (1992) *Trichoderma harzianum* interaction with plants and effect on growth response. Plant Soil 144:267–272

Kleyer H, Tecon R, Or D (2017) Resolving species level changes in a representative soil bacterial community using microfluidic quantitative PCR. Front Microbiol 8:Article 2017

Li S, Lü T, Zhang X, Gu G, Niu Y (2013) Effect of *Trichoderma longbrachiatum* T2 on functional diversity of cucumber rhizomicrobes. J Environ Biol 34:293–299

Lladó S, Baldrian P (2017) Community-level physiological profiling analyses show potential to identify the copiotrophic bacteria present in soil environments. PLoS One 12:e0171638

Lorito M, Woo SL, Harman GE, Monte E (2010) Translational research on *Trichoderma*: from omics to the field. Annu Rev Phytopathol 48:395–417

Louis BP, Maron PA, Menasseri-Aubry S, Sarr A, Lévêque J, Mathieu O, Jolivet C, Leterme P, Viaud V (2016) Microbial diversity indexes can explain soil carbon dynamics as a function of carbon source. PLoS One 11:e0161251

Lumsden RD, Carter JP, Whipps JM, Lynch JM (1990) Comparison of biomass and viable propagule measurements in the antagonism of *Trichoderma harzianum* against *Pythium ultimum*. Soil Biol Biochem 22:187–194

McAllister CB, Garcia-Romera I, Godeas A, Ocampo JA (1994) In vitro interactions between *Trichoderma koningii*, *Fusarium solani* and *Glomus mosseae*. Soil Biol Biochem 26:1369–1374

McLean KL, Dodd SL, Minchin RF, Ohkura M, Bienkowski D, Stewart A (2014) Non-target impacts of the biocontrol agent *Trichoderma atroviride* on plant health and soil microbial communities in two native ecosystems in New Zealand. Australas Plant Pathol 43:33–45

Morán-Diez E, Rubio B, Domínguez S, Hermosa R, Monte E, Nicolás C (2012) Transcriptomic response of *Arabidopsis thaliana* after 24 h incubation with the biocontrol fungus *Trichoderma harzianum*. J Plant Physiol 169:614–620

Mukherjee M, Mukherjee PK, Horwitz BA, Zachow C, Berg G, Zeilinger S (2012) *Trichoderma*–plant–pathogen interactions: advances in genetics of biological control. Indian J Microbiol 52:522–529

Naseby DC, Lynch JM (1998) Impact of wild-type and genetically modified Pseudomonas fluorescens on soil enzyme activities and microbial population structure in the rhizosphere of pea. Mol Ecol 7:617–625

Naseby DC, Pascual JA, Lynch JM (2000) Effect of biocontrol strains of *Trichoderma* on plant growth, *Pythium ultimum* populations, soil microbial communities and soil enzyme activities. J Appl Microbiol 88:161–169

Nieto-Jacobo MF, Steyaert JM, Salazar-Badillo FB, Nguyen DV, Rostás M, Braithwaite M, De Souza JT, Bremont JFJ, Ohkura M, Stewart A, Mendoza-Mendoza A (2017) Environmental growth conditions of *Trichoderma* spp. affects indole acetic acid derivatives, volatile organic compounds, and plant growth promotion. Front Plant Sci 8:Article 102

Ondreičková K, Piliarová M, Bušo R, Hašana R, Schreiber Ľ, Gubiš J, Kraic J (2018) The structure and diversity of bacterial communities in differently managed soils studied by molecular fingerprinting methods. Sustainability 10:1095–1111

Pacwa-Płociniczak M, Płociniczak T, Yu D, Kurola JM, Sinkkonen A, Piotrowska-Seget Z, Romantschuk M (2018) Effect of *Silene vulgaris* and heavy metal pollution on soil microbial diversity in long-term contaminated soil. Water Air Soil Pollut 229:1–13

Pang G, Cai F, Li R, Zhao Z, Li R, Gu X, Shen Q, Chen W (2017) *Trichoderma*-enriched organic fertilizer can mitigate microbiome degeneration of monocropped soil to maintain better plant growth. Plant Soil 416:181–192

Pascual J, Blanco S, Ramos JL, Van Dillewijn P (2018) Responses of bulk and rhizosphere soil microbial communities to thermoclimatic changes in a Mediterranean ecosystem. Soil Biol Biochem 118:130–144

Peñuelas J, Filella I (1998) Visible and near-infrared reflectance techniques for diagnosing plant physiological status. Trends Plant Sci 3:151–156

Ros M, Raut I, Santisima-Trinidad AB, Pascual JA (2017) Relationship of microbial communities and suppressiveness of *Trichoderma* fortified composts for pepper seedlings infected by *Phytophthora nicotianae*. PLoS One 12:e0174069

Sampson PH, Zarco-Tejada PJ, Mohammed GH, Miller JR, Noland TL (2003) Hyperspectral remote sensing of forest condition: estimating chlorophyll content in tolerant hardwoods. For Sci 49:381–391

Saravanakumar K, Li Y, Yu C, Wang QQ, Wang M, Sun J, Gao JX, Chen J (2017) Effect of *Trichoderma harzianum* on maize rhizosphere microbiome and biocontrol of *Fusarium* stalk rot. Sci Rep 7:1771–1783

Shi WL, Chen XL, Wang LX, Gong ZT, Li S, Li CL, Xie BB, Zhang W, Shi M, Li C, Zhang YZ, Song XY (2016) Cellular and molecular insight into the inhibition of primary root growth of Arabidopsis induced by peptaibols, a class of linear peptide antibiotics mainly produced by *Trichoderma* spp. J Exp Bot 67:2191–2205

Shoresh M, Harman GE (2008) The molecular basis of shoot responses of maize seedlings to *Trichoderma harzianum* T22 inoculation of the root: a proteomic approach. Plant Physiol 147:2147–2163

Shrivastava S, Prasad R, Varma A (2014) Anatomy of root from eyes of a microbiologist. In: Morte A, Varma A (eds) Root engineering, vol 40. Springer, Berlin, pp 3–22

Soliman T, Yang SY, Yamazaki T, Jenke-Kodama H (2017) Profiling soil microbial communities with next-generation sequencing: the influence of DNA kit selection and technician technical expertise. Peer J 5:e4178

Szczepaniak Z, Cyplik P, Juzwa W, Czarny J, Staninska J, Piotrowska-Cyplik A (2015) Antibacterial effect of the *Trichoderma viride* fungi on soil microbiome during PAH's biodegradation. Int Biodeterior Biodegrad 104:170–177

Trabelsi D, Mhamdi R (2013) Microbial inoculants and their impact on soil microbial communities: a review. Biomed Res Int 2013:Article ID 863240

Umadevi P, Anandaraj M, Srivastav V, Benjamin S (2018) *Trichoderma harzianum* MTCC 5179 impacts the population and functional dynamics of microbial community in the rhizosphere of black pepper (*Piper nigrum* L.). Braz J Microbiol 49(3):463–470. https://doi.org/10.1016/j.bjm.2017.05.011

Vázquez MM, César S, Azcón R, Barea JM (2000) Interactions between arbuscular mycorrhizal fungi and other microbial inoculants (*Azospirillum, Pseudomonas, Trichoderma*) and their effects on microbial population and enzyme activities in the rhizosphere of maize plants. Appl Soil Ecol 15:261–272

Verma M, Brar SK, Tyagi RD, Surampalli RY, Valero JR (2007) Antagonistic fungi, *Trichoderma* spp.: panoply of biological control. Biochem Eng J 37:1–20

Wang S, Chen X, Gong H, Cai Z (2018) Response of soil microbial abundance and diversity in Sacha Inchi (*Plukenetia volubilis* L.) farms with different land-use histories in a tropical area of Southwestern China. Arch Agron Soil Sci 64:588–596

Wood JL, Zhang C, Mathews ER, Tang C, Franks AE (2016) Microbial community dynamics in the rhizosphere of a cadmium hyper-accumulator. Sci Rep 6:36067

Wu Z, Lin W, Li B, Wu L, Fang C, Zhang Z (2015) Terminal restriction fragment length polymorphism analysis of soil bacterial communities under different vegetation types in subtropical area. PLoS One 10:e0129397

Zhang Z, Qu Y, Li S, Feng K, Wang S, Cai W, Liang Y, Li H, Xu M, Yin H, Deng Y (2017) Soil bacterial quantification approaches coupling with relative abundances reflecting the changes of taxa. Sci Rep 7:4837

Zhu S, Wang Y, Xu X, Liu T, Wu D, Zheng X, Tang S, Dai Q (2018) Potential use of high-throughput sequencing of soil microbial communities for estimating the adverse effects of continuous cropping on ramie (*Boehmeria nivea* L. Gaud). PLoS One 13:e0197095

Chapter 11
Olive Anthracnose and Its Management by Fungal Endophytes: An Overview

Fátima Martins, José Alberto Pereira, and Paula Baptista

Abstract Anthracnose caused by diverse species of fungi belonging to the genus *Colletotrichum* is an important disease of the olive tree, leading to high productivity losses. Their control is very difficult and none of the available control measures are effective enough. Indeed, fungicides are most commonly applied to control this disease, but they are not totally effective, and their use possesses some environmental concerns. In the last decades, the implementation of sustainable production methods has been encouraged with emphasis on the use of living organisms as control agents against plant pests, diseases, and weeds. These control agents comprise a variety of predators, parasitoids, and microorganisms, including fungal endophytes. This review highlights the importance of endophytic fungi for the management of olive anthracnose. These fungi can be effectively used to improve plant performance and plant protection against biotic and abiotic stresses. They produce a wide variety of secondary metabolites that directly or indirectly induced defense responses in the host plant against pathogens. Thus, in this review emphasis is given for the exploitation of fungal endophytes associated to olive tree in the development of new tools/approaches to manage olive anthracnose.

11.1 Olive Anthracnose: A General Overview

The European olive, *Olea europaea* subsp. *europaea* L., is one of the major cultivated species in countries surrounding the Mediterranean Sea. In 2016, approximately 9.2 million ha of land in this region were planted with olive trees (FAOSTAT 2018). Several insect pest and diseases attack the olive crop, reducing its yield both in terms of quantity and quality. Among diseases, anthracnose is the major causes of olive crop damage worldwide (Talhinhas et al. 2018). It was first described in Portugal in 1899 by J.V. d'Almeida (1899) and rapidly expanded to all continents (Cacciola et al. 2012) becoming a serious economic constraints to olive crop production (Mosca et al. 2014; Iliadi et al. 2018). This disease affects different

F. Martins · J. A. Pereira · P. Baptista (✉)
CIMO, Instituto Politécnico de Bragança, Campus de Santa Apolónia, Bragança, Portugal
e-mail: pbaptista@ipb.pt

A. Varma et al. (eds.), *Plant Microbe Interface*,
https://doi.org/10.1007/978-3-030-19831-2_11

Fig. 11.1 Characteristic symptoms of anthracnose on olive tree fruits of cv. Madural (**a**) production of orange/pink sticky masses of conidia in olive surface; (**b**) rot, mummification, and dehydration of fruits; (**c**) symptoms appear mostly in mature fruits, but, in favorable environmental conditions, green fruits may also be infected; (**d**) affected fruits fall prematurely to the ground (photos: Fátima Martins)

organs of the olive tree, including flowers, buds, shoots, leaves, and twigs, being the fruits the most severely affected (Cacciola et al. 2012). Thus, characteristic anthracnose symptoms arise mostly on the fruits, especially when they are nearly ripened. The first symptoms of infected olives are small and brown-colored spots in the epicarp that become later sunken. As the fruits ripe, the center of these sunken spots becomes covered with pink/orange gelatinous masses of conidia that are often produced in a concentric ring pattern (Talhinhas et al. 2011). This causes mummification, rotting, and premature drop of fruits, leading to significant crop losses (Fig. 11.1). The attacks can occur on any part of the fruit, but they are more frequent at the apex, because it stays wet in longer time (Cacciola et al. 2012). In some cases, infected fruits may persist on the tree, becoming an inoculum reservoir of olive anthracnose (Sergeeva 2011a). In the vegetative parts, the symptoms include leaf chlorosis, defoliation, and dieback of shoot and twigs (Cacciola et al. 2012). These effects are due to production of toxins by the pathogen (Cacciola et al. 2012). The infected flowers display blossom blight, dry out, and drop quickly (Moral et al. 2008; Sergeeva et al. 2008). Infections are usually most severe on the lower branches, inside the canopy on the north side, where moisture tends to remain for longer periods of time (de Cantero 1997).

The disease can be devastating, depending on the level of susceptibility of the cultivars, the environmental conditions, the inoculum pressure, and the virulence of the pathogenic strains (Talhinhas et al. 2018). Under favorable conditions, all production can be destroyed. For instance, in some olive-growing countries, the olive anthracnose was described to cause yield losses above 80% (Cacciola et al.

2012). In addition, this disease can reduce the quality of olive oil. The oils' peroxide content and acidity value from anthracnose-infected fruits sometimes can be higher than the maximum legal limit to be considered as virgin olive oil (da Silva 2016). Most of these olive oils show negative sensory and organoleptic characteristics, being classified as lampante (da Silva 2016).

11.2 Anthracnose is Caused by a Complex of *Colletotrichum* Species

Anthracnose in olive tree is associated with at least eight *Colletotrichum* species, belonging to two heterogeneous fungal species complexes, namely, *C. acutatum* sensu lato (s.l.) and *C. gloeosporioides* s.l. (Damm et al. 2012). Of these two complexes, *C. acutatum* s.l. is the most predominant, causing epidemic explosions of anthracnose in most olive-growing countries (Talhinhas et al. 2005). Multilocus molecular phylogenetic analysis revealed that there are six species in the *C. acutatum* complex considered to be causal agents of olive anthracnose, namely, *C. fioriniae*, *C. simmondsii*, *C. nymphaeae*, *C. acutatum* sensu stricto (s.s.), *C. godetiae* (syn. *C. clavatum*), and *C. rhombiforme* (Talhinhas et al. 2018). The same study also revealed that there are two species belonging to the *gloeosporioides* complex, *C. gloeosporioides* s.s. and *C. theobromicola* (Talhinhas et al. 2018). *Colletotrichum boninense* (syn. *C. karstii*) is a third species complex that was recently related with olive anthracnose (Schena et al. 2014). However, this complex does not appear to threaten olive production due to their weakly pathogenicity (Schena et al. 2014). Similarly, other fungal species belonging to *C. gloeosporioides* complex (i.e., *C. aenigma*, *C. queenslandicum*, *C. siamense*, and *C. kahawae* ssp. *cigarro*) were isolated from symptomatic fruits, but their pathogenicity in olives has not yet been confirmed (Schena et al. 2014).

Among all these fungal species identified, *C. acutatum* s.s., *C. godetiae*, and *C. nymphaeae* have been recognized as major causative agents of olive anthracnose in most olive-growing countries (Mosca et al. 2014). For instance, the majority of strains examined from South Africa, Australia, and Tunisia belonged to *C. acutatum* s.s. (Cacciola et al. 2012). In other studies performed in Montenegro, Greece, Italy, and Spain, *C. godetiae* was identified as the most prevalent species (Moral et al. 2008, 2009, 2014). In Portugal, primarily three important species have been related to the olive anthracnose, with *C. godetiae* causing major damage in the northern region, whereas *C. nymphaeae* and *C. acutatum* s.s. have been identified as the most prevalent species in the southern regions (Talhinhas et al. 2009).

Species of *Colletotrichum* have a teleomorph or sexual stage, i.e., *Glomerella* sp. (Wharton and Diéguez-Uribeondo 2004). Nevertheless, in olive crops, the teleomorph of the pathogen has not yet been detected in field conditions (Cacciola et al. 1996), suggesting the imperfect stage, i.e., *Colletotrichum* sp., as the main responsible of olive anthracnose.

11.3 Epidemiology and Life Cycle

Epidemiology and life cycle of olive anthracnose are still poorly understood, especially in what concerns propagation and inoculum maintenance in the olive groves, which required more studies (Moral et al. 2008; Cacciola et al. 2012). In Mediterranean regions, it has been reported that infections begin during the spring in flowers and in young fruits (primary infection; Fig. 11.2) (Moral et al. 2009). The mode of survival and the source of this primary inoculum have yet to be determined (Moral et al. 2009). It is thought that the major primary inoculum reservoirs are mummified fruits that remain on the tree or on the ground, from one season to the next (Moral et al. 2009). It is also plausible that the source of inoculum in spring may originate from fungi that overwinter in woody material and leaves of the tree (Talhinhas et al. 2018). After primary infection, the fungus stops growing and remains dormant until fruit begins to ripen (Moral et al. 2009). At that time, with favorable environmental conditions, sticky masses of spores are produced in acervuli. These spores are then spread by rain splash to newly fruits and other tree parts, giving rise to secondary infections (Moral et al. 2009). The spread of the pathogen and infection of olive tree depend heavily on the climatic conditions (Talhinhas et al. 2015). Olive anthracnose reaches highest disease incidence and severity in areas where relative humidity is highest (over 93%) and the air temperature is warm (ranging from 10 to 30 °C) (Cacciola et al. 2012). The occurrence of precipitation is also crucial for the conidia separation from the gelatinous mass of the acervuli and for their dispersion (Cacciola et al. 2012). Also, the infection of fruits depends on the extent of peel ripeness. Olives at later stages of ripening are more prone to fungal infection than green fruits (da Silva 2016). The severity of symptoms varies widely with the cultivar (i.e., their susceptibility to anthracnose) and the virulence of the strain (Talhinhas et al. 2015). Recent studies showed that, in several olive-growing countries, the pathogen populations are particularly adapted to both environmental conditions and the host, but severe infections occur when only virulent populations of the pathogen are present (Moral et al. 2017).

Usually, penetration and colonization of plant tissues by *Colletotrichum* species comprises a sequential set of stages. Generally, it starts with the fixation and germination of the conidia on the host surface, followed by appressorium development, which facilitates entry through the host epidermis (Wharton and Diéguez-Uribeondo 2004). A detailed study of *C. acutatum* infection on olives showed that after spores' germination, a germ tube is produced and differentiated in an appressorium, which facilitated the penetration of the fungus into the host cells (Gomes et al. 2009). This process occurs within a few hours (48–72 h), and consequently, the infections can occur rapidly under favorable conditions (Gomes et al. 2009). Fungal penetration is also believed to occur through stomas or lenticels as well as wounds caused by insect (e.g., *Bactrocera oleae*) attack (Cacciola et al. 2012).

After penetration on fruit, *Colletotrichum* sp. can follow different infection strategies. These strategies can be range from intracellular hemibiotrophic mode (colonizes living plant tissue and obtains nutrients from living host cells) to the

Fig. 11.2 Diagrammatic representation of disease cycle of olive anthracnose in the Mediterranean region (photos: Fátima Martins)

subcuticular intramural necrotrophic (infects and kills host tissue and extracts nutrients from the dead host cells) mode of nutrition (Gomes et al. 2009), being hemibiotrophic the most common (De Silva et al. 2017). The infection and colonization strategy of *C. acutatum* sp. on olive fruits of both susceptible (cv. *Galega Vulgar*) and tolerant (cv. *Picual*) cultivars was identified as intracellular hemibiotrophic, followed by a necrotrophic phase (Gomes et al. 2009).

11.4 Management Strategies for Olive Anthracnose

Management of olive anthracnose is very difficult, because its spreading and development relies greatly on the climatic conditions. Thus, no effective control measures have been proposed so far for its management. Generally, those measures rely on an integrated approach that combines several means and tools, either to prevent (indirect method) or to protect (direct method) olive crop against anthracnose (Cacciola et al. 2012; Moral et al. 2018).

Indirect or preventive measures of olive anthracnose rely mostly on practices aiming either to reduce the initial levels of inoculum or reduce the rate of spread of the established pathogen. These practices include agronomic techniques such as pruning, drainage and irrigation, fertilization, use of varieties tolerant/resistant to anthracnose, and control of insects that potentially may spread the pathogen, among others. Pruning of olive trees can be an effective way to eliminate sources of fungal inoculum, by removing diseased twigs of infected olive trees. After pruning, the plant material should be removed from the grove and destroyed. Olive pruning also promotes aeration and light penetration in the canopy, helping to reduce the severity of the disease (Sergeeva 2011a). Irrigation management has a strong impact on the olive anthracnose disease severity and epidemic progress rates, since *Colletotrichum* sp. are greatly dependent not only on high humidity levels for all stages of their life cycle but also on available free water for conidia dispersion, which is a process of great epidemiological consequence (Cacciola et al. 2012). Thus, overwatering should be avoided in the grove where anthracnose is present in order to prevent the outbreak of the disease (Sergeeva 2011a). Due to the dependence of *Colletotrichum* sp. to water splash for dispersion, the choice of irrigation method could be extremely important to avoid infections of epidemic-like proportions. Adequate nutrition may also have an important role in reducing the severity of olive anthracnose. Previous studies performed in strawberry showed that the source and level of nitrogen in fertilizers had a great effect on severity of anthracnose (Smith 2009). As far as we know, no studies have been carried out to evaluate the influence of nitrogen fertilization on incidence and development of olive anthracnose. However, a balanced fertilization is frequently recommended for management of olive anthracnose (Sergeeva 2011b). In general, a balanced fertilizer with fairly low nitrogen content will be ideal, since overapplication of nitrogen fertilizers has been reported to increase the incidence of diseases on olive tree canopy (Roca et al. 2018). Use of olive cultivars resistant to the anthracnose pathogens is one of the most successful approaches to the control of this disease (Moral and Trapero 2009). Numerous studies, carried out in several olive-growing countries, have already identified olive tree varieties with different levels of susceptibility to anthracnose, ranging from highly susceptible (e.g., cv. Galega Vulgar) to highly resistant (e.g., cv. Frantoio) (e.g., Talhinhas et al. 2015; Moral et al. 2017). However, response to anthracnose of olive tree cultivars under field conditions has been showed to be dependent on the *Colletotrichum* species (Talhinhas et al. 2015) and on the climatic conditions, in particular of relative humidity (Moral et al. 2014, 2017). Thus, in certain humid olive-growing areas, anthracnose-resistant cultivars can still get infected (Moral et al. 2014, 2017). Control of olive fruit fly attacks, which provides entry points for *Colletotrichum* sp., will limit the surface damage of the fruit and may also be useful to reduce the severity of anthracnose (Malacrinò et al. 2017).

Methods and tools for direct control of olive anthracnose include the use of fungicides and more recently of natural products and biocontrol agents. The fungicides generally recommended for controlling olive anthracnose are protective fungicides based on copper compounds, such as copper oxychloride, copper sulfate, and copper hydroxide (Cacciola et al. 2012). Newer chemicals, such as strobilurins, have also been showed to increase copper-based fungicides effectiveness against olive anthracnose in orchards when used in combination (Moral et al. 2018). Similarly, natural products, like botanicals (i.e., plant extracts) and products of mineral origin (i.e., calcium-rich compounds), have been recently explored in the control of olive anthracnose (Moral et al. 2018). Calcium-rich compounds have been showed to inhibit *Colletotrichum* sp. appressorial formation under in vitro tests, but their field application was not always effective in the control of olive anthracnose (Xaviér 2014). Extract obtained from the peel of pomegranate (*Punica granatum* L.) has proven to be effective against *Colletotrichum* sp. under laboratory conditions and to control olive anthracnose under in field trays (Pangallo et al. 2017). Biological control (BC) is another alternative for olive anthracnose management, although this approach has not been as effective as the chemical control (Holt et al. 2009). The possibilities of using biocontrol agents (BCAs) for controlling the pathogen of olive anthracnose were firstly illustrated by Segura (2003). In artificial inoculations of olives, the microorganisms *Aureobasidium pullulans*, *Curtobacterium flaccumfaciens*, and *Paenibacillus polymyxa* were shown to decrease the severity of the symptoms produced by *C. acutatum* in 76.4, 53.7, and 51.6%, respectively (Segura 2003). Since then, few studies have been done on the BC of olive anthracnose and this strategy has not been used against this disease in field conditions.

Although the several efforts made to better understand the epidemiology and population genetics of the different pathogenic species, the olive anthracnose still remains a "complex disease" to decipher. Indeed, it remains unclear how the pathogen interact with the host plant, which is the variability of *Colletotrichum* species in some olive-growing regions, and which are the best control strategies against this disease. In this regard, the use of fungal endophytes to control olive anthracnose could be a promising approach (Landum et al. 2016; Preto et al. 2017). These microorganisms are able to inhabit the same niche in the same environment that of *Colletotrichum* spp., favoring them as potential biocontrol agents against olive anthracnose.

11.4.1 Fungal Endophytes and Their Potential as Biocontrol Agents Against **Colletotrichum** spp.

Fungal endophytes are microorganisms that inhabit the inner tissue of the plant, at some part or whole of its life cycle, without causing any apparent damage to the hosts (Busby et al. 2016). According to the mechanisms used to colonize the host plant, the fungal endophytes were classified as "obligate" or "facultative" (Andreote et al. 2014). Obligate endophytes are transmitted to other plants by vertical colonization or by vectors and are strictly dependent on host cell metabolism for their survival and replication (Andreote et al. 2014). Facultative endophytes have a free life, living outside of host plant, and during a certain stage of their life cycle, they colonize the plant internally (Andreote et al. 2014).

Overall, most endophytic fungi within plant tissues belong to Ascomycota and Basidiomycota phyla (Arnold and Lutzoni 2007; Selosse et al. 2009). In particular, the composition of fungal endophytic community of olive tree has been only recently analyzed (Martins et al. 2016; Landum et al. 2016; Preto et al. 2017; Gomes et al. 2018). Overall, these studies showed that there is great diversity and abundance of fungal endophytes in several organs of olive tree, including leaves, twigs, fruits, and roots. More than 65 genera from 33 families and 2 phyla of fungal species have been reported to be associated with olive tree (Fig. 11.3). Most of the fungal isolates belong to the phyla *Ascomycota*, accounting to 93% of the total number of fungal isolates, followed by *Basidiomycota* (Martins et al. 2016; Landum et al. 2016; Preto et al. 2017; Gomes et al. 2018). The most abundant fungal families are *Pleosporaceae* (17.1% of the total fungal isolates), *Incertae sedis* (13.7%), and *Nectriaceae* (8.5%). *Alternaria*, *Penicillium*, *Epicoccum*, and *Phomopsis* were identified as the most abundant genera, accounting together 25.5% of total fungal isolates. The various olive tree organs surveyed displayed differences on endophytic fungal composition. Members of *Pleosporaceae* and *Incertae sedis* were the most abundant in leaves and twigs of olive tree, accounting together 90.3% of the total isolates, whereas *Trichocomaceae* and *Nectriaceae* were the most abundant in roots and fruits, respectively (Fig. 11.3) (Martins et al. 2016; Landum et al. 2016; Preto et al. 2017; Gomes et al. 2018). Besides plant organ, host plant geographic location, host genetics (at cultivar level), and season and climatic conditions, such as rainfall and temperature, were also shown to contribute to the shaping of fungal communities in olive tree (Martins et al. 2016; Preto et al. 2017; Gomes et al. 2018). In general, the diversity of fungal endophytes in olive tree leaves and twigs is higher in spring than in autumn (Gomes et al. 2018). The same study also identified differences on fungal composition between spring and autumn. These seasonal shifts were found to be related to climatic factors, especially to rainfall and mean temperature (Gomes et al. 2018). Geographic distance was also found to affect the structure of fungal endophytic communities especially of roots but also of leaves and twigs (Martins et al. 2016). An inverse relationship was noticed between the similarity of endophytic assemblages and their geographic distance (Martins et al. 2016).

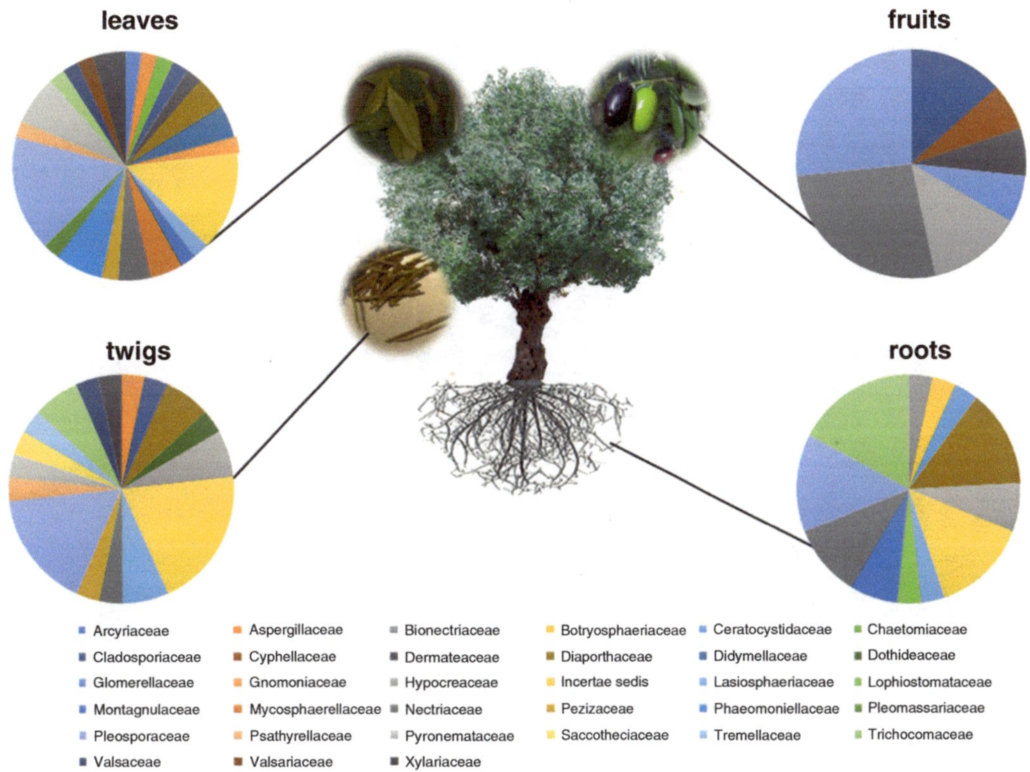

Fig. 11.3 Abundance (number of isolates) of fungal endophytes, at family level, present in leaves, twigs, fruits, and roots of olive tree (*Olea europaea* L.)

There is growing evidence that these endophytic fungi fulfill important functions for plant health and productivity (Khare et al. 2018). Endophytes can, for instance, promote plant nutrition and protection against abiotic (e.g., drought and extreme temperatures) and biotic stresses, such as plant pathogens (Bacon and White Jr 2016). In particular, the mechanisms used by endophytic fungi to protect host plant against pathogens mostly rely on the production of secondary metabolites, such as alkaloids, peptides, steroids, terpenoids, phenols, quinines, flavonoids, siderophores, and volatile organic compounds (Gao et al. 2010; Ownley et al. 2010; Speckbacher and Zeilinger 2018). Most of these classes of compounds comprise phytohormones, mycotoxins, antimicrobial molecules, as well as antibiotics that may reduce pathogen infection directly, through antibiosis, mycoparasitism, and competition, and indirectly by induction of plant resistance response (Lacava and Azevedo 2014). Endophytic fungi are also known to produce cell wall-degrading enzymes (e.g., chitinases, proteases, and glucanases) with the ability to destroy pathogens' cell wall (Lorito et al. 2010; Katoch et al. 2014). The above mechanisms regularly operated simultaneously.

Till date, few studies were conducted to explore the biocontrol activities of endophytes against anthracnose disease caused by *Colletotrichum* species under in vivo conditions (i.e., in detached fruits, field, and/or greenhouse) (Table 11.1). The results obtained up to date appear to be very promising being the level of disease suppression achieved by application of fungal endophytes ranging from 2.5 to 83%, depending on the fungal species (Table 11.1). According to the results shown in Table 11.1, the most promising fungal endophytes to control anthracnose diseases are *Trichoderma* spp., *Nodulisporium* sp., and *Cordana* sp. and also some yeasts belonging to the genera *Debaryomyces* and *Cryptococcus*. These strains were shown to be effective in reducing *Colletotrichum* growth and disease severity in several hosts like papaya (Valenzuela et al. 2015; Hernandez-Montiela et al. 2018), mango (Bautista-Rosales et al. 2014), and wild banana (Nuangmek et al. 2008). Competition for nutrients and space, antibiosis, and mycoparasitism and production of cell wall-degrading enzymes, antibiotics, and volatile organic compounds were the most important modes of action of fungal endophytes for anthracnose disease control (Table 11.1).

11.4.2 Fungal Endophytes on the Control of Olive Anthracnose

Despite the ability of fungal endophytes to control anthracnose disease, there are only limited studies on the use of these fungi against olive anthracnose. In addition, most of these studies were performed under controlled conditions, by using in vitro experiments, being field assays much more limited. Among the various endophytic fungal species tested in in vitro laboratory assays, *Alternaria* sp., *Diaporthe* sp., and *Nigrospora oryzae* isolated from olive tree leaves have been shown to inhibit up to

Table 11.1 Fungal endophytes that have been tested in vivo to control anthracnose disease caused by *Colletotrichum* spp., their possible mechanisms of action, and their efficacy

Antagonistic fungal isolates	Host plant	Assays	Disease agent	Mechanism of action	Efficacy	References
Aureobasidium pullulans	Olive (*Olea europaea* L.)	Field assay	*Colletotrichum* spp.	NA	Reduced both latent infection (14%) and disease severity (40%)	Nigro et al. (2018)
Pichia kudriavzevii *Wickerhamomyces anomalus*	Olive (*Olea europaea* L.)	In vivo (ripe olive fruit)	*C. gloeosporioides*	Competition Antibiotic production Invasive growth	Reduced disease severity (6.99–22.05%)	Pesce et al. (2018)
Debaryomyces hansenii	Papaya (*Carica papaya* L.) var. Maradol	In vivo (fruit)	*C. gloeosporioides*	Volatile organic compounds	Reduced pathogen growth (36%) and disease severity (83%)	Hernandez-Montiela et al. (2018)
Trichoderma spp.	Papaya (*Carica papaya* L.) var. Maradol	In vivo (fruit)	*C. gloeosporioides*	Invasive growth Mycoparasitism	Reduced pathogen growth (50–60%), and disease severity (77.40%)	Valenzuela et al. (2015)
Cryptococcus laurentii	Mango (*Mangifera indica* L.)	In vivo (mango fruit)	*C. gloeosporioides*	Antibiosis Nutrient competition Hydrolytic enzymes	Reduced disease severity (75.88%)	Bautista-Rosales et al. (2014)
Trichoderma viride	Bean (*Phaseolus vulgaris* L.)	In vivo (seeds)	*C. lindemuthianum*	Mycoparasitism Antibiosis	Reduced the growth (59.48%) and the germination (73.60%) of the pathogen, as well as disease severity (32.02%)	Padder and Sharma (2011)
Cordana abramovii *Nodulisporium* sp.	Wild banana (*Musa acuminata* Colla)	In vivo (detached banana)	*C. musae*	Competition Antibiotic production	Reduced the growth (90%) and the germination (91%) of the pathogen, as well as disease severity (53%)	Nuangmek et al. (2008)
Trichoderma viride	Cowpea (*Vigna unguiculata* L.)	In vivo (seedling)	*C. lindemuthianum*	Mycoparasitism Antibiosis	Reduced disease severity (2.5%)	Adebanjo and Bankole (2004)

NA not applicable

26.8% the growth of *C. acutatum* (Landum et al. 2016). This inhibitory effect was ascribed to the production of volatile compounds by the endophyte, in particular of phenylethyl alcohol, 4-methylquinazoline, benzothiazole, benzyl alcohol, lilial, and galaxolide (Landum et al. 2016). Similarly, the endophytic fungal species *Chondrostereum purpureum*, *Chaetomium globosum*, *Aspergillus westerdijkiae*, *Aspergillus* sp. 1, *Quambalaria cyanescens*, *Epicoccum nigrum*, and *Aspergillus brasiliensis*, isolated from olive fruits, have been shown to inhibit *C. acutatum* growth under in vitro conditions, reaching inhibition values of 30.9–71.3% (Preto et al. 2017). Some of these endophytic fungal strains were also shown to induce morphological alterations on pathogen hyphae and to reduce both the production (up to 46%) and germination (up to 21%) of *C. acutatum* spores (Preto et al. 2017). Although the exact mechanism of antagonism displayed by these fungi is not clear, it is hypothesized the involvement of antimicrobial compounds and lytic enzymes, secreted by endophytic isolates, which may act synergistically against the fungal pathogen (Preto et al. 2017). The degree to which fungal endophyte regulates *C. acutatum* infection is dependent on both host plant and the order of arrival of the pathogen and endophyte (Martins et al. 2013). In vitro confrontation assays between the endophyte *Penicillium commune* and *C. acutatum* in the presence of olive leaf (+leaf) revealed a greater inhibitory effect of the endophyte over the pathogen when compared to −leaf treatment (Martins et al. 2013). This result suggests that the plant-endophyte interaction is critical for the biocontrol of the pathogen. The observed inhibitory effect on *C. acutatum* sporulation and germination was strong (around 50 and 60%, respectively) when the endophyte colonized the leaf before the pathogen (Martins et al. 2013).

In olive fruit inoculation assays, the endophytic fungi *Trichoderma koningii* have been shown to reduce significantly ($p < 0.05$) both incidence (AUDPCi) and severity (AUDPCs) of olive anthracnose when compared to control (i.e., in the absence of *T. koningii*), either at 14 or 21 days postinoculation (Fig. 11.4). The effectiveness of this endophyte as a biological control agent against olive anthracnose was most notorious on fruits that start to change skin color (maturation index 2) than on purple or black olives (maturation index 3). The endophyte *T. koningii* also showed the capacity to inhibited significantly the production and germination of spores produced by the pathogen *C. godetiae* in olives, either at maturation index 2 (up to 1.6- and 6.1-fold, respectively) or 3 (up to 2.1- and 5.7-fold, respectively) when compared to control (Martins et al. 2017).

Few studies have determined the efficacy of fungal endophytes against olive anthracnose under field conditions. Only recently, it was reported that the treatment of olive tree with the endophyte *Aureobasidium pullulans* in field trays significantly reduced anthracnose severity by 40% and latent infection by 14% (Nigro et al. 2018).

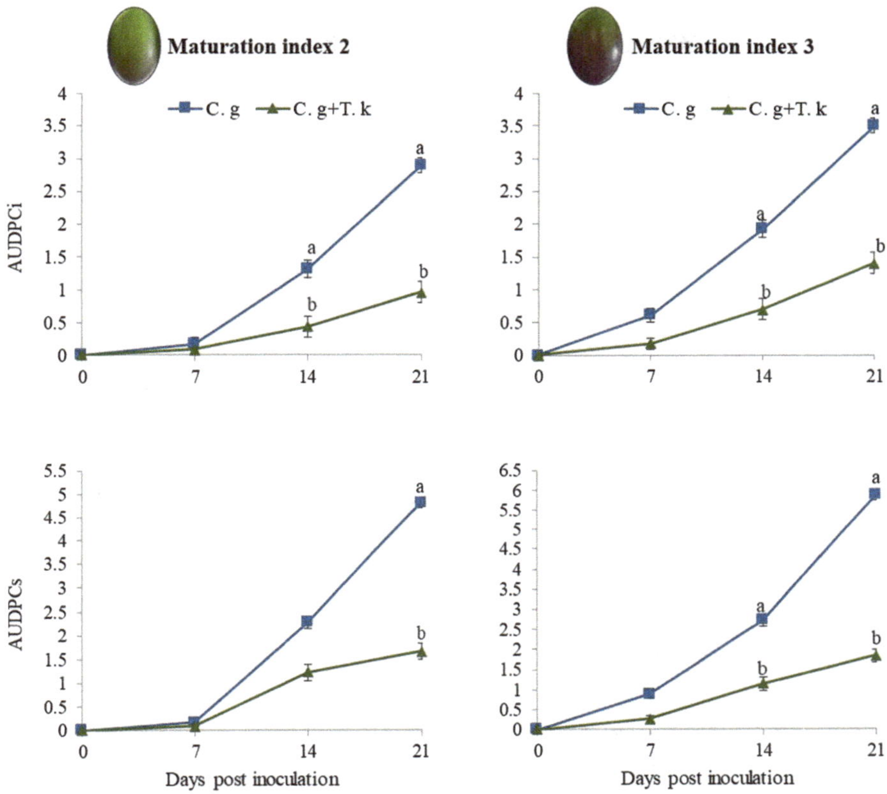

Fig. 11.4 Area under the disease progress curve of incidence (AUDPCi) and severity (AUDPCs) in olive fruits from cv. *Madural*, at maturation index 2 and 3, after 7, 14, and 21 days of inoculation only with *C. godetiae* (C.g) or in combination with the endophyte *T. koningii* (C.g + T.k). In each day, mean values followed by different letters are significantly different ($p < 0.05$)

11.5 Conclusion

The use of endophytic fungi for the biological control of olive anthracnose could be a sustainable alternative to olive crop production (Lugtenberg et al. 2016). Despite no effective biocontrol agents are still available against olive anthracnose, some authors have already described promising results in this area. However, most of these studies have detected the biocontrol activity of the fungal endophyte by using in vitro and in vivo tests on detached fruits, under controlled conditions. They therefore do not replicate the environment in which the biocontrol agent must function. More studies aiming the selection of fungal endophytes as biological control agents against olive anthracnose by using *in planta* assays, either in the field or greenhouses conditions, are required. Similarly, we still have incomplete knowledge on the various relationships that fungal endophytes can establish with their host and with other members of plant-associated microbial community, under natural conditions. Such studies will certainly contribute to enhance the chances to obtain competent endophytic biocontrol agent and therefore develop new successful and sustainable integrated crop protection against olive anthracnose.

Acknowledgments This work was partially funded by European Structural and Investment Funds in the FEDER component, through the Operational Competitiveness and Internationalization Programme (COMPETE 2020); and national funds, through the FCT – Portuguese Foundation for Science and Technology under the project POCI-01-0145-FEDER-031133.

References

Adebanjo A, Bankole SA (2004) Evaluation of some fungi and bacteria for biocontrol of anthracnose disease of cowpea. J Basic Microbiol 44:3–9

Almeida MJV (1899) La gaffa des olives en Portugal. Bulletin de la Société Mycologique de France 15:90–94

Andreote FD, Gumiere T, Durrer A (2014) Exploring interactions of plant microbiomes. Sci Agric 71:528–539

Arnold AE, Lutzoni F (2007) Diversity and host range of foliar fungal endophytes: are tropical leaves biodiversity hotspots? Ecology 88:541–549

Bacon CW, White FE Jr (2016) Functions, mechanisms and regulation of endophytic and epiphytic microbial communities of plants. Symbiosis 68:87–98

Bautista-Rosales PU, Calderon-Santoyoontserrat M, Servín-Villegas R, Ochoa-Álvarez NA, Vázquez-Juárez R, Ragazzo-Sánchez JA (2014) Biocontrol action mechanisms of *Cryptococcus laurentii* on *Colletotrichum gloeosporioides* of mango. Crop Prot 65:194–201

Busby PE, Ridout M, Newcombe G (2016) Fungal endophytes: modifiers of plant disease. Plant Mol Biol 90:645–655

Cacciola SO, Agosteo GE, Pane A, Magnano di San Lio MG (1996) Osservazioni sull'epidemiologia dell'antracnosi dell'olivo in Calabria. Informatore Fitopatologico 46:27–32

Cacciola SA, Faedda R, Sinatra F, Agosteo GE, Schena L, Frisullo S, di San Lio MG (2012) Olive anthracnose. J Plant Pathol 94:29–44

da Silva FCV (2016) Olive anthracnose: passive defense of tolerant and susceptible Portuguese *Olea europaea* L. cultivars and its effect on olive oil quality. Escola Superior Agrária de Bragança Tese de Mestrado, Bragança

Damm U, Cannon PF, Woudenberg JH, Crous PW (2012) The *Colletotrichum acutatum* species complex. Stud Mycol 73:37–113

de Cantero FA (1997) Enfermedades y plagas del olivo, 3ª ed. Riquelme y Varquelme y Vargas Ediciones, S.L., Jaén, 646p

De Silva DD, Crous PW, Ades PK, Hyde KD, Taylor PWJ (2017) Life styles of *Colletotrichum* species and implications for plant biosecurity. Fungal Biol Rev 31:155–168

FAOSTAT (2018) online database, available at link http://faostat.fao.org/. Accessed Nov 2018

Gao FK, Dai CC, Liu XZ (2010) Mechanisms of fungal endophytes in plant protection against pathogens. Afr J Microbiol Res 13:1346–1351

Gomes S, Prieto P, Martins-Lopes P, Carvalho T, Martin A, Guedes-Pinto H (2009) Development of *Colletotrichum acutatum* on tolerant and susceptible *Olea europaea* L. cultivars: a microscopic analysis. Mycopathologia 168:203–211

Gomes T, Pereira JA, Benhadi J, Lino-Neto T, Baptista P (2018) Endophytic and epiphytic phyllosphere fungal communities are shaped by different environmental factors in a Mediterranean ecosystem. Microbiol Ecol 76:668–679

Hernandez-Montiela LG, Gutierrez-Pereza ED, Murillo-Amadora B, Vero S, Chiquito-Contreras RG, Rincon-Enriquez G (2018) Mechanisms employed by *Debaryomyces hansenii* in biological control of anthracnose disease on papaya fruit. Postharvest Biol Technol 139:31–37

Holt KM, George O, Nechols JR, Margolies DC, Williams KA (2009) Comparing chemical and biological control strategies for twospotted spider mites in mixed production of Ivy Geranium and impatiens. J Econ Entomol 102:336–346

Iliadi MK, Tjamos E, Antoniou P, Tsitsigiannis DI (2018) First report of *Colletotrichum acutatum* causing anthracnose on olives in Greece. Plant Dis 102:820–821

Katoch M, Salgotra A, Gurpreet S (2014) Endophytic fungi found in association with Bacopa monnieri as potential producers of industrial enzymes and antimicrobial bioactive compounds. Braz Arch Biol Technol 57(5):714–722

Khare E, Mishra J, Arora NK (2018) Multifaceted interactions between endophytes and plant: developments and prospects. Front Microbiol 9:2732

Lacava PT, Azevedo JL (2014) Biological control of insect-pest and diseases by endophytes. In: Verma VC, Gange AC (eds) Advances in endophytic research part v. Springer India, New Delhi, pp 231–256

Landum MC, Felix MR, Alho J, Garcia R, Cabrita MJ, Rei F, Varanda CMR (2016) Antagonistic activity of fungi of *Olea europaea* L. against *Colletotrichum acutatum*. Microbiol Res 183:100–108

Lorito M, Woo SL, Harman GE, Monte E (2010) Translational research on *Trichoderma*: from 'omics to the field. Annu Rev Phytopathol 48:395–417

Lugtenberg BJJ, Caradus JR, Johnson LJ (2016) Fungal endophytes for sustainable crop production. FEMS Microbiol Ecol 92. https://doi.org/10.1093/femsec/fiw194

Malacrinò A, Schena L, Campolo O, Laudani F, Mosca S, Giunti G, Patricia CS, Vincenzo P (2017) Metabarcoding survey on the fungal microbiota associated to the olive fruit fly. Microb Ecol 73:677–684

Martins F, Pereira JA, Bota P, Bento A, Baptista P (2013) Plant-mediated effects on antagonistic activity of endophytic fungi towards olive fungal diseases. 5th international symposium in plant protection and plant health in Europe/COST action – endophytes for plant protection: the state of the art, Humboldt University, Berlin (26–29 May 2013)

Martins F, Pereira JA, Bota P, Bento A, Baptista P (2016) Fungal endophyte communities in above- and belowground olive tree organs and the effect of season and geographic location on their structures. Fungal Ecol 20:193–201

Martins F, Pereira JA, Bota P, Bento A, Baptista P (2017) Interaction between *Colletotrichum godetiae* and *Trichoderma koningii* and their effect on incidence and severity of olive anthracnose. COST FA1405 annual meeting – "systems biology approaches to identify mechanisms underlying crop-arthropod-microbe interactions", Liubliana (Eslovénia)

Moral J, Trapero A (2009) Assessing the susceptibility of olive cultivars to anthracnose caused by *Colletotrichum acutatum*. Plant Dis 93:1028–1036

Moral J, Bouhmidi K, Trapero A (2008) Influence of fruit maturity, cultivar susceptibility, and inoculation method on infection of olive fruit by *Colletotrichum acutatum*. Plant Dis 92:1421–1426

Moral J, Cherifi F, Muñoz-Dìez C, Xaviér CJ, Taper A (2009) Infection of olive seeds by *Colletotrichum acutatum* and its effect on germination. Phytopathology 99:S88

Moral J, Xaviér C, Roca L, Romero J, Moreda W, Trapero A (2014) La Antracnosis del olivo y su efecto en la calidad del aceite. Grasas Aceites 65(2):e028

Moral J, Xaviér CJ, Viruega JR, Roca LF, Caballero J, Trapero A (2017) Variability in susceptibility to anthracnose in the world collection of olive cultivars of Cordoba (Spain). Front Plant Sci 8:1892. https://doi.org/10.3389/fpls.2017.01892

Moral J, Agustí-Brisach C, Agalliub G, de Oliveira R, Pérez-Rodríguez M, Roca LF, Romero J, Trapero A (2018) Preliminary selection and evaluation of fungicides and natural compounds to control olive anthracnose caused by Colletotrichum species. Crop Prot 114:167–176

Mosca S, Li D, Nicosia MG, Cacciola SO, Schena L (2014) Molecular analysis of *Colletotrichum* species in the carposphere and phyllosphere of olive. PLoS One 9:e114031

Nigro F, Antelmi I, Labarile R, Sion V, Pentimone I (2018) Biological control of olive anthracnose. Acta Hortic 1199. In: Perica S et al (eds) Proceedings of VIII international olive symposium

Nuangmek W, McKenzie EHC, Lumyong S (2008) Endophytic fungi from wild banana (*Musa acuminata* Colla) works against anthracnose disease caused by *Colletotrichum musae*. Res J Microbiol 3:368–374

Ownley BH, Gwinn KD, Vega FE (2010) Endophytic fungal entomopathogens with activity against plant pathogens: ecology and evolution. BioControl 55:113–128

Padder BA, Sharma PN (2011) In vitro and in vivo antagonism of biocontrol agents against *Colletotrichum lindemuthianum* causing bean anthracnose. Arch Phytopathol 44:961–969

Pangallo S, Nicosia MGD, Agosteo GE, Abdelfattah A, Romeo FV, Cacciola SO, Rapisarda P, Schena L (2017) Evaluation of a pomegranate peel extract as an alternative means to control olive anthracnose. Phytopathology 107:1462–1467

Pesce VM, Nally MC, Carrizo GP, Rojo C, Pérez BA, Toro María E, Castellanos de Figueroa LI, Vazquez F (2018) Antifungal activity of native yeasts from different microenvironments against *Colletotrichum gloeosporioides* on ripe olive fruits. Biol Control 120:43–51

Preto G, Martins F, Pereira JA, Baptista P (2017) Fungal community in olive fruits of cultivars with different susceptibilities to anthracnose and selection of isolates to be used as biocontrol agents. Biol Control 110:1–9

Roca LF, Romero J, Bohorquez JM, Alcantara E, Fernandez-Escobar R, Trapero A (2018) A Nitrogen status affects growth, chlorophyll content and infection by *Fusicladium oleagineum* in olive. Crop Prot 109:80–85

Schena L, Mosca S, Cacciola SO, Faedda R, Sanzani SM, Agosteo GE, Sergeeva V, Magnano di San Lio G (2014) Species of the *Colletotrichum gloeosporioides* and *C. boninense* complexes associated with olive anthracnose. Plant Pathol 63:437–446

Segura R (2003) Evaluación de microorganismos antagonistas para el control biológico del Repilo y la Antracnosis del olivo. Tesis doctoral, ETSIAM, Córdoba

Selosse MA, Dubois MP, Alvarez N (2009) Do Sebacinales commonly associate with plant roots as endophytes? Mycol Res 113:1062–1069

Sergeeva V (2011a) Anthracnose in olives symptoms disease cycle and management. 4th Olivebioteq international conference for olive products, vol 1, pp 269–274

Sergeeva V (2011b) Balanced plant nutrition may help reduce anthracnose. http://olivediseases.com/media/VeraTOPJan11.pdf

Sergeeva V, Spooner-Hart R, Nair N (2008) Evidence of early flower infection in olives (*Olea europaea*) by *Colletotrichum acutatum* and *C. gloeosporioides* causing anthracnose disease. Aust Plant Dis Notes 3:81–82

Smith BJ (2009) Influence of nitrogen, phosphorus and potassium on the severity of strawberry anthracnose crown rot. Acta Hortic 842:235–237

Speckbacher V, Zeilinger S (2018) Secondary metabolites of mycoparasitic fungi, secondary metabolites, Ramasamy Vijayakumar and Suresh S.S. Raja, IntechOpen. https://doi.org/10.5772/intechopen.75133

Talhinhas P, Sreenivasaprasad S, Neves-Martins J, Oliveira H (2005) Molecular and phenotypic analyses reveal the association of diverse *Colletotrichum acutatum* groups and a low level of *C. gloeosporioides* with olive anthracnose. Appl Environ Microbiol 71:2987–2998

Talhinhas P, Neves-Martins J, Oliveira H, Sreenivasaprasad S (2009) The distinctive population structure of *Colletotrichum* species associated with olive anthracnose in the Algarve region of Portugal reflects a host–pathogen diversity hot spot. FEMS Microbiol Lett 296:31–38

Talhinhas P, Mota-Capitão C, Martins S, Ramos AP, Neves-Martins J, Guerra-Guimarães L, Várzea V, Silva MC, Sreenivasaprasad S, Oliveira H (2011) Epidemiology, histopathology and aetiology of olive anthracnose caused by *Colletotrichum acutatum* and *C. gloeosporioides* in Portugal. Plant Pathol 60:483–495

Talhinhas P, Gonçalves E, Sreenivasaprasad S, Oliveira H (2015) Virulence diversity of anthracnose pathogens (*Colletotrichumacutatum* and *C. gloeosporioides* species complexes) on eight olive cultivars commonly grown in Portugal. Eur J Plant Pathol 142:73–83

Talhinhas P, Loureiro A, Oliveira H (2018) Olive anthracnose: a yield- and oil quality-degrading disease caused by several species of *Colletotrichum* that differ in virulence, host preference and geographical distribution. Mol Plant Pathol 19:1797–1180

Valenzuela LN, Angel DN, Ortiz DT, Rosas RA, García CFO, Santos MO (2015) Biological control of anthracnose by postharvest application of Trichoderma spp. on maradol papaya fruit. Biol Control 91:88–93

Wharton PS, Diéguez-Uribeondo J (2004) The biology of *Colletotrichum acutatum.* Anales del Jardín Botánico de Madrid 61:3–22

Xaviér JC (2014) Resistencia y control quimico en la antracnosis del olive por *Colletotrichum* spp. Universidade de Cordoba Tesis Doctoral, Cordoba

Chapter 12
Metagenomics as a Tool to Explore New Insights from Plant-Microbe Interface

Prachi Bhargava, Mahejibin Khan, Ankit Verma, Anushka Singh, Sukriti Singh, Siddharth Vats, and Reeta Goel

Abstract Microbial communities colonizing in and around the plants are essential for their survival and act as key determinants for plant's holistic health to make the dynamic plant microbiome. The microbiome comprises of trillions of bacteria, fungi, viruses and other microorganisms interacting with each other as well as with the plants. Metagenomics is a powerful tool that enables rapid analysis of microbial heterogenicity, thus helping us to understand the association of microorganisms within their environment and the overall functioning of microbiome. Herein, an overview of culture-independent methods to explore the unculturable/yet to culture microbial diversity of plant microbiome is addressed. This chapter focuses on the different constituents of plant-microbe interface and the metagenomic studies related to them.

12.1 Introduction: Plant Microbiome

A vast diversity of microorganisms present in nearby microenvironment of either outside or inside of plants constitute the plant microbiome. Plant microbiome consists of almost all groups of microbes including virus, archaea, bacteria, oomycetes and fungi; in spite of many decades of long studies, details of the composition and their interrelationship related to microbial diversity and richness of species that comprise the plant microbiome are not yet fully explored (Raaijmakers et al. 2009; Singh et al. 2019). To understand the dynamics and functioning of microbiome of a plant, it is divided in three different fractions,

P. Bhargava · A. Verma · A. Singh · S. Singh · S. Vats
Institute of Biosciences and Technology, Shri Ramswaroop Memorial University, Barabanki, Uttar Pradesh, India

M. Khan
Central Food and Technological Research Institute, Lucknow, Uttar Pradesh, India

R. Goel (✉)
Department of Microbiology, College of Basic Sciences and Humanities, GBPUA&T, Pantnagar, Uttarakhand, India

A. Varma et al. (eds.), *Plant Microbe Interface*,
https://doi.org/10.1007/978-3-030-19831-2_12

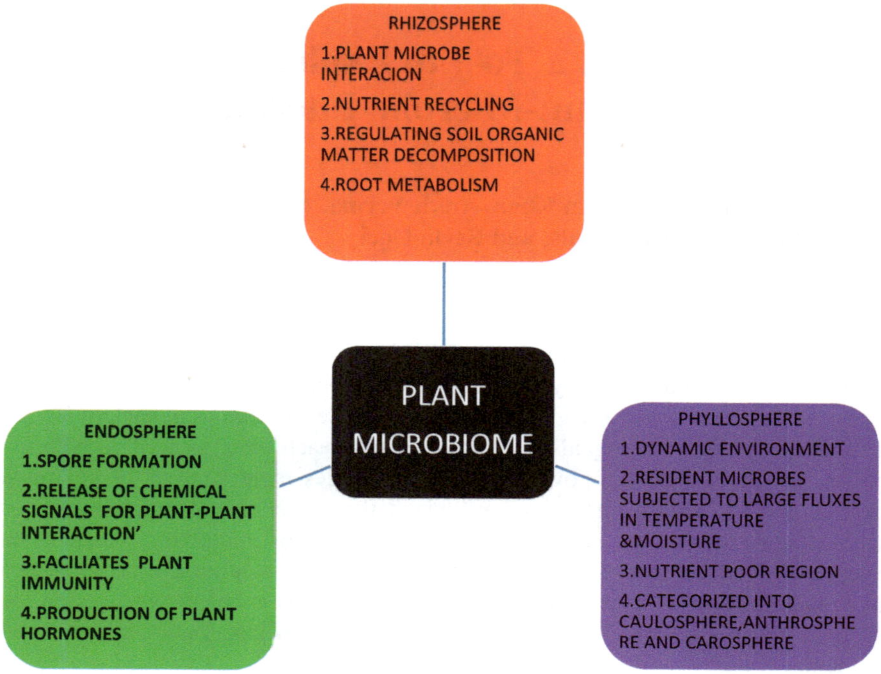

Fig. 12.1 Composition of plant microbiome and their specific functions

i.e. (1) microorganisms that reside outside the plants in rhizosphere (soil surrounding the roots and root surface); (2) endophytes that exist inside the plants tissues, and (3) epiphytes, colonizing the outer surface of plants mostly on phyllosphere (the aerial parts of plants). It is assumed that thousands of epiphytes and endophytic species exist on a single plant species.

Microorganisms constituting the plant microbiome are part of a complex food web; they utilize the nutrients released from plants and in return help in nutrient cycling and detoxification of harmful compounds and induce resistance in plants against both types of stresses, be it abiotic or biotic, and protect plants from plant pathogens, thus imposing a significant impact on plant productivity (Yadav et al. 2018). Exudates released from both plants and microorganisms are reported to act as inducer molecules which play a major role in signalling each other through which plants and microorganism communicate. Due to the interaction of plants with microbe, the whole plant microbiome is referred as extension of genetic compendium of plants and coined as plant's 'second genome' (Berendsen et al. 2012). Microbiome regulates several physiological processes in the host. Figure 12.1 explains the composition of plant microbiome and their specific functions.

Rhizosphere acts as a strong componential pillar of the plant microbiome. The complex microbial diversity in rhizosphere is influenced by climatic conditions, viz. temperature, salinity etc., of soil and physical factors including presence of metal ions and organic compounds, as well as biotic factors. Extensive researches reveal that plants also contribute to design their own designer microbiome as per their requirement. Exudates released from plant roots are composed of carbohydrates, proteins, lipids, phenolic compounds, organic acid and enzymes. These molecules are utilized by selective groups of microorganisms, thus proving that the composition of microbial community in any given plant microbiome is highly influenced by the released root exudates from the existing plant species. The presence of nitrogen-fixing bacteria and phosphate-solubilizing bacteria and production of plant-growth hormones are few examples of selective induction of microbial diversity in plant vicinity (Mendes et al. 2011). Similarly, natural suppressive soils are also an example of plant-driven stimulation of antibiotic-producing bacteria. Hence it can be said that plants have evolved themselves in such a way that they know how to build their designer rhizosphere communities which aid in protection from various stresses.

Endophytes constitute the other important component of plant microbiome. In the ancient time, microorganisms existing inside the plants were considered as only disease agents. The revelation of the presence of non-pathogenic microorganisms inside the plant led to the concept that microbes can colonize inside the plants as nonpathogens without posing any threat to their health. Later on these endophytes were studied by several groups, and it was found that no plant is free from endophytes (Rosenblueth and Martínez-Romero 2006). They help host plants in managing their pathogens and also promote plant growth. Endophytic *Burkholderia* spp. is known to control the growth of the pathogen *Fusarium moniliforme*. Endophytic diazotrophs from sugarcane roots produce amino acids and other plant growth-promoting substances which aid in improving their health (Suman et al. 2001). Genomic studies show the vast diversity in endophytic community and suggest that the ecology and genome size of entophytic population depend on environmental conditions.

Epiphytes or microorganism colonizing the outer surface of plants is the third important component of plant microbiome. It comprises of bacteria, fungi and algae. Protozoa and nematodes have also been reported at lower frequencies in as epiphytes (Lindow and Brandl 2003). As per rough estimate, plant leaves surface could harbour approximately 10^{26} bacterial cells (Vorholt 2012). The structure and composition of epiphytic microbial diversity is largely influenced by the nutritional heterogeneity of plant surfaces and also with the environmental interaction with plants. Epiphytes play a major role in plant development by acting as soldiers combating against invading pathogens; some of the epiphytes also help in phytohormones biosynthesis, performing nitrogen fixation, etc. (Padhi et al. 2013). Phyllosphere-associated fungi also interact with pathogenic fungi and help to control pathogenic invasion on leaves and leaf litter degradation.

12.2 Metagenomics: Effective Tool to Explore Plant Host Interface

Plant microbiome comprises of relatively diverse yet under-characterized microbial community. Exploring it can potentially enrich our understanding of plant-microbial ecology and their interaction within the community. Phylogenetic surveys show that the unknown prokaryotic microbial species outnumber the known cultured prokaryotes in any single plant microbiome. In recent past decades, several studies have compared the phylogenetic presence and abundance of different microbes in the phyllosphere region of various plants like spinach, apple, lettuce, rice and *Arabidopsis* by traditional and culture-independent methods revealing that the phyllosphere comprised more of the unknown uncultivable microbial population which tallied to a percentage of 90–99 (Rastogi et al. 2012).

Fortunately, the recent developments in metagenomics, viz. next-generation sequencing technologies and other culture-independent approaches, have enabled the investigations of the functional genetic diversity of various microorganisms without the inherent biases of manual cultivation, competition amongst microbes and plants, parasitism and other biotic/abiotic stress (e.g. salinity, temperature, humidity, etc.) and have helped us to have a deeper knowledge of microbial ecology (Oulas et al. 2015). The term metagenomics is based on the concept of meta-analysis (the statistical process of combining separate analyses) and genomics (the comprehensive analysis of an organism's genetic material). Figure 12.2 clearly depicts the work methodology of metagenomic analysis from any environmental sample be it from rhizosphere, endosphere or phyllosphere. Metagenomics is a combination of all modern techniques of the field of genomics that have metamorphosized our understanding of the microbial population and their interactions with the environment. It has opened a magnificent door to the biotechnology field especially based on the exploitation of uncultivated microbial species. Initially, metagenomic studies were focused on only uncultured microflora and ancient DNA findings, but nowadays the technology has reached to another level and is applied to study the whole plant microbiome and gastrointestinal ecosystems of human and animals as well (Müller and Ruppel 2014).

Although the new metagenomic techniques allow us to conclude changes in microbial communities at the genetic level, few challenges have to be fought like heterogeneity of the scales used for sampling and the connectivity between those scales. Selecting a good site for sampling and methods used thereafter are important factors to contemplate when beginning a metagenomic analysis of a microbiome. Microbial activity and population are affected by its physical, chemical and biological properties. The minute changes in any condition which affects plant growth such as increased or decreased nutrient concentration or major changes like drought all have profound impacts on the structure and functions of epiphytic and endophytic microbial communities.

Metagenomic methodology starts with isolation of environmental DNA. A library of clones is constructed and screened followed by sequencing and analysis of

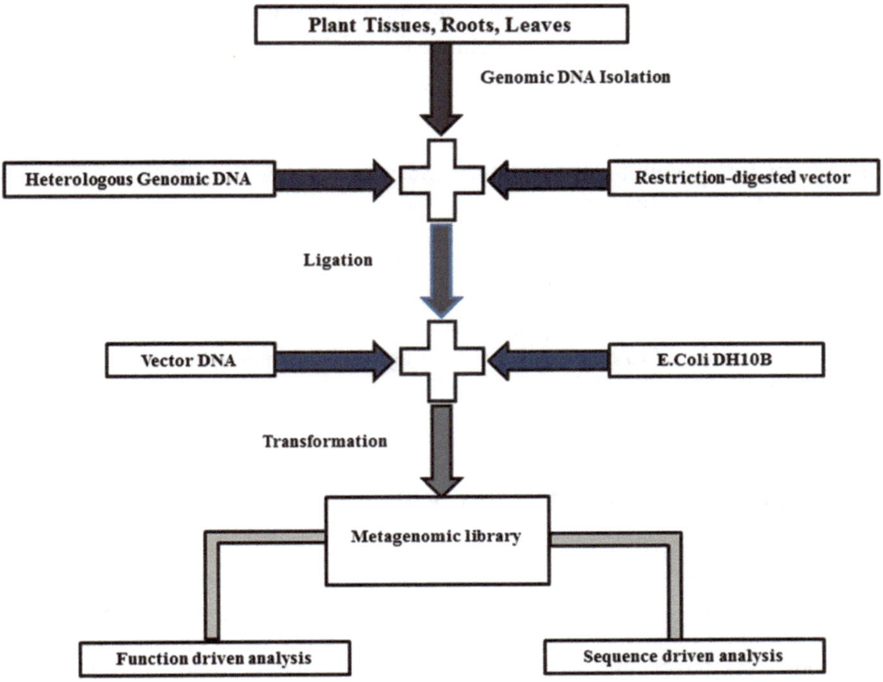

Fig. 12.2 Brief methodology of the metagenomic process related to all the components of microbiome

isolated metagenomic DNA which renders informative data on various aspects of the studied sample, allowing to typify the microbial life in any given environment extensively. It not only identifies the species of the microbiome but also provides a glimpse about the metabolomic activities related to the functional aspects of the cultivable and unculturable microbes of a given population (Langille et al. 2013).

12.3 Rhizosphere and Its Components

The physically, chemically and biologically agile zone of the soil around the plant roots is referred to as rhizosphere. It comprises of the soil adhering to plant roots which has great importance and exhibits complex interrelationships between microbial confraternity and their mutual interactions with the plant. The diversity and complexity shown by any rhizospheric microbial community is greatly influenced by root exudation and physiochemical properties reflected by soil owing to the agronomic operations and practices used (Shrivastava et al. 2014). It harbours all the soil-borne microbes including protozoans, fungi, archaea and bacteria, which have a great impact on the roots and its exudates through their biological, physical and chemical interactions. Hence, it is imperative to study the interactions between plants and these soil microorganisms for cognizing various plant-related processes (Amann et al. 1995). Most of the microorganisms residing in the soil are not culturable in the standard laboratory conditions. Different plant species may be biased in supporting various microbial communities in their rhizospheric zones owing to their root exudates. The plant rhizosphere profusely secretes a large number of compounds that are utilized by the soil microbial communities in many different ways. These phytochemical exudates act as selective nutritional sources for stimulation and enrichment of specific groups of soil microorganisms which in turn help in the growth and development of the plants (Larkin et al. 1993; Mendes et al. 2013). This hot spot is considered to be one of the most aggressively enterprising interfaces on earth (Philippot et al. 2013).

12.4 Metagenomics to Explore Rhizosphere Environment

Rhizosphere is the most active interface in which plants and microorganisms establish a complex and varied molecular dialogue, involving nutrient transfer as well as specific interactions mediated by the release of signalling molecules from plant roots, thereby resulting in enhanced plant productivity (Prasad et al. 2015). The rhizosphere microbiome is a dynamic blend of beneficial and pathogenic (plant, human) microorganisms. Kumar et al. (2018) studied the rhizosphere of alfalfa and explored the structure and diversity of microbial community using 16S rRNA metagenome analysis. Metabolic network approaches also find their usage in exploring the associations between structure and functions of environment in complex microbial rhizosphere microbiome. The DNA data from two agricultural crops, viz., wheat and cucumber, were extracted using the same techniques (Ofaim et al. 2017).

Hypergeometric enrichment tests have been used to study enriched pathways (metabolites/enzymes) and possible functional significance for observed co-occurrence patterns of various taxonomic combinations and their complementary metabolite profiles. The soil plays a prominent role in the structural composition of microbial communities residing in it. Many novel members of *Crenarchaeota* group displaying resistance to different metals were discovered using these techniques from Tinto river (Mirete et al. 2007). The microbially incongruent communities in the rhizosphere showed more complexity than those in the river and mainly represented heterotrophic acidophiles suggesting that the soil composition rendered the diverse resistance to the microbiome.

Results in the effects of fertilizers and other agricultural practices on rhizosphere microbiome have been studied revealing that some genes for phytic acid utilization were upregulated after the incorporation of fertilizers. Plants start harbouring those microbes which are beneficial for their growth in varied conditions. There was a clear demarcation in the taxonomic profiles of the samples collected from rhizosphere and bulk soil again suggesting a role of plants and soil environment on the microbial species present there (Uroz et al. 2010). Similarly some crops like soybean have been shown to allow some selected microbial communities to inhabit their rhizosphere based on beneficial functional traits aiding in their growth promotion and nutrition. Techniques like stable-isotope probing (SIP) along with metagenomics of fungal communities have led to the discovery of many new OTUs belonging to *Basidiomycota* and *Ascomycota* (Gkarmiri et al. 2017). Metagenomic analysis of many citrus rhizospheres have concluded that functional properties involved in host-microbe interactions are significantly critical for the microbiome-inhabiting plant root surfaces and are influenced remarkably by the availability of plant exudates. These rhizoplane-enriched functional properties are advantageous to the plant host. Thereby, determining genetic and microbial intricacy in the citrus rhizoplane microbiome compared to that in the rhizosphere communities, indicating the filter effect of plant hosts on the closely associated rhizoplane microbiome assembly (Zhang et al. 2017). Pyrosequencing analysis has been used to analyse the shift in microbial communities as an effect of addition of various fertilizers (Li et al. 2016). Table 12.1 enlists the dominating phylum's explored metagenomically in different rhizospheres.

12.5 Endosphere

12.5.1 Endosphere and Its Components

Endosphere is defined as the region present inside the plant. Within the endosphere, microbes inhabit various microenvironments like the intra- and intercellular spaces inside the plant body, and each microenvironment presents an unmatched and diverse biochemical profile. Plants as metaorganisms populate microbes showing heterogeneity residing in different habitats (such as endosphere, rhizosphere and phyllosphere), situated inside or on the surface of vegetative parts (roots, stems and leaves) and reproductive parts (flowers, fruits and seeds) of the host plant (Truyens et al. 2015). Endophytic microbes refer to microbial population that reside within the tissues of plants without resulting in any visible adverse effects on their host (Knief 2014). Hence, endophytic microbes are mostly facultative rhizospheric microorganisms and/or accidental passengers in the root, suggesting that the overall composition of the various taxonomic and phylogenetic profiles of the dominant residing microbes will have homology and similarity. Bacterial colonies extensively colonize the internal plant tissues, found in almost every plant worldwide. Both culture-dependent and culture-independent techniques have shown the diversity of endophytic bacteria that include various bacterial taxa across a broad range of different plant species. Studies suggest that endophytes originated from the rhizosphere (soil) and/or are maternally transferred to future generations (vertical transmission through seeds). Several microbes residing in endosphere have the potential to affect plant growth either directly or indirectly by helping them in the production of siderophores and procurement of different macronutrients by mineral phosphate solubilization

Table 12.1 List of dominating phyla explored metagenomically in different rhizospheres

Plant	Area	Dominating Phylum	Approach	References
Oilseed Rape	Potato, maize, rice, grassland	*Chloroflexi* *Proteobacteria* *Bacteroidetes*	16S rRNA reference taxonomy	Peiffer et al. (2013), Fierer et al. (2012)
Mangroves	Diazotrophic communities	*Fusobacteria* *Cyanobacteria* *Betaproteobacteria*	16S rRNA and *nif*H genes	Varon-lopez et al. (2014)
Populus tremula *Populus alba*	Rhizosphere soil, root	*Alphaproteobacteria* *Firmicutes* *Betaproteobacteria*	16S rRNA amplicon pyrosequencing	Edwards et al. (2015), Shakya et al. (2013)
Maize	Maize genotype and bulk soil	*Verrucomicrobia* *Planctomycetes*	16S rRNA gene sequence	Janssen (2006)
Willows	Contaminated and non-contaminated soils	*Deltaproteobacteria* *Gammaproteobacteria* *Firmicutes*	16S rRNA gene sequencing	Lundberg et al. (2012), Dennis et al. (2010)
Barley	Root samples	*Bacteroidetes* *Proteobacteria* *Actinobacteria*	16S rRNA gene ribotyping	Schlaeppi et al. (2014)
Grey mangroves	Red sea coast	*Microsporidia* *Blastocladiomycota* *Ascomycota*		Lauber et al. (2008)
Chamaecyparis obtusa *Pinus densiflora*	Temperate forest	*Glomeromycota* *Ascomycota* *Basidiomycota*	Forward primer ITS1F-KYO1	Toju et al. (2012)
Zostera japonica *Zostera marina*	Mixed beds of Netarts Bay, United States	*Bacteroidetes* *Deltaproteobacteria* *Spirochaetes*	16S rRNA gene amplicon sequencing, Metatranscriptomics	Crump and Koch (2008), Fahimipour et al. (2017)
Wheat	Turkey	*Acidobacteria* *Actinobacteria* *Bacteroidetes*	16S rDNA rhizosphere sequences	Donn et al. (2015), Corneo et al. (2016)

and biological nitrogen fixation. The direct routes also involve production of phytohormones such as auxin (indole-3-acetic acid, IAA) by microbes which stimulates plant growth, especially of roots (Bulgarelli et al. 2013). The mechanisms of promoting plant growth used by endophytic bacteria are similar to the mechanisms used by rhizospheric bacteria. Similar to rhizospheric plant growth-promoting bacteria, endophytic bacteria can also act to facilitate plant growth in horticulture, agriculture and silviculture as well as in strategies for environmental cleanup (viz. phytoremediation). Understanding these mechanisms is crucial to determine the principles governing structure, function and robustness of microbial community. Bacterial endophytes may have a benefit over bacteria inhabiting the rhizosphere, since living inside plant's tissues gives them a chance to always be in direct contact with the plant's cells, and thus, they can more easily exert an enhanced beneficial effect. Bacteria residing within the rhizosphere also have potential chances to enter and colonize the plant roots. This microecosystem is one most commonly studied primary route of endophytic colonization (Hallmann et al. 1997). More and more extensive studies suggest that endophytic bacterial diversity can be considered a subset of the rhizosphere and/or root-associated bacterial population, and rhizospheric and endophytic bacterial communities sometimes exhibit different overall patterns of relative sufficiency of the major groups at the phylum level (Kent and Triplett 2002; Cocking 2003). Fungi with different morphological characteristics were isolated from both rhizosphere and endosphere fungi of *C. japonicum*. The genus *Trichoderma* is most often isolated and deeply studied endospheric fungi, and the distribution of fungi is similar between rhizosphere and endosphere.

12.5.2 Metagenomics of Endosphere Environment

It has been surmised that endophytic root bacterial communities comprehend a subset of colonists originating from the encircling rhizosphere soil (Cocking 2003), and the resulting community framework is affected by the surrounding soil and environmental properties. The term metagenomics incorporates the analysis of an assemblage of similar but non-identical items (Glass 1976). It basically involves isolating DNA from an environmental sample which is called a metagenome, cloning the environmental DNA into a suitable vector, transforming the clones into a host bacterium and screening the resulting transformants. The resultant clones are then screened for phylogenetic markers or 'anchors', for example, conserved sequences, viz. 16S rRNA and recA, or any other conserved genes by hybridization or multiplex PCR or for any function like expression of specific traits like enzyme activity or antibiotic production (Courtois et al. 2003) or they can be sequenced randomly. Traditional and metagenomic approach both have certain benefits over each other along with certain limitations. Therefore it is advisable to use these approaches together to enrich the understanding of the uncultured world, providing insight into the vast microbial population that is still unrevealed and entirely unknown. Metagenomic analysis has unveiled substantial microheterogeneity in apparently uniform populations where the challenge lies in linking the genomic information with the organism or ecosystem from which the DNA was isolated. Culture-based techniques allow to study isolated microbes in depth, and the modern molecular techniques like metagenomics and metabolomics help to explore the unidentified microbial communities in situ. These studies hold an important place in core areas like plant breeding and microbiology apart from allied field of agriculture and healthcare system as plant microbiome is a decisive determinant of plant health and productivity and has received substantial attention in recent years (Bulgarelli et al. 2013). Comparisons between endogenome and rhizogenome with emphasis on plant growth-promoting bacteria have disclosed potential genetic factors involved in an endophytic lifestyle, which facilitates a better cognizance of the functioning of bacterial endophytes. Competition for resources among community members is based on the usage of diverse survival tactics, like antagonism and mutualism among the members. Metagenomics has redefined the concept of a genome which has enhanced the prospects of solving many problems and given a momentum to the rate of gene discovery. The potential for application of metagenomics to biotechnology seems endless. Table 12.2 shows the list of microbes metagenomically isolated and identified from endosphere region.

Usually, endophytic bacteria are known to be non-pathogenic, causing no visible symptoms, but sometimes they may include latent pathogens that may cause disease depending on the availability of favourable environmental conditions and host genotypes. Model organisms like *Burkholderia*, *Herbaspirillum* and *Azoarcus* spp., residing in the non-leguminous plants, mainly grasses, have been extensively studied for extracting information about the taxonomic diversity and mechanisms of infection and colonization of endophytic microbes within plant system (Thomas 2017). Culture-independent methods, such as analyses of 16S rRNA and nifH transcripts and metagenome analyses, have paved a way for exploring huge melange of endophytes in the economically important crops like sugarcane and rice. The studies suggest that rhizobia (and other α-Proteobacteria) are very common endophytes, as are β-proteobacteria, γ-proteobacteria and *Firmicutes*. The core endophytic bacterial microbiome of *A. thaliana* was studied using high-throughput sequencing (HTS) of 16S rRNA. These studies showed that although various soil types altered the bacterial endophyte microbiome, some species of prokaryotes were persistently present in endosphere as compared with the rhizosphere environment and included *Actinobacteria* and some families from the *Proteobacteria*.

12.6 Phyllosphere

12.6.1 Phyllosphere and Its Components

Phyllosphere represents the microbial flora and fauna dwelling on and in aerial plant organs, which constitute the total part of living plant above the ground (Newton et al.

Table 12.2 List of microbes metagenomically isolated and identified from endosphere region

Species	Area	Techniques used	Frequency	References
Firmicutes	Nea Apollonia (NAP) geothermal, Greece	DNA extraction and quantitative PCR	41.70%	Filippidou et al. (2015)
C. spinosa	Saint Catherine Mountain	Terminal restriction fragment length polymorphism (TRFLP)	3.5%	Compant et al. (2010)
Cypripedium japonicum	Korea peninsula	CTAB method	215 isolates	Gang et al. (2017)
Aquificae	Geothermal Hot Springs of Manikaran	Power soil DNA isolation method	64%	Bhatia et al. (2015)
Chiliadenus iphionoides	Sinai, Egypt	Polymerase chain reaction	35%	El-Badry (2016)
B. amyloliquefaciens	Country Value Seeds, UK	Amplification and high-throughput sequencing	49%	Gadhave et al. (2018)
Discodermia calyx	Shikine-jima Island, Tokyo	PGM sequencing	250,000 colonies	Nakashima et al. (2016)
P. deltoids	Caney Fork River	Bacterial and fungal ribosomal PCR amplification and sequencing	97%	Gottel et al. (2011)
Hordeum vulgare L.	Different locations in Germany	Fluorescent in situ hybridization and confocal laser scanning microscopy (FISH-CLSM)	39.2%	Rahman et al. (2018)

2010). It is further categorised into caulosphere (stems), carosphere (fruits) and anthosphere (flowers) (Berlec 2012). The major part of this surface is provided by green leaves, and it is believed to represent one of the largest dwelling sites on earth. There exist little information about the bacterial communities which reside in the above said categories apart from leaves; therefore, the maximum information about phyllosphere consists of the knowledge pertaining to leaves.

Recent cultivation-independent studies have helped us to examine the composition of microbial phyllosphere communities in a better way. It is evident that these communities do not represent random assemblies of microorganisms, but instead undergo selection that results, at least partially, in predictable microbial communities with few dominant phyla and their subgroups. Diverse communities of microorganisms including bacteria, fungi, archaea and protists are known to exist in harmony in the phyllosphere region. *Actinobacteria*, *Firmicutes*, *Bacteroidetes* and *Proteobacteria* dominate the phyllosphere community along with few bacterial genera including *Bacillus*, *Pseudomonas*, *Massilia*, *Sphingomonas*, *Arthrobacter*,

Methylobacterium and *Pantoea*, which reside as the core phyllosphere microbial communities (Delmotte et al. 2009).

Most studies done on the abundance of organisms in the phyllosphere region have focused on bacteria and a lesser range to fungi as archaea are apparently not abundant in the phyllosphere (Knief et al. 2012; Finkel et al. 2011). Most bacteria on leaf surfaces do not occur as single cells or small groups of cells, as fungi tend to, but form larger assemblages which are particularly common at the depressions formed at the boundary of epidermal cells, along the veins and at the bases of trichomes, and in these depressions, they are generally lodged within extracellular polymeric substances (EPS). The EPS helps in providing a hydrated area to the bacterial surrounding and also concentrates detoxifying enzymes (Baldotto and Olivares 2008; Lindow and Brandl 2003).

The microbial communities of phyllosphere play a vital role in remediation of pesticides, hydrocarbon pollutants from atmosphere and cycling of nutrients as saprophytes, which are important for plant growth and healthy development serving as phytostimulators, biofertilizers and biopesticides to combat plant pathogens (Zhou et al. 2011; Ali et al. 2012) and affect global carbon and nitrogen cycles (Whipps et al. 2008)

The phyllosphere is an ephemeral or short-lived environment as compared to the rhizosphere, as the annual plants complete their life cycle within a single growth season, whereas perennial deciduous plants spontaneously form and shed leaves every year and evergreen plants do so sequentially throughout the year. Successful phyllosphere inhabitants can be expected to multiply and occupy newly formed niches while the leaves are expanding. Moreover, the waxy cuticle covering the plant epidermal cells is hydrophobic and reduces evaporation of water as well as leaching of plant metabolites, thus resulting in an oligotrophic environment.

Microorganisms dwelling within the phyllosphere live as commensals on their hosts; they can either be endophytic or epiphytic. Presently the extent to which plants are benefited by colonization of these commensal microbiota in their aerial parts is almost unknown (Innerebner et al. 2011; Knief et al. 2012). Although the exact extent of benefits which the plants receive from the endophytic microbes is not fully explored, the presence of surface appendages, comprising of trichomes and hydathodes, veins and stomata alter nutrient availability in addition to the environmental factors which affect them such as fluctuations in UV, temperature, humidity, water availability and light irradiation (Innerebner et al. 2011; Knief et al. 2012).

Consequently, the frequency of occurrence and multiplication of these microorganisms is uneven over the leaf surface (Remus-Emsermann et al. 2012) owing to the environmental variabilities and their encounter to the antimicrobial compounds produced by plants and other microbes. Trees adapted to xerophytic conditions have the capacity to secrete some soluble compounds which consequently result in alkaline and saline leaf surfaces, thus leading to saline or alkaline stress of phyllosphere microbes (Finkel et al. 2011). Figure 12.3 shows the interactions of the factors governing the different components of phyllosphere.

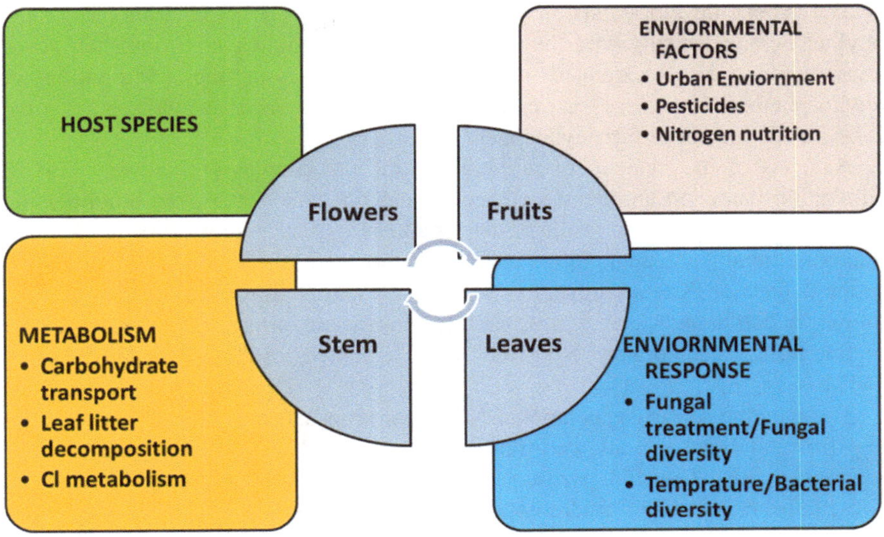

Fig. 12.3 Interactions of the factors governing the different components of phyllosphere

12.6.2 Metagenomics of Phyllosphere

The need for exploration of microbial life within the phyllosphere is crucially important for two reasons—first, understanding the survival strategies of disease-causing pathogens and developing methods to prevent their spread, thereby improving plant health to improve biomass production and prevent biomass losses. Second, there is an alarming rise in the food poisoning cases associated with vegetables, fruits and salads contaminated with food-borne pathogenic microbes especially bacteria, *Salmonella enterica* and *Escherichia coli* (Teplitski et al. 2011). Proper safety methods and decontamination strategies are important to prevent any outbreak affecting public health. Another interesting area of potential is phytoremediation, using microorganisms for removal of volatile pollutants such as phenol or benzene from the air using phyllosphere also called phylloremediation (De Kempeneer et al. 2004; Sandhu et al. 2007).

The triangular relationship between host, environment and pathogenic phyllospheric microbiota can give valuable insights into the population biology and genetics of phylloplane pathogens leading to more effective and sustainable disease management practices (Montarry et al. 2008).

The realization that a huge percentage of the microorganisms associated with plants, as those in other natural environments too, is metabolically active, but nonculturable in commonly used media and culture conditions, has had important accompaniment for plant microbiology and has brought about the beginning of culture-independent detection methods into phyllosphere research.

The recent developments in the area of exploration of microbiome in the phyllosphere, especially with the advances in metagenomics, environmental genomics, have greatly extended our understanding about the contribution of phyllosphere microbiome in plant-environment interactions along with the ecosystemic impact of the phyllosphere.

Analysis of the makeup of microbiome in leaf samples without any bias of cultivation based on amplicon sequencing and the 16S rRNA gene amplification has given many milestone results. There is a benefit of accessing a broader range of microbial inhabitants than culture techniques; however, the shortcomings comprise of the defects of PCR amplification, lack of quantitative information, sensitivity to inhibitory compounds, primer mismatch sensitivity and, primarily, the amplification of interfering plant organelle-derived RNA sequences (Saito et al. 2007; Berlec 2012).

The oncome of next-generation DNA sequencing has significantly reduced the experimental costs and allowed multiplexing of hundreds of samples in a single sequencing run. The 454 pyrosequencing platform was among the 'first' to be commonly executed in microbiota analysis through rRNA or whole-genome sequencing, shotgun metagenomics, ITS amplicon sequencing and transcriptional profiling (Delmotte et al. 2009; Rastogi et al. 2012). Ultra-high-throughput sequencing of microbial communities by 'second' next-generation sequencing technology like the Illumina platform (Degnan and Ochman 2012) yields amounts of sequence data that are of several order magnitude higher than generated by other techniques. Proteogenomics represents another important technical advancement which summates the application of metagenomic with metaproteomic analysis (Delmotte et al. 2009). Combined together, these technological revolutions are nobly helpful in the relative ecological analyses and help provide new introspections into the structure, function and heterogeneity of microbiome in the phyllosphere and different environments. In Table 12.3, many recent examples of phyllosphere studies that used high-throughput molecular methods are listed.

12.7 Conclusion

Plant microbiome is a composite ecosystem that hosts a number of interactions at 'microbe-soil-microbe-plant-microbe interface'. Earlier it was difficult to study and understand the plant microbiome as a whole due to the unculturability of majority of microorganisms. Advances in latest molecular technologies, culture-independent methodology and next-generation sequencing have rapidly expanded the research in the area of microbial ecology of a particular niche and provided an in-depth knowledge of various genes present within the microbiome. Several studies have proved that the microbes are an integral part of plant genome, but their population is highly diverse and varies with the environmental as well as the biotic elements. Horizontal gene transfer and plant-based selection add to the plant microbiome diversity. Although, in the past decade, understanding of microbial ecology has

Table 12.3 List of dominating phyla and studies applying high-throughput molecular approaches to phyllosphere communities

Plant	Dominating phyla	Molecular approach	References
Soybean, clover, *Arabidopsis*	*Actinobacteria* *Bacteroidetes* *Proteobacteria*	16S rRNA gene pyrosequencing, metaproteogenomics	Delmotte et al. (2009)
Lettuce	*Bacteroidetes* *Firmicutes* *Proteobacteria*	16S rRNA gene pyrosequencing	Rastogi et al. (2012)
Rice	*Actinobacteria* *Proteobacteria*	Metaproteogenomics	Knief et al. (2012)
Salt cedar	*Actinobacteria* *Firmicutes* *Proteobacteria*	16/18S rRNA gene pyrosequencing	Finkel et al. (2011)
Spinach	*Actinobacteria* *Proteobacteria*	16S rRNA gene pyrosequencing	Lopez-Velasco et al. (2011)
Grapes	*Firmicutes*, *Bacteroidetes*, *Actinobacteria*	16S rRNA gene pyrosequencing	Leveau and Tech (2010)
Oak	*Alternaria*, *Epicoccum*, *Erysiphe*	ITS pyrosequencing	Jumpponen and Jones (2009)

grown very rapidly but to predict the ecophysical behaviour and to improve the plant productivity using custom-made microbiome, still more research is required.

Acknowledgement PB thanks DST-SERB: SB/YS/LS-213/2013 for financial support.

References

Ali N, Sorkhoh N, Salamah S, Eliyas M, Radwan S (2012) The potential of epiphytic hydrocarbon-utilizing bacteria on legume leaves for attenuation of atmospheric hydrocarbon pollutants. J Environ Manag 93:113–120

Amann RI, Ludwig W, Schleifer KH (1995) Phylogenetic identification and in situ detection of individual microbial cells without cultivation. Microbiol Rev 59:143–169

Baldotto LEB, Olivares FL (2008) Phylloepiphytic interaction between bacteria and different plant species in a tropical agricultural system. Can J Microbiol 54:918–931

Berendsen RL, Pieterse CM, Bakker PA (2012) The rhizosphere microbiome and plant health. Trends Plant Sci 17:478–486

Berlec A (2012) Novel techniques and findings in the study of plant microbiota: search for plant probiotics. Plant Sci 193:96–102

Bhatia S, Batra N, Pathak A, Green SJ, Joshi A, Chauhan A (2015) Metagenomic evaluation of bacterial and archaeal diversity in the geothermal hot springs of Manikaran, India. Genome Announc 3:e01544–e01514

Bulgarelli D, Schlaeppi K, Spaepen S, van Themaat EVL, Schulze-Lefert P (2013) Structure and functions of the bacterial microbiota of plants. Annu Rev Plant Biol 64:807–838

Cocking EC (2003) Endophytic colonization of plant roots by nitrogen-fixing bacteria. Plant Soil 252:169–175

Compant S, Clément C, Sessitsch A (2010) Plant growth-promoting bacteria in the rhizo- and endosphere of plants: their role, colonization, mechanisms involved and prospects for utilization. Soil Biol Biochem 42:669–678

Corneo PE, Suenaga H, Kertesz MA, Dijkstra FA (2016) Effect of twenty four wheat genotypes on soil biochemical and microbial properties. Plant Soil 404(1–2):141–155

Courtois S, Cappellano CM, Ball M, Francou FX, Normand P, Helynck G et al (2003) Recombinant environmental libraries provide access to microbial diversity for drug discovery from natural products. Appl Environ Microbiol 69:49–55

Crump BC, Koch EW (2008) Attached bacterial populations shared by four species of aquatic angiosperms. Appl Environ Microbiol 74:5948–5957

De Kempeneer L, Sercu B, Vanbrabant W, Van Langenhove H, Verstraete W (2004) Bioaugmentation of the phyllosphere for the removal of toluene from indoor air. Appl Microbiol Biotechnol 64:284–288

Degnan PH, Ochman H (2012) Illumina-based analysis of microbial community diversity. ISME J 6 (1):183

Delmotte N, Knief C, Chaffron S, Innerebner G, Roschitzki B, Schlapbach R, Vorholt JA (2009) Community proteogenomics reveals insights into the physiology of phyllosphere bacteria. Proc Natl Acad Sci U S A 106:16428–16433

Dennis PG, Miller AJ, Hirsch PR (2010) Are root exudates more important than other sources of rhizodeposits in structuring rhizosphere bacterial communities? FEMS Microbiol Ecol 72:313–327

Donn S, Kirkegaard JA, Perera G, Richardson AE, Watt M (2015) Evolution of bacterial communities in the wheat crop rhizosphere. Environ Microbiol 17:610–621

Edwards J, Johnson C, Santos-Medellín C, Lurie E, Podishetty NK, Bhatnagar S, Eisen JA, Sundaresan V (2015) Structure, variation, and assembly of the root-associated microbiomes of rice. Proc Natl Acad Sci U S A 112:E911–E920

El-Badry MA (2016) Bacterial community metagenomic and variation of some medicinal plant rhizosphere collected form Sinai. SCIREA J Agric 1(1):16

Fahimipour AK, Kardish MR, Lang JM, Green JL, Eisen JA, Stachowicz JJ (2017) Global-scale structure of the eelgrass microbiome. Appl Environ Microbiol. https://doi.org/10.1128/AEM.03391-16

Fierer N, Lauber CL, Ramirez KS, Zaneveld J, Bradford MA, Knight R (2012) Comparative metagenomic, phylogenetic and physiological analyses of soil microbial communities across nitrogen gradients. ISME J 6(5):1007–1017

Filippidou S, Junier T, Wunderlin T, Lo CC, Li PE, Chain PS, Junier P (2015) Under-detection of endospore-forming firmicutes in metagenomic data. Comput Struct Biotechnol J 13:299–306

Finkel OM, Burch AY, Lindow SE, Post AF, Belkin S (2011) Geographical location determines the population structure in phyllosphere microbial communities of a salt-excreting desert tree. Appl Environ Microbiol 77:7647–7655

Gadhave KR, Devlin PF, Ebertz A, Ross A, Gange AC (2018) Soil inoculation with Bacillus spp. modifies root endophytic bacterial diversity, evenness, and community composition in a context-specific manner. Microb Ecol 76:741–750

Gang GH, Cho G, Kwak YS, Park EH (2017) Distribution of rhizosphere and endosphere fungi on the first-class endangered plant *Cypripedium japonicum*. Mycobiology 45:97–100

Gkarmiri K, Mahmood S, Ekblad A, Alström S, Högberg N, Finlay R (2017) Identifying the active microbiome associated with roots and rhizosphere soil of oilseed rape. Appl Environ Microbiol. https://doi.org/10.1128/AEM.01938-17

Glass GV (1976) Primary, secondary, and meta-analysis of research. Educ Res 5:3–8

Gottel NR, Castro HF, Kerley M, Yang Z, Pelletier DA, Podar M et al (2011) Distinct microbial communities within the endosphere and rhizosphere of *Populus deltoides* roots across contrasting soil types. Appl Environ Microbiol 77:5934–5944

Hallmann J, Quadt-Hallmann A, Mahaffee WF, Kloepper JW (1997) Bacterial endophytes in agricultural crops. Can J Microbiol 43:895–914

Innerebner G, Knief C, Vorholt JA (2011) Protection of *Arabidopsis thaliana* against leaf-pathogenic *Pseudomonas syringae* by Sphingomonas strains in a controlled model system. Appl Environ Microbiol 77:3202–3210

Janssen PH (2006) Identifying the dominant soil bacterial taxa in libraries of 16S rRNA and 16S rRNA genes. Appl Environ Microbiol 72:1719–1728

Jumpponen A, Jones KL (2009) Massively parallel 454 sequencing indicates hyperdiverse fungal communities in temperate *Quercus macrocarpa* phyllosphere. New Phytol 184:438–448

Kent AD, Triplett EW (2002) Microbial communities and their interactions in soil and rhizosphere ecosystems. Annu Rev Microbiol 56:211–236

Knief C (2014) Analysis of plant microbe interactions in the era of next generation sequencing technologies. Front Plant Sci 5:216–239

Knief C, Delmotte N, Chaffron S, Stark M, Innerebner G, Wassmann R, Von Mering C, Vorholt JA (2012) Metaproteogenomic analysis of microbial communities in the phyllosphere and rhizosphere of rice. ISME J 6:1378–1390

Kumar V, AlMomin S, Al-Aqeel H, Al-Salameen F, Nair S, Shajan A (2018) Metagenomic analysis of rhizosphere microflora of oil-contaminated soil planted with barley and alfalfa. PLoS One 13: e0202127

Langille MG, Zaneveld J, Caporaso JG, McDonald D, Knights D, Reyes JA, Clemente JC, Burkepile DE, Thurber RL, Knight R, Beiko RG (2013) Predictive functional profiling of microbial communities using 16S rRNA marker gene sequences. Nat Biotechnol 31:814–821

Larkin RP, Hopkins DL, Martin FN (1993) Effect of successive watermelon plantings on *Fusarium oxysporum* and other microorganisms in soils suppressive and conducive to fusarium wilt of watermelon. Phytopathology 83:1097–1105

Lauber CL, Strickland MS, Bradford MA, Fierer N (2008) The influence of soil properties on the structure of bacterial and fungal communities across land-use types. Soil Biol Biochem 40:2407–2415

Leveau JHJ, Tech JJ (2010) Grapevine microbiomics: bacterial diversity on grape leaves and berries revealed by high-throughput sequence analysis of 16S rRNA amplicons. In: International symposium on biological control of postharvest diseases: challenges and opportunities, vol 90, pp 31–42

Li JG, Shen MC, Hou JF, Li L, Wu JX, Dong YH (2016) Effect of different levels of nitrogen on rhizosphere bacterial community structure in intensive monoculture of greenhouse lettuce. Sci Rep 6:25305–25314

Lindow SE, Brandl MT (2003) Microbiology of the phyllosphere. Appl Environ Microbiol 69:1875–1883

Lopez-Velasco G, Welbaum GE, Boyer RR, Mane SP, Ponder MA (2011) Changes in spinach phylloepiphytic bacteria communities following minimal processing and refrigerated storage described using pyrosequencing of 16S rRNA amplicons. J Appl Microbiol 110:1203–1214

Lundberg DS, Lebeis SL, Paredes SH, Yourstone S, Gehring J, Malfatti S, Tremblay J, Engelbrektson A, Kunin V, Del Rio TG, Edgar RC (2012) Defining the core *Arabidopsis thaliana* root microbiome. Nature 488:86–90

Mendes R, Kruijt M, De Bruijn I, Dekkers E, van der Voort M, Schneider JH, Piceno YM, DeSantis TZ, Andersen GL, Bakker PA, Raaijmakers JM (2011) Deciphering the rhizosphere microbiome for disease-suppressive bacteria. Science 332(6033):1097–1100

Mendes R, Garbeva P, Raaijmakers JM (2013) The rhizosphere microbiome: significance of plant beneficial, plant pathogenic, and human pathogenic microorganisms. FEMS Microbiol Rev 37:634–663

Mirete S, De Figueras CG, González-Pastor JE (2007) Novel nickel resistance genes from the rhizosphere metagenome of plants adapted to acid mine drainage. Appl Environ Microbiol 73:6001–6011

Montarry J, Cartolaro P, Delmotte F, Jolivet J, Willocquet L (2008) Genetic structure and aggressiveness of Erysiphe necator populations during grapevine powdery mildew epidemics. Appl Environ Microbiol 74:6327–6332

Müller T, Ruppel S (2014) Progress in cultivation-independent phyllosphere microbiology. FEMS Microbiol Ecol 87:2–17

Nakashima Y, Egami Y, Kimura M, Wakimoto T, Abe I (2016) Metagenomic analysis of the sponge Discodermia reveals the production of the cyanobacterial natural product kasumigamide by 'Entotheonella'. PLoS One 11:e0164468

Newton AC, Gravouil C, Fountaine JM (2010) Managing the ecology of foliar pathogens: ecological tolerance in crops. Ann Appl Biol 157:343–359

Ofaim S, Ofek-Lalzar M, Sela N, Jinag J, Kashi Y, Minz D, Freilich S (2017) Analysis of microbial functions in the rhizosphere using a metabolic-network based framework for metagenomics interpretation. Front Microbiol 8:1606–1620

Oulas A, Pavloudi C, Polymenakou P, Pavlopoulos GA, Papanikolaou N, Kotoulas G, Arvanitidis C, Iliopoulos L (2015) Metagenomics: tools and insights for analyzing next-generation sequencing data derived from biodiversity studies. Bioinf Biol Insights 9:75–88

Padhi L, Mohanta YK, Panda SK (2013) Endophytic fungi with great promises: a review. J Adv Pharm Educ Res 3:152–170

Peiffer JA, Spor A, Koren O, Jin Z, Tringe SG, Dangl JL, Buckler ES, Ley RE (2013) Diversity and heritability of the maize rhizosphere microbiome under field conditions. Proc Natl Acad Sci U S A 110(16):6548–6553

Philippot L, Raaijmakers JM, Lemanceau P, Van Der Putten WH (2013) Going back to the roots: the microbial ecology of the rhizosphere. Nat Rev Microbiol 11:789–799

Prasad R, Kumar M, Varma A (2015) Role of PGPR in soil fertility and plant health. In: Egamberdieva D, Shrivastava S, Varma A (eds) Plant growth-promoting rhizobacteria (PGPR) and medicinal plants. Springer International Publishing, Cham, pp 247–260

Raaijmakers JM, Paulitz TC, Steinberg C, Alabouvette C, Moënne-Loccoz Y (2009) The rhizosphere: a playground and battlefield for soilborne pathogens and beneficial microorganisms. Plant Soil 321:341–361

Rahman MM, Flory E, Koyro HW, Abideen Z, Schikora A, Suarez C, Schnell S, Cardinale M (2018) Consistent associations with beneficial bacteria in the seed endosphere of barley (*Hordeum vulgare* L.). Syst Appl Microbiol 41:386–398

Rastogi G, Sbodio A, Tech JJ, Suslow TV, Coaker GL, Leveau JH (2012) Leaf microbiota in an agroecosystem: spatiotemporal variation in bacterial community composition on field-grown lettuce. ISME J 6:1812–1822

Remus-Emsermann MN, Tecon R, Kowalchuk GA, Leveau JH (2012) Variation in local carrying capacity and the individual fate of bacterial colonizers in the phyllosphere. ISME J 6:756–765

Rosenblueth M, Martínez-Romero E (2006) Bacterial endophytes and their interactions with hosts. Mol Plant-Microbe Interact 19:827–837

Saito A, Ikeda S, Ezura H, Minamisawa K (2007) Microbial community analysis of the phytosphere using culture-independent methodologies. Microbes Environ 22:93–105

Sandhu A, Halverson LJ, Beattie GA (2007) Bacterial degradation of airborne phenol in the phyllosphere. Environ Microbiol 9:383–392

Schlaeppi K, Dombrowski N, Oter RG, van Themaat EVL, Schulze-Lefert P (2014) Quantitative divergence of the bacterial root microbiota in *Arabidopsis thaliana* relatives. Proc Natl Acad Sci U S A 111:585–592

Shakya M, Gottel N, Castro H, Yang ZK, Gunter L, Labbé J et al (2013) A multifactor analysis of fungal and bacterial community structure in the root microbiome of mature *Populus deltoides* trees. PLoS One 8:e76382

Shrivastava S, Prasad R, Varma A (2014) Anatomy of root from eyes of a microbiologist. In: Morte A, Varma A (eds) Root engineering, vol 40. Springer, Berlin, pp 3–22

Singh D, Raina TK, Kumar A, Singh J, Prasad R (2019) Plant microbiome: a reservoir of novel genes and metabolites. Plant Gene. https://doi.org/10.1016/j.plgene.2019.100177

Suman A, Shasany AK, Singh M, Shahi HN, Gaur A, Khanuja SPS (2001) Molecular assessment of diversity among endophytic diazotrophs isolated from subtropical Indian sugarcane. World J Microbiol Biotechnol 17:39–45

Teplitski M, Warriner K, Bartz J, Schneider KR (2011) Untangling metabolic and communication networks: interactions of enterics with phytobacteria and their implications in produce safety. Trends Microbiol 19:121–127

Thomas P (2017) Potential applications of endophytic microorganisms in agriculture. Biotechnol Dev 19:3–23

Toju H, Tanabe AS, Yamamoto S, Sato H (2012) High-coverage ITS primers for the DNA-based identification of ascomycetes and basidiomycetes in environmental samples. PLoS One 7: e40863

Truyens S, Weyens N, Cuypers A, Vangronsveld J (2015) Bacterial seed endophytes: genera, vertical transmission and interaction with plants. Environ Microbiol Rep 7:40–50

Uroz S, Buée M, Murat C, Frey-Klett P, Martin F (2010) Pyrosequencing reveals a contrasted bacterial diversity between oak rhizosphere and surrounding soil. Environ Microbiol Rep 2:281–288

Varon-Lopez M, Dias ACF, Fasanella CC, Durrer A, Melo IS, Kuramae EE, Andreote FD (2014) Sulphur-oxidizing and sulphate-reducing communities in Brazilian mangrove sediments. Environ Microbiol 16:845–855

Vorholt JA (2012) Microbial life in the phyllosphere. Nat Rev Microbiol 10:828–840

Whipps J, Hand P, Pink D, Bending GD (2008) Phyllosphere microbiology with special reference to diversity and plant genotype. J Appl Microbiol 105:1744–1755

Yadav AN, Kumar V, Dhaliwal HS, Prasad R, Saxena AK (2018) Microbiome in crops: diversity, distribution, and potential role in crop improvements. In: Prasad R, Gill SS, Tuteja N (eds) Crop improvement through microbial biotechnology. Elsevier, pp 303–327

Zhang Y, Xu J, Riera N, Jin T, Li J, Wang N (2017) Huanglongbing impairs the rhizosphere-to-rhizoplane enrichment process of the citrus root-associated microbiome. Microbiome 5:97–114

Zhou Y, Qiao X, Li W, Xu J, Wang W, Chen X (2011) Phyllosphere bacterial communities associated with the degradation of acetamiprid in *Phaseolus vulgaris*. Afr J Biotechnol 10:3809–3817

Chapter 13
A Concise Compilation of the Diverse Detection Methods to Study Plant-Microbe Interfaces at the Cellular and Molecular Level: The Past, Present and Future

Regina Sharmila Dass and Rathijit Mallick

Abstract Plant-microbe interaction can be classified under certain distinctive categories like synergistic, associative or pathogenic. The degree of friendly or hostile consortium depends on the kinds and species of microorganisms involved. These interactions are observed at various physiological planes of the host plant which in turn build the basis of molecular and genetic modifications. These changes then direct the path for biochemical reactions which occur between the plants and microbes. As a result of which, nutrient sequestration, mineral solubilization, nitrogen fixation, etc. are embarked upon by the plants, and in exchange the microbes get building blocks for energy conservation in their system. Due to this coevolutionary existence in the same niche, both have acquired mechanisms to defend each other's non-complementary company as well. Before the scientific advent of modern molecular instruments and technologies, traditional methods such as culturing on solid media, light and electron microscopic observations and biochemical tests provided initial insight into a broader realization of how these two beings communicate. As years passed, the dire need of new, effortless techniques with contemporary serological and molecular-based methodologies like isozyme assays, polymerase chain reaction (PCR), enzyme-linked assays (ELISA, RIA), microarrays (lab-on-chip), nucleic acid-based techniques (next-generation sequencing, whole genome sequencing, etc.) surfaced. Also, the invention of particular high-resolution microscopic techniques like video microscopy, confocal laser scanning microscopy (CLSM) and fluorescence microscopy brought a whole set of new information at cellular level. Apart from these, high spectral imaging also proved to be efficient enough to detect the disease symptoms at an early stage based on volatile organic compound profiling. A compilation is presented.

R. S. Dass (✉)
Department of Microbiology, School of Life Sciences, Pondicherry University, Pondicherry, India

R. Mallick
Pondicherry University, Pondicherry, India

A. Varma et al. (eds.), *Plant Microbe Interface*,
https://doi.org/10.1007/978-3-030-19831-2_13

13.1 Introduction

The term 'interaction' simply draws a picture in one's mind of active involvement of two entities under given circumstances. When we try to explore the humongous impact of the living micros on a crucial group of eukaryotic organisms, i.e. the plants, we must not avoid the most prodigious issue of all, the type of plant-microbe interactions taking place in nature every single microsecond. A precisely elucidated definition of plant-microbe interaction would not be able to accommodate the vastness of several kinds of relationships that a plant builds with the associated microscopic creatures. These organisms can communicate with a variety of plants in either of the following ways or in conjunction with phylloplane and phyllosphere (leaf surfaces and surrounding zone), rhizoplane (root surfaces), rhizosphere (an expansive zone around the roots in soil) and finally vascular system of plants.

In mutualism the interaction benefits both the host and the infecting microbe by allowance of increased nutrient availability, susceptibility against drought, heightened immunity against pathogens, usable forms of minerals from soil, production of plant hormones, nitrogen fixation, etc. In return, the exudates secreted by plants such as amino acids, sugars, vitamins and certain growth factors enhance the colonization of the microbes more efficiently. This can also be denoted as a synergistic relationship (Mitchell and Gu 2010; Madigan et al. 2012; Talaro and Chess 2012).

Commensalism is a type of symbiotic relationship where the *commensal* generally benefits, whereas the other partner is neither harmed nor benefitted by the association. Usually, a different form of the ingested nutrient by the plant is by-passed to the microorganism living in close contact with its host (Willey et al. 2009). Amensalism states the reciprocal action where the adverse effect of one organism's activity affects the other in a negative manner (Willey et al. 2009; Madigan et al. 2012). The sole purpose of these interactions, in most cases, is to confiscate nutrients which are otherwise difficult to utilize, for both the plants and the microbes. Also, it adds up some beneficial features to the plants which make them resistant to grazing, flood conditions so that they can hold on to the soil tightly (e.g. dense mycorrhizal network), infection by other group of pathogenic microorganisms (e.g. bacteria, fungi, nematode, algae, etc.) by increasing immunity, being devoured by herbivores, etc. Parasitism is a type of long-term, co-existing relationship that evolves over a period of time to favour the parasite to feed onto its host figuratively. This intimacy gives beneficial gains to the predator at the expense of the host member. Usually, if the host defence mechanism is too strong, then the interaction equilibrium will favour the host over the parasite, but in the opposite scenario, the host organism becomes ill and may eventually die depending on certain factors, like the infecting species. One remarkable aspect about parasitism is that the parasite is 'expectant' of the host to remain alive as long as possible to draw maximum nutrition to prolong the reproduction process as a mode of continuum for infecting new hosts (Willey et al. 2009).

13.2 Definition of Plant-Microbe Interaction

Acquisition of nutritional substances from environment is indulged by both parties involved in a synergistic plant-microbe relationship. These nutrients are essential for all metabolic activities that take place in vivo (Talaro and Chess 2012). Plant's exudative properties and rhizoplane-allied microbial communities depend on certain factors such as nutritional standing of the plant, species and, most importantly, the developmental stage of the plant. Root microbiome, in return, helps in carbon and nitrogen sequestration, solubilization of macro- and micronutrients and synthesis of growth factors and protective substances (Mitchell and Gu 2010; Talaro and Chess 2012). This co-operative behaviour is encouraging for both plants and the microbes. Although, there are a handful of examples present which negate this collaboration.

13.2.1 Differentiation Between Beneficial and Harmful Effects

Few of the outstanding examples of plant-microbe interactions in bacteria and fungi are symbiotic nitrogen-fixing bacteria and the mycorrhizal fungi, respectively (Bever et al. 2001; Buscot 2015). Examples of parasitic associations are numerous and have been elaborately outlined in various studies. Many leguminous plants and some others have acquired the propensity to fix atmospheric dinitrogen by forming associations with bacteria. This feature is widely distributed among prokaryotes, but is unknown and undefined in eukaryotes. The bacteria *Rhizobium* and *Bradyrhizobium* (rhizobia) and the actinomycetes *Frankia* form nodules on plant roots and are major contributors to symbiotic nitrogen fixation. During the course of evolution, the leguminous plants developed a couple of systems to attain mutual symbioses with rhizobia and mycorrhizae. The genetic requirements for rhizobial and mycorrhizal interactions in plants overlap in a common symbiosis pathway (CSP) leading to successful, mutually beneficial associations (Imaizumi-Anraku et al. 2005; Kouchi et al. 2010).

13.3 Plant-Bacteria Interactions

When a host plant is threatened by any microbe, due to the process of co-evolution, a cascade of defence strategies are turned on in both. *Phytoanticipins* and *phytoalexins*, two low molecular antibiotic compounds, are products of such defence mechanism. The former is usually present in plants inherently, but the latter is synthesized as soon as the microbes confront plants, as a mode of recognition (VanEtten et al. 1994). Studies suggest that the production of extracellular matrix (ECM) is a more precise and irreversible signal definitive of cellular specification once the initial yet firm root surface adhesion takes place (Allard-Massicotte et al. 2016). This ECM principally consists of exo-polysaccharides, nucleic acids and diverse groups of proteins which are the basis of biofilm formation (Branda et al. 2005). With the help of microfluidics concomitant with laser confocal microscopy, the distribution of bacterial cells such as *Bacillus subtilis* (soil-dwelling microorganism) and *Escherichia coli* (common contaminant of manure fertilizer) in distinctive parts of the root has been well studied. It has been found out that root exudates play a major role as chemical attractants in the colonization of growing root tips by bacteria, which, after a certain point, is well distributed throughout the root surface as the cell density decreases and eventually forms a bacterial plug (Englert et al. 2010; Grossmann et al. 2011; Mendes et al. 2011; Parashar and Pandey 2011; Nezhad 2014; Jiang et al. 2014; Panke-Buisse et al. 2015; Shapiro et al. 2016).

Bacteria are ubiquitous. They are capable of occupying ecological niches as well as colonizing hotspots like plant rhizosphere and rhizoplane. By the process of rhizodeposition, plants release up to 40% of their photosynthetically fixed carbon through the roots into the vicinity (Barber and Martin 1976; Lynch and Whipps 1991; Marschner 1995; Hütsch et al. 2002). The colonization of the root itself (rhizoplane) and the surrounding soil zone (rhizosphere) directs towards a crucial link between the plant roots and soil zone (Lenc et al. 2011; Bulgarelli et al. 2013; Reinhold-Hurek et al. 2015). Bacteria utilize this constant flow of organic plant-based substrates and in return promote plant growth by mobilizing and providing inorganic nutrients and plant growth-promoting substances (Spaink et al. 1998; Brimecombe et al. 2007; Nannipieri et al. 2007; Compant et al. 2010). Investigation of such a continual stream of affairs in a complex and happening habitat is a major challenge.

Biochemical studies of tissue extracts are not very ideal to study plant-microbe interactions, because they are initiated at the level of the cell and need high-resolution studies of cellular responses. The potential to visualize or detect such interactions at a single-cell level becomes particularly important. More than about a decade and a half ago, imaging techniques have addressed the question to a certain extent. Video microscopy (Inoue and Spring 1997), confocal laser scanning microscopy (CLSM), laser trapping, image processing using a wide variety of commercially available software programs (Russ 1999) and fluorescence microscopy have been especially used and applied to a few plant-microbe systems (Heath 2000).

Some examples of bacterial phytopathogens that have been documented over the years include comprehension of the biology underlying disease initiation and progression in *Erwinia chrysanthemi* (Collmer and Keen 1986; Hugouvieux-Cotte-Pattat et al. 1996; Bouchart et al. 2007), *Pseudomonas syringae* (Loubens et al. 1993; Bohin and Lacroix 2006), *Xanthomonas campestris* (Minsavage et al. 2004) and *Xylella fastidiosa* known to cause citrus variegated chlorosis (Chang et al. 1993), Pierce's disease in grapevine (Davis et al. 1978) and leaf scorch in oleander, mulberry, coffee, almond and plum (Purcell and Hopkins 1996).

13.3.1 Phylloplane Interactions

Distribution of bacteria on all over the plant foliage is heterogeneous, and in turn their occurrence is affected by factors like orientation of the leaves within the herbage, climatic conditions, chemical composition of the cuticles and competition with other group of microbes, plant-eaters, etc. (Bodenhausen et al. 2013). Sometimes a co-existing foundation is observed between bacteria and fungi, where they form aggregates as they myriad spread over the leaves' surface at the depressions formed at the joints of epidermal cells and stomata, alongside the veins and at the base of trichomes (Remus-Emsermann et al. 2014). These structural features of the leaves also enhance the nutrient availability by facilitating percolation of photoassimilates onto the leaf surface (Leveau and Lindow 2001; Vorholt 2012).

13.4 Plant-Virus Interaction

Unlike many mutually beneficial acts of bacterial and fungal cells with plants, viruses always play a catastrophic role when they are in close contact with their hosts. This intimate involvement ultimately leads to infection, resulting in plant diseases by causing certain physiological changes. Technically, plants are not affected by viruses via the so-called receptor-binding mechanism due to the lack of it and rather are threatened when there is a mechanical damage by either vectors or environmental causes. The molecular basis of this interaction is dependent on the formation of multi-structural complexes between the host and virus proteins. Almost around 80% of the viruses (Mandahar 2006) have single-stranded RNA (mRNA) molecules (either singly or in multiple copies) as their genetic material (rice stripe virus, maize stripe virus, tomato spotted wilt virus, etc.), but few examples have also been documented on DNA (ssDNA and dsDNA) viruses (cauliflower mosaic virus, soybean crinkle leaf virus, para-retroviruses, single-stranded Gemini viruses).

Spread of infection depends on viral factors and host component interactions. This is specified by targeting viral RNA to plasmodesmata (PD) and increasing the PD pore size to allow the viral ribonucleoprotein (RNP) complex (i.e. viral genetic material/infectious viral RNA particle along with the movement proteins). For promoting the diffusion of these virus particles, the size exclusion limit of PD is exploited by the movement proteins (MP) which otherwise limits transport of macromolecules between cells. This natural phenomenon is also called 'PD gating' (Wolf et al. 1989).

It is to be noted that the whole process sometimes takes place by diffusion of the aforementioned substances through endoplasmic reticulum (ER)-Golgi complex secretory pathway or simply in association with the ER, independent of the above route. Varied set of reactions are generated in a cascade during plant-virus interactions at every stage of the infection favouring either the host or the infectious agent. Depending on the type of interaction mentioned earlier, viral replication and their movement are halted as they are tightly coupled (Heinlein 2015). A more stable interaction is formed if the virus particles are not recognized by the host plant, but the obverse is true if the viral particles are detected. This follows an incompatible match for both and is unfavourable for the virus (Hammond-Kosack and Jones 2000). In order to counteract this ordeal, some major defence mechanisms have been imposed to plants by nature. One of these is RNA silencing, in which Dicer or an RNA-induced silencing complex (RISC) recognizes and degrades the virus genome (Soosaar et al. 2005; Dunoyer and Voinnet 2005; Li and Ding 2006; Valli et al. 2009; Shiekh 2014). But, these nonliving particles have learned to deal with that too by producing suppressor genes against the defence (Incarbone and Dunoyer 2013) in the host and accelerate replication by coinfecting viruses to possess multiple infections (Pruss et al. 1997; Fukuzawa et al. 2010; Syller 2012).

13.5 Plant-Fungal Interactions

Plant health can in part be attributed to their ability to team up with filamentous microorganisms. The primary plant part for nutrient and water uptake is the root system, which is inhabited and surrounded by a complex microbial community referred to as the root microbiome (Hacquard et al. 2015). It may be of special importance to mycologists and plant pathologists to comprehend the underlying mechanisms. Microbiologists have been drawn to this field of research mainly because of the need for identification of microbial agents responsible for causing infectious diseases in economically important crop plants (Montesines 2000).

Plant pathogenic fungi fall into one of the three categories on the basis of their growth within the tissues of their hosts, namely, necrotrophs (perthotrophs), biotrophs and hemibiotrophs. The enzymes and toxins are used by necrotrophs, which kill the host cells, in advance of their hyphal proliferation. Eventually they grow between and into dead and dying cells. On the other hand, biotrophs obtain nutrients from living host cells and serve as typical examples of fungi exhibiting a parasitic mode of nutrition. Hemibiotrophs like the anthracnose fungus *Colletotrichum lindemuthianum* initially require living host cells but soon cause their destruction like necrotrophs (Alexopoulos et al. 2010). The culprit which caused the Irish famine, namely, *Phytophthora infestans*, continues to cause dramatic yield losses in crops such as potato and tomato (Fry 2008). There is an increasing impact of plant diseases in crop plants; plant-pathogen research has resulted in extensive documentation on plant defence mechanisms.

There is considerable evidence of fossilized fungal structures inside plant cells (Remy et al. 1994), and further nearly about 80% of all existing higher plants are colonized by arbuscular mycorrhizal (AM) fungi (Wang and Qiu 2006; Prasad et al. 2017). Plants also engage in beneficial root associations, namely, endophytic mycorrhizal interactions (Parniske 2008). The fungal partner is known to provide mineral nutrients such as phosphorus, etc. Conversely, plants supply carbohydrates generated via photosynthesis. It is also to be understood that plant carbohydrates may also serve as attractant molecules to root-infecting filamentous pathogens, namely, fungi and oomycetes. Of particular importance are the arbuscular mycorrhizal (AM) and ectomycorrhizal (ECM) fungi because they actively participate in plant development (Smith and Read 2008).

13.6 Methods Used to Detect Plant-Microbe Interactions: Conventional and Molecular Methods

Conventional techniques in the detection of phytopathogenic microorganisms involve the use of culturing methodologies on specific media with subsequent morphological and biochemical characterization (Lopez et al. 2003). These traditional methods are extremely time-consuming and laborious and require expert and skilled personnel for the purpose of identification. In the 1970s, phytopathogenic viruses have been detected based on isolation, electron microscopy, electrophoresis, biological indexing and serological techniques. Meanwhile the advent of ELISA and PCR have revolutionized phytopathogen detection, especially in the last quarter of the twentieth century, and have become commonplace, being routinely used by plant pathologists (Lopez et al. 2003).

Erwinia chrysanthemi is a Gram-negative enterobacterium that is a causative agent of soft rot diseases in ornamentals and vegetables. Plant hosts become vulnerable to this bacterium because of its ability to secrete a number of enzymes that are responsible for degrading plant cell wall components. Techniques such as 2-DE and MALDI/TOF MS have been carried out to characterize the secretome and compared with mutants of *Erwinia chrysanthemi*. The secretome of *Xanthomonas* pv. *campestris* was analysed by another research group elsewhere (Watt et al. 2005) using the aforementioned techniques.

In fact, the first phytopathogenic organism to be sequenced completely was *Xylella fastidiosa* following which the annotation of its 2849 genes was found in the chromosome and two extrachromosomal plasmids (Simpson et al. 2000). Consequent to the determination of the complete genome sequence, efforts were made to analyse the whole bacterial cell proteome as well as the secreted protein profile (*secretome*), which led to the identification of 142 different proteins including some that were homologous to proteins involved in different cellular adhesion systems (Smolka et al. 2003).

The genetic basis of this robust specificity of plant-bacteria interaction can be described by gene-for-gene elicitor-receptor model (Flor 1955; Baker et al. 1997). It goes in a way where the avirulence (*avr*) gene acts like a resistance (R) gene counterpart in hosts. A complementary amalgamation of the *avr* and R genes ends up in unsuitable plant-pathogen interaction, and plant defences are triggered, but in a compatible reaction, infection takes place (Bent 1996; Ellis et al. 2000; Hammond-Kosack and Jones 2000). A hypersensitivity response (HR) is provoked in host plants due to the presence of these HR genes and pathogenicity genes (a set of genes consisting of these two). With the help of these genes, viruses can also elicit HR in non-host plants (Lindgren 1997; Nakahara and Masuta 2014; Rodriguez et al. 2015) (same is applicable for viral infection as well). In contrast to the positive side of the relationship, a set of genes are also involved in pathogen-host relation.

While studying the principles of colonization, it is disadvantageous to separate plant pathology and symbiosis systems on different plant species. *Arabidopsis thaliana*, the choicest plant for numerous plant-pathogen interactions, has not been found to be ideal to study the feeding structure formation by endomycorrhizal fungi and has thus found limited use. On the other hand, *Phytophthora* and beneficial AM fungi in legumes follow analogous steps to establish an interaction.

Gehrig et al. (1996) are of the opinion that the establishment of early land plants was facilitated by the interaction with symbiotic fungal association, ever since the evolution of land plants more than 700 million years ago as suggested by molecular clock estimates (Heckman et al. 2001). The exposition of the plants to microbes is a continual and ongoing process. To exhibit pathogenesis, most microbes must gain entry and access to the plant interior. The entry process may be direct by the formation of specialized structures (e.g. in bacteria, haustoria and appressoria in fungi) or indirectly through wounds or natural openings such as stomata, thus overcoming the first line of host defence. Digit-like haustoria formed by *P. palmivora* in *Nicotiana benthamiana* roots have been documented using green fluorescent protein (GFP) (Rey and Schornack 2013). Subsequently, the microbes are required to conquer the cellulose-based support, namely, the rigid cell wall and the host plasma membrane, where they encounter extracellular surface receptors that recognize pathogen-associated molecular patterns (PAMPs). During the course of evolution, microbes have found ways and means to suppress the PAMP-triggered immunity (PTI) that supposedly alter resistance signalling and responses in plants (Chisholm et al. 2006).

13.6.1 Conventional Methods

Traditional methods involve the isolation of the fungal pathogens onto suitable standard mycological agar media (general purpose, routine, semisynthetic, synthetic, semi-selective, selective, specialized media), studying the cultural characteristics such as colony colour, obverse and reverse colony morphology and micromorphological characteristics like sporangiospores, conidiophores, conidiospores, chlamydospores, etc., identifying sexual structures such as ascospores, asci, cleistothecia, perithecia and miscellaneous structures like Hulle cells, etc. However, culture-dependent techniques allow the phenotypic analyses of only culturable strains and limit its use in the case of fungi that cannot be isolated on artificial media. Temporary mounts using cellophane and agar plug techniques have also been successfully used to study fungal propagules. Study of *Fusarium* species has been extensively elaborated by many researchers using standard protocols (Booth 1971; Nelson et al. 1983; Leslie and Summerell 2006). Other examples of fungi include

Phytophthora bischeri species (Erwin and Ribeiro 1996; Abad et al. 2008), *Botrytis* species (Mirzaei et al. 2008) being some of them.

13.6.2 Biochemical Methods

13.6.2.1 Detection of GUS Activity

Gene reporters enable valuable insight into gene expression. The GUS gene reporter system is one of the popular and common plant reporter systems to establish the cause of certain diseases. Fungal transformant strains of *Cladosporium fulvum* infecting tomatoes and *Leptosphaeria maculans* infecting brassica crops expressing β-glucuronidase activity have been produced and used to histochemically detect the presence of the hyphae in infected host plant tissues. Oliver et al. (1993) also reported that this activity could be used as a measure of fungal biomass in the cotyledons of infected tomato seedlings.

13.6.2.2 Isozyme Analysis

Isozyme analysis is a powerful biochemical analysis tool whose usefulness is obvious for the detection, differentiation and identification of morphologically similar or closely related species, varieties and *formae speciales*. This in turn helps in the 'fingerprinting' of strains by protein profiling. This technique which was used more than three decades ago is still relevant and can be applied to study fungal interactions with plants. Isozyme analysis has been performed with *Peronosclerospora* spp. from maize for distinguishing *P. sorghi*, *P. sacchari* and *P. sacchari-phillipinensis* complex (Bonde et al. 1984) and in other studies (Bonde et al. 1985, 1989, 1993; Pan et al. 1991; Kaufmann and Weidemann 1996). In a particular study, uniform isozyme patterns were noted from different *Fusarium* species, independent of the geographical origin and hosts from which they were isolated. The different strains studied include *F. cerealis*, *F. culmorum*, *F. graminearum* and *F. pseudograminearum* (Láday and Szécsi 2001, 2002).

13.6.3 Immunoassays and Nucleic Acid-Based Assays

Immunodiagnostic methods fall into two broad groups of (a) direct and (b) labelled methods. While immunoprecipitation, immunoagglutination and immunodiffusion are examples of direct methods, enzyme immunoassay (EIA), radioimmunoassay (RIA) and immunofluorescence are categorized as labelled methods. The legacy dates back when Tempel (1959) developed a gel diffusion test for the detection and differentiation of *formae speciales* of *Fusarium oxysporum*. Later on, there have been arrays of studies which different research groups have undertaken in different parts of the world. Some of the assays include detection of *Botrytis cinerea* using RIA (Savage and Sall 1981), *Phytophthora cinnamomi* by ELISA (Hardham et al. 1986) and turf disease causers, namely, *Pythium* species, *Rhizoctonia solani* and *Sclerotina homeocarpa* (Rittenburg et al. 1988), using commercial kits of ELISA and its variants. Similar studies based on antigen-antibody-based detection has been carried out (Kitagawa et al. 1989; Sundaram et al. 1991; Plantiño-Álvarez et al. 1999) in the detection of *F. oxysporum* f. sp. *cucumerinum*, *Verticillium dahliae* and *F. oxysporum* f. sp. *radicis-lycopersici*, respectively. ELISA tests have been used for the detection of *Phytophthora* at the generic level and *P. ramorum* at the species level using the diagnostic method in combination with TaqMan PCR (Kox et al. 2007). Karthikeyan et al. (2006) have used ELISA and PCR for the detection of *Ganoderma lucidum* known to cause *Ganoderma* disease in coconuts.

13.7 Assays for Virus Detection

Identification of viral phytopathogens by traditional methods like culturing is not easily achievable. For that reason, serological methods and advanced molecular methods have been employed to guide their detection regime methodically. ELISA, being a highly sensitive virus detection tool, uses targeted epitopes which bind specifically with a desired antibody conjugated to an enzyme. Upon irreversible binding, a colour reaction (Clark and Adams 1977) is generated as a suitable substrate is added to the reaction mixture. This binding specificity can in turn be enhanced by using monoclonal (using hybridoma technology against single epitope on a particular cell line) (Holzloehner et al. 2013) or recombinant antibodies (Gorris et al. 1994; López et al. 2001). Initially, PCR was invented for the robust detection of bacterial and viral pathogens (Cai et al. 2014).

Polymerase chain reaction and its modifications have successfully been used to enormously detect plant pathogens over the years. RT-PCR (mostly used for its high sensitivity to detect RNA viruses) (Lopez et al. 2003), multiplex PCR (simultaneous detection of different DNA and RNA in a single reaction) (Osiowy 1998; Pallisgaard et al. 1998; James 1999; Williams et al. 1999; Nassuth et al. 2000), real-time PCR (for viral DNA identification in real time for improved diagnosis of diseases and on-site as well) (Schaad and Frederick 2002; Lievens et al. 2006). Nucleic acid sequence-based amplification (NASBA) is another example of excellent technique for amplification and identification of RNA sequence containing viruses (Klerks et al. 2001; Olmos et al. 2005), the predominant category responsible for viral plant diseases. Loop-mediated isothermal amplification has been extensively in use to detect plant virus such as plum pox virus (PPV) (Varga and James 2006). This method is very simple to use without incorporating the need of an expensive thermocycler, where the amplicons are detected by observing solution turbidity through photometric analysis (Mori et al. 2001). SYBR green is usually used as a colour detector. Microarray is another molecular method where the RNA of interest (mRNA) is isolated from cells and is reverse transcribed to its complementary DNA. After labelling, they are hybridized with prefixed probes on the chip and detected for light signals.

13.8 Challenges and Lacunae in the Detection of Microbes

The advancement in the field of plant pathology over years has made it easier for the scientists to detect and identify several different kinds of diseases and the etiological agents related to them. Also, this progress has led farmers and growers achieve a contamination-free cultivation with minimum economic loss. Apart from being extremely advantageous, these unique detection methods also have certain downsides. Among the direct methods for detection, PCR (normal PCR, QPCR, qRT-PCR, etc.) is one such molecular method with higher ability, specificity and sensitivity for a particular pathogen. The technique is however challenged at the point, where it demands species-specific probe design to amplify target DNA which is cost-prohibitive and time-consuming and can be applied for high target value analytes. Alongside this, parameters like buffer concentration of nucleotide solution, polymerase activity, etc. may compromise with the quality of result expected.

Another issue to be addressed is that the infection can be detected only on the onset of disease, i.e. when the host starts expressing disease symptoms. As soon as the infective viral particle (RNA/DNA) is found, it undergoes a tiresome sample processing for detection, which includes isolation and purification of the genetic material, complementary DNA synthesis and amplification. The whole method makes it less of a choice for large-scale preventive measure. ELISA is another serological technique which is highly sensitive for detecting viral plant diseases but fails to do so in case of bacterial infections as it is poorly sensitive. Next-generation sequencing of the disease-causing pathogens is also a robust method, but the high analysis cost per sample, duration of sample processing, complexity of the data analysis and compromisation for low-titre virus samples make it unfit economically. Certain protocols are available as commercial kits (fluorescent in situ hybridization, immunofluorescence) which require skilled personnel to operate and prepare samples and to decode the data.

Volatile organic compounds (VOCs) are released by plants as a mode of metabolic activities which have a distinct pattern and vary from each species of plant to the other and also under stressful conditions. This change can be measured with using gas chromatography (GC), but the limitations to this method are as follows:

1. Time taken to process the diseased plant parts which release the VOCs.
2. Differentiate between the variation in VOCs released naturally which might cover up the ones secreted as an outcome of environmental stress or diseased condition. This requires the usage of characteristic volatile markers specific for a plant and a diseased condition that will be different from when its produced under environmental or nutritional stress.

3. Performing instrumentation and analysis of the complicated data which demand skill and expertise.

Another crucial point to be mentioned in this context is that all these direct measures of detection restrict their application in cultivation field which is the biggest drawback. This problem could partly be solved with the help of thermography or high spectral imaging, but again their object instability to environmental changes and sensitivity towards climate change is extreme which makes them unsuitable in field detection. A plausible way to prevail over this situation would be to gather more knowledge about a particular wavelength range which can be sensitized for a particular plant disease and at the same time would also be least affected by the changes in surrounding.

The void that has been left unfulfilled by all these preventive measures could now be resolved by the application of nanotechnology as is evident in a number of research studies. Nanoparticles like nanorods, nanocrystals, nanotubes and microcapsules can effectively carry chemical and biological pesticides, host immune factors, defensive molecules against pathogens, etc. and can release them as per host's requirement and will contain the infection. The other potential next-generation large-scale field detection method would be biotrophic sensors and VOC sensors which are yet to be brought into limelight. These overwhelming technologies could detect plant pathogens at the earliest stages of infection with robustness before symptoms appear. Interestingly it could serve as an edge over other molecular or serological methods, way before by detecting early induced volatiles when pathogens make their way through the host system.

13.9 Future Prospects and Conclusions

The need to detect, identify and eliminate all primary sources of inocula in disease production in plants through microbial interactions is of great consequence. This is of vital importance to prevent infection and spread of microbial plant pathogens which can cause major economic losses in crop plants. While there are beneficial aspects also ruling these interactions, the need of the hour is the use of methods to distinguish pathogens and closely related species and strains. Conventional and rapid assays have always proved handy for this purpose. However, the former techniques are quite laborious, time-consuming and incommodious. Rapid assays such as immunoassays and nucleic acid-based methods are not only fast but also specific and reliable. Further, these methods are distinctly advantageous over immunoassays because microbial antigens or propagules are complex in nature, especially varying depending on the stages in their life cycle. On the contrary, the nature of the genomic elements remains constant thus enhancing specificity and sensitivity with PCR-based detection being one of the answers to problems faced while studying such interactions. In nature, diverse microorganisms reside in, on and around plants as endophytes and epiphytes (Hallmann et al. 1997; Mano et al. 2007; Whipps et al.

2008). Many questions underlying these plant-microbe relationships remain unanswered (Saito et al. 2007; Hardoim et al. 2008). Both beneficial and harmful interactions exist, which are a resultant of relationships between plants and bacteria, fungi, viruses, viroids, etc. Symbiosis is the living together of two or more organisms involved in this association (Ogle and Brown 1997). All the organisms involved in this association derive benefits. Parasitic interactions on the other hand lead to derivation of benefit from one of the partners, while the other associate could face highly detrimental effects, and the association ceases to exist with the passage of time. The challenges for the present and the next decades include understanding the complex behaviour of microbes in their natural habitats. Plant-microbe interactions have been extensively studied and researched upon in habitats such as the rhizosphere and phyllosphere. The underlying principles are microbes, which like human beings, want to survive, even if conditions are inhospitable or hostile, want to feed on nutrients, grow, multiply and proliferate on the onset of favourable environments.

References

Abad ZG, Abad JA, Cofffey MD, Oudemans PV, Veld WA, de Gruyter H, Cunnington J, Louws FJ (2008) *Phytophthora bischeria* sp. nov., a new species identified in isolates from the rosaceous raspberry, rose and strawberry in three continents. Mycologia 100:99–110

Alexopoulos CJ, Mims CW, Blackwell M (2010) Introductory mycology, 4th edn. Wiley, Noida

Allard-Massicotte R, Tessier L, Lecuyer F, Lakshmanan V, Lucier JF, Garneau D, Caudwell L, Vlamakis H, Bais HP, Beauregard PB (2016) *Bacillus subtilis* early colonization of *Arabidopsis thaliana* roots involves multiple chemotaxis receptors. MBio 7:e01664-16

Baker B, Zambryski P, Staskawicz B, Dinesh-Kumar SP (1997) Signalling in plant-microbe interactions. Science 276:726–733

Barber DA, Martin JK (1976) The release of organic substances by cereal roots into soil. New Phytol 76:69–80

Bent AF (1996) Plant disease resistance genes: function meets structure. Plant Cell 8:1757–1771

Bever JD, Schultz PA, Pringle A, Mortan JB (2001) Arbuscular mycorrhizal fungi: more diverse than meets the eye, and the ecological tale of why. Bioscience 51:923–931

Bodenhausen N, Horton MW, Bergelson J (2013) Bacterial communities associated with the leaves and the roots of *Arabidopsis thaliana*. PLoS One 8:e56329

Bohin JP, Lacroix JM (2006) Osmoregulation in the periplasm. In: Ehrmann M (ed) The Periplasm. American Society for Microbiology, Washington, DC, pp 325–341

Bonde MR, Peterson GL, Dowler WM, May B (1984) Isozyme analysis to differentiate species of *Peronosclerospora* causing downy mildews for maize. Phytopathology 74:1278

Bonde MR, Peterson GL, Dowler WM, May B (1985) Comparison of *Tilletia indica* isolates from India and Mexico by isozyme analysis. Phytopathology 75:1309

Bonde MR, Peterson GL, Matsumoto TT (1989) The use of isozymes to identify teliospores of *Tilletia indica*. Phytopathology 79:596–599

Bonde MR, Micales JA, Peterson GL (1993) The use of isozyme analysis for identification of plant pathogenic fungi. Plant Dis 77:961–968

Booth C (1971) The genus *fusarium* commonwealth mycological institute. CAB International, Kew, Surrey

Bouchart F, Delangle A, Lemoine J, Bohin JP, Lacroix JM (2007) Proteomic analysis of a non-virulent mutant of the phytopathogenic bacterium *Erwinia chrysanthemi* deficient in

osmoregulated periplasmic glucans: change in protein expression is not restricted to the envelope, but affects general metabolism. Microbiology 153:760–767
Branda SS, Vik S, Friedman L, Kolter R (2005) Biofilms: the matrix revisited. Trends Microbiol 13:20–26
Brimecombe MJ, De Leij FAAM, Lynch JM (2007) Rhizodeposition and microbial populations. In: Pinton R, Varanini Z, Nannipieri P (eds) The rhizosphere: biochemistry and organic substances at the soil-plant interface. Marcel Dekker, New York, NY, pp 41–93
Bulgarelli D, Schlaeppi K, Spaeten S, Ver Loren van Themaat E, Schulze-Lefert P (2013) Structure and functions of the bacterial microbiota of plants. Annu Rev Plant Biol 64:807–838
Buscot F (2015) Implication of evolution and diversity in arbuscular and ectomycorrhizal symbioses. J Plant Physiol 172:55–61
Cai H, Caswell J, Prescott J (2014) Nonculture molecular techniques for diagnosis of bacterial disease in animals a diagnostic laboratory perspective. Vet Pathol 51:341–350
Chang CJ, Garnier M, Zreik L, Rossetti V, Bove JM (1993) Culture and serological detection of the xylem-limited bacterium causing citrus variegated chlorosis and its identification as a strain of *Xylella fastidiosa*. Curr Microbiol 27:137–142
Chisholm ST, Coaker G, Day B, Staskawicz BJ (2006) Host-microbe interaction: shaping the evolution of the plant immune response. Cell 124(4):803–814
Clark MF, Adams A (1977) Characteristics of the microplate method of enzyme-linked immunosorbent assay for the detection of plant viruses. J Gen Virol 34:475–483
Collmer A, Keen NT (1986) The role of pectic enzymes in plant pathogenesis. Annu Rev Phytopathol 24:383–409
Compant S, Clement C, Sessitsch A (2010) Plant growth promoting bacteria in the rhizo- and endosphere of plants: their role, colonization, mechanisms involved and prospects for utilization. Soil Biol Biochem 42:669–678
Davis MJ, Purcell AH, Thomson SV (1978) Pierce's disease of grapevines: isolation of the causal bacterium. Science 199:75–77
Dunoyer P, Voinnet O (2005) The complex interplay between plant viruses and host RNA-silencing pathways. Curr Opin Plant Biol 8:415–423
Ellis J, Dodds P, Pryor T (2000) The generation of plant disease resistance gene specificities. Trends Plant Sci 5:373–379
Englert DL, Manson MD, Jayaraman A (2010) A microfluidic device for quantifying bacterial chemotaxis in stable concentration gradients. J Vis Exp 38:2–7
Erwin DC, Ribeiro OK (1996) *Phytophthora* diseases worldwide. The American Phytopathology Society Press, St. Paul, MN
Flor AH (1955) Host-parasite interactions in flax rust – its genetics and other implications. Phytopathology 45:680–685
Fry W (2008) *Phytophthora infestans*: the plant (and R gene) destroyer. Mol Plant Pathol 9:385–402
Fukuzawa N, Itchoda N, Ishihara T, Goto K, Masuta C, Matsumura T (2010) HC-pro, a potyvirus RNA silencing suppressor, cancels cycling of cucumber mosaic virus in *Nicotiana benthamiana* plants. Virus Genes 40:440–446
Gehrig H, Schussler A, Kluge M (1996) Geosiphon pyriforme, a fungus forming endocytobiosis with Nostoc (cyanobacteria), is an ancestral member of the Glomales: evidence by SSU rRNA analysis. J Mol Evol 43:71–81
Gorris MT, Alarcon B, Lopez M, Cambra M (1994) Characterization of monoclonal antibodies specific for *Erwinia carotovora* subsp. *atroseptica* and comparison of serological methods for its sensitive detection on potato tubers. Appl Environ Microbiol 60:2076–2085
Grossmann G, Guo WJ, Ehrhardt DW, Frommer WB, Sit RV, Quake SR, Meier M (2011) The root chip: an integrated microfluidic chip for plant science. Plant Cell 23:4234–4240
Hacquard S, Garrido-Oter R, Gonza'lez A, Spaepen S, Ackermann G, Lebeis S, AC MH, Dangl JL, Knight R, Ley R, Schulze-Lefert P (2015) Microbiota and host nutrition across plant and animal kingdoms. Cell Host Microbe 17(5):603–616

Hallmann J, Quadt-Hallmann A, Mahaffee WF, Kloepper JW (1997) Bacterial endophytes in agricultural crops. Can J Microbiol 43:895–914

Hammond-Kosack K, Jones JDG (2000) Responses to plant pathogens. American Society of Plant Physiology, Rockville, pp 1102–1156

Hardham AR, Suzaki E, Perkin JL (1986) Monoclonal antibodies to isolate- species- and genus specific components on the surface of zoospores and cysts of the genus *Phytophthora cinnamomi*. Canad J Bot 64:311–321

Hardoim PR, van Overbeek LS, van Elsas JD (2008) Properties of bacterial endophytes and their proposed role in plant growth. Trends Microbiol 16:463–471

Heath CM (2000) Advances in imaging the cell biology of plant-microbe interactions. Annu Rev Phytopathol 38:443–549

Heckman DS, Geiser DM, Eidell BR, Stauffer RL, Kardos NL, Hedges SB (2001) Molecular evidence for the early colonization of land by fungi and plants. Science 293:1129–1133

Heinlein M (2015) Plant virus replication and movement. Virology 479–480:657–671

Holzloehner P, Schliebs E, Maier N, Füner J, Micheel B, Heilmann K (2013) Production of monoclonal camelid antibodies by means of hybridoma technology. J Immunol 190:135.14

Hugouvieux-Cotte-Pattat N, Condemine G, Nasser W, Reverchon S (1996) Regulation of pectinolysis in *Erwinia chrysanthemi*. Annu Rev Microbiol 50:213–257

Hütsch BW, Augustin J, Merbach W (2002) Plant rhizodeposition – an important source for carbon turnover in soils. J Plant Nutr Soil Sci 165:397–407

Imaizumi-Anraku H, Takeda N, Charpentier M, Perry J, Miwa H, Umehara Y, Kouchi H, Murakami Y, Mulder L, Vickers K, Pike J, Downie JA, Wang T, Sato S, Asamizu E, Tabata S, Yoshikawa M, Murooka Y, Wu GJ, Kawaguchi M, Kawasaki S, Parniske M, Hayashi M (2005) Plastid proteins crucial for symbiotic fungal and bacterial entry into plant roots. Nature 433:527–531

Incarbone MP, Dunoyer P (2013) RNA silencing and its suppression: novel insights from in planta analyses. Trends Plant Sci 18:382–392

Inoue S, Spring KR (1997) Video microscopy: the fundamentals, 2nd edn. Plenum, New York/London, p 741

James D (1999) A simple and reliable protocol for the detection of apple stem grooving virus by RT-PCR and in a multiplex PCR assay. J Virol Methods 83:1–9

Jiang H, Xu Z, Aluru MR, Dong L (2014) Plant chip for high-throughput phenotyping of *Arabidopsis*. Lab Chip 14:1281–1293

Karthikeyan M, Radhika K, Bhaskaran R, Mathiyazhagan S, Samiyappan R, Velazahahan R (2006) Rapid detection of *Ganoderma* disease of coconut and assessment of inhibition effect of various control measures by immunoassay and PCR. Plant Protect Sci 42:49–57

Kaufmann PJ, Weidemann GJ (1996) Isozyme analysis of *Colletotrichum gloeosporioides* from five host genera. Plant Dis 80:1289–1293

Kitagawa T, Sakamoto Y, Furumi K, Ogura H (1989) Novel enzyme immunoassays for specific detection of *Fusarium oxysporum* f.sp. *cucumerinum* and for general detection of various *Fusarium* species. Phytopathology 79:162–165

Klerks MM, Leone G, Lindner JL, Schoen CD, van den Heuvel JFJM (2001) Rapid and sensitive detection of apple stem pitting virus in apple trees through RNA amplification and probing with fluorescent molecular beacons. Phytopathology 91:1085–1091

Kouchi H, Imaizumi-Anraku H, Hayashi M, Hakoyama T, Nakagawa T, Umehara Y, Suganama N, Kawaguchi M (2010) How many peas in a pod? Legume genes responsible for mutualistic symbioses underground. Plant Cell Physiol 51:1381–1397

Kox LFF, van Brouwershaven IR, van de Vossenberg BTLH, van den Beld HE, Bonants PJM, de Gruyter J (2007) Diagnostic values and utility of immunological, morphological and molecular methods for *in planta* detection of *Phytophthora ramorum*. Phytopathology 97:1119–1129

Láday M, Szécsi Á (2001) Distinct electrophoretic isozyme profiles of *Fusarium graminearum* and closely related species. System Appl Microbiol 24:67–75

Láday M, Szécsi Á (2002) Identification of *Fusarium* species by isozyme analysis. Acta Microbiol Immunol Hung 49:321–330

Lenc L, Kwasna H, Sadowski C (2011) Dynamics of the root/soil pathogens and antagonists in organic and integrated production of potato. Eur J Plant Pathol 131:603–620

Leslie JF, Summerell BA (2006) The *Fusarium* laboratory manual, 1st edn. Blackwell, Oxford/London

Leveau JHJ, Lindow SE (2001) Appetite of an epiphyte: quantitative monitoring of bacterial sugar consumption in the phyllosphere. Proc Natl Acad Sci U S A 98:3446–3453

Li F, Ding SW (2006) Virus counterdefense: diverse strategies for evading the RNA-silencing immunity. Annu Rev Microbiol 60:503–531

Lievens B, Brouwer M, Vanachter ACRC, Cammue BPA, Thomma BPHJ (2006) Real-time PCR for detection and quantification of fungal and oomycete tomato pathogens in plant and soil samples. Plant Sci 171:155–165

Lindgren PB (1997) The role of *hrp* genes during plant-bacterial interactions. Annu Rev Phytopathol 35:129–152

López MM, Llop P, Cubero J, Penyalver R, Caruso P, Bertolini E, Penalver J, Gorris MT, Cambra M (2001) Strategies for improving serological and molecular detection of plant pathogenic bacteria. In: Plant pathogenic bacteria. Springer, Cham, pp 83–86

Lopez MM, Bertolini E, Olmos A, Caruso P, Gorris MT, Llop P, Penyalver R, Cambra M (2003) Innovative tools for the detection of plant pathogenic viruses and bacteria. Int Microbiol 6:233–243

Loubens I, Debarbieux L, Bohin A, Lacroix JM, Bohin JP (1993) Homology between a genetic locus (mdoA) involved in the osmoregulated biosynthesis of periplasmic glucans in *Escherichia coli* and a genetic locus (hrpM) controlling pathogenicity of *Pseudomonas syringae*. Mol Microbiol 10:329–340

Lynch JM, Whipps JM (1991) In: Keister DL, Cregan B (eds) Substrate flow in the rhizosphere, in the rhizosphere and plant growth. Kluwer Academic, Dordrecht, pp 15–24

Madigan TM, Martinko MJ, Stahl AD, Clark PD (2012) Brock biology of microorganisms, 13th edn. Pearson, London

Mandahar CL (2006) Multiplication of RNA plant viruses. Springer, Dordrecht

Mano H, Tanaka F, Nakamura C, Kaga H, Morisaki H (2007) Culturable endophytic bacterial flora of the maturing leaves and roots of rice plants (*Oryza sativa*) cultivated in a paddy soil. Microbes Environ 22:175–185

Marschner H (1995) Mineral nutrition of higher plants. Ann Bot 78:527–528

Mendes R, Kruijt M, Bruijn I, Dekkers E, Voort M, Schneider JHM, Piceno YM, Desantis TZ, Andersen GL, Bakker PAHM, Raaijmakers JM (2011) Deciphering the rhizosphere microbiome for disease-suppressive bacteria. Science 332:1097–1100

Minsavage GV, Mudgett MB, Stall RE, Jones JB (2004) Importance of opgHXcv of *Xanthomonas campestris* pv. *vesicatoria* in host parasite interactions. Mol Plant-Microbe Interact 17:152–161

Mirzaei S, Goltapeh EM, Shams-Bakhsh M, Safaie N (2008) Identification of *Botrytis* spp. on plants grown in Iran. J Phytopathol 156:21–28

Mitchell R, Gu J (2010) Environmental microbiology, 2nd edn. Wiley-Blackwell, Hoboken, NJ

Montesines E (2000) Pathogenic plant-microbial interactions. What we know and how we benefit? (editorial). Intl Microbiol 3:69–70

Mori Y, Nagamine K, Tomita N, Notomi T (2001) Detection of loopmediated isothermal amplification reaction by turbidity derived from magnesium pyrophosphate formation. Biochem Biophys Res Commun 289:150–154

Nakahara KS, Masuta C (2014) Interaction between viral RNA silencing suppressors and host factors in plant immunity. Curr Opin Plant Biol 20:88–95

Nannipieri P, Ascher J, Ceccherini MT, Landi L, Pietramellara G, Renella G (2007) Microbial diversity and microbial activity in the rhizosphere. Ciencia del Suelo (Argentina) 25:89–97

Nassuth A, Pollari E, Helmeczy K, Stewart S, Kofalvi SA (2000) Improved RNA extraction and one tube RT-PCR assay for simultaneous detection of control plant RNA plus several viruses in plant extracts. J Virol Methods 90:37–49

Nelson PE, Toussoun TA, Marasas WFO (1983) *Fusarium* species: an illustrated manual for identification. Pennsylvania State University Press, University Park, PA

Nezhad AS (2014) Microfluidic platforms for plant cells studies. Lab Chip 14:3262–3274

Ogle HJ, Brown JF (eds) (1997) Plant pathogens and plant diseases. Rockvale Publications, Berringer, NSW

Oliver RP, Farman ML, Jones JDG, Harmon-Kosack KE (1993) Use of fungal transformants expressing ß-glucuronidase activity to detect infection and measure hyphal biomass in infected plant tissues. Mol Plant-Microbe Interact 6:521–525

Olmos A, Bertolini E, Gil M, Cambra M (2005) Real-time assay for quantitative detection of non persistently transmitted Plumpox virus RNA targets in a single aphids. J Virol Methods 128:151–155

Osiowy C (1998) Direct detection of respiratory syncytial virus, parainfluenza virus, and adenovirus in clinical respiratory specimens by a multiplex reverse transcription-PCR assay. J Clin Microbiol 36:3149–3154

Pallisgaard N, Hokland P, Riishoj DC, Pedersen B, Jørgensen P (1998) Multiplex reverse transcription polymerase chain reaction for simultaneous screening of 29 translocations and chromosomal aberrations in acute leukemia. Blood 92:574–588

Pan SQ, Ye XS, Kuc J (1991) A technique for detection of chitinase ß-1,3-glucanase and protein patterns after a single separation using polyacrylamide gel electrophoresis or isoelectrofocusing. Phytopathology 81:970–974

Panke-Buisse K, Poole AC, Goodrich JK, Ley RE, Kao-Kniffin J (2015) Selection on soil microbiomes reveals reproducible impacts on plant function. ISME J 9:980–989

Parashar A, Pandey S (2011) Plant-in-chip: microfluidic system for studying root growth and pathogenic interactions in *Arabidopsis*. Appl Phys Lett 98:263703

Parniske M (2008) Arbuscular mycorrhiza: the mother of plant root endosymbioses. Nat Rev Microbiol 6:763–775

Plantiño-Álvarez B, Rodríguez-Cámara MC, Rodríguez Fernandez T, González-Jaen MT, Vázquez Estévez C (1999) Immunodetection of an exo-polygalacturonase in tomato plants infected with *Fusarium oxysporum* f.sp. *radicis-lycopersici*. Bol Sandad Vegetal Plagas 25:529–536

Prasad R, Bhola D, Akdi K, Cruz C, Sairam KVSS, Tuteja N, Varma A (2017) Introduction to mycorrhiza: historical development. In: Varma A, Prasad R, Tuteja N (eds) Mycorrhiza. Springer, Cham, pp 1–7

Pruss G, Ge X, Shi XM, Carrington JC, Bowman Vance V (1997) Plant viral synergism: the potyviral genome encodes a broadrange pathogenicity enhancer that transactivates replication of heterologous viruses. Plant Cell 9:859–868

Purcell AH, Hopkins DL (1996) Fastidious xylem-limited bacterial plant pathogens. Annu Rev Phytopathol 34:131–151

Reinhold-Hurek B, Bünger W, Burbano CS, Sabale M, Hurek T (2015) Roots shaping their microbiome: global hotspots for microbial activity. Annu Rev Phytopathol 53:403–424

Remus-Emsermann MNP, Lücker S, Müller DB, Potthoff E, Daims H, Vorholt JA (2014) Spatial distribution analyses of natural phyllosphere-colonizing bacteria on *Arabidopsis thaliana* revealed by fluorescence in situ hybridization. Environ Microbiol 16:2329–2340

Remy W, Taylor TN, Hass H, Kerp H (1994) Four hundred-million-year-old vesicular arbuscular mycorrhizae. Proc Natl Acad Sci U S A 91:11841–11843

Rey T, Schornack S (2013) Interactions of beneficial and detrimental root-colonizing filamentous microbes with plant hosts. Genome Biol 14:121

Rittenburg JH, Petersen FP, Grothaus GD, Miller SA (1988) Development of a rapid, field-usable immunoassay format, for detection and quantitation of *Pythium*, *Rhizoctonia* and *Sclerotinia* spp. in plant tissue. Phytopathology 78:1516

Rodriguez EE, Ghoul H, Mundy J, Petersen M (2015) Making sense of plant autoimmunity and 'negative regulators'. FEBS J 283(8):1385–1391

Russ JC (1999) The image processing handbook, 3rd edn. CRC, Boca Raton, FL, 771 p

Saito A, Ikeda S, Ezura H, Minamisawa K (2007) Microbial community analysis of the phytosphere using culture-independent methodologies. Microbes Environ 22:93–105

Savage SD, Sall MA (1981) Radioimmunosorbent assay for *Botrytis cinerea*. Phytopathology 71:411–415

Schaad NW, Frederick RD (2002) Real-time PCR and its application for rapid plant disease diagnostics. Can J Plant Pathol 24:250–258

Shapiro OH, Kramarsky-Winter E, Gavish AR, Stocker R, Vardi A (2016) A coral-on-a-chip microfluidic platform enabling live-imaging microscopy of reef-building corals. Nat Commun 7:10860

Shiekh MA (2014) Plant defense and the counter defense by viruses. Global J Res Rev 1:37–44

Simpson AJG, Reinach FC, Arruda P, Abreu FA, Acencio M, Alvarenga R, Alves LM, Araya JE et al (2000) The genome sequence of the plant pathogen *Xylella fastidiosa*. The *Xylella fastidiosa* consortium of the organization for nucleotide sequencing and analysis. Nature 406:151–159

Smith SE, Read DJ (2008) Mycorrhizal symbiosis, 3rd edn. Academic, New York

Smolka MB, Martins D, Winck FV, Santoro CE, Castellari RR, Ferrari F, Brum IJ, Galembeck E (2003) Proteome analysis of the plant pathogen *Xylella fastidiosa* reveals major cellular and extracellular proteins and a peculiar codon bias distribution. Proteomics 3:224–237

Soosaar JLM, Burch-Smith TM, Dinesh-Kumar SP (2005) Mechanisms of plant resistance to viruses. Nat Rev Microbiol 3:789–798

Spaink HP, Kondorosi A, Hooykaas PJJ (1998) The rhizobiaeceae. Kluwer Academic, Dordrecht

Sundaram S, Plascencia S, Banttari EE (1991) Enzyme-linked immunosorbent assay for detection of *Verticillium* spp. using antisera produced to *V. dahliae* from potato. Phytopathology 81:1485–1489

Syller J (2012) Facilitative and antagonistic interactions between plant viruses in mixed infections. Mol Plant Pathol 13:204–216

Talaro KP, Chess B (2012) Foundations in microbiology, 8th edn. McGraw-Hill, New York

Tempel A (1959) Serological investigations in *Fusarium oxysporum*. Meded No. 138 Lanbouwhogeschool, Wageningen 59: 1–60

Valli A, Lopez-Moya JJ, Garcia JA (2009) RNA silencing and its suppressors in the plant-virus interplay. Wiley, Chichester

VanEtten HD, Mansfield JW, Bailey JA, Farmer EE (1994) Two classes of plant antibiotics: phytoalexins versus "phytoanticipins". Plant Cell 6:1191–1192

Varga A, James D (2006) Use of reverse transcription loop-mediated isothermal amplification for the detection of plum pox virus. J Virol Methods 138:184–190

Vorholt JA (2012) Microbial life in the phyllosphere. Nat Rev Microbiol 10:828–840

Wang B, Qiu YL (2006) Phylogenetic distribution and evolution of mycorrhizas in land plants. Mycorrhiza 16:299–363

Watt SA, Wilke A, Patschkowski T, Niehaus K (2005) Comprehensive analysis of the extracellular proteins from *Xanthomonas campestris* pv. *campestris* B100. Proteomics 5:153–167

Whipps JM, Hand P, Pink D, Bending GD (2008) Phyllosphere microbiology with special reference to diversity and plant genotype. J Appl Microbiol 105:1744–1755

Willey MJ, Sherwood ML, Woolverton JC (2009) Prescott's principles of microbiology, 1st edn. Mc-Graw-Hill, New York

Williams K, Blake S, Sweeney A, Singer JT, Nicholson BL (1999) Multiplex reverse transcriptase PCR assay for simultaneous detection of three fish viruses. J Clin Microbiol 37:4139–4141

Wolf S, Lucas JW, Deom CM, Beachy RN (1989) Movement protein of tobacco mosaic virus modifies plasmodesmatal size exclusion limit. Science 246:377–379

Chapter 14
Anthosphere Microbiome and Their Associated Interactions at the Aromatic Interface

Nagarathinam Arunkumar, Suchitra Rakesh, Kaushik Rajaram, Narayanasamy Ravi Kumar, and Siva Sundara Kumar Durairajan

Abstract The American quote "snug as a bug in a rug" (which means very comfortable and everyone has their own tastes) fits perfectly for the relation between plant and microbes with their associated interactions. Microbes interact at anthosphere, caulosphere, carposphere, phyllophane, rhizosphere, and spermosphere regions of the plants, and the plant-microbe interface acts as a medium of communication between these two diversified living systems. The interface is influenced by an extensive variety of biotic and abiotic determinants responsible for shaping plant-associated habitats, considerably modifying the active composition of the microbial communities, which alter themselves according to the environment for beneficial interactions. The microbiome of root and leaf interactions is most studied as evident from the availability of humongous literature; however, even a small microhabitat such as the anthosphere has its own group of associated microbes obtained from autochthonous or allochthonous. In addition, these microhabitats are contiguous with mutualistic pollinators, florivores, and nectar robbers, which alter the dynamic microbial inhabitants of these aromatic interfaces. To attain sustainability in plant conservation, food, and agriculture, an in-depth understanding of the entire plant-microbe environment is crucial. This chapter was written to provide an overview of the different interfaces, in particular, the anthosphere region of the phyllosphere.

14.1 Introduction to the Plant-Microbe Interface

The emergence, structural formation, and development of biological systems depend on complex signal exchanges between the systems over space and time (Adam et al. 2018). Plant-microbe interface (PMI) is a point where the two diversified living systems, plants and microbes, meet or more specifically the dynamic environment in

N. Arunkumar · S. Rakesh · K. Rajaram · N. R. Kumar · S. S. K. Durairajan (✉)
Department of Microbiology, School of Basic and Applied Sciences, Central University of Tamil Nadu, Thiruvarur, India
e-mail: d.sivasundarakumar@cutn.ac.in

A. Varma et al. (eds.), *Plant Microbe Interface*,
https://doi.org/10.1007/978-3-030-19831-2_14

which plants and microbes interact. A detailed signal exchange exists between the microbe and plant even before they engage in any physical contact, and PMI acts as a medium of long-distance chemical communication. The knowledge on inherent molecular, chemical, and physical processes occurring at the interface has facilitated the understanding of diversified microbe-plant interactions. The interface is considerable because microorganisms have evolved over time for possible microbial interactions based on the environment (Felestrino et al. 2017). Colonizing microorganisms have obtained most of their housekeeping genes from plants through horizontal gene transfer, and the most promising explanation for this evolution is to form a successful plant-microbe interaction (Kaneko et al. 2010) and hence gaining the importance toward PMI studies. The interface mainly occurs in the rhizosphere (a narrow zone influenced by plant roots), spermosphere (a zone or habitat surrounding the seeds where the soil, germinating seeds, and the microbial communities interact), phyllosphere (the total aerial aboveground plant surfaces, particularly the leaves), and anthosphere (an adjacent zone around the flowers, a subdivision of phyllosphere) (Fürnkranz et al. 2012; Remus-Emsermann and Schlechter 2018).

The interface is influenced by an extensive variety of biotic (Table 14.1) and abiotic (Fig. 14.1) factors responsible for shaping plant-associated habitats, considerably modifying the active composition of the microbial communities, which alter themselves according to the environment for beneficial interactions (Aleklett et al. 2014; Ushio et al. 2015; Schiltz et al. 2015; Santoyo et al. 2017; Bumroongsook 2018). The interfaces may remain disconnected or sometimes interconnected to each other and can also considerably influence the interactions of microbial communities. The below- and aboveground parts of plants create specific habitats for various microbial communities without interference, which was confirmed using a metaproteogenomic approach on the phyllosphere and rhizosphere microbiota of domestic cultivars (Knief et al. 2012). Substantial research has proved that variance in environmental factors across any plant surfaces can affect the distribution of microbial communities around the other plant parts (Aleklett et al. 2014), thereby influencing the plant ecosystem functions under different environmental conditions (Kembel et al. 2014). On the basis of the pertaining studies on these interaction habitats, this chapter aimed to provide an overview of the different interfaces, in particular, the anthosphere region of the phyllosphere.

14.2 Rhizosphere Interface

The immediate thin layer of soil surrounding plant roots is the rhizosphere and rhizoplane (the zone on the root surface) (Shrivastava et al. 2014; Prasad et al. 2015). Microorganisms attach to these rhizo-zones for possible interactions using the special appendages (fimbriae and flagella) and secretions (surface polysaccharides) (Mwajita et al. 2013). This continuum (layers separated by an extremely slim boundary) of rhizosphere and rhizoplane layers is extensively studied because of

Table 14.1 Biotic factors governing the plant-microbe interactions at the plant-microbe interface

S. no.	Interface	Role player	Factors involved	References
1.	Rhizosphere	Plant	Root length/density/depth, root exudates, foliar leaching, MAMP (microbe-associated molecular pattern), chemoattractants, carbon sources, defense metabolites, and enzyme secretion	Mommer et al. (2016), Garcia and Kao-Kniffin (2018)
		Microbes	Quorum-sensing-involved biofilm formation, phytohormone biosynthesis, virulence of pathogenic bacteria, production of antimicrobial compounds, microbial effectors, soil, and the rhizosphere microbiome	Gianfreda (2015), Santoyo et al. (2017)
2.	Phyllosphere	Plant	Phenological stage of the plant, plant phenotype and genotype, biochemical secretions, leaf characteristics, leaf food resources, phytohormones, green leaf volatiles, and plant traits	Whipps et al. (2008), Kembel et al. (2014), Hacquard et al. (2017)
		Microbes	PPFM (pink pigmented facultative methylotrops) characteristics, microbial fitness (surfactants and extracellular polysaccharides), metabolic response, phyllosphere, and allochthonous microbiome through insect-, atmosphere-, seed-, or even animal-borne sources	El-Gawad et al. (2015), Remus-Emsermann and Schlechter (2018)
3.	Spermosphere	Plant	Seed exudations (nature and composition), seed genotype, seed carbon deposition, germination pattern, seed-borne pathogens, host-dependent microbiome	Schiltz et al. (2015), Chohan et al. (2017)
		Microbes	Chemotaxis, tropic and signal-mediated interactions, evolutionary traits for colonization, and spermosphere-dependent microbiome	Tian et al. (2015), Lemanceau et al. (2017)

its importance in root activity and metabolism. A diversified group of microorganisms co-occurs and multiplies in the rhizosphere where bacteria are abundant (Saharan and Nehra 2011). The rhizosphere interface is pooled with biochemical secretions backed by molecular pattern mechanisms for possible interaction with the soil microorganisms. Root-soil interface is constantly exposed to a vast array of stresses, and the interface responds to these abiotic and biotic stresses by secreting an admixture of root exudates to enhance positive interactions and protect the interface

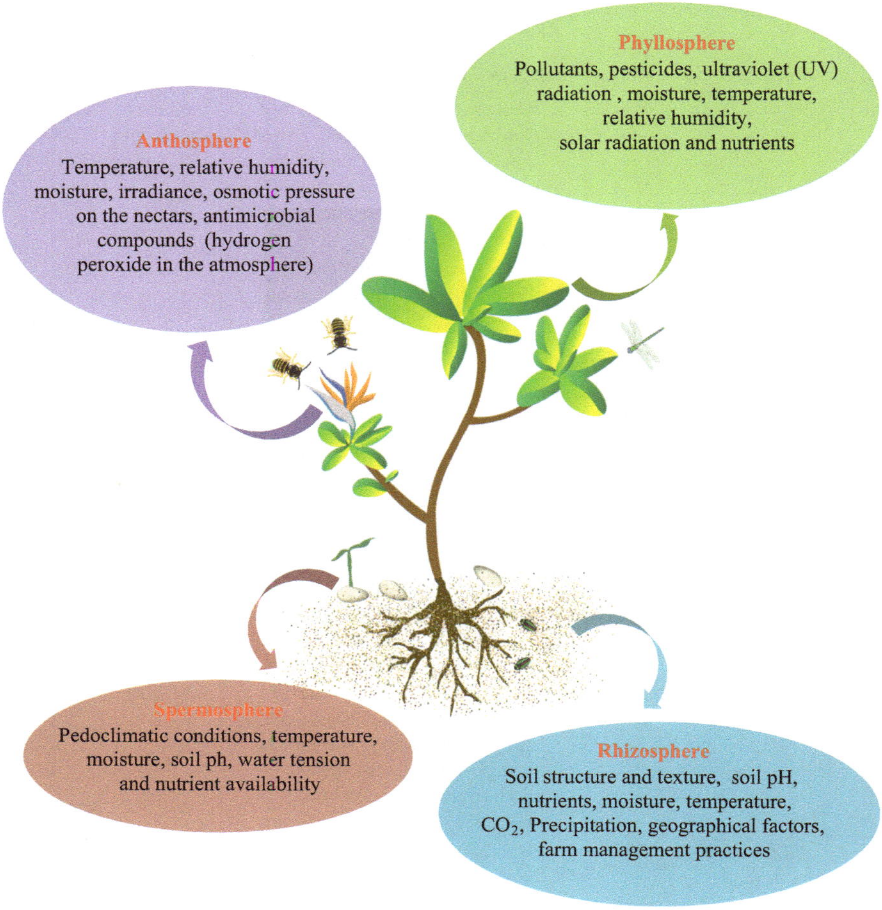

Fig. 14.1 Major abiotic factors affecting plant-microbe interactions at the plant-microbe interface

against harmful negative influences (Badri and Vivanco 2009). In addition, abiotic factors such as light, soil structure or texture, temperature, soil moisture, and soil pH highly modulate root exudation (a biotic factor), which in turn modulates the microbial interactions. For example, neutral soils tend to show a greater microbial diversity by providing an environment for controlled root exudation, whereas acidic soils show lower diversity indices (Rousk et al. 2010). Likewise, the genetic diversity of nitrogen-fixing rhizobacterium is influenced by soil type and other geographical factors as reported by Santoyo et al. (2017). The bioavailability of soil nutrients at the rhizosphere interface has both direct (toxic effects) and indirect (plant exudates) effects on the abundance and diversity of the rhizosphere microbiome by reducing the possibility of interactions (Berendsen et al. 2012).

The biotic factors include the secretion of root exudates, enzymes for metabolism, antimicrobial compounds, phytotoxic chemicals, and processes including molecular plant-pathogen detection at the interface (Field et al. 2006). At the plant-pathogen interface, either at rhizo- or phyllosphere, the first line of defense is the secretion of biochemical compounds (antimicrobial or phytotoxic compounds), failure of which leads to the pathogen-associated molecular pattern (PAMP)-triggered immunity (PTI). Plants can identify microbe-associated molecular patterns (MAMPs) or PAMPs through the effectors, which are pathogen-specific signatures. Plants maintain a complex genetic system for recognizing effectors (flagella proteins, elongation factor Tu, peptidoglycan, and lipopolysaccharides) as signals of invasion leading to PTI either at the interface or initial interaction (Newman et al. 2013). However, in case of positive beneficial interactions, microbial effectors form a biotrophic interface complex, which acts as suppressors of plant defense machinery (Pellegrin et al. 2015). In addition, a combined effect of living and nonliving factors is reported; for example, the biomolecules and clay minerals form natural nanoprecipitates at the soil-root interface, acting as an active microsite for favorable root-microbe interactions (Violante and Caporale 2015).

14.3 Spermosphere Interface

The spermosphere represents the soil adjacent to a germinating seed: a habitat with a short span but a microbiologically dynamic and rapidly changing zone. When the germination starts, various carbon compounds are released as seed exudations (fatty acids, carbohydrates, amino acids, and organic acids) into the soil. These exudations alter and control the microbial activities that occur in the spermosphere interface, and also these microbial interactions can continue for a short time or the whole life cycle of the plant (Nelson 2004). This interconnection may occur during the seed development in the fruit, dormancy, or germination, and these interrelations are rather planted species specific and microorganism specific. These interrelations may be casual and nonspecific, but in most cases, they are beneficial (root nodulation) or pathogenic (seed-borne fungal diseases) (Chohan et al. 2017). The dynamic associations between plant-microbe interactions around a germinating seed are governed by certain extrinsic (temperature, moisture, and biotic habitat of the soil) and intrinsic factors (plant genotype and phenotype) prevailing at the interface (Simon et al. 2001). Schiltz et al. (2015) reported that the nature and quantity of the seed exudates are dependent on the plant species and abiotic factors such as soil pH, temperature, and type, thereby indirectly influencing the microbial community of the spermosphere.

Microbial attachment to seeds is mainly through biofilm formation or by bacterial adhesins and is regulated by quorum sensing (Tian et al. 2015). Biofilm formations by the colonizers provide resistance to various antimicrobial compounds produced by the seed during germination (Nelson 2004). In the spermosphere, the hydration state of the seed during germination acts as a leading factor behind plant-microbe interactions. In addition, seed carbon deposition and germination of dicot (epigeal germination) and monocot seeds (i.e., hypogeal germination) influence the microbial behavior in the seed habitat (Lemanceau et al. 2017). Only microorganisms with specific traits (trophic and signal-mediated interactions) can succeed in colonizing germinating seeds because of the high competition prevailing within the spermosphere for resources and space. These competitive trophic interactions include the judicious consumption of available resources, chemotaxis motility toward seed exudates (amino acids and organics acids), and versatile metabolic potential to colonize and sustain spermosphere competence (Barret et al. 2015); spermosphere bacterial taxa including *Pseudomonas*, *Bacillus*, and *Rhizobium* with specific evolutionary traits for colonization were found to be abundant in the spermosphere during germination (Lemanceau et al. 2017). Hence, the biochemical secretions of the seeds triggered by their internal genomic and external environmental factors coupled with the microbial evolutionary characters at the interface confirm the nature and survival of the interactions (Truyens et al. 2015).

14.4 Phyllosphere Interface

The total aerial aboveground plant surfaces or phyllosphere is an ubiquitous environment for harboring diversified microscopic living organisms. In agricultural and native plants, phyllosphere is dominated by the bacterial members of the phyla *Firmicutes*, *Proteobacteria*, and *Actinobacteria* (Williams and Marco 2014). The quantitative microbial proposition of each taxon may vary depending on the geographical location, plant phenotype and genotype, human intervention, and seasonal variation (Maignien et al. 2014), which make the phyllosphere interface crucial for qualitative and quantitative plant-microbe interactions. Lettuce plants grown indoors confirmed that the leaf microbiota of laboratory-grown plants is distinct and low compared with that of field-grown plants (Williams and Marco 2014). In addition, similar studies on indoor-grown plants have reported low cell numbers with little diversity under laboratory conditions, confirming the intervention of natural factors before and after interaction (Reisberg et al. 2012). Recent genomics and metabolomics studies have characterized the interaction and survival of phyllosphere microbial communities with regard to the ecological, utilitarian, and structural properties of host plants and environment properties, such as changing climate dynamics and composition of trace gases in the surrounding atmosphere (Bringel and Couée 2015). Functional plant traits such as plant stature, leaf dry matter content, leaf mass per area, height and diameter, wood density, relative growth, and mortality rate have a functional role in plant-microbe interactions (Kembel et al. 2014).

Green leaf volatiles, that is, small organic compounds such as methane to higher complex compounds (fatty acid derivative and sesquiterpenes), are formed at the interface for the recruitment of microbes (Matsui 2006). Pink-pigmented facultative methylotrophic bacteria are capable of growing on small carbon compounds such as formate, formaldehyde, and methanol, which are constantly available on the leaf surface, which in turn renders systemic resistance against diseases, produce plant growth regulators, and offer drought resistance to young leaves (El-Gawad et al. 2015). Field and greenhouse experiments performed by plant pathologists confirmed that fungal species found on a leaf, such as *Cladosporium*, are potential antagonist against *Septoria nodorum*, *Alternaria zinniae*, *Cochliobolus sativus*, and *Botrytis cinerea* (Rodríguez et al. 2001). Certain green leaf volatile organic compounds (VOCs) have growth-inhibiting effects on some microbial strains; however, leaf-colonizing bacteria have developed adaptation mechanisms for their survival. Likewise, some coniferous species prevent the establishment of airborne bacteria on needles through volatile compounds, and the denseness of microbes in the surrounding atmosphere of the conifer stands was considerably reduced by VOCs in addition to some epiphytic communities (Gao et al. 2005). Similarly, in order to colonize the plant internal tissues, plant growth-promoting bacterial endophytes have developed a complex genome compared with other microbes; however, the molecular basis of endophytic microbes to overcome the plant defense is still not well understood, but it is well confirmed that the sensing of MAMPs by pattern recognition receptors in plants control the endophytic load on the leaves (Hacquard et al. 2017).

14.5 Anthosphere Interface

The aboveground portions of phyllosphere include the carposphere (fruits), anthosphere (flowers), phyllophane (surface of leaves), and caulosphere (stems), which withhold several peculiar microbial life interactions. In particular, the anthosphere region around flowers is colonized by a vast diversity of microorganisms, which are flower specific; however, some members of the genera *Pseudomonas* and *Acinetobacter* (*Proteobacteria*), *Metschnikowia* (*Ascomycota*), and *Cryptococcus* (*Basidiomycota*) are consistent members of the floral microbiome across many agricultural and ornamental plants. Another most notable feature of this interface is the permanence; this habitat has a shorter life span when compared with other spheres. The studies on apple floral microbiome confirmed the presence of fast-growing bacterial communities (Shade et al. 2013). Novel studies on flower-associated microbial communities have highlighted that the fungal population is the highest in the anthosphere, followed by bacteria (Ushio et al. 2015). The floral components, namely, pollen, nectar, sepals, petals, stamens, style, ovary, and stigmas, act as short-span microsites for the colonization of microorganisms (Fridman et al. 2012). As flowers are dependent on other biotic and abiotic factors to assist in pollination (for fertilization), seed dispersal, and germination, they have every opportunity for inhabitation by harmful, beneficial, and commensal microorganisms. The horizontal transmission of the floral microbiome occurs through the wind in case of wind-pollinated species and mostly through the frequent pollinators in non-wind-pollinated plants (Frank et al. 2017). In addition, anthosphere plays an important role in the biographic interactions among other interfaces (Huisman et al. 2015); hence, the knowledge on possible environmental factors, biochemical compounds, and molecular patterns formed at this interface is of a great importance.

14.5.1 Anthosphere Microbiome and Their Composition

Advanced sequencing techniques offer in-depth knowledge on the composition and diversity of microbial communities in the anthosphere, and these techniques circumvent both laboratory-grown and non-laboratory-grown bacterial species and provide accurate identification up to the genera level (Samuni-Blank et al. 2014). Flowers are always considered whole structures; however, the interactions at this interface are also more organ-specific because the microbiome diversity varies with the floral components. For example, studies on culturable yeast species across different floral parts (nectar, pollen, and inner and outer corollae) found that the floral surface organs were abundant with basidiomycetous yeasts, whereas nectar and pollen were filled with ascomycetous yeast species and nectar was found to host most fungi compared with other floral parts (Pozo et al. 2012), indicating a far down intense separation in the microbial world within the flower.

In pollen, the cultivable bacterial count is abundant, ranging between 10^6 and 10^9, and the diversity and composition vary from species to species because of the difference in nutrient composition, pollen viability, pollen structure, pollen coat antimicrobial peptides, moisture, and the special attachment of the bacterial cells (Frank et al. 2017). The epiphytic bacteria in the anthosphere exist either single or in clusters with the special formation of thin biofilms in certain habitats. The anthosphere microbiome of various plants at different floral components is listed in Table 14.2. Anthosphere interactions are, however, restricted to epiphytic microbes, but novel and biologically active endophytic microbes with potential sources of useful metabolites were also documented among the floral microbiome. Therapeutic metabolites producing endophytic fungi identified as *Pestalotiopsis disseminate*, *Phomopsis* sp., and *Coelomycete* sp. were reported from *Tripterygium wilfordii* flowers, a traditional Chinese medicinal plant that proves the role of endophytes in this unique interface (Kumar and Hyde 2004).

14.5.2 Abiotic and Biotic Determinants of Microbial Colonization

In any living system, both biotic and abiotic factors work in combination to alter species distributions and abundance (Bumroongsook 2018). Compared with other phyllosphere components, floral surfaces provide some unique conditions, such as elevated levels of humidity and moisture, increased irradiance, low pH and high alcohol concentrations (fermentation of nectar sugars by microbes), osmotic pressure, antimicrobial compounds, and some protection against extreme weather conditions similar to some leaf structures (Aleklett et al. 2014). Flowers are exposed to various abiotic stresses such as seasonal variations, geographical factors, rainfall, temperature, and humidity, which alter the microclimate around the flowers (Vega and Marques 2015), directly affecting the flower longevity, pollen viability, and

Table 14.2 List of anthosphere microorganisms of various domestic and wild plants at different floral components

S. no.	Plant host	Flower part	Microbial group	Most abundant microbes	References
1.	*Epilobium canum*	Nectar	Bacteria	*Neokomagataea* sp.	Rering et al. (2017)
2.	*Mimulus aurantiacus*	Nectar	Bacteria	*Asaia astilbes*	Rering et al. (2017)
3.	*Linaria vulgaris*	Nectar	Yeast	*Metschnikowia reukaufii*	Bartlewicz et al. (2016)
4.	*Linaria vulgaris*	Nectar	Bacteria	*Acinetobacter nectaris*	Bartlewicz et al. (2016)
5.	*Malus domestica*	Flower	Archaea	*Deinococcus* sp.	Shade et al. (2013)
6.	*Atropa baetica*	Nectar	Fungi	*Coniochaeta* sp.	Pozo et al. (2012)
7.	*Digitalis obscura*	Pollen	Yeast	*Metschnikowia* sp.	Pozo et al. (2012)
8.	*Amygdalus communis*	Nectar	Bacteria	*Phaseolibacter* sp.	Aizenberg-Gershtein et al. (2013)
9.	*Phleum* sp.	Pollen	Fungi	*Botrytis* sp.	Heydenreich et al. (2012)
10.	*Pulmonaria officinalis*	Nectar	Bacteria	*Rhodococcus* sp.	Jacquemyn et al. (2013)
11.	*Eugeissona tristis*	Nectar	Fungi	*Trichomonascus* sp.	Wiens et al. (2008)
12.	*Atropa baetica*	Nectar	Fungi	*Coniochaeta* sp.	Pozo et al. (2012)
13.	*Helianthus annuus*	Flower	Fungi	*Sclerotinia sclerotiorum*	Rodríguez et al. (2001)
14.	*Epipactis palustris*	Nectar	Bacteria	*Rosenbergiella nectarea*	Lenaerts et al. (2014)
15.	*Protea subvestita*	Nectar	Bacteria	*Tatumella citrea*	Lenaerts et al. (2014)
16.	*Delphinium nuttallianum*	Nectar	Yeast	*Metschnikowia reukaufii*	Schaeffer and Irwin (2014)
17.	*Silene latifolia*	Nectar	Yeast	*Microbotryum violaceum*	Golonka and Vilgalys (2013)
18.	*Mimulus aurantiacus*	Nectar	Bacteria	*Gluconobacter* sp.	Vannette et al. (2013)
19.	*Nicotiana glauca*	Nectar	Bacteria	*Erwinia amylovora*	Fridman et al. (2012)
20.	*Citrus paradisi*	Nectar	Bacteria	*Acinetobacter gerneri*	Fridman et al. (2012)
21.	*Echium leucophaeum*	Nectar	Yeast	*Cryptococcus carnescens*	Mittelbach et al. (2015)
22.	*Helleborus foetidus*	Nectar	Yeast	*Metschnikowia reukaufii*	Pozo et al. (2014)
23.	*Epilobium canum*	Nectar	Endo yeast	*Aureobasidium pullulans*	Rering et al. (2017)
24.	*Iris xiphium*	Nectar	Bacteria	*Rosenbergiella australoborealis*	Álvarez-Pérez and Herrera (2013)

nectar viscosity in turn indirectly modifying the flower-inhabiting microbes. The studies on urban and rural habitat-grown *Linaria vulgaris* (yellow toadflax-late-flowering herb) suggested that environmental changes related to urbanization (landscapes, pollution, and special isolation) may impact inhabiting yeasts in the floral nectar of plants (Bartlewicz et al. 2016). Likewise, the role of rainfall in indirectly shaping the microbial community is evident. The studies on Hindu lotus and East Indian lotus flowers revealed that the thrips population was more abundant in summer than in the rainy season as insect vectors play a major role in the microbiome alteration of the flowers (Bumroongsook 2018). The specific interaction studies on the flowers of the *Rosaceae* family with the blight pathogen, *Erwinia amylovora*, confirmed that a specific combination of relative humidity and temperature on flower surfaces can exert extremely strong specific pressure on the flower microbiome (Aleklett et al. 2014).

The microbial model system studies on bacterial communities confirmed that the floral microbes are shaped not only by the abiotic factors in the flower niche but also by the plant genotype and phenotype, nectar allure, pollen surface structure, volatile and nonvolatile organic compounds, microbe-microbe interactions, temporal dynamics of flower communities, visiting pollinators, and insect vectors (Samuni-Blank et al. 2014). The high levels of hydrogen peroxide generated by the nectar proteins (nectarins) and the saponins, alkaloids, terpenoids, and phenolics of the flower keep the colonizing microbes under control (Kessler and Baldwin 2007). In addition, it has been suggested that pollen odors provide defense against pathogens (Basim et al. 2006). The greenhouse and field experiments performed on the fungal pathogens of *Helianthus annuus* (sunflower) anthosphere have confirmed that pollens reduce injury by pathogens, increase the colonization of beneficial microbes, and provide additional protection from pathogens because of competition toward limited anthosphere resource (Rodríguez et al. 2001). In addition, biochemical studies showed a marked difference between the secondary metabolite profiles of the anthosphere and phyllosphere through diverse biochemical pathways such as aliphatics, aromatics, and terpenoids; the flowers maintain distinct organic metabolites in their environment (Knudsen et al. 2006). Nonvolatile metabolites such as proteins, alkaloids and phenolics, and volatile organic compounds such as phenylacetonitrile, 2-phenylethyl alcohol, and sesquiterpene synthesized by the flowers vary among species and even among the flowers of the same plant. Most of these volatile and nonvolatile compounds have specific antimicrobial properties through which the recruitment of microorganisms is thoroughly filtered, and they also offer protection to the nectar-inhabiting indigenous microbes (Junker and Tholl 2013).

14.5.3 Plant-Microbe-Pollinator Triangle

Nonpollinating floral herbivores (florivores) and pollinators frequently visit the flowers (de Vega and Herrera 2013), and these visitors carry their own internal

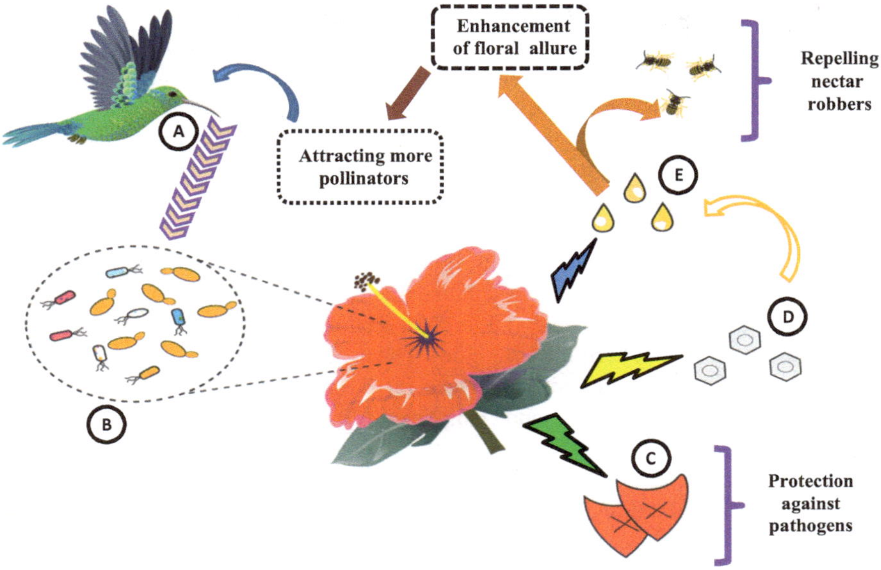

Fig. 14.2 Role of pollinator-transferred microbes in altering the anthosphere; (**a**) pollinators carrying microbes to the flower, (**b**) altering the anthosphere microbiome, (**c**) secretion of antimicrobial compounds, (**d**) altering the secondary metabolites, and (**e**) changing the nectar composition and aroma

and external surface microbiota that could be shared with the anthosphere (Jacquemyn et al. 2013). The microbial load varies with the pollinators and the flowers they visit. The studies through direct visualization and microbial fingerprint approach on different insect pollinators, such as *Bombus ardens ardens* (bumblebees), *Xylocopa appendiculata circumvolans* (carpenter bees), and *Apis cerana japonica* (honeybees), under field conditions suggested that an average of 12.2×10^5 microbial cells were harbored by individual insects on its surface (Ushio et al. 2015). Reports are available showing that, between flowers, the transport of yeast is facilitated by the pollinators (Belisle et al. 2012) and the diversified yeast speciation and their composition in the nectar are because of the outcome of these diversified vectors. Herrera et al. (2009) highlighted the association between *Zygosaccharomyces* and *Debaryomyces* yeasts in plants and bumblebee mutualism (Brysch-Herzberg 2004). In addition, after colonization, microbial communities play their role in affecting the quality and composition of the nectar, which alters nectar attractiveness to active pollinators (Fig. 14.2); hence, the reproductive success of the flower is directly affected by the floral microbiota (de Vega and Herrera 2013).

In several plant species, flower-insect-yeast interactions have been observed, and insect-mediated microbial dispersal and alteration in the microbiome composition of the anthosphere were inseparable. Nectar-inhabiting microorganisms such as the yeast *Metschnikowia reukaufii* sourced by pollinators produce volatile compounds, which differentially affect the preference of honey bee as reported by Rering et al. (2017). A study involving the shrub *Mimulus aurantiacus* and the nectar yeast, *Metschnikowia reukaufii*, confirmed that the nectar attractiveness (for the pollinator humming bird) was enhanced because of the mutualism, whereas the same plant with bacterial genus *Gluconobacter* sp. decreased the nectar attractiveness by altering the sugar composition (Aleklett et al. 2014). The various volatile and nonvolatile metabolites of the flowers either attract or repel the pollinators, thus preventing the entry of pathogens from infected vectors (Junker and Tholl 2013). At present, researchers have started using the help of pollinators to dispense microbial biocontrol agents for sustainable agriculture production. Recently, entomovectoring (the practice of using bees to spread microbial biocontrol agents) is gaining momentum to reduce the use of harmful fungicides. These new techniques can be successful only by maintaining an attractive interface for the pollinators and their associated microbes (Menzler-Hokkanen and Hokkanen 2017). Likewise, insect species-specific microbes can be transferred from the insect body surface to a floral surface, and these insect-specific microbes can act as a fingerprint of the specified insect, particularly for large-bodied insects (Ushio et al. 2015).

14.6 Interactions Among the Interfaces

The interfaces are connected internally through a series of pathways in the plant life cycle, for example, through foliar leaching, residue decomposition, volatilization, and debris incorporation, the phyllosphere alter the rhizosphere habitat leading to changes in the microbial diversity. Through the floral pathway, the microorganisms are transmitted to the seed compartments and seed coats, which in turn enter the spermosphere region and then to the surrounding soil environment (Singh and Mathur 2004). Evidence suggests that the plant-soil interface is often the preferable site for horizontal gene transfer processes from plant to soil microbes as a result of plant biomass decomposition (Heuer and Smalla 2012). Likewise, the rhizosphere microbes also find their way through dust, water droplets, and agricultural equipment into the other interfaces for possible colonization. For example, *Verticillium dahliae*, a broad host-fungal pathogen, infects the flowers of agricultural crops but always maintains its local reservoir inoculums in spinach seed stock and soil (Maruthachalam et al. 2013). The volatiles phenyl acetonitrile and 2-phenylethyl alcohol emitted by floral structures have strong growth-inhibiting effects on phyllosphere bacterial strains (Junker et al. 2011), proving the interconnection between PMI.

14.7 Inference on the Interface

Our knowledge on the interaction of these microbial communities with the plant kingdom is still insufficient; the hidden process and pathways are still a challenge with an exciting new frontier. The knowledge of interpreting these unanswered questions will help to attain sustainability in various fields such as agriculture, genetic conservation, food safety and security, and the development of genetically modified organisms. The PMI is a region of discrete transience with a mixture of various microbial consortia; in addition, the investigation of these models will provide insights into the general ecology and evolution of various species. However, the archaeal and viral communities of the interface should be classified for their possible role in the interactions; likewise, the role of florivores (flower-feeding nonpollinating herbivores) in the flower microbiota is still unclear and needs elaborate research.

References

Adam ZR, Fahrenbach AC, Kacar B, Aono M (2018) Prebiotic geochemical automata at the intersection of radiolytic chemistry, physical complexity, and systems biology. Complexity 2018:1. https://doi.org/10.1155/2018/9376183

Aizenberg-Gershtein Y, Izhaki I, Halpern M (2013) Do honeybees shape the bacterial community composition in floral nectar? PLoS One 8(7):e67556

Aleklett K, Hart M, Shade A (2014) The microbial ecology of flowers: an emerging frontier in phyllosphere research. Botany 92(4):253–266

Álvarez-Pérez S, Herrera CM (2013) Composition, richness and nonrandom assembly of culturable bacterial–microfungal communities in floral nectar of Mediterranean plants. FEMS Microbiol Ecol 83(3):685–699

Badri DV, Vivanco JM (2009) Regulation and function of root exudates. Plant Cell Environ 32 (6):666–681

Barret M, Briand M, Bonneau S, Préveaux A, Valière S, Bouchez O, Jacques MA (2015) Emergence shapes the structure of the seed microbiota. Appl Environ Microbiol 81 (4):1257–1266

Bartlewicz J, Lievens B, Honnay O, Jacquemyn H (2016) Microbial diversity in the floral nectar of *Linaria vulgaris* along an urbanization gradient. BMC Ecol 16(1):18

Basim E, Basim H, Özcan M (2006) Antibacterial activities of Turkish pollen and propolis extracts against plant bacterial pathogens. J Food Eng 77(4):992–996

Belisle M, Peay KG, Fukami T (2012) Flowers as islands: spatial distribution of nectar-inhabiting microfungi among plants of *Mimulus aurantiacus*, a hummingbird-pollinated shrub. Microb Ecol 63(4):711–718

Berendsen RL, Pieterse CM, Bakker PA (2012) The rhizosphere microbiome and plant health. Trends Plant Sci 17(8):478–486

Bringel F, Couée I (2015) Pivotal roles of phyllosphere microorganisms at the interface between plant functioning and atmospheric trace gas dynamics. Front Microbiol 6:486

Brysch-Herzberg M (2004) Ecology of yeasts in plant–bumblebee mutualism in Central Europe. FEMS Microbiol Ecol 50(2):87–100

Bumroongsook S (2018) Abiotic and biotic factors affecting the occurrence of thrips on lotus flowers. Appl Ecol Environ Res 16(3):2827–2836

Chohan S, Perveen R, Abid M, Naqvi AH, Naz S (2017) Management of seed borne fungal diseases of tomato: a review. Pak J Phytopathol 29(1):193–200

de Vega C, Herrera CM (2013) Microorganisms transported by ants induce changes in floral nectar composition of an ant-pollinated plant. Am J Bot 100(4):792–800

El-Gawad HA, Ibrahim MFM, El-Hafez AA, El-Yazied AA (2015) Contribution of pink pigmented facultative methylotrophic bacteria in promoting antioxidant enzymes, growth and yield of snap bean. Synthesis 2:4

Felestrino ÉB, Santiago IF, Freitas LDS, Rosa LH, Ribeiro SP, Moreira LM (2017) Plant growth promoting bacteria associated with *Langsdorffia hypogaea*-rhizosphere-host biological interface: a neglected model of bacterial prospection. Front Microbiol 8:172

Field B, Jordán F, Osbourn A (2006) First encounters–deployment of defence-related natural products by plants. New Phytol 172(2):193–207

Frank AC, Saldierna Guzmán JP, Shay JE (2017) Transmission of bacterial endophytes. Microorganisms 5(4):70

Fridman S, Izhaki I, Gerchman Y, Halpern M (2012) Bacterial communities in floral nectar. Environ Microbiol Rep 4(1):97–104

Fürnkranz M, Lukesch B, Müller H, Huss H, Grube M, Berg G (2012) Microbial diversity inside pumpkins: microhabitat-specific communities display a high antagonistic potential against phytopathogens. Microb Ecol 63(2):418–428

Gao Y, Jin YJ, Li HD, Chen HJ (2005) Volatile organic compounds and their roles in bacteriostasis in five conifer species. J Integr Plant Biol 47(4):499–507

Garcia J, Kao-Kniffin J (2018) Microbial group dynamics in plant rhizospheres and their implications on nutrient cycling. Front Microbiol 9. https://doi.org/10.3389/fmicb.2018.01516

Gianfreda L (2015) Enzymes of importance to rhizosphere processes. J Soil Sci Plant Nutr 15 (2):283–306

Golonka AM, Vilgalys R (2013) Nectar inhabiting yeasts in Virginian populations of *Silene latifolia* (Caryophyllaceae) and coflowering species. Am Midl Nat 169:235–258

Hacquard S, Spaepen S, Garrido-Oter R, Schulze-Lefert P (2017) Interplay between innate immunity and the plant microbiota. Annu Rev Phytopathol 55:565–589

Herrera CM, de Vega C, Canto A, Pozo MI (2009) Yeasts in floral nectar: a quantitative survey. Ann Bot 103(9):1415–1423

Heuer H, Smalla K (2012) Plasmids foster diversification and adaptation of bacterial populations in soil. FEMS Microbiol Rev 36(6):1083–1104

Heydenreich B, Bellinghausen I, König B, Becker WM, Grabbe S, Petersen A, Saloga J (2012) Gram-positive bacteria on grass pollen exhibit adjuvant activity inducing inflammatory T cell responses. Clin Exp Allergy 42(1):76–84

Huisman R, Bouwmeester K, Brattinga M, Govers F, Bisseling T, Limpens E (2015) Haustorium formation in *Medicago truncatula* roots infected by *Phytophthora palmivora* does not involve the common endosymbiotic program shared by arbuscular mycorrhizal fungi and rhizobia. Mol Plant-Microbe Interact 28(12):1271–1280

Jacquemyn H, Lenaerts M, Brys R, Willems K, Honnay O, Lievens B (2013) Among-population variation in microbial community structure in the floral nectar of the bee-pollinated forest herb *Pulmonaria officinalis* L. PLoS One 8(3):e56917

Junker RR, Tholl D (2013) Volatile organic compound mediated interactions at the plant-microbe interface. J Chem Ecol 39(7):810–825

Junker RR, Loewel C, Gross R, Dötterl S, Keller A, Blüthgen N (2011) Composition of epiphytic bacterial communities differs on petals and leaves. Plant Biol 13(6):918–924

Kaneko T, Minamisawa K, Isawa T, Nakatsukasa HIROKI, Mitsui H, Kawaharada YASU, Shimizu Y (2010) Complete genomic structure of the cultivated rice endophyte *Azospirillum* sp. B510. DNA Res 17(1):37–50

Kembel SW, O'Connor TK, Arnold HK, Hubbell SP, Wright SJ, Green JL (2014) Relationships between phyllosphere bacterial communities and plant functional traits in a neotropical forest. Proc Natl Acad Sci U S A 111(38):13715–13720

Kessler D, Baldwin IT (2007) Making sense of nectar scents: the effects of nectar secondary metabolites on floral visitors of *Nicotiana attenuata*. Plant J 49(5):840–854

Knief C, Delmotte N, Chaffron S, Stark M, Innerebner G, Wassmann R, Vorholt JA (2012) Metaproteogenomic analysis of microbial communities in the phyllosphere and rhizosphere of rice. ISME J 6(7):1378

Knudsen JT, Eriksson R, Gershenzon J, Ståhl B (2006) Diversity and distribution of floral scent. Bot Rev 72(1):1

Kumar DSS, Hyde KD (2004) Biodiversity and tissue-recurrence of endophytic fungi in Tripterygium wilfordii. Fungal Divers 17:69–90

Lemanceau P, Barret M, Mazurier S, Mondy S, Pivato B, Fort T, Vacher C (2017) Plant communication with associated microbiota in the spermosphere, rhizosphere and phyllosphere. In: Advances in botanical research, vol 82. Academic, Cambridge, MA, pp 101–133

Lenaerts M, Álvarez-Pérez S, De Vega C, Van Assche A, Johnson SD, Willems KA, Lievens B (2014) *Rosenbergiella australoborealis* sp. nov., *Rosenbergiella collisarenosi* sp. nov. and *Rosenbergiella epipactidis* sp. nov., three novel bacterial species isolated from floral nectar. Syst Appl Microbiol 37(6):402–411

Maignien L, DeForce EA, Chafee ME, Eren AM, Simmons SL (2014) Ecological succession and stochastic variation in the assembly of *Arabidopsis thaliana* phyllosphere communities. MBio 5 (1):e00682–e00613

Maruthachalam K, Klosterman SJ, Anchieta A, Mou B, Subbarao KV (2013) Colonization of spinach by *Verticillium dahliae* and effects of pathogen localization on the efficacy of seed treatments. Phytopathology 103(3):268–280

Matsui K (2006) Green leaf volatiles: hydroperoxide lyase pathway of oxylipin metabolism. Curr Opin Plant Biol 9(3):274–280

Menzler-Hokkanen I, Hokkanen HM (2017) Entomovectoring: an agroecological practice of using bees for biocontrol. Agroecol Pract Sustain Agric Princ Appl Mak Transit. https://doi.org/10.1142/9781786343062_0007

Mittelbach M, Yurkov AM, Nocentini D, Nepi M, Weigend M, Begerow D (2015) Nectar sugars and bird visitation define a floral niche for basidiomycetous yeast on the Canary Islands. BMC Ecol 15(1):2

Mommer L, Hinsinger P, Prigent-Combaret C, Visser EJ (2016) Advances in the rhizosphere: stretching the interface of life. Plant Soil 407(1–2):1–8. https://doi.org/10.1007/s11104-016-3040-9

Mwajita MR, Murage H, Tani A, Kahangi EM (2013) Evaluation of rhizosphere, rhizoplane and phyllosphere bacteria and fungi isolated from rice in Kenya for plant growth promoters. SpringerPlus 2(1):606

Nelson EB (2004) Microbial dynamics and interactions in the spermosphere. Annu Rev Phytopathol 42:271–309

Newman MA, Sundelin T, Nielsen JT, Erbs G (2013) MAMP (microbe-associated molecular pattern) triggered immunity in plants. Front Plant Sci 4:139

Pellegrin C, Morin E, Martin FM, Veneault-Fourrey C (2015) Comparative analysis of secretomes from ectomycorrhizal fungi with an emphasis on small-secreted proteins. Front Microbiol 6:1278

Pozo MI, Lachance MA, Herrera CM (2012) Nectar yeasts of two southern Spanish plants: the roles of immigration and physiological traits in community assembly. FEMS Microbiol Ecol 80 (2):281–293

Pozo MI, Herrera CM, Alonso C (2014) Spatial and temporal distribution patterns of nectar-inhabiting yeasts: how different floral microenvironments arise in winter-blooming Helleborus foetidus. Fungal Ecol 11:173–180

Prasad R, Kumar M, Varma A (2015) Role of PGPR in soil fertility and plant health. In: Egamberdieva D, Shrivastava S, Varma A (eds) Plant growth-promoting rhizobacteria (PGPR) and medicinal plants. Springer International Publishing, Switzerland, pp 247–260

Reisberg EE, Hildebrandt U, Riederer M, Hentschel U (2012) Phyllosphere bacterial communities of trichome-bearing and trichomeless *Arabidopsis thaliana* leaves. Antonie Van Leeuwenhoek 101(3):551–560

Remus-Emsermann MN, Schlechter RO (2018) Phyllosphere microbiology: at the interface between microbial individuals and the plant host. New Phytologist 218(4):1327–1333
Rering CC, Beck JJ, Hall GW, McCartney MM, Vannette RL (2017) Nectar-inhabiting microorganisms influence nectar volatile composition and attractiveness to a generalist pollinator. New Phytol 220(3):750–759
Rodríguez MA, Venedikian N, Godeas A (2001) Fungal populations on sunflower (*Helianthus annuus*) anthosphere and their relation to susceptibility or tolerance to *Sclerotinia sclerotiorum* attack. Mycopathologia 150(3):143
Rousk J, Baath E, Brookes PC, Lauber CL, Lozupone C, Caporaso JG, Fierer N (2010) Soil bacterial and fungal communities across a pH gradient in an arable soil. ISME J 4(10):1340
Saharan BS, Nehra V (2011) Plant growth promoting rhizobacteria: a critical review. Int J Life Sci Med Res 21(1):30
Samuni-Blank M, Izhaki I, Laviad S, Bar-Massada A, Gerchman Y, Halpern M (2014) The role of abiotic environmental conditions and herbivory in shaping bacterial community composition in floral nectar. PLoS One 9(6):e99107
Santoyo G, Pacheco CH, Salmerón JH, León RH (2017) The role of abiotic factors modulating the plant-microbe-soil interactions: toward sustainable agriculture. A review. Span J Agric Res 15 (1):13
Schaeffer RN, Irwin RE (2014) Yeasts in nectar enhance male fitness in a montane perennial herb. Ecology 95(7):1792–1798
Schiltz S, Gaillard I, Pawlicki-Jullian N, Thiombiano B, Mesnard F, Gontier E (2015) A review: what is the spermosphere and how can it be studied? J Appl Microbiol 119(6):1467–1481
Shade A, McManus PS, Handelsman J (2013) Unexpected diversity during community succession in the apple flower microbiome. MBio 4(2):e00602–e00612
Shrivastava S, Prasad R, Varma A (2014) Anatomy of root from eyes of a microbiologist. In: Morte A, Varma A (eds) Root engineering, vol 40. Springer, Berlin, pp 3–22
Simon HM, Smith KP, Dodsworth JA, Guenthner B, Handelsman J, Goodman RM (2001) Influence of tomato genotype on growth of inoculated and indigenous bacteria in the spermosphere. Appl Environ Microbiol 67(2):514–520
Singh D, Mathur SB (2004) Location of fungal hyphae in seeds. In: Histopathology of seed-borne infections. CRC Press, Boca Raton, FL, pp 101–168
Tian Y, Zhao Y, Wu X, Liu F, Hu B, Walcott RR (2015) The type VI protein secretion system contributes to biofilm formation and seed-to-seedling transmission of *Acidovorax citrulli* on melon. Mol Plant Pathol 16(1):38–47
Truyens S, Weyens N, Cuypers A, Vangronsveld J (2015) Bacterial seed endophytes: genera, vertical transmission and interaction with plants. Environ Microbiol Rep 7(1):40–50
Ushio M, Yamasaki E, Takasu H, Nagano AJ, Fujinaga S, Honjo MN, Kudoh H (2015) Microbial communities on flower surfaces act as signatures of pollinator visitation. Sci Rep 5:8695
Vannette RL, Gauthier MPL, Fukami T (2013) Nectar bacteria, but not yeast, weaken a plant–pollinator mutualism. Proc R Soc Lond B Biol Sci 280(1752):20122601
Vega Y, Marques I (2015) Both biotic and abiotic factors influence floral longevity in three species of *Epidendrum* (Orchidaceae). Plant Species Biol 30(3):184–192
Violante A, Caporale AG (2015) Biogeochemical processes at soil-root interface. J Soil Sci Plant Nutr 15(2):422–448
Whipps J, Hand P, Pink D, Bending GD (2008) Phyllosphere microbiology with special reference to diversity and plant genotype. J Appl Microbiol 105(6):1744–1755
Wiens F, Zitzmann A, Lachance MA, Yegles M, Pragst F, Wurst FM, Spanagel R (2008) Chronic intake of fermented floral nectar by wild treeshrews. Proc Natl Acad Sci U S A 105 (30):10426–10431
Williams TR, Marco ML (2014) Phyllosphere microbiota composition and microbial community transplantation on lettuce plants grown indoors. MBio 5(4):e01564–e01514

Chapter 15
Efficiency of Soil, Plant and Microbes for Healthy Plant Immunity and Sustainable Agricultural System

Gufran Ahmad, Yasar Nishat, Mohammad Haris, Mohammad Danish, and Touseef Hussain

Abstract For the vital functioning of soil ecosystem, microbes have always been the superior force in driving many processes. These microorganisms are the main key facilitators in nutrient cycles associated with plant root system by delivering nutrients and suppressing pathogens, thereby sustaining plant health. Their amazing activity and biochemical versatility, especially the roots of growing plants, show great potential for beneficial microorganisms, for the development of biotechnology applications, for the control of plants of wild plants and for increased food crops. In this chapter we review the existing literature on the interaction between plants, microorganisms and soil. The rhizosphere is an arena where the complex rhizosphere community, which includes both microflora and microfauna, communicates with pathogens and influences the outcome of pathogen infection. A number of microorganisms are advantageous to the plants which include nitrogen-fixing bacteria, endo- and ectomycorrhizal fungi and plant growth-promoting bacteria and fungi. Some of the activities include complex systems of communication, in case of symbiosis such as arbuscular microscopic symbiosis, many millions of years old, while others include exudates from the root and other products of the rhizodeposition which are used as substrates for soil microorganisms. Since degradation of organic compounds in the rhizosphere is encouraged by the release of root expressions and enzymes in plants, therefore, biodegradation plays an important role, depending on the contact between the soil and the contaminated substances surrounding the plants. There is a considerable potential in the expanded area of microorganisms to replace synthetic biological chemistry. Since microbial activities are an important and sensitive component of soil, they are also good indicators of soil disorder and ecosystem. Still, an extended use of microorganisms for bioindication purposes and sustainable means of soil management depends on advances in understanding microbial ecology, especially on a field scale. As a result, to enhance the regenerative capacity of soil ecosystems for sustainable agriculture, it is best to understand how to

G. Ahmad · Y. Nishat · M. Haris · M. Danish · T. Hussain (✉)
Deptartment of Botany, Aligarh Muslim University, Aligarh, Uttar Pradesh, India

A. Varma et al. (eds.), *Plant Microbe Interface*,
https://doi.org/10.1007/978-3-030-19831-2_15

increase the dynamics and potential of soil biology. This will allow new applications of knowledge to address the challenges of pest and diseases and increase global food production and sustainable farming.

15.1 Introduction

This chapter gives a general outline of the underground processes related with the plant, microbes and soil. It covers the interactions among plants and microbes in the soil, their compound communications, the natural procedures they sustain and the expenses and advantages to plants related with these collaborations. One of the significant difficulties for the twenty-first century will be protection of environment and sustainable development in agricultural crop. An improved production is important to give adequate nourishment to the expanding human (population). Current generation techniques in agribusiness, e.g. the ill-advised utilization of synthetic chemicals in the form of pesticides and fertilizers, create pollution in environment and also effect on human health (Gunnell et al. 2007; Leach and Mumford 2008). There is a developing interest for environmentally compatible strategies in agriculture. Plant breeding technology improves the variety of crops which enhanced disease resistance; provide better growth against abiotic stress. Plants grow in close relationship with the microorganisms that inhabit the soil in which plants live. Soil microbes provide beneficial environment for plants' roots. On the other hand, pathogenic microbes are harmful for the plants; microbes and plants' interaction may be beneficial, neutral or harmful to the plant; and it's totally dependent upon the particular microorganisms and plants' environment condition (Bais et al. 2006). These may be bacteria, fungi, protozoans, nematodes and algae (Raaijmakers and Weller 2001). All plants correlated with microenvironments, particularly the rhizosphere, are found in high plenitudes by microorganism (Berg et al. 2005). Research showed that microbes and parasites have a personal cooperation with their host plants and can advance plant development and vanquish plant pathogens (Whipps 2001; Thakore 2006; Ehlers 2006). Plant rhizosphere is the soil closest to the plant root framework where roots discharge absorbe number of metabolites from root hairs or stringy root system. These metabolites go about as compound signs for motile microscopic organisms to move to the root surface yet additionally speak to the fundamental supplement sources accessible to help development and growth in the rhizosphere (Venant Nihorimbere et al. 2011). The loss of organic material from the roots gives the vitality to the improvement of dynamic microbial populations in the rhizosphere near the root (Whipps 2001). Sustainable agriculture includes physical, chemical, biological, ecological, and economic sequence in a way to develop latest agricultural practices that are safe and nonhazardous for environment (Lichtfouse et al. 2009). Plants grow in soil and get all ground requirements which they want. When rhizosphere is not compatible for root then plant immunity is affected and interrupt metabolic activity aproduced by microorganisms present already in rhizosphere that provides a complete and healthy environment for the planthealthy root system. Plant growth rate and soil have been showing the composition of rhizosphere microbial communities (Broeckling 2008). Organic farming is one of the best cultural practices that protects the crop and also maintains the diversity of the soil (Megali et al. 2013). To address the expanding yield demand which is diminished in available agricultural land, conventional agriculture have been constrained to apply consistently higher doses of destructive synthetic chemical and pesticides (Foley et al. 2011). Using of these toxic chemical and pesticides, plant disease is diagnosed, but their pollution affects upon the soil, environment, human and other plants' health. Organic amendment, biofertilizer and organic manure all these things keep the soil condition healthy in all respect. Micro and macro-biota rich soil and fill the nutrient value by nitrogen fixation, phosphate and potassium solubilization or mineralization, help in the release of plant growth regulating compounds, producing anti infection agent (antibiotics) and biodegradable organic substance in the soil. Microbial biotechnology and its applications are getting better attention in the sustainable development of microbial biotechnology and agriculture and environmental health (Mosttafiz et al. 2012). A broad understanding of the mechanisms controlling the selection and activity of microbial communities by the roots of plants will provide new opportunities for increase crop production in sustainable agricultural system.

15.2 Biogeochemical Process and Microbial Interaction in the Mycorrhizosphere

Soil microbiota have a significant influence on soil fertility and plant health (Gianinazzi and Schuepp 1994). Microbial communities in the mycorrhizosphere (i.e. in the small volume of soil immediately surrounding a mycorrhizal root) actively interact with the establishment and functioning of the mycorrhizal symbiosis (Giri et al. 2005). Symbiotic mycorrhizal fungi, such as arbuscular mycorrhizal (AM) fungi, increase the absorptive surface area of their host plant root systems. The hyphae of these symbiotic fungi provide an increased area for interactions with many different microbial populations which directly or indirectly influence plant growth and uptake of nutrients (Johansson et al. 2004). The significance of biological interactions in the mycorrhizosphere is considered from the standpoints of plant ecology and of practical application.

15.2.1 Diversity in the Mycorrhizosphere

A wide diversity of microorganism lives in the surrounding zone of roots or mycorrhizas, taking advantage the influence of the different organic compounds released by the plant (Garbaye 1991). These organisms include members of most taxonomic groups of aerobic and anaerobic heterotrophic organisms, from bacteria through fungi and protozoa to animals. These are purely saprophytic organisms, specialized mycorrhizosphere organisms, root pathogens, root symbionts and predators.

15.2.2 Biochemical Interaction in Mycorrhizosphere

Root exudates used by the fungus modify root functions; microbial communities in the mycorrhizosphere differ from those in the rhizosphere and in the soil. Microbes suppress the infection of roots by mycorrhizal organisms. The specificity of mycorrhizosphere microorganisms has been shown under many conditions (Bianciotto et al. 1996; Carpenter-Boggs et al. 1995; Citernesi et al. 1996).

Saprophytic microbiota which live in the close vicinity of roots or mycorrhizas are able to use a number of different complex organic molecules such as lignin, proteins, glycoproteins, cellulose and other polysaccharides (Garbaye 1991). They do not differ very much from those living in the soil far away from the roots: they are able to live on dead organic material. However, they are more plentiful in the mycorrhizosphere, where large amounts of their substrates are produced such as sloughed root cells, dead rootlets, decaying mycorrhizal fungi, polysaccharides and glycoproteins exudates or secreted by the root or the symbionts, etc.

Specialized mycorrhyzospheric organism depends completely on the simple organic molecules such as sugars, amino acids and organic acids, released by the plant and its fungal symbiont. They cannot multiply significantly in the non-rhizospheric soil because of the lack of these short-lived substances. As a consequence, they compete strongly with each other as well as with the symbionts and pathogens in their root surface colonization (Garbaye 1991).

15.3 Interactions of Plants and Microbes for the Phytoremediation of Air Pollutants

Air pollution is one of the major problems and a global cause of concern. There are two types of air pollutants composing of a primary and secondary air pollutants, primarily the NOx, SO_2, CO_2, O_3 (inorganic pollutants), BTEX (benzene, toluene, ethylbenzene, xylene) [volatile organic compounds (VOCs)], particulate matter (PM) and PAHs (polyaromatic hydrocarbons). Despite the complexity in composition, phytoremediation has been defined as the in situ use of plants to stabilize, remediate and reduce or restore contaminated soil and eco-friendly plant-based biotechnology for the reduction and detoxification of indoor and outdoor air pollutants. There are millions of microorganisms like fungi and bacteria that are associated with plants; that's why the plants never live alone. Moreover, with the association of microorganisms, the plants are known to remove the significant amount of air pollutants and are metabolized (Nowak et al. 2006; Brack 2002). Plants are supporting to handle the living and nonliving stresses, and they produce siderophores (a type of plant hormones) and some allelochemicals which are inhibitory by assisting their host for the uptake of water and nutrients (Bulgarelli et al. 2013; Weyens et al. 2009a, b). Generally it is known that the interaction of plants and microbes plays a vital role in the process of phytoremediation by detoxification or sequestration, by degradation of air pollutants and by promoting plant growth (Weyens et al. 2009a, b).

In this part of the chapter, the available knowledge about the above-described plant–microbe interactions during phytoremediation of air pollution is summarized for the main air pollutants (particulate matter, volatile organic compounds and inorganic pollutants). Moreover, a concise overview of the specific contributions of the plant and its microbiome has been discussed.

The outer surface of leaves and stems is known to adsorb significant amounts of air pollutants; bacteria and fungi living on the surface of leaves are known as the phyllospheric bacteria and fungi; that's why it is of too much importance. On the other hand, the parts of air pollutants, adsorbed by the plant through the leaves, are known as leaf endophytes. By means of degradation, transformation or sequestration, the endospheric and phyllospheric bacteria and fungi can detoxify the part of the air pollutants. After these the air pollutants mix in the rhizospheric soil through the rainfall.

15.4 Plant Associated with Microorganisms and PM Phytoremediation

The mixture of liquid and solid materials with various sizes and shapes and mainly the chemical composition and their different origins are known as particulate matter (PM) (Pastuszka 2007). It is mainly generated outside by anthropogenic activity like exhaust from vehicles, dust of roads, burning of fossil fuels, and by industrial activities also (Amato et al. 2002), indoors mainly by heating, cleaning and cooking activities (Myers and Maynard 2005; Amato et al. 2002; WHO 2000), naturally by eruption of volcanoes, forest fire, breezing of ocean, erosion of rocks and sandstorms. There are many plants having the ability to remove the amounts of particulate matter, especially in city areas which are close to roads, by adsorbing PM on the foliage (sPM) or stabilizing them in waxes (wPM) (Popek et al. 2012; Saebo et al. 2012; Dzierzanowski et al. 2011; Beckett et al. 2000). A large number of trees and shrubs reduce PM that is accumulated by the foliage by the process of biofilter (Popek et al. 2015). For the removal of pollutants, the most important model for the explanation of forest structure and ecosystem is the "i-Tree model" which was developed by Nowak et al. (2008). In this model, the trees and plants in Beijing

(China) in the city centre removed 772 tons of PM10 on a yearly basis (Yang et al. 2005); in West Midlands (UK), trees reduce 26% of PM10 in air; in Chicago (USA) 234 tons of PM10 are reduced by the trees; and in the USA, a whole trees and shrubs reduce 215,000 tons of PM10 annually (Mcdonald et al. 2007; Nowak 2015; Nowak et al. 2006). Plant-associated microorganisms are known to play an important role during plant growth and development by increasing nutrient availability by the production of organic acids, siderophores (plant hormones) and plant growth hormones (indole acetic acid), and they help the plant to reduce the abiotic and biotic stresses by the production of 1-aminocyclopropane-1-carboxylate (ACC) deaminase (Bulgarelli et al. 2013; Arshad et al. 2007; Weyens et al. 2009a, b). The process of biological nitrogen fixation by rhizobia in nodulated leguminous plants incorporates C and N into soil, which, besides increasing nutrient uptake capacity, also improves their tolerance to environmental stresses (Vessey 2003) and promotes the growth of plants by producing the growth regulators such as auxins [e.g. IAA, cytokinins (Glick 2010)]. Plant-associated microorganisms might also play a role in the detoxification of the PM absorbed by their host plant and the process of degradation, metabolic capabilities resulting in phytotoxicity and evapotranspiration (Weyens et al. 2009a, b). The toxic trace elements which were found in the soil can be remediated by the root endophytes (Rajkumar et al. 2012). Hence it might be proved that the plants associated with microbes may support to remove the PM-bounded contaminants and improve the ability of phytoremediation.

15.5 Plant Associated with Microorganisms and Volatile Organic Compounds (VOCs) Phytoremediation

An organic compound having a 20 °C vapour pressure of 0.01 kPa or more or having a corresponding volatility under the particular conditions of use is known as volatile organic compounds (VOCs) (Directive 1999). They depend on their composition of the range of number of carbon and also on nature like their chemical and physical composition. On the basis of the emissions of solvents, it was defined as "The presence of VOCs is negatively affecting outdoor as well as indoor air quality. It is the main component for the ozone (O3) formation by the reaction of the oxides of nitrogen and sunlight (Calfapietra et al. 2013; Arneth et al. 2010). There are two types of VOCs like AVOCs [anthropogenic (transportation and industrial process, e.g. benzene, toluene, ethylbenzene, xylene (BTEX)), polyaromatic hydrocarbons (PAHs) and formaldehyde] and BVOCs [biogenic (trees and other plants, e.g. chloromethane, isoprene and monoterpenes), indoors (carpets, wallpaper, curtains, paper products, office chairs and electronic equipment)] (Salthammer and Uhde 2009; Yu and Crump 1998; Harrison et al. 2012). Plants are important source of VOCs as well as have a capability of coping the VOCs from the air. As per Karl et al. (2010), Yang et al. (2009), Kim et al. (2008), Agrawal et al. (2003), Giese et al. (1994), Wolverton and Wolverton (1993), Ren et al. (2014) and the plants which are emitting the VOCs very low used for the phytoremediation. Opening of the stomata in light is greater in the removal of VOCs as compared to darkness because the pollutants (VOCs) enter through the leaf stomata. In some plants the stomata are open in night and close at day time, for example, CAM plants, and they take VOCs along with CO_2 and help in phytoremediation (Winter and Holtum 2014). There are some plants which are recommended for the remediation of indoor air pollutants like *Zamioculcas zamiifolia*, *Agave* and cactus by the uptake BTEX (Sriprapat and Thiravertyan 2013). There are several species of plants that have a capability to host the several microorganisms, and they help in the degradation of VOCs. Phyllosphere is the surface of leaf, and it is the best habitat for the survival of microbes. Globally the bacterial population present in the surface of leaves are 1026 cells, fungi are less numerous, and archaea are minor (Vorholt 2012; Voriskova and Baldrian 2013; Baldotto and Olivares 2008; Lindow and Brandl 2003; Knief et al. 2012; Delmotte et al. 2009). BVOCs and AVOCs are degraded by the phyllospheric microorganisms and also help in the cleaning of indoor air pollution like fungi and bacteria. The endophytic and rhizospheric bacteria have an ability to remove toxic compounds from soil, and which are associated with plants and the mycorrhizal fungi help in the mineralization of pollutants (McGuinness and Dowling 2009; Gao et al. 2010; Mohsenzadeh et al. 2010; Bouwer and Zehnder 1993).

15.6 Role of Plant-Associated Microorganisms During IAP Phytoremediation

The oxides of sulphur (SO_2, SO_3), oxides of nitrogen (NOx), carbon monoxide (CO), CO_2 and O_3 are considered the principal and common inorganic air pollutants. CO_2 contributes the major amount among all greenhouse gasses. It is released through the anthropogenic activities like burning of fossil fuels and automobile exhausts. The rising concentration of carbon dioxide in the atmosphere is the main cause of global warming including rising of surface temperatures, melting ice and snow, rising sea levels and increasing climate variability. The main sources of sulphur dioxide (SO_2) are burning of coal in thermal power plants for the generation of electricity and other fuels which are used in industry and domestic purposes. Nitric oxide (NO) and nitrogen dioxide (NO_2) are the main gasses which come under the oxides of nitrogen (NOx). Automobile exhausts are the main man-made source of NOx (Parrish and Zhu 2009; Fowler et al. 1998). NO_2 becomes toxic at high concentration and plays an important role in the formation of ozone (O_3) by photochemical oxidation reaction cycle, concerning with the health of humans (Samoli et al. 2006; Ostro et al. 2006). Ozone (O3) is formed by the induction of photochemical complex in the presence of sunlight (UV radiation), reacted with NOx, VOCs and CO in troposphere. There are many disease related to humans by the tropospheric ozone like respiratory diseases, asthma and premature mortality

(Ostro et al. 2006; Vagaggini et al. 2002; Borrego-Hernandez et al. 2014). Urban trees cop the significant amount of air pollution and improve the air quality by the reduction of ozone because several species of plants accumulate the ozone in their cells and tissues (Nowak et al. 2000; Taha 1996; Cardelino and Chameides 1990). In plants SO_2 enters into the leaves through the way of stomata, detoxifies and is used in reductive sulphur cycle and forms sulphur-containing amino acids for growth and development (Gheorghe and Ion 2015). The process of uptake and storage of CO_2 in plants is known as carbon sequestration and removes CO_2 from the atmosphere (Sedjo and Sohngen 2012; Scheller et al. 2011). Plants use NO_2 as a fertilizer which helps in the reduction of air pollution (Welburn 1998). On the other side, the adsorption of atmospheric NO_2 in leaves and roots also plays a vital role in the reduction of nitrogen dioxide (Welburn 1990). It is the main source of photochemical reaction by entering into the plants and metabolized organics like amino acid through the assimilation of nitrogen. Phytoremediation through the microorganisms which are associated with plants is very limited in the sense of inorganic air pollutants. The metabolic process of N and S, which shows the existence of microorganisms, helps to seize the oxides of nitrogen and sulphur. The chemolithoautotrophic bacteria are found to be the deposition of NO_2 on the surface of leaves (Papen et al. 2002). CO_2 is used by autotrophic microorganisms as a source of carbon in the sequestration of carbon and affects humus formation and composition (Langley and Hungate 2003). On the other side, the mycorrhizal fungi are also used in carbon sequestration in soil organic matter (SOM), and large amount of C is stored in oil organic matter in a boreal forest system originated from roots and fungi (Clemmensen et al. 2015; Lesaulnier et al. 2008). The elevated CO_2 concentration affects the soil microbial population associated with aspen (Lesaulnier et al. 2008). Ozone helps in the generation ROS associated with bacteria which have high antioxidative properties; that's why it is known as an antimicrobial agent, and the contribution of the microbiome during ozone phytoremediation will be limited to decline the toxicity (Wu et al. 2014; Van sluys et al. 2002). Moreover, the bacteria can play a role in ROS detoxification. Generally, the plant-associated microbiome promotes the growth and development of plants upon exposure to inorganic air pollutants.

15.7 Microbial Bio Control in the Rhizosphere, Its Effect on Root Health

15.7.1 What Completes a Plant Require from the Rhizosphere?

The response to this inquiry is just that a plant gets nearly everything specifically from the soil to help development, with the striking exemptions of carbon dioxide, oxygen and light. The soil already have a structure that is physically fit for supporting the over the ground half of the plant through its creating root framework as it develops. Furthermore, the soil should be kept up at a fitting pH, give insurance from harmful substances and pathogens and contain reasonable levels of water. Past this, all the fundamental mineral components that a plant requires are acquired from the soil. No less than 17 components are fundamental for plant development and multiplication (Marschner 1995). Fourteen of these components are procured essentially from the soil arrangement. These incorporate six macronutrients (N, K, P, S, Mg and Ca) and eight micronutrients (B, Cl, Cu, Fe, Mn, Mo, Ni and Zn). A large portion of these components are for the most part taken up from the soil, arrangement in their ionic frame (White 2003). The interactions between plant roots and microbes inside their rhizosphere assist them by acquiring essential mineral supplements and keep out the aggregation of toxic components. The release of carbon compounds from plants into the soil brings about more prominent microbial population in the rhizosphere in respect to the mass soil and expanded microbial biomass and their activity (Lynch 1987; Bending 2003). Plants can have unique functions with mycorrhizal and nodulation-based associations.

15.7.2 The Rhizosphere Microsphere

The developing plant and residing microorganisms which are present in their surrounding soil interact with each other when the seed germinates and seedlings grow. The assorted variety of microorganisms related with plant roots is tremendous; in the request of a huge number of species, this complex plant-related microbial network, additionally alluded to as the second genome of the plant, is pivotal for plant's wellbeing. Recent study has shown that in plant–microbe interactions, plants are capable to make their rhizosphere niche.

Rhizosphere includes microflora like fungi, bacteria, protozoa, nematodes and algae (Raaijmakers and Weller 2001). The rhizosphere consists three distinct components: the rhizosphere, rhizoplane and plant root. The rhizosphere is in this manner the zone of soil impacted by roots through the arrival of substrates that influence microbial action. The rhizoplane is the root surface, including the strongly adhering root particles. The root itself is a piece of the framework, on the grounds that specific endophytic microorganisms can colonize inward root tissues (Bowen and Rovira 1999; Prasad et al. 2015).

15.7.3 Microbial Biocontrol in the Rhizosphere

A large number of microorganisms including fungi and bacteria show biocontrol activity in the rhizosphere region. Their mode of action in different biocontrol includes antagonism, parasitism, rivalry for supplements and space and initiation of plant defence (Whipps 2001). Different types of microbial biocontrol agents work against the root-feeding insects, and other root pathogens have also promoted plant growth from their biocontrol nature. The plant development impact of the BCA *Trichoderma harzianum* (Harman et al. 2004a, b) has been proposed to be founded on the making of adversarial mixes against root pathogens, which are also used as plant hormones, which can promote root development (Vinale et al. 2008). The soil is a natural habitat of microorganism and pathogen. Assorted scientific categorizations of pathogenic microorganisms isolated from soil include nematodes, fungi, viruses, protozoa and bacteria (Jackson and Glare 1992). In the soil condition, abiotic factors assume a part in the frequency, constancy and spread of microorganisms and nematodes (Studdert and Kaya 1990; Ignoffo 1992). Considerable interaction between parasitic microorganisms has been observed in the rhizosphere because of their performance depends upon different trophic level in soil but is still not known. In the other side, some microorganisms like bacteria and fungi are used as biocontrol agents, and they are also used in combination, bacteria–fungus, fungus–fungus and bacteria–bacteria. Different combinations of microbes has shown good activity against plant pathogens and also showed plant growth promoter activity too (Goettel 2008; Vega 2008). Bacteria *Paecilomyces lilacinus* and fungi *Trichoderma harzianum* combined treatment show the effective control of root-knot nematodes population (Hafeez Ullah Khan et al. 2001).

15.7.4 Root Health

Some plant pathogens are root pests include protozoans, fungi, nematode, oomycetes and root-feeding insects, which pathogens complete their basal need like food, shelter obtain from the plants root with different vital strategy of pathogen. In many cases, root disease symptoms particularly from oomycete and nematode pathogens can be mistaken for supplement and water insufficiency on the grounds that the root system cannot help the plant with adequate supplements and water. Root-feeding insects cause same supplement and water lack side effects since root herbivory harms the root framework and, in this manner, its absorptive limit (Johnson et al. 2013). Another critical perspective to consider is that the supplement absorptive capacity of the plant likewise relies upon the outer mycelium of root possessing arbuscular mycorrhiza. AM fungi have been appeared to reduce plant stress from root parasites, which might be identified with enlistment of resilience indemnity by enhancing host plant supplement and water progression (Whipps 2004; St. Arnaud and Vujanovic 2007). Microorganism in the rhizosphere is very important for the root health except pathogen.

15.8 Rhizodeposition

Rhizodeposition is a major process affecting nutritional availability through the stimulation of microbial activity in the rhizosphere. The microbial community structure and function in the rhizosphere is thought to be the result of the difference in quantity and quality of rhizodeposition. Rhizosphere is also defined as the amount of soil under the influence of the root as well as the root itself. Depending on distance to the root, Marilley et al. (1998) described three different experimental fractions (Fig. 15.1): the bulk soil (BS) or non-rhizosphere soil (NRS), corresponding to soil that is not influenced by the roots; the rhizosphere soil (RS), defined by the soil that adheres to the root when the root system is shaken manually; and the rhizoplane–endorhizosphere (RE), corresponding to the surface and the interior of the root. The soil area near the roots is only 2 to 3% of the total soil volume (Lee and Pankhurst 1992) but, in the case of microbial population and community mobility, displays the most complex area of soil. The impact on crop productivity is the most relevant for discussion of sustainable agriculture and soil conservation efforts (Lynch 1986).

The word rhizosphere was first introduced by Hiltner (1904) which is related to the excitement created by microbial growth around the roots of the legumes. Immediately after this, the concept was applied to all crops as well as in terrestrial plants (Lynch 1986). Many reviews discussed about the rhizosphere and its effects to the health and growth of the plant but also for quality of soil (Lynch 1990; Pinton et al. 2001). There is an agreement that the rhizosphere effect is a complex and dynamic process that is started by the original exhaust and is influenced by the factors of plants such as plants' age and phase of development, rather than the factors

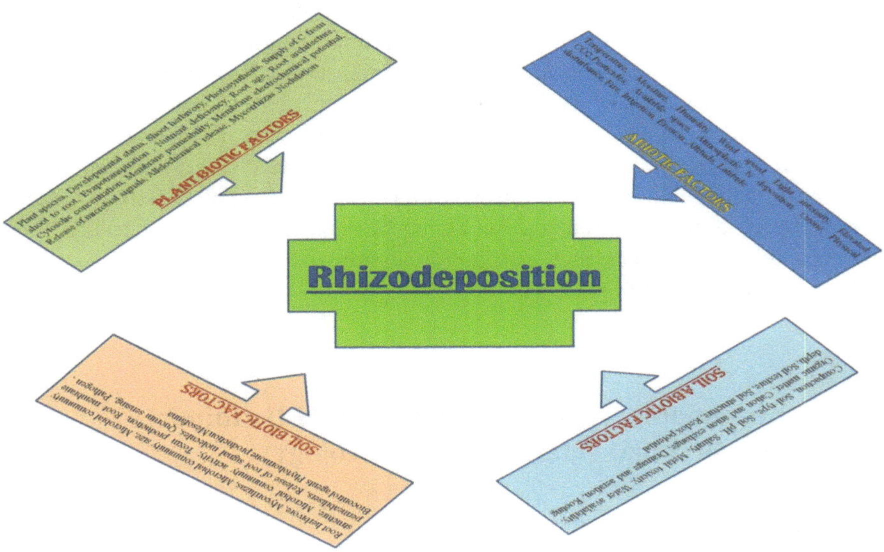

Fig. 15.1 Schematic representation of factors that influence rhizodeposition

of soil; environmental conditions and cultural practices are also influenced by it (Linderman 1988). In many ways of the rhizosphere effect, this route is the stimulus of microbial growth and activity by the released compounds which always gets and still focuses because the sustainable crop growth promotion and soil for the development of effective rhizosphere microbial technologies biological treatment due to scope. Nowadays, because of rising efforts to implement plant-aided soil remediation techniques and to assess the possible risks of soil contaminants for entering field crops and groundwater, research increasingly focuses also on direct interactions between root-released compounds and soil chemicals. Rhizodeposits, on the contrary, have large direct and indirect effects on soil organisms, soil environment and neighbouring plants. Root exhaust plants can affect the availability of nutrients directly or indirectly.

15.9 Biological Control by Microbes in the Rhizosphere Region

Various microorganisms, including bacteria and fungi, complete the activities of biocontrol in the rhizosphere. Many of the biocontrol modes of action include parasitism, rivalry, nutrients and competition for space and protection of plants (Whipps 2001). Many hidden nontargeted effects of biocontrol agents on other soil microbes and their functions are largely overlooked (Brimner and Boland 2003), which are obstructing successful integration of beneficial microorganisms of these plants in agricultural plants. Many of the microbial biocontrol agents against root pathogens have been shown to promote the development of plants independently from biological control facilities and to promote the good health effect of biocontrol agent *Trichoderma harzianum* (Harman et al. 2004a). Root pathogens have been suggested to be based on the production of rival compounds, which increase root growth and also act as hormones in the plant (Vinale et al. 2008). Soil is the pathogenic natural reservoirs present on insects. Various taxonomic groups of pathogenic microorganisms differ from many soil-dwelling insects: viruses, bacteria, fungus, protozoa and nematodes (Jackson and Glare 1992). They are capable of epidemics and causing immature and adult insect phases with pathogenic agricultural importance.

15.10 Sustainable Agriculture System and Its Impact on Microbial Interaction and Ecology

Microbes are always better at running important processes for the functioning and production of soil ecosystem. Microorganisms are important aids in major nutrient cycles, and they distribute nutrients to the plants and are associated with suppressing the pathogens, thereby maintaining plant's and animal's good health, and thus there is life on earth. Their amazing activity and biochemical versatility, especially around the roots of growing plants, show the potential of beneficial microorganisms for the development of applications of biotechnology, which is responsible for the control of weed and plant pathogens, food and forest increasing yields of crops and improved durable outlook of soil (Vessey 2003; Prasad et al. 2017).

Since microbial activities are very important and sensitive components of soil, they are also good indicator of soil defects and ecosystem functioning. Nevertheless, an extended use of microorganisms for biological purposes and durable management of soil management depends on progress in understanding microbial ecosystem especially on regional scale. During the last century, soil research has focused largely on chemical and physical factors of soil with regard to biological neglect. As a result, there is limited understanding about dynamics and capitalization on the possibilities of soil biology to increase the Renaissance potential of soil ecosystem for durable agriculture (Mosttafiz et al. 2012).

Plant–microbe–soil interactions actively co-ordinate C, N and P mobility in the soil. However, the traditional rhizosphere concept focuses primarily on root–microbe interactions with plant nutrition (mycorrhiza and rhizobium) and microbe–microbe interactions with respect to plant-to-plant health (biocontrol of root pathogens) (John 2001). Conversation between soil fauna and rhizosphere microbes on the interaction between beneficial rhizosphere microbes from different functional groups remains still relatively overlooked (John 2001).

15.10.1 Benefits to Humans from an Improved Understanding of the Rhizosphere

The rhizosphere is the area of soil around the root of the plants, which is affected by the root secretion and becomes the centre of the attraction of microorganisms. Algae, archaea, arthropod, bacteria, fungi, nematodes, oomycetes, protozoa and viruses are all microorganisms found in the rhizosphere region. Microorganism and their population activities occur by exudates release from the root because food is the main source of microorganism, so rhizosphere acts as a hot spot of microbial interactions. In natural ecosystems the composition and biomass of various plants can also directly and indirectly affect by the rhizospheric microbes (Karami et al. 2012; Schneider et al. 2013).

According to various studies, many rhizospheric microorganisms are beneficial for the development and health of plants, for instance, nitrogen-fixing bacteria, protozoa, biocontrol microorganisms and symbiotic association. In the rhizosphere many beneficial microorganisms improve the plant growth, crop productivity, soil quality, microbial interaction and physiochemical properties of soil (Morris and Blackwood 2015). The effects of rhizospheric microbiota make substantial improvements in plant nutrients (Fig. 15.2). A popular example is nitrogen-fixing bacteria

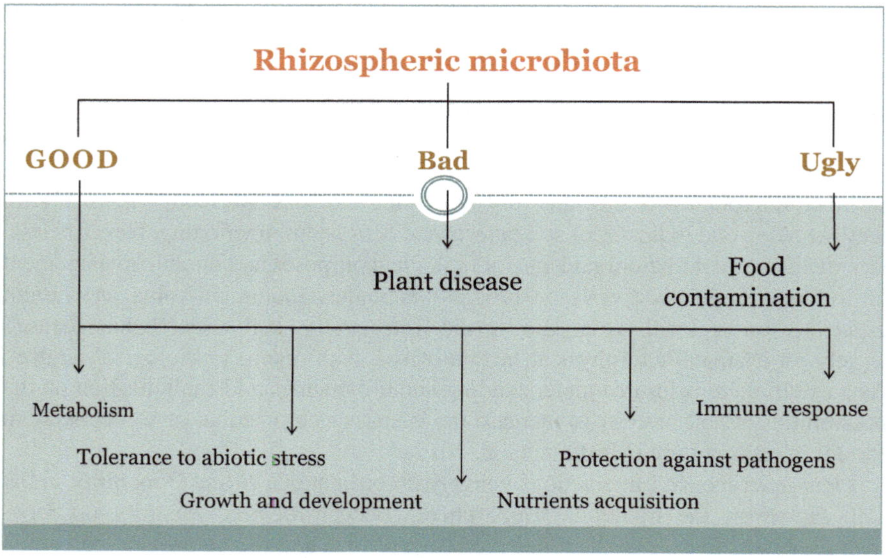

Fig. 15.2 Schematic diagram showing the different functions of rhizospheric microbiota on the host plants. The words 'good', 'bad' and 'ugly' are known as microbial species, and based on its abundance, it may be beneficial and pathogenic (Maurhofer et al. 1992)

(rhizobia) and mycorrhizal fungus which clear the way for the phosphorus absorption (Hawkins et al. 2000; Richardson et al. 2009; Miransari 2011). The approaches of rhizospheric zone play an important role in the sustainable soil property and are beneficial for crop production which ultimately help to human health and also economically. To full fill human need, plants are the mainly source of food, medicine, reducing carbon dioxide levels, oxygen, fuel, fibre, residential woody material and also use for the phytoremediation.

There are multiple benefits to humans with the concurrence of rhizosphere. It is difficult for humans to measure the benefits by better understanding of the rhizosphere and its processes. Plants survive better in normal soil without any involvement of humans in nature, but in case of agricultural field, many problems take place for the growth of plants. Due to this harmful contaminant present in environment in the form of particles containing heavy metals and many of the manure made up of harmful agents added in the soil to hamper the plant growth. Due to the presence of toxic substances (heavy metals) are sensitive to rhizobium species and roots are not able to nodulation (McGrath et al. 1995). A understanding of underground processes can help us to meet the repercussion of problems like global warming and climate change and agriculture, but it is difficult to determine the benefits of this knowledge.

In this world Zero Hunger Challenge is the main challenge for achieving in the present for human lifetimes. This achievement can be oriented with sustainable development in relation to agricultural productivity and the remaining demands of the growing population for the human economy. Increasing the food items from less and less holdings of fields is the demand of human economy. Currently we are touching the boundaries of land ready for farming which is harmful to reduce the size of the agricultural sector due to the spread of industrialization, highways and habitat areas. So, we need to maintain and improve the soil fertility with involvement of microorganisms. In the soil, microorganisms represent a good relation to prolong soil fertility. Natural and anthropogenic agricultural soils maintained by the presence of rhizospheric microbiota in relation with plant beneficial rhizospheric microbes (PBRMs) play an important role to enhance soil quality (Bharti et al. 2016) (Table 15.1).

Some microorganisms like bacteria and fungi secrete brown-coloured substances during decomposition of organic matter which can physically and chemically bury soil particles in subtle formulations. Soil particles can cross-link by the network of fungal hyphae that benefit in the formation and maintenance of soil association (Lehmann et al. 2017).

Origin of new technique has attracted considerable attention and plays a pivotal role to help the improvement in growth and development of plants by bacteria and fungi. In rhizosphere, microorganism *Trichoderma* spp. belongs to such group of microorganisms which is responsible for improving the growth and development of plants (Harman et al. 2004a, b; Qi and Zhao 2013).

15.11 Conclusions

The study of plant microbes has benefited from the overall ecological study on one side and mechanical discoveries which reduce the other. Both schools of thought are providing intense insights into ecological processes which control plants and microorganisms along with the molecular mechanism that makes them comfortable. The study of large separate collections and synthetic microbial communities in conjunction with plant genetic resources will allow us to conduct hypothesis-driven studies that bridge this gap and rapidly decrease complex ecological references to field trials. In these progresses, nature and agriculture have the potential to change our understanding of plant–microorganisms interactions and will make significant contributions to the next green revolution. Once the species of plants is selected naturally with microorganisms, the next step is to achieve the most promising bioagentated plant microbe for the most promising microorganisms (declines, changes, sequencing, detoxification and growth of plants). In relation to capacity it will be their promotion. Since the Philosphere is reducing the major part of air pollution, in this case, philosphere is recommended instead of the rhizosphere inoculation. Plants were brought in contact with toluene, and compared to non-inoculated control plants, there was a considerable increase in the rate of toluene removal after phyllosphere insulation. In addition, plants with their respective microorganisms play a leading role in maintaining ecological sustainability of biodiversity and urban green infrastructure, and the basic knowledge of this symbiosis is of high importance

Table 15.1 Significant effects of various PBRMs of vegetables and grains

Plants	Rhizospheric microbes	Significant effects
Vegetables		
Brinjal	*Azotobacter* spp., *Azospirillum* spp.	Biofertilizer
	Pseudomonas fluorescens	Biocontrol for bacterial wilt
Broccoli	*Bacillus cereus*, *Brevibacillus reuszeri*, *Rhizobium rubi*	Biofertilizer
Cabbage	*Bacillus subtilis*, *B. megaterium*, *Pantoea agglomerans*	Enhance nutrient uptake and ISR[a]
Chilli	*Bacillus* spp., *Arthrobacter* and *Serratia* spp.	ISR against the disease of anthracnose in *Colletotrichum gloeosporioides* and endophytic bacteria
Cucumber	*Bacillus pumilus*	Biocontrol for angular leaf spot by *Pseudomonas syringae*
	Bacillus spp.	Biocontrol for root rot by *Pythium* sp.
Okra	*Bacillus firmus*, *B. subtilis*	ISR against root-knot nematodes
Potato	*Bacillus subtilis*, *B. cereus*, *Azotobacter* spp.	Biofertilizer, ISR for late blight
Radish	*Pseudomonas fluorescens*, *Bacillus subtilis*	ISR
Spinach	*Bacillus megaterium*, *B. subtilis*, *B. licheniformis* *B. cereus*, *Paenibacillus polymyxa*, *Pseudomonas putida*	ISR against *Rhizoctonia*, *Pythium*, *Fusarium* and *Phytopthora* spp.
Tomato	*Bacillus subtilis*, *B. cereus*, *Arthrobacter* spp., *Trichoderma viride*, *Pseudomonas fluorescens*, *P. putida*, *P. marginalis*, *P. syringae*	ISR and biocontrol for late blight
Food grains		
Bean	*Rhizobium tropici*, *Azotobacter brasilence*, *Glomus sinuosum*, *Glomus intraradices*, *Gigaspora albida*, *Pseudomonas fluorescens*	Enhance biofertilization (N and P) Biocontrol against *Rhizoctonia solani*
Maize	*Serratia liquefaciens*, *Bacillus* Sp. *Pseudomonas* spp., *Azospirillum brasilense*, *Bradyrhizobium japonicum*	Biofertilization, biocontrol for root pathogens and phytostimulation

(continued)

Table 15.1 (continued)

Plants	Rhizospheric microbes	Significant effects
Rice	*Pseudomonas fluorescence* Aur6	Biocontrol against *Magnaporthe grisea*
	Chryseobacterium balustinum, *BGA*	Salinity and N fixation
Soybean	*Bradyrhizobium japonicum*, *Bacillus subtilis*, *Bacillus Thuringiensis*, *Pseudomonas flaviporus*, *Bradyrhizobium japonicum*	Enhance biofertilization and phytostimulation
Wheat	*Bacillus circulans*, *Glomus* spp., *Cladosporium herbarum*, *Calothrix* sp., *Anabaena* sp., *Arthrobacter* sp., *Bacillus Subtilis*	Biofertilization and biocontrol against salinity

[a]ISR-Induce systemic resistance

for human health and environmental sustainability. Air pollution affects the ecosystem in many ways, and the effects should be quantified in a series of ecosystem service types to provide a more holistic approach to the effects.

References

Agrawal M, Singh B, Rajput M, Marshall F, Bell JN (2003) Effect of air pollution on peri-urban agriculture: a case study. Environ Pollut 126:323–329

Amato G, Liccardi GD, Amato M, Cazzola M (2002) Outdoor air pollution, climate changes and allergic bronchial asthma. Eur Respir J 20:763–776

Arneth A, Harisson S, Zaehle S, Tsigaridis K, Menon S, Bartlein PJ, Feichter J, Korhola A, Kulmala M, O'Donnell D et al (2010) Terrestrial biogeochemical feedbacks in the climate system. Nat Geosci 3:525–532

Arshad M, Saleem M, Hussain S (2007) Perspectives of bacterial ACC deaminase in phytoremediation. Trends Biotechnol 25:356–362

Bais HP, Weir TL, Perry LG, Gilroy S, Vivanco JM (2006) The role of root exudates in rhizosphere interactions with plants and other organisms. Annu Rev Plant Biol 57:233–266

Baldotto LEB, Olivares FL (2008) Phylloepiphytic interaction between bacteria and different plant species in a tropical agricultural system. Can J Microbiol 54:918–931

Beckett KP, Freer-Smith P, Taylor G (2000) Effective tree species for local air quality management. J Arboric 26:12–19

Bending GD (2003) The rhizosphere and its microorganisms. In: Thomas B, Murphy DJ, Murray BG (eds) Encyclopaedia of applied plant sciences. Academic, London, pp 1123–1129

Berg G, Krechel A, Ditz M, Faupel A, Ulrich A, Hallmann J (2005) Endophytic and ectophytic potato-associated bacterial communities differ in structure and antagonistic function against plant pathogenic fungi. FEMS Microbiol Ecol 51:215–229

Bharti N, Barnawal D, Wasnik K, Tewari SK, Kalra A (2016) Co-inoculation of Dietzia natronolimnaea and Glomus intraradiences with vermi compost positively influences Ocimum basilicum growth and resident microbial community structure in salt affected low fertility soils. Appl Soil Ecol 100:211–225

Bianciotto V, Minerdi D, Perotto S, Bonfante P (1996) Cellular interactions between arbuscular mycorrhizal fungi and rhizosphere bacteria. Protoplasma 193:123–131

Borrego-Hernandez O, Garcia-Reynoso JA, Ojeda-Ramirez MM, Suarez-Lastra M (2014) Retrospective health impact assessment for ozone pollution in Mexico city from 1991 to 2011. Atmosfera 27:261–271

Bouwer EJ, Zehnder AJB (1993) Bioremediation of organic compounds-putting microbial metabolisms to work. Trends Biotechnol 11:360–367

Bowen G, Rovira A (1999) The rhizosphere and its management to improve plant growth. Adv Agron 66:1–102

Brack CL (2002) Pollution mitigation and carbon sequestration by an urban forest. Environ Pollut 116:195–200

Brimner TA, Boland GJ (2003) A review of the non-target effects of fungi used to biologically control plant diseases. Agric Ecosyst Environ 100:3–16

Broeckling CD (2008) Root exudates regulate soil fungal community composition and diversity. Appl Environ Microbiol 74:738–744

Bulgarelli D, Schlaeppi K, Spaepen S, Ver Loren van Themaat E, Schulze-Lefert P (2013) Structure and functions of the bacterial microbiota of plants. Annu Rev Plant Biol 64:807–838

Calfapietra C, Fares S, Manes F, Morani A, Sgrigna G, Loreto F (2013) Role of biogenic volatile organic compounds (BVOC) emitted by urban trees on ozone concentration in cities: a review. Environ Pollut 183:71–80

Cardelino CA, Chameides WL (1990) Natural hydrocarbons, urbanization, and urban ozone. J Geophys Res 95:13971–13979

Carpenter-Boggs L, Loynachan TE, Stahl PD (1995) Spore germination of Gigaspora margarita stimulated by volatiles of soil-isolated actinomycetes. Soil Biol Biochem 27(11):1445–1451

Citernesi AS, Fortuna P, Filippi C, Bagnoli G, Giovannetti M (1996) The occurrence of antagonistic bacteria in Glomus mosseae pot cultures. Agronomie 16(10):671–677

Clemmensen KE, Finlay RD, Dahlberg A, Stenlid J, Wardle DA, Lindahl BD (2015) Carbon sequestration is related to mycorrhizal fungal community shifts during long-term succession in boreal forests. New Phytol 205:1525–1536

Delmotte N, Knief C, Chaffron S, Innerebner G, Roschitzki B, Schlapbach R, Von Mering C, Vorholt JA (2009) Community proteogenomics reveals insights into the physiology of phyllosphere bacteria. Proc Natl Acad Sci U S A 106:16428–16433

Directive 1999/13/EC. http://eur-lex.europa.eu/legal-content/EN/TXT/PDF/?uri=CELEX:01999L0013-20101201&qid=1406008935749&from=ES. Accessed 31 Aug 2015

Dzierzanowski K, Popek R, Gawronska H, Saebo A, Gawronski SW (2011) Deposition of particulate matter of different size fractions on leaf surface and waxes of urban forest species. Int J Phytorem 13:1037–1046

Ehlers RU (2006) Einsatz der Biotechnologie im biologischen Pflanzenschutz. Schrreihe Dtsch Phytomed Ges 8:17–31

Foley JA, Ramankutty N, Brauman KA, Cassidy ES, Gerber JS, Johnston M, Mueller ND, O'Connell C, Ray DK, West PC, Balzer C, Bennett EM, Carpenter SR, Hill J, Monfreda C, Polasky S, Rockstrom J, Sheehan J, Siebert S, Tilman D, Zaks DPM (2011) Solutions for a cultivated planet. Nature 478(7369):337–342

Fowler D, Flechard C, Skiba U, Coyle M, Cape JN (1998) The atmospheric budget of oxidized nitrogen and its role in ozone formation and deposition. New Phytol 139:11–23

Gao Y, Cheng Z, Ling W, Huang J (2010) Arbuscular mycorrhizal fungal hyphae contribute to the uptake of polycyclic aromatic hydrocarbons by plant roots. Bioresour Technol 101:6895–6901

Garbaye J (1991) Biological interactions in the mycorrhizosphere. Experientia 47(4):370–375

Gheorghe IF, Ion B (2015) The effects of air pollutants on vegetation and the role of vegetation in reducing atmospheric pollution. http://cdn.intechopen.com/pdfs-wm/18642.pdf. Accessed 22 Oct 2015

Gianinazzi S, Schuepp H (1994) Impact of arbuscular mycorrhizas on sustainable agriculture and natural ecosystems. Birkhäuser Verlag, Basel, p 226

Giese M, Baue-Doranth LC, Sandermann H Jr (1994) Detoxification of formaldehyde by the spider plant (Chlorophytum comosum L.) and by soybean (Glycine max L.) cell-suspension cultures. Plant Physiol 104:1301–1309

Giri B, Giang PH, Kumari R, Prasad R, Sachdev M, Garg AP, Oelmuller R, Varma A (2005) Mycorrhizosphere: strategies and functions. In: Buscot F, Varma A (eds) Microorganisms in soils: roles in genesis and functions, vol 3. Springer, Berlin, pp 213–252

Glick BR (2010) Using soil bacteria to facilitate phytoremediation. Biotechnol Adv 28:367–374

Goettel MS (2008) Are entomopathogenic fungi only entomopathogens? A preamble. J Invertebr Pathol 98:255

Gunnell D, Eddleston M, Phillips MR, Konradsen F (2007) The global distribution of fatal pesticide self-poisoning: systematic review. BMC Public Health 21(7):357

Harman EG, Howell CR, Viterbo A, Chet I, Lorito M (2004a) Trichoderma species opportunistic, avirulent plant symbionts. Nat Rev Microbiol 2:43–56

Harman WL, Gerik TJ, Williams JR (2004b) CROPMAN: a crop production/environmental simulation tool to enhance agricultural sustainability. In: Dilawari VK, Brar LS, Jalota SK (eds) Proceedings of sustaining agriculture: problems and prospects. Punjab Agricultural University, Ludhiana, pp 22–30

Harrison SP, Morfopoulos C, Dani KGS, Prentice IC, Arneth A, Atwell BJ, Barkley MP, Leishman MR, Loreto F, Medlyn BE et al (2012) Volatile isoprenoid emissions from plastid to planet. New Phytol 197:49–57

Hawkins HJ, Johansen A, George E (2000) Uptake and transport of organic and inorganic nitrogen by arbuscular mycorrhizal fungi. Plant Soil 226:275–285

Hiltner L (1904) Über neuere Erfahrungen und Probleme auf dem Gebiete der Bodenbakteriologie unter besonderer Berücksichtigung der Gründungung und Brache. ArbDtsch Landwirtsch Ges 98:59–78

Ignoffo CM (1992) Environmental factors affecting persistence of entomopathogens. Fla Entomol 75:516–525

Jackson T, Glare TR (1992) Use of pathogens in scarab pest management. Intercept, Andover

Johansson JF, Paul LR, Finlay RD (2004) Microbial interactions in the mycorrhizosphere and their significance for sustainable agriculture. FEMS Microbiol Ecol 48(1):1–13

John MW (2001) Microbial interactions and biocontrol in the rhizosphere. J Exp Bot 52:487–511

Johnson SC, Mitchell C, McNicol JW, Thompson J, Karley AJ (2013) Downstairs drivers - root herbivores shape communities of above-ground herbivores and natural enemies via changes in plant nutrients. J Anim Ecol 82:1021–1030

Karami M, Homaee S, Afzalinia H, Ruhipour S (2012) Organic resource management: impacts on soil aggregate stability and other soil physico-chemical properties. Agric Ecosyst Environ 148:22–28

Karl T, Harley P, Emmons L, Thornton B, Guenther A, Basu C, Turnipseed A, Jardine K (2010) Efficient atmospheric cleansing of oxidized organic trace gases by vegetation. Science 330:816–819

Khan HU, Ahmad R, Ahmed W, Khan SM, Aslam Khan M (2001) Evaluation of the combined effects of Paecilomyces lilacinus and Trichoderma harzianum against root-knot disease of tomato. J Biol Sci 1(3):139–142

Kim KJ, Kil MJ, Song JS, Yoo EH (2008) Efficiency of volatile formaldehyde removal by indoor plants: contribution of aerial plant parts versus the root zone. Hortscience 133:521–526

Knief C, Delmotte N, Chaffron S, Stark M, Innerebner G, Wassman R, von Mering C, Vorholt JA (2012) Metaproteogenomic analysis of microbial communities in the phyllosphere and rhizosphere of rice. ISME J 6:1378–1390

Langley JA, Hungate BA (2003) Mycorrhizal controls on belowground litter quality. Ecology 84:2302–2312

Leach AW, Mumford JD (2008) Pesticide environmental accounting: a method for assessing the external costs of individual pesticide applications. Environ Pollut 151:139–147

Lee KE, Pankhurst CE (1992) Soil organisms and sustainable productivity. Aust J Soil Res 30 (6):855–892

Lehmann A, Leifheit EF, Rillling MC (2017) Mycorrhiza and soil aggregation. Mycorrhizal mediation of soil fertility, structure and carbon storage, Chapter 14, pp 241–262

Lesaulnier C, Papamichail D, McCorkle S, Olivier B, Skiena S, Taghavi S, Zak D, van der Lelie D (2008) Elevated atmospheric CO_2 affects soil microbial diversity associated with trembling aspen. Environ Microbiol 10:926–941

Lichtfouse E, Navarrete M, Debaeke P, Souchère V, Alberola C, Ménassieu J (2009) Agronomy for sustainable agriculture: a review. In: Sustainable agriculture. Springer, Dordrecht, pp 1–7

Linderman RG (1988) Mycorrhizal interactions with the rhizosphere microflora: the Mycorrhizosphere effect. Phytopathology 78:366–370

Lindow SE, Brandl MT (2003) Microbiology of the phyllosphere. Appl Environ Microbiol 69:1875–1883

Lynch JM (1986) Rhizosphere microbiology and its manipulation. Biol Agric Hortic 3:143–152

Lynch JM (ed) (1987) The rhizosphere. Wiley, Chichester

Lynch JM (1990) The Rhizosphere. Wiley, Chichester, 458 p

Marilley L, Vogt G, Blanc M, Aragno M (1998) Bacterial diversity in the bulk soil and rhizosphere fractions of *Lolium perenne* and *Trifolium repens* as revealed by PCR restriction analysis of 16S rDNA. Plant Soil 198:219–224

Marschner H (1995) Mineral nutrition of higher plants, 2nd edn. Academic, London

Maurhofer M, Keel C, Schnider U, Voisard C, Haas D, Defago G (1992) Influence of enhanced antibiotic production in Pseudomonas fluorescens strain CHA0 on its disease suppressive capacity. Phytopathology 82:190–195

McDonald AG, Bealey WJ, Fowler D, Dragosits U, Skiba U, Smith RI, Donovan RG, Brett HE, Hewitt CN, Nemitz E (2007) Quantifying the effect of urban tree planting on concentrations and depositions of PM10 in two UK conurbations. Atmos Environ 41:8455–8467

McGrath SP, Chaudri AM, Giller KE (1995) Long-term effects of metals in sewage-sludge on soils, microorganisms and plants. J Ind Microbiol 14:94–104

McGuinness M, Dowling D (2009) Plant-associated bacterial degradation of toxic organic compounds in soil. Int J Environ Res Public Health 6:2226–2247

Megali L, Glauser G, Rasmann S (2013) Fertilization with beneficial microorganisms decrease tomato defenses against insect pests. Agron Sustain Dev 34:649–656

Miransari M (2011) Arbuscular mycorrhizal fungi and nitrogen uptake. Arch Microbiol 193:77–81

Mohsenzadeh F, Nasseri S, Mesdaghinia A, Nabizadeh R, Zafari D, Khodakaramian G, Chehregani A (2010) Phytoremediation of petroleum-polluted soils: application of Polygonum aviculare and its root-associated (penetrated) fungal strains for bioremediation of petroleum-polluted soils. Ecotoxicol Environ Saf 73:613–619

Morris SJ, Blackwood CB (2015) The ecology of soil biota and their function, Chapter 10. In: Soil microbiology, ecology and biodiversity, 4th edn. Academic, London, pp 273–309

Mosttafiz S, Rahman M, Rahman M (2012) Biotechnology: role of microbes in sustainable agriculture and environmental health. Internet J Microbiol 101:1937–8289

Myers I, Maynard RL (2005) Polluted air-outdoors and indoors. Occup Med 55:432–438

Nihorimbere V, Ongena M, Smargiassi M, Thonart P (2011) Beneficial effect of the rhizosphere microbial community for plant growth and health. Biotechnol Agron Soc Environ 15 (2):327–337

Nowak DJ (2015) Air pollution removal by Chicago's urban forest. http://www.nrs.fs.fed.us/pubs/gtr/gtr_ne186.pdf. Accessed 22 Oct 2015

Nowak DJ, Civerloo KL, Rao ST, Sistla G, Luley CJ, Crane DEA (2000) Modeling study of the impact of urban trees on ozone. Atmos Environ 34:1610–1613

Nowak DJ, Crane DE, Stevens JC (2006) Air pollution removal by urban trees and shrubs in the United States. Urban For Urban Green 4:115–123

Nowak DJ, Crane DE, Stevens JC, Hoehn RE, Walton JT, Bond JA (2008) Ground-based method of assessing urban forest structure and ecosystem services. Arboricult Urban For 34:347–358

Ostro BD, Tran H, Levy JI (2006) The health benefits of reduced tropospheric ozone in California. J Air Waste Manag Assoc 56:1007–1021

Papen H, Gebler A, Zumbusch E, Rennenberg H (2002) Chemolithoautotrophic nitrifiers in the phyllosphere of a spruce ecosystem receiving high atmospheric nitrogen input. Curr Microbiol 44:56–60

Parrish DD, Zhu T (2009) Clean air for megacities. Science 326:674–675

Pastuszka JS (2007) Effect of particulate aerosols on air quality—methods of identification and assessment (Wplyw aerozoli ziarnistych na jakosc powietrza—Metody identyfikacji I oceny. Ekoprofit. Finanse Nauka Technol. Prawo). Ekoprofit Financ Sci Technol Law 2:7–15

Pinton R, Varanini Z, Nannipieri P (2001) The rhizosphere: biochemistry and organic substances at the soil-plant interface. Marcel Dekker, New York, NY, 424 p

Popek R, Gawronska H, Wrochna M, Gawronski SW, Saebo A (2012) Particulate matter on foliage of 13 woody species: deposition on surfaces and phytostabilization in waxes- a 3-year study. Int J Phytorem 15:245–256

Popek R, Gawrońska H, Gawroński SW (2015) The level of particulate matter on foliage depends on the distance from the source of emission. Int J Phytorem 17:1262–1268

Prasad R, Kumar M, Varma A (2015) Role of PGPR in soil fertility and plant health. In: Egamberdieva D, Shrivastava S, Varma A (eds) Plant growth-promoting rhizobacteria (PGPR) and medicinal plants. Springer International Publishing, Switzerland, pp 247–260

Prasad R, Gupta N, Satti U, Wang S, Ahmed ISA, Varma A (2017) Management of fungal pathogens by mycorrhiza. In: Varma A, Prasad R, Tuteja N (eds) Mycorrhiza. Springer International Publishing AG, pp 179–194

Qi WZ, Zhao L (2013) Study of the siderophore-producing Trichoderma asperellum Q1 on cucumber growth promotion under salt stress. J Basic Microbiol 53:355–364

Raaijmakers JM, Weller DM (2001) Exploiting genotypic diversity of 2,4 diacetylphloroglucinol producing pseudomonas spp.: characterization of superior root-colonizing P. fluorescens strain Q8r1-96. Appl Environ Microbiol 67:2545–2554

Rajkumar M, Sandhya S, Prasad MNV (2012) Freitas, H. perspectives of plant-associated microbes in heavy metal phytoremediation. Biotechnol Adv 1:1562–1574

Ren Y, Ge Y, Gu B, Min Y, Tani A, Chang J (2014) Role of management strategies and environmental factors in determining the emission of biogenic volatile organic compounds from urban greenspaces. Environ Sci Technol 48:6237–6246

Richardson AE, Barea JM, McNeill AM, Prigent-Combaret C (2009) Acquisition of phosphorus and nitrogen in the rhizosphere and plant growth promotion by microorganisms. Plant Soil 321:305–339

Saebo A, Popek R, Nawrot B, Hanslin HM, Gawronska H, Gawronski SW (2012) Plant species differences in particulate matter accumulation on leaf surfaces. Sci Total Environ 427-428:347–354

Salthammer T, Uhde E (2009) Organic indoor air pollutants: occurrence-measurement-evaluation, 2nd edn. Wiley, New York, NY, pp 1–464

Samoli E, Aga E, Touloumi G, Nisiotis K, Forsberg B, Lefranc A, Pekkanen J, Wojtyniak B, Schindler C, Niciu E et al (2006) Short-term effects of nitrogen dioxide on mortality: an analysis within the APHEA project. Eur Respir J 27:1129–1137

Scheller RM, Van Tuyl S, Clark KL, Hom J, La Puma I (2011) Carbon sequestration in New Jersey pine barrens under different scenarios of fire management. Ecosystems 2011(14):987–1004

Schneider A, Tesileanu R, Charles R, Sinaj S (2013) Kinetics of soil potassium sorption-desorption and fixation. Commun Soil Sci Plant Anal 44:837–849

Sedjo R, Sohngen B (2012) Carbon sequestration in forests and soils. Annu Rev Resour Econ 4:127–144

Sriprapat W, Thiravertyan P (2013) Phytoremediation of BTEX from indoor air by Zamioculcas zamiifolia. Water Air Soil Pollut 224:1482

St-Arnaud M, Vujanovic V (2007) Effect of the arbuscular mycorrhizal symbiosis on plant diseases and pests. In: Hamel C, Plenchette C (eds) Mycorrhizae in crop production. Haworth Food & Agricultural Products Press, Binghampton, NY, pp 67–122

Studdert JP, Kaya HK (1990) Water potential, temperature, and soil type on the formation of Beauveria bassiana soil colonies. J Invertebr Pathol 56:380–386

Taha H (1996) Modeling impacts of increased urban vegetation on ozone air quality in the south coast Air Basin. Atmos Environ 30:3423–3430

Thakore Y (2006) The biopesticide market for global agricultural use. Ind Biotechnol 2:194–208

Vagaggini B, Taccola M, Cianchetti S, Carnevali S, Bartoli ML, Bacci E, Dente FL, Di Franco A, Giannini D, Paggiaro PI (2002) Ozone exposure increases eosinophilic airway response induced by previous allergen challenge. Am J Respir Crit Care Med 166:1073–1077

Van Sluys MA, Monteiro-Vitorello CB, Camargo LE, Menck CF, da Silva AC, Ferro JA, Oliveira MC, Setubal JC, Kitajima JP, Simpson AJ (2002) Comparative genome analysis of plant-associated bacteria. Annu Rev Phytopathol 40:169–189

Vega FE (2008) Insect pathology and fungal endophytes. J Invertebr Pathol 98:277–279

Vessey JK (2003) Plant growth promoting rhizobacteria as biofertilizers. Plant Soil 2003 (255):571–586

Vinale F, Sivasithamparam K, Ghisalberti EL, Marra R, Barbetti MJ, Lim H, Woo SL, Lorito M (2008) A novel role for Trichoderma secondary metabolites in the interactions with plants. Physiol Mol Plant Pathol 72:80–86

Vorholt JA (2012) Microbial life in the phyllosphere. Nat Rev Microbiol 10:828–840

Voriskova J, Baldrian P (2013) Fungal community on decomposing leaf litter undergoes rapid successional changes. ISME J 7:477–486

Welburn A (1990) Why are atmospheric oxides of nitrogen usually phytotoxic and not alternative fertilizers? New Phytol 115:395–429

Welburn A (1998) Atmospheric nitrogenous compounds and ozone-Is NOx fixation by plants a possible solution? New Phytol 139:5–9

Weyens N, van der Lelie D, Taghavi S, Newman L, Vangronsveld J (2009a) Exploiting plant–microbe partnerships for improving biomass production and remediation. Trends Biotechnol 27:591–598

Weyens N, van der Lelie D, Taghavi S, Vangronsveld J (2009b) Phytoremediation: plant-endophyte partnerships take the challenge. Curr Opin Biotechnol 20:248–254

Whipps J (2001) Microbial interactions and biocontrol in the rhizosphere. J Exp Bot 52:487–511

Whipps JM (2004) Prospects and limitations for mycorrhizas in biocontrol of root pathogens. Can J Bot 82:1198–1227

White PJ (2003) Ion transport. In: Thomas B, Murphy DJ, Murray BG (eds) Encyclopaedia of applied plant sciences. Academic, London, pp 625–634

Winter K, Holtum JAM (2014) Facultative crassulacean acid metabolism (CAM) plants: powerful tools for unravelling the functional elements of CAM photosynthesis. J Exp Bot 65:425–3441

Wolverton BC, Wolverton JD (1993) Plants and soil microorganisms: removal of formaldehyde, ethylbenzene, and ammonia from the indoor environment. J Miss Acad Sci 38:11–15

World Health Organization (WHO) (2000) The right to healthy indoor air—report on a WHO meeting. World Health Organization (WHO), Bilthoven

Wu D, Sun MZ, Zhang C, Xin Y (2014) Antioxidant properties of lactobacillus and its protecting effect to oxidative stress Caco-2 cells. J Anim Plant Sci 24:1766–1771

Yang J, McBride J, Zhou J, Sun Z (2005) The urban forest in Beijing and its role in air pollution reduction. Urban For Urban Green 3:65–68

Yang DS, Pennisi SV, Son KC, Kays SJ (2009) Screening indoor plants for volatile organic pollutant removal efficiency. Horscience 44:1377–1381

Yu C, Crump D (1998) A review of the emission of VOCs from polymeric materials used in buildings. Build Environ 33:357–337

Chapter 16
Biological Control of Soft-Rot of Ginger: Current Trends and Future Prospects

Mahendra Rai, Patrycja Golińska, Sudhir Shende, Priti Paralikar, Pramod Ingle, and Avinash P. Ingle

Abstract Ginger (*Zingiber officinale* Roscoe) is an important crop having various medicinal, nutritional, and ethnomedicinal properties cultivated all over the world. *Pythium* and *Fusarium* spp. are pathogens responsible for the deteriorating disease in ginger known as soft- or rhizome-rot, causing more than 50% loss of ginger crop worldwide. The application of chemical fungicides is a promising method for control of soft-rot in ginger. But use of such fungicides is harmful to both environment and human health. Thus, there is an obligatory need for the search of an eco-friendly and economic approach for the control of soft-rot in ginger. Various physical, chemical, and biological methods have already been in practice since many years for managing soft-rot in ginger. This chapter primarily focuses on the advantages of biological control over chemical methods of *Pythium* and *Fusarium* spp. management using antagonistic fungi, bacteria, actinomycetes, and plant extracts. These biocontrol agents offer the best opportunity in control of diseases and also help to maintain the quality and crop yield. Moreover, the emerging role of nanotechnology in the management of these pathogens is also briefly discussed.

16.1 Introduction

Ginger (*Zingiber officinale* Roscoe) is an important plant crop cultivated all over the world for its promising medicinal properties (Rai et al. 2018). However, India is among the most leading producers and exporter of ginger (Anisha and Radhakrishnan 2015; Gupta and Kaushal 2017). Due to potential medicinal, nutritional, and ethnomedicinal properties, ginger is widely used as a spice, flavoring

M. Rai (✉) · S. Shende · P. Paralikar · P. Ingle
Department of Biotechnology, SGB Amravati University, Amravati, Maharashtra, India
e-mail: mahendrarai@sgbau.ac.in

P. Golińska
Department of Microbiology, Nicolaus Copernicus University, Torun, Poland

A. P. Ingle
Department of Biotechnology, Engineering School of Lorena, University of Sao Paulo, Lorena, SP, Brazil

A. Varma et al. (eds.), *Plant Microbe Interface*,
https://doi.org/10.1007/978-3-030-19831-2_16

agent, and herbal medicines (Dhanik et al. 2017). The ginger crop is susceptible to various diseases caused by bacteria, fungi, or viruses which mainly include soft-rot, yellows, *Phyllosticta* leaf spot, storage-rot, bacterial wilt, mosaic, chlorotic fleck, etc. These diseases reduce potential yields drastically (Gupta and Kaushal 2017; Rai et al. 2018).

Soft-rot (rhizome-rot) is one of the most common and destructive diseases of ginger caused by various species of *Pythium* (mainly by *P. aphanidermatum* (Edson) Fitz, *P. deliense* Meurs, *P. graminicola* Subram, *P. myriotylum* Drechsler, *P. spinosum* Sawada, *P. splendens* Braun, *P. ultimum* Trow, *P. vexans* de Bary, and *P. zingiberis* Takahashi), *Fusarium* spp. (mainly by *F. oxysporum* f. sp. *zingiberi*), and bacteria (e.g., *Ralstonia* spp.) (Le et al. 2014, 2016; Gupta and Kaushal 2017; Rai et al. 2018). The disease is both seed and soilborne, and its development depends on moisture and temperature conditions of soil (Gupta and Kaushal 2017). Soft-rot caused by *Pythium* spp. is carried over and maintained through diseased rhizomes as oospores in scales and soil. Fungal pathogens have ability to survive as saprophytes in plant debris, which may contain a large number of oospores and thus acts as a source of primary inoculum (Gupta and Kaushal 2017).

Soft-rot is considered as a complex disease condition. There are various conventional strategies, namely, cultural practices and biological and chemical agents, commonly used for disease management. The application of these practices in ginger fields helps in controlling the diseases and also restricts the dissemination of fungal pathogens (Le et al. 2014; Gupta and Kaushal 2017; Rai et al. 2018). It is demonstrated that the management of soft-rot is difficult by using any one conventional approach. Therefore, combination of more than one approaches has been found to be more satisfactory in the control of this disease (Dohroo et al. 2015; Gupta and Kaushal 2017).

Cultural practices like seed selection, crop rotation, organic amendment, drainage and quarantine, and chemical fungicides are most frequently used to control soft-rot of ginger. There are two types of chemical fungicides: one is applied to soil (zineb, captafol, methyl bromide, mercuric chloride, thiram, phenylmercury acetate, copper oxide, mancozeb, and many more), and another is commonly used for seed treatment (Ridomil MZ, Fytolan, Bavistin, Thimet, etc.). However, these fungicides are more effective when used in combination (Mathur et al. 2002; Rajan et al. 2002; Singh 2011; Smith and Abbas 2011; Le et al. 2014). Although chemical fungicides can be effectively used for the control of soft-rot of ginger, their continuous use may cause harm to both environmental and human health (Rai et al. 2018). Moreover, the frequent use of chemical fungicides leads to increase in resistance of fungi toward such fungicides and also reduces soil fertility (Ponmurugan et al. 2016). Therefore, search for novel eco-friendly agents such as biological agents (microorganisms, plants) for control of soft-rot is essentially required. Among microorganisms, *Trichoderma* spp. are the most widely used biocontrol agents of rhizome-rot of ginger caused by *Fusarium* and *Pythium* spp. (Selvakumar et al. 2013; Shanmugam et al. 2013a, b). However, combined applications of bioagents such as *Trichoderma harzianum*, *Pseudomonas fluorescens*, and *Bacillus subtilis* were found to be more effective when compared to the individual treatments (Dohroo and Gupta 2014). Moreover, many other bacteria (*Enterobacter* sp., *Rhizobium* sp.), actinobacteria (*Nocardiopsis* sp., *Streptomyces* sp., *Micromonospora* sp.), and plants (*Lawsonia inermis, Nigella sativa, Azadirachta indica, Zingiber zerumbet*, etc.) have also been used as biocontrol agents against disease caused by *Pythium* (El-Tarabily et al. 1997; Chin-A-Woeng et al. 2003; Bardin et al. 2004; Bhai et al. 2005; Gupta et al. 2013; Loliam et al. 2013; Parveen and Sharma 2014; Ravi et al. 2017).

In the present chapter, we focused on biological control strategy used for the control of soft-rot disease of ginger caused by fungal pathogens. In addition, advantages of biocontrol methods over chemical control have also been discussed.

16.2 Soft-Rot: Causal Organisms (Mainly Different Species of *Pythium*)

Several species of *Pythium* have been reported from various parts of world which have ability to cause soft-rot disease in ginger. *Pythium gracile* (de Bary) was reported for the first time to cause rhizome-rot of ginger by Butler (1907) from Surat (Gujarat), and in Bengal, from Kerala by Sen (1930), and from Assam and Fiji by Parham (1935) in India. Apart from these, various other *Pythium* species have been found associated with soft-rot of ginger. For instance, *P. aphanidermatum* (Edson) Fitz. in Pusa (Bihar) (Mitra and Subramaniam 1928), in Nagpur (Maharashtra) (Sahare and Asthana 1962), in Madhya Pradesh (Haware and Joshi 1974), and in Kerala (Sarma et al. 1979). Similarly, 11 species of *Pythium* were recovered from infected rhizome of ginger showing symptoms of soft-rot, and it was reported that *P. aphanidermatum* and *P. myriotylum* were the most prevalent species (Dohroo 2005). *P. butleri* Subram. exists from 1918 in the Malabar and South Kanara district of South India (Thomas 1938) as a causative agent of rhizome-rot, and it was reported afterward in Ceylon (Park 1934). *P. complectans* Braun was isolated from infected rotted part of ginger in Ceylon (Park 1934). *P. graminicolum* Subram was reported from Ceylon (Park 1935). *P. delense* Meurs was described from Madhya Pradesh (Haware and Joshi 1974).

In Bombay, Ceylon, Hong Kong, Kerala, Nagpur, Poona, and Taiwan, *P. myriotylum* Drechsler was found to be the main causative agent that affected the ginger rhizome (Park 1937; Uppal 1940; Bertus 1942; Patel et al. 1949; Sahare and Asthana 1962; Lin et al. 1971; Dake and Edison 1989). *P. zingiberum* was reported from Osaka (Japan) and Korea (Takahashi 1954; Yang et al. 1988). *P. pleroticum* T Ito causes disease in Solan of Himachal Pradesh (Sharma and Dohroo 1982). *P. ultimum* affected the rhizomes of ginger in Himachal Pradesh (Dohroo 1987). In Rajasthan *P. myriotylum* was found in association with *Fusarium solani* causing soft-rot of ginger (Mathur et al. 1984; Drojee 1986). According to Le et al. (2014), more than 15 *Pythium* species may cause soft-rot in ginger, and *P. aphanidermatum* causes about 60% of yield loss. In another study, they have recovered 11 different species of *Pythium* from infected rhizome of ginger from farms in Queensland, Australia, and assessed them for their pathogenicity on ginger (Le et al. 2016). Out of these *Pythium* isolates, *P. aphanidermatum*, *P. deliense*, *P. myriotylum*, *P. splendens*, *P. spinosum*, and *P. ultimum* were found to be most pathogenic.

However, several other species of *Pythium*, viz., *P. myriotylum* Drechsler (Wang et al. 2003) and *P. aphanidermatum* (Edson) Fitzpatrick (Kavita and Thomas 2008), were reported from various countries such as Taiwan, Malaysia, the USA, Japan (Moreira et al. 2013), India (Ravindran and Babu 2005), Australia, and Fiji (Stirling et al. 2009).

16.3 Current Physical and Chemical Methods to Control Infection

As discussed earlier, various physical, chemical, and biological methods are commonly used for the management of soft-rot of ginger; all these methods are briefly described below.

16.3.1 Physical Methods

One of the most important criteria to avoid the soft-rot is the selection of healthy and disease-free seeds or rhizomes; such selection helps to minimize the probability of contamination by *Pythium* spp. (Dake 1995). In order to obtain good quality of seeds, there are various approaches of seed treatment like seed fortification (using biological or physical approaches or their combinations), seed disinfestations (to kill the pathogens present on the seed surface), and seed disinfection (using various disinfection agents kill the pathogens present in the cells). All these approaches are promisingly helpful in the management of pathogen without causing any harm to embryo or potential of seed germination (Bennett et al. 1991; Rai et al. 2018). Prevalence of the pathogen in soil is also responsible for setting the infections, if a particular crop or any other crop which acts as host for the same pathogen is cultivated every year. In this context, there is necessity to cultivate the different crops, i.e., crop alternation or rotation can be the prominent approach which avoids the recurrence of pathogen in the subsequent harvesting. It was suggested that crops like corn and rice can be used as alternate crops after cultivation of ginger in the same field because corn and rice are tolerant to pathogens of ginger (Pordesimo and Raymundo 1963; Quimio and Chan 1979; Bennett et al. 1991).

In conventional agricultural practices, application of suppressive soil for fastidious pathogens is another approach for better crop protection. Lee et al. (1990) proposed that soil with higher clay content and lower pH is suitable for ginger cultivation as it suppress the growth of *Pythium zingiberum* and *F. oxysporum* f. sp. *zingiberi* as compared to conductive soils. Soil solarization is another important approach which helps to destruct molds present in the soil for better growth and health of crop. The heating of soil covered with plastic films using solar energy in the summer season for 1–2 months reduces load of pathogens and various other pests and weeds. In addition to soil solarization, use of biocontrol agents is advantageous to growth of plants and to restrict the growth of a variety of pathogens. Soil solarization is considered as one of the most suitable approaches for home gardens, nurseries, landscaping, and greenhouses due to its low-cost and long-term benefits (Dake 1995; Stapleton and Devay 1986). Moreover, use of silicon (Si) (may be in the form of potassium silicate) as supplement in the soil is reported to enhance the plant growth and also inhibits the growth of *P. aphanidermatum* (Chérif et al. 1994). Routine phytosanitation is recommended as soon as disease symptoms appear in the field to decrease its spread to the other healthy plants. Similarly, rouging of diseased plants and demolishing them followed by disinfection of tools used for phytosanitation to avoid transfer of inocula to healthy plants is an essential practice (Dake 1995).

16.3.2 Chemical Methods

A variety of fungicides are commonly used around the globe for controlling postharvest diseases in ginger since 1940. *Pythium* spp. have the ability to survive in the soil for years together once introduced (Hoppe 1966), and hence the management of soft-rot is more difficult. Till date, a large number of chemical fungicides have been discovered and routinely used worldwide. Some of the important fungicides include mancozeb, ziram, guazatine, propineb, and copper oxychloride. These fungicides are considered as most promising in the effective control of soft-rot (Dohroo and Sharma 1986; Thakore et al. 1988). In addition, metalaxyl (fosetyl-aluminum/Ridomil) is one of the most commonly used chemical fungicides. This fungicide is useful in both soil application and also as drench alone or in combination with other fungicides for the significant control of soft-rot caused by *Pythium* (Chase et al. 1985; Ramachandran et al. 1989; Dake 1995; Hwang et al. 2001; Luong et al. 2010). Singh (2011) performed a comparative study on seed treatment with Ridomil MZ (1.25 g/L) and hot water (51 °C for 30 min) in a naturally contaminated field with *P. aphanidermatum* in Raigarh, India, and reported 30% more survival of rhizomes treated with Ridomil MZ. Similarly, in a pot trial experiment, it was observed that seed coated with Fytolan (copper oxychloride 0.2%) + Ridomil (500 ppm) + Bavistin (carbendazim 0.2%) + Thimet keep ginger rhizomes free from soft-rot (Rajan et al. 2002). In addition, seed treatment with Smith and Abbas (2011) proposed that fungicides like metalaxyl, Ridomil, Maxam XL (fludioxonil) and Proplant (propyl carbamate hydrochloride) considerably helps in the management of soft-rot caused by *P. myriotylum* than sole carbendazim seed treatment in a pot trial.

Various other antifungal agents like zineb, captafol, methyl bromide, mercuric chloride, thiram, phenyl mercury acetate, copper oxide, mancozeb, etc. reported to have effective antifungal activity against different *Pythium* species (Doshi and Mathur 1987). Dohroo et al. (1984) reported significant efficacy of metalaxyl in the control of rhizome-rot. Similarly, treatment of seed (1 day before) and soil drenching (3 months after planting) with the mixture of metalaxyl and captafol effectively controlled the soft-rot of ginger (Rathaiah 1987). Apart from these, fosetyl-Al, metalaxyl, oxadixyl, propamocarb and ethazole (epidiazole) were also evaluated against *P. aphanidermatum*. Among these, metalaxyl formulations (Ridomil 5G and Apron 35 WS) were found to be most effective when used in soil and seed treatments (Ramachandran et al. 1989). Srivastava (1994) effectively controlled the soft-rot of ginger by inhibiting growth of causative agent (*P. aphanidermatum*) in Sikkim by drenching the soil with zineb or mancozeb following rhizome treatment with carbendazim and incorporating Thiodan dust into the soil to control insect invasion. Rhizome fly is a common insect pest found in association with rhizome-rot of ginger caused by *Pythium* sp. Gautam and Mainali (2016) demonstrated that the combination of Chlorpyrifos 20 EC (insecticide) + Dithane M-45 (Mancozeb 80 WP) (pesticides) and Bavistin (Carbendazim 50 DF) (pesticides) was significantly effective against rhizome fly and rhizome-rot (Gautam and Mainali 2016).

16.3.3 Biological Control of Pythium *spp.*

Eco-friendly methods of disease management are being practiced nowadays. The increasing use of hazardous fungicides in agriculture has been growing cause worldwide concern. Therefore, increased concern for the hazards associated with the use of synthetic pesticides and the use of biological agents for control of plant pathogens during the past 20 years has been driven in part by trends in agriculture toward greater sustainability. The biological control is defined as the reduction in disease producing ability or density of microbial inoculum by one or more organism accomplished naturally in its active state or through manipulation of the environment, by mass introduction of antagonists (Agrios 2005; Heydari and Pessarakli 2010). The new insights into the underlying mechanisms by which biocontrol agents function can be evolved by technologies from molecular biology and genetics which allowed the evaluation of the behavior of microbial inoculants in natural environments to a degree not previously possible (Thomashow and Weller 1996).

Presently, a variety of organisms, mainly bacteria and fungi that counteract important agronomical pests and diseases, have been described. These include *Trichoderma* species (Harman 2006; Rai et al. 2018); mycoparasitic (where a fungus directly attacks and feeds on other fungi, resulting in the direct destruction or lysis of propagules and structures) members of the genus *Verticillium* (Gajera et al. 2013); *Pseudomonas*, *Bacillus*, and *Streptomyces* (Ashwini and Srividya 2012; Beneduzi et al. 2012; Fróes et al. 2012; Sivasakthi et al. 2014; Menendez and Garcia-Fraile

2017); and *Lecanicillium* species (Fenice and Gooday 2006). Fungal species against plant pathogens have attracted a great deal of attention from the researchers around the globe as potential biocontrol agents in many crops, of which one of the most well-studied fungal genera is *Trichoderma/Hypocrea* (Gajera et al. 2013; Yacoub et al. 2017; Rai et al. 2018). Biocontrol agents may also induce plant physiological processes that lead to plant defense mechanism activation such as production of phytoalexins, the hypersensitive response, or synthesis of chitinase and glucanase (lytic enzymes) (Thakur and Sohal 2013).

Biological control of *Pythium* species is considerably difficult because of immediate infection of sporangia in seed or root the ability to cause long-term root rots (Whipps and Lumsden 1991). In spite of these constraints, many important diseases have been controlled with antagonistic fungi, bacteria, and actinomycetes (Nayak et al. 2017; Rai et al. 2018). In vitro tests using *T. viride*, *T. harzianum*, and *T. hamatum* against *P. aphanidermatum*, *F. equiseti*, and *F. solani* showed inhibitory effect (Dohroo et al. 2012; Shanmugam et al. 2013a, b; Hudge 2015; Mudyiwa et al. 2016). Good control of storage-rot caused by *P. aphanidermatum* and *F. equiseti* was obtained when *T. viride* and *T. hamatum* were applied to rhizomes either by socking them in smear or spore suspension with the antagonists (Khatso and Tiameren Ao 2013; Hudge 2015; Mudyiwa et al. 2016). Effective suppression of soft-rot of ginger was reported under field condition, when *T. viride* and *T. harzianum* was applied to soil in combination with sawdust (Kulkarni and Hegde 2002). The effective control of *P. aphanidermatum* causing soft-rot of ginger was observed when *T. harzianum* or *T. hamatum* was applied to soil along with neem oil cake (Abbasi et al. 2005). The potential inhibition of growth of *Fusarium oxysporum f.* sp. *zingiberi* and *P. aphanidermatum* causing yellows and rhizome-rot of ginger was observed after use of *T. harzianum*, *T. virde*, *Azadirachta indica* Juss, and *Agave americana* L. as biocontrol agents (Rajan et al. 2002; Singh 2011; Gupta et al. 2013; Parveen and Sharma 2014; Gupta and Kaushal 2017). Ram et al. (2000) demonstrated the role of different biocontrol agents like *T. harzianum*, *T. aureoviride*, and *T. virens* in the control of ginger rhizome-rot. It was reported that all the abovementioned biocontrol agents significantly reduced the population density of both *F. solani* and *P. aphanidermatum*. Similarly, when *T. harzianum* was applied in the soil for the management of rhizome-rot of turmeric (*F. solani*), this resulted in reduced disease incidence and increased yield (Reddy et al. 2003).

16.3.4 Bacteria and Actinomycetes as a Biocontrol Agents

The bacteria are potentially used as biocontrol agent due to its ability to produce important metabolites like lipopeptides which possess strong antifungal activity (Meena and Kanwar 2015; Fira et al. 2018). Among the bacteria and actinomycetes, fluorescent pseudomonads, *Bacillus* spp., and *Streptomyces* received maximum attention because these microorganisms can be grown easily in large-scale and applied to the both seed and soil. Similarly, the fluorescent *Pseudomonas* includes the species of *P. fluorescens*, *P. putida*, *P. aeruginosa*, *P. chlororaphis*, *P. aureofaciens*, and *P. syringae* (Hagedorn et al. 1990; Howie and Suslow 1991; Zheng and Sinclair 2000; Naseby et al. 2001). These species produces various secondary metabolites with antagonistic characteristics most of which are nitrogen containing heterocyclic compounds or unusual amino acids and peptides. However, the members of the genus *Trichoderma* are most widely used biocontrol agents all over the world.

Recently, Zouari et al. (2016) reported broad-spectrum antifungal potential shown by *Bacillus amyloliquefaciens* strain CEIZ-11 against various plant pathogens especially *P. aphanidermatum.* Sellem et al. (2017) demonstrated the potential activity of actinomycetes, *Streptomyces* strain TN258 isolated form Tunisian Sahara soil against *P. ultimum* responsible for potato tubers leak. The results suggest that mycelial growth of *P. ultimum* was completely inhibited by total destruction of hyphae after application of *Streptomyces* strain TN258 extract. Further, author reported that there was significant decrease in pathogen penetration activity was observed on treatment of *Streptomyces* strain TN258 filtrate to potato tubers. *Stenotrophomonas maltophilia*, *Lysobacter enzymogenes*, *Paenibacilli*, *Serratia entomophila*, *E. faecalis*, and *Streptomyces rubrolavendulae* were reported for their ability to control different diseases caused by several *Pythium* species including *P. ultimum*, *P. aphanidermatum*, etc. (Palumbo et al. 2005; El-Tarabily 2006; Chairat and Pasura 2013; Loliam et al. 2013; Fira et al. 2018).

16.3.4.1 Fungi as Biocontrol Agent

The use of endophytic *Trichoderma* as a biocontrol agent is widely applicable for control of various *Pythium* species. The endophytic *Trichoderma* exhibits various significant activities like production of cell wall degrading activity, production of hydrogen cyanide and indole acetic acid, solubilization of phosphate, etc. These activities are important in destruction of cell wall of oomycetes of *Pythium* spp. (Mishra 2010; Vinayarani and Prakash 2018). Recently, Vinayarani and Prakash (2018) reported the control of soft-rot disease in turmeric plant caused by *P. aphanidermatum*. The study revealed that endophytic *T. harzianum* showed significant inhibition of mycelial growth of causative agent of rhizome-rot disease in turmeric. The preemergence of diseases caused by *Pythium* species can be achieved by coating the seeds with fungal extract. El-Katatny et al. (2001) described the preemergence of damping-off induced by *Pythium* species in radish and pea seeds by coating them with *T. harzianum* and *T. koningii* as a biocontrol agent. Besides, control of *Pythium* spp., by application of various *Trichoderma* spp., in cauliflower, sugar beet, tobacco, chili, cucumber, and tomato has been reported (Das et al. 2002; Jayaraj et al. 2006; Muthukumar et al. 2011; Mbarga et al. 2012; Kipngeno et al. 2015).

16.3.5 Plants as a Biocontrol Agent

Use of various plants as a biocontrol agent is eco-friendly and cost-effective approach for the management of plant diseases caused by *Pythium* spp. Gholve et al. (2016) demonstrated the antifungal potential of different plant extracts, namely, *Ocimum sanctum* (tulsi), *Parthenium hysterophorus* (*Parthenium*), *Lawsonia inermis* (mehndi), *Datura metel* (*Datura*), *Zingiber officinale* (ginger), *Azadirachta indica* (neem), *Asparagus racemosus* (shatawari), *Allium sativum* (garlic), *Curcuma longa* (turmeric), etc. against *P. ultimum* causing damping-off disease in Brinjal. The study revealed that all tested plant extract showed significant antifungal potential against *P. ultimum* by inhibiting mycelial growth. Similar studies on effective management of *Pythium* spp. causing damping-off disease by using botanical extracts were reported by Muthukumar et al. (2010) and Ambikapathy et al. (2011). Pandey et al. (2016) also reported the efficacy of different plant extracts, namely, *Azadirachta indica*, *Eucalyptus globulus*, *Catharanthus roseus*, *Lawsonia inermis*, *Ocimum sanctum*, *Murraya koenigii*, and *Lantana camara*, against *P. aphanidermatum* causing damping-off disease in chili.

Previously, extracts from leaves, stem, and flowers of *Euphorbia macroclada* were found to be effective against *Pythium* spp. (Al-Mughrabi 2003). Uma et al. (2012) reported the antifungal potential of *C. papaya*, *P. granatum*, *V. vinifera*, *A. zapota*, *A. squamosa*, and plant extracts against *Pythium capsici*, and *T. indica*, *C. papaya*, *P. granatum*, *V. vinifera*, *C. colocynthis*, and *A. zapota* plant extracts showed antifungal efficacy against *P. aphanidermatum*. Vinayaka et al. (2014) reported the inhibitory activity of *Usnea pictoides* against *P. aphanidermatum* which causes rhizome-rot disease of ginger. Bahraminejad (2012) reported the antifungal activity of Iranian plants' methanolic and aqueous extract against *Pythium* sp. Kim et al. (2000) reported antifungal potential of *Xanthium strumarium* and *Cinnamomum zelanicum* against *Pythium drechsleri*. Tahira and Sharma (2014) stated the antifungal activity of crude aqueous, alcoholic, and partial hydroalcoholic extracts of *Cassia fistula*, *Clitoria ternatea*, *Eucalyptus globulus*, *Jacaranda mimosifolia*, *Azadirachta indica*, *Aegle marmelos*, *Polyalthia longifolia*, *Tecomella undulata*, and *Terminalia arjuna* against *P. aphanidermatum* and *P. myriotylum*. Hence, according to abovementioned studies, the management of *Pythium* causing different plant diseases can be controlled by use of biocontrol agents, which offers the best opportunity in control of diseases and also helps to maintain the quality and crop yield (Table 16.1).

Table 16.1 Biological control agents of *Pythium* species

Biocontrol agent	Phytopathogens	References
Fungi		
Trichoderma hamatum	*Pythium* sp.	Bhardwaj et al. (1988), Hudge (2015), Mudyiwa et al. (2016)
Trichoderma sp., *Gliocladium* sp.	*Pythium* sp.	Howell and Stipanovic (1983), Fravel (2005)
Trichoderma spp.	*Fusarium* sp.	Selvakumar et al. (2013)
Gliocladium virens, *Glomus* sp.	*Pythium ultimum*	Lumsden and Locke (1989)
Trichoderma harzianum	*P. aphanidermatum* *Pythium* sp. *Fusarium* sp.	Dohroo et al. (2012), Singh (2011), Rajan et al. (2002), Khatso and Tiameren Ao (2013)
Trichoderma viride	*Fusarium* sp.	Khatso and Tiameren Ao (2013)
Bacteria		
Pseudomonas sp. *Enterobacter* *Erwinia* *Bacillus* *Burkholderia* *Stenotrophomonas* *Rhizobium*	*Pythium* sp.	Chin-A-Woeng et al. (2003), Bardin et al. (2004)
Pseudomonas fluorescens *Bacillus* sp. *B. lentus* *B. polymyxa* *Enterobacter agglomerans* *Glomus* sp.	*Pythium myriotylum*	Bhai et al. (2005)
Bacillus mycoides	*Pythium aphanidermatum*	Peng et al. (2017)
Pseudomonas fluorescens	*Pythium ultimum*	Callan et al. (1991)
Rhizobium japonicum	*Fusarium solani*	Smitha and Singh (2014), Al-Ani et al. (2012)
Bacillus subtilis	*Pythium ultimum*, *Fusarium solani*	Mohammady and Abbas (2017)
Actinomycetes		
Streptomyces, *Actinoplanes*, *Micromonospora*	*Pythium coloratum*	El-Tarabily et al. (1997)
Streptomyces rubrolavendulae S4	*Pythium aphanidermatum*	
Nocardiopsis sp.	*Pythium myriotylum*	Sabu et al. (2017)
Plants		
Jacaranda mimosifolia, *Moringa oleifera* *Polyalthia longifolia*,	*Pythium aphanidermatum*	Parveen and Sharma (2014), Gupta et al. (2013)

(continued)

Table 16.1 (continued)

Biocontrol agent	Phytopathogens	References
Terminalia arjuna *Lawsonia inermis* *Aegle marmelos* *Nigella sativa* *Azadirachta indica*		
Zingiber zerumbet (wild ginger)	*Pythium myriotylum*	Ravi et al. (2017)

16.4 Emerging Nanotechnological Strategies for the Management of *Pythium* sp. and *Fusarium*

Nanotechnology applied to agriculture for the effective management of plant diseases is an eco-friendly and outstanding tool over the conventional approaches, which are toxic and hazardous to the environment (Ismail et al. 2017; Abd-Elsalam and Prasad 2018). Ultimately, the nanotechnology will help in minimizing the use of synthetic chemical compounds used in the control of plants diseases (Gogos et al. 2012). The nanoparticles which can be used in agriculture include copper, titanium, zinc, silica, aluminum, chitosan, sulfur, silver, and gold. The broad-spectrum antimicrobial activity of nanoparticles can sustainably replace the existing ecotoxic chemicals commonly used in agriculture (Sabir et al. 2014; Fraceto et al. 2016; Banker et al. 2017). The nanoparticles are responsible for the suppression of the augmenting pathogens; at the same time, they promote the growth of plants by maintaining NPK content of the soil (Ponmurugan et al. 2016; El-Argawy et al. 2017). Recently, the market value of nanoparticles as antifungal and plant growth promoters is increasing enormously. Nanomaterials are beneficial in a controlled delivery of nutrients in agriculture with minimum nutrient loss during application (Prasad et al. 2014, 2017). The nanoparticles have small size, large surface area, greater stability, and easier availability to plants, imparting them property of delivering active ingredients or nutrients in a controlled manner and serving as a great fungicide delivery system in agriculture (Sekhon 2014; Manjuntha et al. 2016; Bhattacharyya et al. 2016; Gupta et al. 2018). Rai and Ingle (2012) have suggested the development of nano-based biosensors and kits for the detection and control of fungal pathogens in agriculture thus flourishing the agriculture-based nanotechnology industry. This will lend a hand in changing the present status of food and agriculture industries worldwide. The chemical fertilizers and fungicides have deleterious effect on human health and environment, mainly on endangered species posing high risk of their extinction. This may lead to imbalance of biodiversity and ultimately disturbing the ecosystem. The soft-rot of ginger is caused by fungal pathogen, i.e., *Pythium* spp., causing huge loss of yield around the world. Hence, there is an imperative need of a nanotechnological strategy for the management of soft-rot of ginger, to overcome the hazardous impact of traditional practices (Patel et al. 2014; Mishra et al. 2014).

Although there is no report on management of *Pythium* spp. and *Fusarium* spp. infection in ginger by nanotechnological approach, in vitro antifungal activity of some plant-mediated nanoparticles has been reported. Nanoparticles are found to be potential fungicidal agents against phytopathogens. Copper nanoparticles are shown to have antifungal activity against many fungal pathogens like *Alternaria alternata*, *Rhizopus stolonifer*, *F. oxysporum*, and *Mucor plumbeus* (Wani and Shah 2012; Viet et al. 2016; Shende et al. 2016; Brahmanwade et al. 2016). It has been found that CTAB (cetyltrimethylammonium bromide)-mediated copper nanoparticles have potential antifungal activity against *F. oxysporum*, *Curvularia lunata*, *A. alternata*, and *Phoma destructiva* as compared to the commercially used fungicide Bavistin (Kanhed et al. 2014). Ponmurugan et al. (2016) showed antifungal activity of biosynthesized *Streptomyces griseus*-mediated copper nanoparticles against root-rot causing soil pathogen *Poria hypolateritia* in tea plant. The morphology-dependent antifungal activity of CuS nanoparticles against *Mucor*, *Rhizopus*, *F. oxysporum*, *Alternaria* spp., and *Helminthosporium* was reported by Chakraborty et al. (2016).

Sulfur nanoparticles have demonstrated higher antifungal activity against pathogenic fungi of fruits like grape, strawberry, vegetables, and many other crops (Deshpande et al. 2008; Suleiman et al. 2013; Llorens et al. 2017). The pathogenic *Fusarium solani* and *Venturia inaequalis* causing wilt and apple scab disease, respectively, are found to susceptible to sulfur nanoparticles, efficiently inhibiting the cell wall of fungi (Rao and Paria 2013). Chitosan nanoparticles also have shown antifungal activity against *Pyricularia grisea*, *A. solani*, and *F. oxysporum* and growth promotion in chickpea seedlings contributing to the increased seed vigor index, enhanced germination, and increase in biomass of seeds (Sathiyabama and Parthsarthy 2016). Chitin and chitosan nanoparticles are found to enhanced defense in host plants against microbial attack by increasing the synthesis of defense proteins, proteinase inhibitors, and phytoalexins, protecting the host plant from fungal pathogens (Sharan et al. 2015; Ahmed and Lee 2015).

Silver nanoparticles are one of the widely studied antifungal agents. But there are scanty reports against *Pythium* spp. causing soft-rot in ginger. Kasprowicz et al. (2010) studied antifungal efficacy of silver nanoparticles against *F. culmorum*, a plant pathogenic fungus. Oh et al. (2006) observed the total inhibition of phytopathogenic fungi, namely, *P. ultimum*, *R. solani*, *M. grisea*, *Colletotrichum gloeosporioides*, and *Botrytis cinerea*, by silica-silver nanoparticles at 10 ppm concentration. Inhibition of *R. solani*, *B. cinerea*, *A. alternata*, *M. phaseolina*, *Sclerotinia sclerotiorum*, and *C. lunata* by silver nanoparticles at 15 mg was reported by Krishnaraj et al. (2012). Silver nanoparticles cause damage to the fungal hyphae, conidial germination, decrease fungal growth, and interfere with microbial absorption (Woo et al. 2009; Jo et al. 2009). In vitro antifungal activity of zinc nanoparticles showing 77% mycelia inhibition in *F. oxysporum* and *P. expansum*, at concentration of 12 mg/L was shown by Ramy and Osama (2013). Parizi et al. (2014) reported antifungal activity of magnesium nanoparticles against tomato wilt causing *F. oxysporum* f. sp. lycopersici.

16.5 Conclusion

Ginger is a cash crop cultivated around the globe for its various novel properties which mainly include culinary and medicinal properties. This crop is susceptible to the attack of various microbes including most important *Pythium* and *Fusarium*, and there is a huge economic loss owing to diseases caused by these fungi. Various physical and chemical methods are in practice since long for the effective control of soft-rot disease of ginger. Unfortunately, the fungal pathogens have developed resistance to the fungicides, and therefore, the management of these pathogenic fungi seems difficult. Moreover, there are increasing concerns of toxicity of these fungicides to humans and environment. Hence, now the biological methods or biocontrol methods are being used for the management of this disease as these methods are eco-friendly and economically viable. A variety of microbes especially bacteria and fungi found to have promising activity against different pathogenic fungi. The members of genus *Trichoderma* such as *T. viride*, *T. aureoviride*, *T. virens*, and *T. hamatum* have been reported as potential biocontrol agents. Among the bacteria and actinomycetes, fluorescent pseudomonads, *Bacillus* spp., and *Streptomyces* have demonstrated their potential against *Pythium* spp. In addition, extract of different plants also showed high potential against *Pythium* spp. Although several efforts have been made for biological control of *Pythium* spp., a little success has been achieved. In future, there is huge possibility of use of nanotechnology-based methods to control soft-rot disease.

Acknowledgement MR is thankful to University Grants Commission, New Delhi for award of BSR faculty fellowship.

References

Abbasi PA, Riga E, Conn KL, Lazarovits G (2005) Effect of neem cake soil amendment on reduction of damping-off severity and population densities of plant-parasitic nematodes and soil borne plant pathogens. Can J Plant Pathol 27:38e45

Abd-Elsalam KA, Prasad R (2018) Nanobiotechnology applications in plant protection. Springer, Cham, (ISBN 978-3-319-91161-8)

Agrios GN (2005) Plant pathology, 5th edn. Elsevier\Academic, Burlington, MA

Ahmed IS, Lee YS (2015) Nanoparticles as alternative pesticides: concept manufacturing and activities. Kor J Mycol 43:207–215

Al-Ani RA, Adhab MA, Mahdi MH, Abood HM (2012) *Rhizobium japonicum* as a biocontrol agent of soybean root rot disease caused by *Fusarium solani* and *Macrophomina phaseolina*. Plant Protect Sci 48:149–155

Al-Mughrabi KI (2003) Antimicrobial activity of extracts from leaves, stems and flowers of *Euphorbia macroclada* against plant pathogenic fungi. Phytopathol Mediterr 42:245–250

Ambikapathy V, Gomathi S, Panneerselva MA (2011) Effect of antifungal activity of some medicinal plants against *Pythium debaryanum* (Hesse). Asian J Plant Sci Res 1(3):131–134

Anisha C, Radhakrishnan EK (2015) Gliotoxin-producing endophytic *Acremonium* sp. from *Zingiber officinale* found antagonistic to soft-rot pathogen *Pythium myriotylum*. Appl Microbiol Biotechnol 175:3458–3467

Ashwini N, Srividya S (2012) Optimization of chitinase produced by a biocontrol strain of *B. subtilis* using Plackett-Burman design. Eur J Exp Biol 2:861–865

Bahraminejad S (2012) *In vitro* and *in vivo* antifungal activities of Iranian plant species against *Pythium aphanidermatum*. Ann Biol Res 3:2134–2145

Banker S, Volova T, Prudnikova SV, Satish S, Prasad N (2017) Nanoagroparticles emerging trends and future prospect in modern agriculture system. Environ Toxicol Pharmacol 53:10–17

Bardin SD, Huang HC, Pinto J, Amundsen EJ, Erickson RS (2004) Biological control of *Pythium* damping-off of pea and sugar beet by *Rhizobium leguminosarum* bv Viceae. Can J Bot 82:291–296

Beneduzi A, Ambrosini A, Passaglia LMP (2012) Plant growth-promoting rhizobacteria (PGPR): Their potential as antagonists and biocontrol agents. Genet Mol Biol 35(4 Suppl):1044–1051

Bennett MA, Callan NW, Fritz VA (1991) Seed treatments for disease control. Hortic Technol 1:84–87

Bertus LS (1942) Plant pathology. Administration report, Directorate of Agriculture, Ceylon, 1941, p. 5

Bhai RS, Kishore VK, Kumar A, Anandaraj M, Eapen SJ (2005) Screening of rhizobacterial isolates against soft-rot disease of ginger (*Zingiber officinale* Rosc). J Spices Aromat Crops 14:130–136

Bhardwaj SS, Gupta PK, Dohroo NP, Shyam KR (1988) Biological control of rhizome rot of ginger in storage. Ind J Plant Pathol 6:56–58

Bhattacharyya A, Duraisamy P, Govindarajan M, Buhroo AA, Prasad R (2016) Nano-biofungicides: emerging trend in insect pest control. In: Prasad R (ed) Advances and applications through fungal nanobiotechnology. Springer, Cham, pp 307–319

Brahmanwade K, Shende S, Bonde S, Gade A, Rai M (2016) Fungicidal activity of cu nanoparticles against Fusarium causing crop disease. Environ Chem Lett 14:229–235

Butler EJ (1907) An account of the genus *Pythium* and some Chytridiaceae. Memoirs of the Department of Agriculture, India. Bot Ser 1:1–162

Callan NW, Mathre DE, Miller JB (1991) Field performance of sweet corn seed bio-primed and coated with *Pseudomonas fluorescens* AB254. Hotrscience 26(9):1163–1165

Chairat Y, Pasura A (2013) Isolation and identification of rhizobacteria having inhibitory capability on pathogenic fungi, *Pythium* sp. J Sci Technol Human 11:117–127

Chakraborty P, Adikary J, Chatterjee S, Biswas B, Chattopadhyay T (2016) Facile synthesis of copper nanoparticles antibacterial and antifungal activity study. Rasayan J Chem 9:77–83

Chase AR, Brunk DD, Tepper BL (1985) Fosetyl aluminum fungicide for controlling *Pythium* root rot of foliage plants. Proc Florida State Hortic Soc 95:119–122

Chérif M, Menzies JG, Ehret DL, Bogdanoff C (1994) Yield of cucumber infected with *Pythium aphanidermatum* when grown with soluble silicon. Hortic Sci 29:896–897

Chin-A-Woeng TF, Bloemberg GV, Lugtenberg BJ (2003) Phenazines and their role in biocontrol by *Pseudomonas* bacteria. New Phytol 157:503–523

Dake GN (1995) Diseases of ginger (*Zingiber officinale* rose) and their management. J Spices Aromat Crops 4:40–48

Dake GN, Edison S (1989) Association of pathogen with rhizome rot of ginger in Kerala. Indian Phytopath 42:116–119

Das S, Biswapati M, Maity D, Raj SK (2002) Different techniques of seed treatment in the management of seedling disease of sugar beet. J Mycopathol Res 40:175–178

Deshpande AS, Rhomane RB, Vaidya BK, JoshiRM HAS, Kulkarni BD (2008) Sulphur nanoparticles synthesis and characterization from H_2S gas using novel biodegradable iron chelates in W/O micro emulsion. Nanoscale Res Lett 3:221–229

Dhanik J, Arya N, Nand V (2017) A review on *Zingiber officinale*. J Pharmacogn Phytother 6:174–184

Dohroo NP (1987) *Pythium ultimum* on ginger. Indian Phytopath 40:275
Dohroo NP (2005) Diseases of ginger. In: Ravindran PN, Babu KN (eds) Ginger, the genus *Zingiber*. CRC Press, Boca Raton, FL, pp 305–340
Dohroo NP, Gupta M (2014) Effect of bioagents on management of rhizome diseases, plant growth parameters and nematode population in ginger. Agric Sci Digest 34:41–44
Dohroo NP, Sharma SL (1986) Evaluation of fungicides for the control of rhizome rot of ginger in storage. Indian Phytopath 36:691–693
Dohroo NP, Sharma SL, Bhardwaj SS (1984) Efficacy of soil applied fungitoxicants against rhizome rot of ginger. Indian J Plant Prot 12:59–60
Dohroo NP, Kansal S, Mehta P, Ahluwalia N (2012) Evaluation of eco-friendly disease management practices against soft-rot of ginger caused by *Pythium aphanidermatum*. Plant Dis Res 27:1e5
Dohroo NP, Kansal S, Ahluwalia N (2015) Studies on eco-farmer-friendly practices for management of soft-rot of ginger (*Zingiber officinale*). Indian Phytopathol 68:93–96
Doshi A, Mathur S (1987) Symptomatology, interaction and management of rhizome rot of ginger. Xenobiotics 26:261–265
Drojee SL (1986) Studies on rhizome rot of ginger in Udaipur. M.Sc. thesis, Sukhadia University, Udaipur, p 98
El-Argawy E, Rahhal M, El-Korany K, Elshabrawy E, Eitahan R (2017) Efficacy of some nanoparticles to control dampping off and root rot of sugar beet in El-Behiera. Asian J Plant Pathol 11:35–37
El-Katatny MH, Gudelj M, Robra KH, Elnaghy MA, Gubitz GM (2001) Characterization of a chitinase and an endo-b-1,3- glucanase from *Trichoderma harzianum* Rifai T24 involved in control of the phytopathogen *Sclerotium rolfsii*. Appl Microbiol Biotechnol 56:137–143
El-Tarabily KA (2006) Rhizosphere-competent isolates of streptomycete and non-streptomycete actinomycetes capable of producing cell-wall degrading enzymes to control *Pythium aphanidermatum* damping-off disease of cucumber. Can J Bot 84:211–222
El-Tarabily KA, St J, Hardy GE, Sivasithamparam K, Hussein AM, Kurtböke DI (1997) The potential for the biological control of cavity-spot disease of carrots, caused by *Pythium coloratum*, by streptomycete and non-streptomycete actinomycetes. New Phytol 137:495–507
Fenice M, Gooday GW (2006) Mycoparasitic actions against fungi and oomycetes by a strain (CCFEE 5003) of the fungus *Lecanicillium muscarium* isolated in Continental Antarctica. Ann Microbiol 56(1):1–6
Fira D, Dimkić I, Berić T, Lozo J, Stanković S (2018) Biological control of plant pathogens by Bacillus species. J Biotechnol 285:44–55
Fraceto LF, Grillo R, de Medeiros GA, Scognamiglio V (2016) Nanotechnology in agriculture: which innovation potential does it have? Front Environ Sci 4:20. https://doi.org/10.3389/fenvs2016000204
Fravel DR (2005) Commercialization and implementation of biocontrol. Annu Rev Phytopathol 43:337–359
Fróes A, Macrae A, Rosa J, Franco M, Souza R, Soares R, Coelho R (2012) Selection of a Streptomyces strain able to produce cell wall degrading enzymes and active against *Sclerotinia sclerotiorum*. J Microbiol 50:798–806
Gajera H, Domadiya R, Patel S, Kapopara M, Golakiya B (2013) Molecular mechanism of Trichoderma as bio-control agents against phytopathogen system – a review. Curr Res Microbiol Biotechnol 1:133–142
Gautam J, Mainali RP (2016) Management of ginger rhizome fly (*Calobata* sp) and associated rhizome rot (*Pythium* sp.). World J Agric Res 4:128–131
Gholve VM, Tatikundalwar VR, Suryawanshi AP, Dey U (2016) Effect of fungicides, plant extracts/botanicals and bioagents against damping off in brinjal. Afr J Microbiol Res 8:2835–2848
Gogos A, Knauer K, Bucheli TD (2012) Nanomaterials in plant protection and fertilization: current state foreseen applications and research priorities. J Agric Food Chem 60:9781–9792

Gupta M, Kaushal M (2017) Diseases infecting ginger (*Zingiber officinale* Roscoe): a review. Agric Rev 38(1):15–28. https://doi.org/10.18805/ag.v0iOF.7305

Gupta SL, Paijwar MS, Rizvi G (2013) Biological management of rot disease of ginger (*Zingiber officinale* Rosc). Trends Biosci 6:302

Gupta N, Upadhyaya CP, Singh A, Abd-Elsalam KA, Prasad R (2018) Applications of silver nanoparticles in plant protection. In: Abd-Elsalam K, Prasad R (eds) Nanobiotechnology applications in plant protection. Springer, Cham, pp 247–266

Hagedorn C, Nelson N, Skwara JE (1990) Evaluation of a *Pseudomonas fluorescens* strain for repression of seedling disease. Virginia J Sci 41(4B):492–500

Harman GE (2006) Overview of mechanisms and uses of *Trichoderma* spp. Phytopathology 96:190–194

Haware MP, Joshi LK (1974) Studies on soft-rot of ginger from Madhya Pradesh. Indian Phytopathology 27:158–161

Heydari A, Pessarakli M (2010) A review on biological control of fungal plant pathogens using microbial antagonists. J Biol Sci 10:273–290

Hoppe PE (1966) *Pythium* species still viable after 12 years in air-dried muck soil. Phytopathology 56:1411

Howell CR, Stipanovic RD (1983) Gliovirin, a new antibiotic from *Gliocladium virens* and its role in the biological control of *Pythium ultimum*. Canadian J Micro 29:321–324

Howie WJ, Suslow TV (1991) Role of antibiotic biosynthesis in the inhibition of *Pythium ultimum* in the cotton spermosphere and rhizosphere by *Pseudomonas fluorescens*. Mol Plant Microb Interact 4:393–399

Hudge BV (2015) Management of damping-off disease of soybean caused by *Pythium ultimum* Trow. Int J Curr Microbiol App Sci 4:799–808

Hwang SF, Gossen BD, Chang KF, Turnbull GD, Howard RJ (2001) Effect of seed damage and metalaxyl seed treatment on *Pythium* seedling blight and seed yield of field pea. Can J Plant Sci 81:509–517

Ismail M, Prasad R, Ibrahim AIM, Ahmed ISA (2017) Modern prospects of nanotechnology in plant pathology. In: Prasad R, Kumar M, Kumar V (eds) Nanotechnology. Springer, Singapore, pp 305–317

Jayaraj J, Radhakrishnan NV, Velazhahan R (2006) Development of formulations of *Trichoderma harzianum* strain M1 for control of damping-off of tomato caused by *Pythium aphanidermatum*. Arch Phytopathol Plant Prot 39:1–8

Jo YK, Kim BH, Jung G (2009) Antifungal activity of silver ions and nanoparticles on phytopathogenic fungi. Plant Dis 93:1037–1043

Kanhed P, Birla S, Gaikwad S, Gade A, Seabra AB, Rubilar O, Duran N, Rai M (2014) *In vitro* antifungal efficacy of copper nanoparticles against selected crop pathogenic fungi. Mater Lett 15:13–17

Kasprowicz MJ, Kozioł M, Gorczyca A (2010) The effect of silver nanoparticles on phytopathogenic spores of *Fusarium culmorum*. Can J Microbiol 56:247–253

Kavita PG, Thomas G (2008) Population genetic structure of the clonal plant *Zingiber zerumbet* (L) smith (Zingiberaceae) a wild relative of cultivated ginger and its response to *Pythium aphanidermatum*. Euphytica 160:89–100

Khatso K, Tiameren Ao N (2013) Biocontrol of rhizome rot disease of ginger (*Zingiber officinale* Rosc.). Int J Bio-Resour Stress Manag 4:317–321

Kim JS, Kwon CS, Son KH (2000) Inhibition of α-glucosidase and amylase by luteolin, a flavonoid. Biosci Biotechnol Biochem 64:2458–2461

Kipngeno P, Losenge T, Maina N, Kahangi E, Juma P (2015) Efficacy of *Bacillus subtilis* and *Trichoderma asperellum* against *Pythium aphanidermatum* in tomatoes. Biol Control 90:92–95

Krishnaraj C, Ramachandran R, Mohan K, Kalaichelvan PT (2012) Optimization for rapid synthesis of silver nanoparticles and its effect on phytopathogenic fungi. Spectrochim Acta A 93:95–99

Kulkarni S, Hegde YR (2002) Diseases of plantation crops and their management. Agrotech Publishing Academy, Udaipur, p 131

Le DP, Smith M, Hudler GW, Aitken E (2014) *Pythium* soft-rot of ginger: detection and identification of the causal pathogens and their control. Crop Prot 65:153–167

Le DP, Smith MK, Aitken E (2016) An assessment of *Pythium* spp. associated with soft-rot disease of ginger (*Zingiber officinale*) in Queensland Australia. Australas Plant Pathol 45:377–387

Lee WH, Cheong SS, So IY (1990) Properties of suppressive and conducive soils to ginger rhizome rot. Kor J Plant Pathol 6:338–342

Lin LT, Chang SS, Leu LS (1971) Soft-rot of ginger. Plant Protect Bull Taiwan 13:54–67

Llorens E, Agustí-Brisach C, González-Hernández AI, Troncho P, Vicedo B, Yuste T, Orero M, Ledó C, García-Agustín P, Lapeña L (2017) Bioassimilable sulphur provides effective control of *Oidium neolycopersici* in tomato enhancing the plant immune system. Pest Manag Sci 73 (5):1017–1023

Loliam B, Morinaga T, Chaiyanan S (2013) Biocontrol of *Pythium aphanidermatum* by the cellulolytic actinomycetes *Streptomyces rubrolavendulae* S4. Sci Asia 39:584–590

Lumsden RD, Locke JC (1989) Biological control of damping-off caused by *Pythium ultimum* and *Rhizoctonia solani* with *Gliocladium virens* in soilless mix. Phytopathology 79:361–366

Luong TM, Huynh LMT, Hoang HMT, Tesoriero LA, Burgess LW, Phan GHT, Davies P (2010) First report of *Pythium* root rot of chrysanthemum in Vietnam and control with metalaxyl drench. Aust Plant Dis Notes 5:51–54

Manjuntha SB, Biradar DP, Alandkantti YR (2016) Nanotechnology and its applications in agriculture. J Pharmacol Sci 29:1–13

Mathur S, Thakore BL, Singh RB (1984) Effect of different fungicides on ginger rhizome rot pathogen and their effect on germination and rotting of rhizome. Indian J Mycol Plant Path 14:155–157

Mathur K, Ram D, Poonia J, Lodha BC (2002) Integration of soil solarization and pesticides for management of rhizome rot of ginger. Indian Phytopath 55:345–347

Mbarga JB, Hoopen GMT, Kuate J, Adiobo A, Ngonkeu MEL, Ambang Z, Akoa A, Tondje PR, Begoude BAD (2012) *Trichoderma asperellum*: a potential biocontrol agent for *Pythium myriotylum*, causal agent of cocoyam (*Xanthosoma sagittifolium*) root rot disease in Cameroon. Crop Protect 36:18e22

Meena KR, Kanwar SS (2015) Lipopeptides as the antifungal and antibacterial agents: applications in food safety and therapeutics. BioMed Res Int 2015:473050. https://doi.org/10.1155/2015/473050

Menendez E, Garcia-Fraile P (2017) Plant probiotic bacteria: solutions to feed the world. AIMS Microbiol 3:502–524

Mishra VK (2010) *In vitro* antagonism of *Trichoderma* species against *Pythium aphanidermatum*. J Phytol 2:28–35

Mishra S, Singh BR, Singh A, Keswani C, Naqvi AH, Singh HB (2014) Biofabricated silver nanoparticles act as a strong fungicide against *Bipolaris sorokiniana* causing spot blotch disease in wheat. PLoS One 9:978–981

Mitra M, Subramaniam LS (1928) Fruit rot diseases of cultivated Cucurbitaceous caused by *Pythium aphanidermatum* (Eds.) Fitz. Memorial Department of Agriculture India Botany 15:79–84

Mohammady A, Abbas A (2017) Biological control of *Pythium ultimum* and Fusarium solani by indigenous strains *Bacillus subtilis*. Boil J microorg. https://www.academia.edu/32658977/Biological_control_of_Pythium_ultimum_and_Fusarium_solani_by_indigenous_strains_Bacillus_subtilis

Moreira SI, Dutra DC, Rodrigues AC, de Oliveira JR, Dhingra OD, Pereira OL (2013) Fungi and bacteria associated with post-harvest rot of ginger rhizomes in Espírito Santo Brazil. Trop Plant Pathol 38:218–226

Mudyiwa RM, Chaibva P, Takawira M (2016) Evaluation of *Trichoderma harzianum* in controlling damping-off (*Pythium* spp.) on tomato (*Solanum lycopersicum*) seedling varieties. Ann Biol Res 7(6):6–11

Muthukumar A, Eswaran A, Nakkeeran S, Sangeetha G (2010) Efficacy of plant extracts and biocontrol agents against *Pythium aphanidermatum* inciting chilli damping off. Crop Prot 29:1483–1488

Muthukumar A, Eswaran A, Sanjeevkumas K (2011) Exploitation of Trichoderma species on the growth of *Pythium aphanidermatum* in chilli. Braz J Microbiol 42:1598–1607

Naseby DC, Way JA, Bainton NJ, Lynch JM (2001) Biocontrol of *Pythium* in the pea rhizosphere by antifungal metabolite producing and non-producing *Pseudomonas* strains. J App Micro 90 (3):421–429. https://doi.org/10.1046/j.1365-2672.2001.01260.x

Nayak SK, Nayak S, Mishra BB (2017) Antimycotic role of soil bacillus sp. against rice pathogens: a biocontrol prospective. In: Patra J, Vishnuprasad C, Das G (eds) Microbial biotechnology. Springer, Singapore

Oh SD, Lee S, Choi SH, Lee IS, Lee YM, Chun JH, Park HJ (2006) Synthesis of Ag and Ag-SiO_2 nanoparticles by γ-irradiation and their antibacterial and antifungal efficiency against Salmonella enteric serovar Typhimurium and *Botrytis cinerea*. Colloids Surf A Physicochem Eng Asp 275:228–233

Palumbo JD, Yuen GY, Jochum CC, Tatum K, Kobayashi DY (2005) Mutagenesis of beta-1, 3-glucanase genes in *Lysobacter enzymogenes* strain C3 results in reduced biological control activity toward Bipolaris leaf spot of tall fescue and Pythium damping-off of sugar beet. Phytopathology 95:701–707

Pandey M, Ahmad S, Khan KZ (2016) *In Vitro* evaluation of some antagonists and plant extracts against *Pythium Aphanidermatum* causes damping off of Chilli. Bioscan 11:879–884

Parham BEV (1935) Annual report of general mycological and botanical work for 1934. Annual Bulletin Department of Agriculture, Fiji, 1935, pp 55–56

Parizi MA, Moradpour Y, Roostaei A, KhaniM NM, Rahimi G (2014) Evaluation of the antifungal effect of magnesium oxide nanoparticles on *Fusarium oxysporum* f. sp. lycopersici pathogenic agent of tomato. Eur J Expt Biol 4:151–156

Park M (1934) Report of the work of the mycological division. Annual report of Director of Agriculture, Ceylon, pp 126–133

Park M (1935) Report of the work of the mycological division. Annual report of Director of Agriculture, Ceylon, pp 124–131

Park M (1937) Report of the work of the mycological division. Annual report of Director of Agriculture, Ceylon, pp 20–21

Parveen T, Sharma K (2014) Management of "Soft-rot" of ginger by botanicals. Int J Pharm Life Sci 5:3478–3484

Patel MK, Kamat MN, Bhide VP (1949) Fungi of Bombay. Indian Phytopath 2:142–155

Patel N, Desai P, Patel N, Jha A, Gautam HK (2014) Agro nanotechnology for plant fungal disease management a review. Int J Microbiol Appl Sci 3:71–84

Peng YH, Chou YJ, Liu YC et al (2017) Inhibition of cucumber *Pythium* damping-off pathogen with zoosporicidal biosurfactants produced by *Bacillus mycoides*. J Plant Dis Prot 124:481. https://doi.org/10.1007/s41348-017-0110-z

Ponmurugan P, Manjukarunagmbika K, Elango V, Gnamamangai BM (2016) Antifungal activity of biosynthesized copper nanoparticles evaluated against red root rot disease in tea plants. J Exp Nanosci 11:1019–1031

Pordesimo AN, Raymundo SA (1963) Rhizome rot of ginger and its control. Coffee Cocoa Res J 5:240

Prasad R, Kumar V, Prasad KS (2014) Nanotechnology in sustainable agriculture: present concerns and future aspects. Afr J Biotechnol 13:705–713

Prasad R, Bhattacharyya A, Nguyen QD (2017) Nanotechnology in sustainable agriculture: recent developments challenges and perspectives. Front Microbiol 8:1014. https://doi.org/10.3389/fmicb201701014

Quimio AJ, Chan HH (1979) Survival of *Pseudomonas solanacearum* in the rhizosphere of some weed and economic plant species. Philipp Phytopathol 15:108–121
Rai M, Ingle A (2012) Role of nanotechnology in agriculture with special reference to management of insect's pests. Appl Microbiol Biotechnol 94:287–293
Rai M, Ingle A, Paralikar P, Anasane N, Gade R, Ingle P (2018) Effective management of soft-rot of ginger caused by *Pythium* spp. and *Fusarium* spp.: emerging role of nanotechnology. Appl Microbiol Biotechnol 102:1–13. https://doi.org/10.1007/s00253-018-9145-8
Rajan PP, Gupta SR, Sarma YR, Jackson GVH (2002) Diseases of ginger and their control with *Trichoderma harzianum*. Indian Phytopathol 55:173–177
Ram D, Mathur K, Lodha BC, Webster J (2000) Evaluation of resident biocontrol agents as seed treatments against ginger rhizome rot. Indian Phytopathol 53:450–454
Ramachandran N, Dake GN, Sarma YR (1989) Effect of systemic fungicides on in vitro growth of *Pythium aphanidermatum*, the rhizome rot pathogen of ginger. Indian Phytopathol 42:463–465
Ramy SY, Osama FA (2013) In vitro study of the antifungal efficacy of zinc oxide nanoparticles against *Fusarium oxysporum* and *Penicillium expansum*. Afr J Microbiol Res 7:1917–1923
Rao KJ, Paria S (2013) Use of sulfur nanoparticles as a green pesticide on *Fusarium solani* and *Venturina equalis*. Phytopathogens 3:10471–10478
Rathaiah Y (1987) Control of soft-rot of ginger with Ridomil. Pesticides 21:29–30
Ravi A, Varghese S, Krishnankutty RE (2017) Biocontrol activity of the extract prepared from *Zingiber zerumbet* for the management of rhizome rot in *Zingiber officinale* caused by *Pythium myriotylum*. Arch Phytopathol Plant Protect 50:555–567
Ravindran PN, Babu N (eds) (2005) Ginger—the genus *Zingiber*. CRC Press, Boca Raton, p 310
Reddy MN, Charitha Devi M, Sreedevi NV (2003) Biological control of rhizome rot of turmeric (*Curcuma longa* L.) caused by *Fusarium solani*. J Biol Control 17:193–195
Sabir S, Arshad M, Chaudhari SK (2014) Zinc oxide nanoparticles for revolutionizing agriculture synthesis and applications. Sci World J 2014:925494. https://doi.org/10.1155/2014/925494
Sabu R, Soumya KR, Radhakrishnan EK (2017) Endophytic *Nocardiopsis* sp. from *Zingiber officinale* with both antiphytopathogenic mechanisms and antibiofilm activity against clinical isolates. 3 Biotech 7(2). https://doi.org/10.1007/s13205-017-0735-4
Sahare KC, Asthana RP (1962) Rhizome rot of ginger and its control. Indian Phytopathol 15:77–78
Sarma YR, Nambiar KKN, Brahma RN (1979) Studies on rhizome rot of ginger and its control. In Venkataram CS (ed) Proceedings of PLACROSYM_II. 1979, Indian Society of Plantation Crops, Central Plantation Crops Research Institute, Kasargod, Kerala, pp 386–397
Sathiyabama M, Parthsarthy R (2016) Biological preparation of chitosan nanoparticale and its *in vitro* antifungal efficacy against some phytopathogenic fungi. Carbohydr Polym 151:321–325
Sekhon BP (2014) Nanotechnology in agri-food production: an overview. Nanotechnol Sci Appl 7:31–53
Sellem I, Triki MA, Elleuch L, Cheffi M, Chakchouk A, Smaoui S, Mellouli L (2017) The use of newly isolated Streptomyces strain TN258 as potential biocontrol agent of potato tubers leak caused by *Pythium ultimum*. J Basic Microbiol 57:393–401
Selvakumar R, Singh AK, Raja P (2013) Identification of potential *Trichoderma* sp. for management of rhizome rot of ginger in Arunachal Pradesh. Environ Ecol 31:67–70
Sen TN (1930) Appendix I: IV Mycology: annual report of the Department of Agriculture. Assam 1929–39:57–59
Shanmugam V, Gupta S, Dohroo NP (2013a) Selection of a compatible biocontrol strain mixture based on co-cultivation to control rhizome rot of ginger. Crop Prot 43:119–127
Shanmugam V, Thakur H, Kaur J, Gupta S, Rajkumar S, Dohroo NP (2013b) Genetic diversity of *Fusarium* spp. inciting rhizome rot of ginger and its management by PGPR consortium in the western Himalayas. Biol Conl 66:1–7
Sharan V, Sharma G, Yadav M, Choudhary MK, Shrma SS, Pal A, Raliya R, Biswas P (2015) Synthesis and *in vitro* antifungal efficacy of Cu-chitosan nanoparticles against pathogenic fungi of tomato. Int J Biol Macromol 75:346–353

Sharma SL, Dohroo NP (1982) Efficacy of chemicals in controlling rhizome rot of ginger (*Zingiber officinale* Rosc.). In: Nair MK, Premkumar T, Ravindran PN, Sarma YR (eds) Proceedings of national seminar on Ginger and Turmeric Central Plantation Crops Research Institute, Kasargod, Kerala, 8–9 April 1980, Calicut, pp 120–122

Shende SS, Gaikwad ND, Bansod SD (2016) Synthesis and evaluation of antimicrobial potential of copper nanoparticle against agriculturally important phytopathogens. Int J Biol Res 1(4):41–47

Singh AK (2011) Management of rhizome rot caused by *Pythium fusarium* and *Ralstonia* spp. in ginger (*Zingiber officinale*) under natural field conditions. Indian J Agric Sci 81:268–270

Sivasakthi S, Usharani G, Saranraj P (2014) Biocontrol potentiality of plant growth promoting bacteria (PGPR)-*Pseudomonas fluorescens* and *Bacillus subtilis*: A Review. Afr J Agric Res 9:1265–1277

Smith M, Abbas R (2011) Controlling *Pythium* and associated pests in ginger. RIRDC publication No. 11/128, Canberra

Smitha M, Singh R (2014) Biocontrol of phytopathogenic fungi using mycolytic enzymes produced by rhizospheric bacteria of *Cicer arietinum*. Indian J Agric Biochem 27:215–218

Srivastava LS (1994) Management of soft-rot of ginger in Sikkim. Plant Dis Res 9:146–149

Stapleton JJ, Devay JE (1986) Soil solarization: a non-chemical approach for management of plant pathogens and pests. Crop Prot 5:190–198

Stirling GR, Turaganivalu U, Stirling AM, Lomavatu MF, Smith MK (2009) Rhizome rot of ginger (*Zingiber officinale*) caused by *Pythium myriotylum* in Fiji and Australia. Australas Plant Pathol 38:453–460

Suleiman M, Al Ali A, Hussein A, Hammouti B, Hadda TB, Warad I (2013) Sulfur nanoparticles: synthesis characterizations and their applications. J Mater Environ Sci 4:1029–1033

Tahira P, Sharma K (2014) Phytochemical profiling of leaves and stem bark of *Terminalia arjuna* and *Tecomella undulata*. Int J Pharma Biosci 1:1–7

Takahashi (1954) *Pythium zingiberum*. Ann Phytopathol Soc Jpn 18:115

Thakore BBL, Mathur S, Singh RB (1988) Effect of rhizome treatment with fungicides for economic control of rot. J Phytol Res 1:83–84

Thakur M, Sohal BS (2013) Role of elicitors in inducing resistance in plants against pathogen infection: a review. ISRN Biochem 2013:762412. https://doi.org/10.1155/2013/762412

Thomas KM (1938) Administration report of the government mycologist, Madras, 1937-38

Thomashow L, Weller DM (1996) Current concepts in the use of introduced bacteria for biological disease control: mechanisms and antifungal metabolites. In: Plant-microbe interactions. Springer, Boston, pp 187–235. https://doi.org/10.1007/978-1-4613-1213-0_6

Uma T, Mannam S, Lahoti J, Devi K, Kale RD, Bagyaraj DJ (2012) Biocidal activity of seed extracts of fruits against soil borne bacterial and fungal plant pathogens. J Biopest 5:103–105

Uppal BN (1940) Report of the department of agriculture. Bombay 1938–39:203–211

Viet PV, Nguyen HT, Cao TM, Hieu LV (2016) Fusarium antifungal activities of copper nanoparticles synthesized by a chemical reduction method. J Nanomater 2016:1957612. https://doi.org/10.1155/2016/1957612

Vinayaka KS, Prashitha Kekuda TR, Noor Nawaz AS, Junaid S, Dileep N, Rakesh KN (2014) Inhibitory Activity of Usnea pictoides G.Awasthi (Parmeliaceae) Against *Fusarium oxysporum* F. Sp. Zingiberi and *Pythium aphanidermatum* isolated from Rhizome Rot of Ginger. Life Sci Leafl 49:17–22

Vinayarani G, Prakash HS (2018) Fungal endophytes of turmeric (*Curcuma longa* L.) and their biocontrol potential against pathogens *Pythium aphanidermatum* and *Rhizoctonia solani*. World J Microbiol Biotechnol 34:49. https://doi.org/10.1007/s11274-018-2431-x

Wang PH, Chung CY, Lin YS, Yeh Y (2003) Use of polymerase chain reaction to detect the soft-rot pathogen *Pythium myriotylum* in infected ginger rhizomes. Lett Appl Microbiol 36:116–120

Wani AH, Shah MA (2012) A unique and profound effect of MgO and ZnO nanoparticles on some plant pathogenic fungi. J Appl Pharm Sci 02:40–44

Whipps JM, Lumsden R (1991) Biological control of Pythium species. Biocontrol Sci Technol 1:75–90. https://doi.org/10.1080/09583159109355188

Woo KS, Kim KS, Lamsal K, Kim YJ, Kim SB, Jung M, Sim SJ, Kim HS, Chang SJ, Kim JK, Lee YS (2009) An *in vitro* study of the antifungal effect of silver nanoparticles on oak wilt pathogen *Raffaelea* sp. J Microbiol Biotechnol 19:760–764

Yacoub A, Gerbore J, Magnin N, Haidar R, Compant S, Rey P (2017) Transcriptional analysis of the interaction between the oomycete biocontrol agent, *Pythium oligandrum*, and the roots of *Vitis vinifera* L. Biol Control 120:26–35. https://doi.org/10.1016/j.biocontrol.2017.02.007

Yang KD, Kim HM, Lee WH, So IN (1988) Studies on rhizome rot of ginger caused by *Fusarium oxysporum* f. sp. *zingiberi* and *Pythium zingiberum*. Korean J Plant Pathol 4:271–277

Zheng XY, Sinclair JB (2000) The effects of traits of *Bacillus megaterium* on seed and root colonization and their correlation with the suppression of *Rhizoctonia* root rot of soybean. BioControl 45:223. https://doi.org/10.1023/A:1009998304177

Zouari I, Jlaiel L, Tounsi S, Trigui M (2016) Biocontrol activity of the endophytic *Bacillus amyloliquefaciens* strain CEIZ-11 against *Pythium aphanidermatum* and purification of its bioactive compounds. Biol Control 100:54–62

Chapter 17
Biological Effects of Uranium and Its Decay Products on Soil Microbes, Plants, and Humans

Chanda Siddoo-Atwal

Abstract In this atomic age, exposure to toxins in the environment ranges from radioactive heavy metals to chemical pesticides, and detoxification has become an issue of considerable importance. Recently, many parts of the world have been contaminated with radioactive waste from depleted uranium bombs and projectiles including the Arabian Gulf, Iraq, Syria, Bosnia, Serbia, and Afghanistan. In addition, other areas of the globe have been contaminated by nuclear testing sites and accidents at nuclear power plants involving radioactive uranium and its decay products. There are three naturally occurring uranium isotopes that are of major significance with regard to mining of this element and the nuclear industry. These include uranium-238 (U-238), which comprises the majority of this element in the Earth's crust, uranium-235 (U-235), and uranium-234 (U-234), which together comprise a much smaller portion. The half-lives of these isotopes are approximately 4500 million years, 703 million years, and 246,000 years, respectively. The transfer of radionuclides of the uranium decay series through the environment is important for assessing the impact of nuclear weapons use, nuclear power plant leaks, and the mining and milling of uranium ores. The pathway from soil through plants to humans contributes significantly to the overall radiation dose. The transfer of mobilized radionuclides within the environment is determined by weathering rate, which, in turn, depends on particle composition and chemical conditions such as pH of the soil after deposition. Specific geographical sites that have been contaminated with uranium attract specific bacterial species that display resistance to the metal. Moreover, various plant species exhibit substantial differences in the soil-plant transfer factor for uranium and other related radionuclides. The biological effects of environmental radionuclides in humans (particularly depleted uranium) have been documented as part of the Gulf War syndrome and Balkan syndrome and comprise a complex set of seemingly unrelated symptoms. Some of these include incapacitating fatigue, musculoskeletal and joint pains, headaches, neuropsychiatric disorders, confusion, visual problems, changes of gait, loss of memory, lymphadenopathies, respiratory impairment, impotence, and urinary tract morphological and functional alterations. Moreover, the overall incidence of breast and lung cancer, leukemia, and

C. Siddoo-Atwal (✉)
Moondust Cosmetics Ltd., West Vancouver, BC, Canada

A. Varma et al. (eds.), *Plant Microbe Interface*,
https://doi.org/10.1007/978-3-030-19831-2_17

lymphoma has doubled or tripled in certain areas of Iraq contaminated with depleted uranium during the Gulf War. An association with lung cancer has also been found in uranium miners. Thus, soil remediation, plant selection, phytoremediation, and human detoxification are the main issues to be considered in relation to environmental contamination with uranium and its decay products.

17.1 Introduction

In this atomic age, exposure to toxins in the environment ranges from radioactive heavy metals to chemical pesticides, and detoxification has become an issue of considerable importance. It is something of note particularly for official governmental agencies dealing with the cleanup of environmental pollution and setting regulatory health standards for such hazards in the surrounding environment. Recently, many parts of the world have been contaminated with radioactive waste from depleted uranium bombs and projectiles including the Arabian Gulf, Iraq, Bosnia and Serbia, Afghanistan, and Syria (Durakovic 2001, 2003; Obralic et al. 2004; Briner 2006; Middle East Eye). Depleted uranium (DU) is a by-product of the uranium enrichment process employed in nuclear reactors. It has been used to make depleted uranium bombs and to coat bullets, which are effective armor-piercing projectiles, since the 1990s. Other areas of the globe have also been contaminated by nuclear testing, such as Nevada and the Bikini Islands, and accidents at nuclear power plants, such as Three Mile Island, Chernobyl, and Fukushima. Heavy metal exposure can occur in humans potentially through the air, water, and soil via food chain (Figs. 17.1, 17.2, and 17.3). However, depleted uranium exposure is most likely to occur via inhalation (Bleise et al. 2003).

Natural uranium consists of three isotopes that are of significance with regard to mining of this element and the nuclear industry. These include uranium-238 (U-238), which comprises approximately 99.3% of this element by mass; uranium-235 (U-235), which comprises 0.72% by mass; and uranium-234 (U-234), which comprises 0.006% by mass. The half-life of U-238 is approximately 4.5 billion years, while those of U-235 and U-234 are estimated at 703 million and 246,000 years, respectively (Priest 2001). U-235 is fissile, and, therefore, it is used to trigger nuclear fission in atomic bombs and to power steam turbines which generate electricity at nuclear power plants (World Nuclear Association 2017). The process of uranium enrichment alters the relative percentages of the three isotopes to produce more U-235, and depleted uranium with a higher percentage of U-238 is a by-product. The decay products of U-238 form the "uranium-radium decay chain" and include uranium, protactinium, thorium, actinium, radium, francium, radon, astatine, polonium, bismuth, lead, and thallium. The decay products of U-235 form the similar, but distinct, "actinium decay chain." Lead (Pb) is the end product for both the uranium and actinium decay chains (Pb-206 and Pb-207, respectively) (Thoennessen 2016).

Fig. 17.1 Bosnia-Herzegovina is an example of a country that has been bombed with depleted uranium (DU) weapons (photo of Trebinje, Bosnia and Herzegovina)

Fig. 17.2 It is quite possible that the soil in Bosnia is contaminated with radioactive uranium isotopes and these may accumulate in plants (photo of an olive and olive tree in Trebinje, Bosnia)

The transfer of radionuclides of the uranium decay series through the environment is important for assessing the impact of nuclear weapons use, nuclear power plant leaks, and the mining and milling of uranium ores. The pathway from soil through plants to humans contributes significantly to the overall radiation dose received (Sheppard and Evenden 1988). Size distribution pattern, radionuclide and matrix composition, morphology, and structure are all important factors in studying weathering, mobilization, and biological uptake of radionuclides when radioactive particles are released into the environment. The particle characteristics are dependent on the radiation source which will determine release, dispersion, and deposition. For example, high-temperature releases like Chernobyl can result in varying

Fig. 17.3 The water sources in Bosnia may also be contaminated with uranium, and these may affect the local wildlife (photo of waterfowl in Trebinje, Bosnia)

composition, morphology, and structure of uranium fuel particles, while low-temperature releases like Windscale may result in flake-like uranium fuel particles that are significantly different. In fact, the release of radioactive particles from nuclear sources occurs more frequently than expected. In order to predict transport, ecosystem transfer, and radiation dose, experimental information is also required with respect to factors such as radionuclide speciation and association with other particles and colloids that influence mobility and biological uptake (Salbu and Krekling 1998).

The transfer of mobilized radionuclides within the environment is determined by weathering rate, which, in turn, depends on particle composition and chemical conditions such as pH of the soil after deposition. It has been found that soil pH is highly linearly correlated with log sorption ratios as a probable consequence of different uranium chemical complexes as a function of soil pH. For example, there is such a great difference in sorption behavior between uranium carbonate complexes that other effects of soil properties on uranium sorption are hidden (Echevarria et al. 2001).

Prokaryotes and eukaryotes contribute actively to geological phenomena including the transformation of metals. Microbes have a variety of properties that can effect changes in metal speciation, toxicity, and mobility, as well as mineral formation or deterioration. Such mechanisms are important components of natural biogeochemical cycles for metals as well as associated elements. Bacteria and fungi are the most important organisms for reclamation, immobilization, or detoxification of metallic and radionuclide pollutants (Gadd 2010).

Geographical sites that have been contaminated with uranium attract specific bacterial species that display resistance to the metal. This may manifest as radiation resistance as in the case of *Deinococcus radiodurans*, the most radioresistant organism known, or chemical resistance as in the case of *Arthrobacter*. As an example, one acidic uranium-contaminated site was found to contain heterotrophic *Arthrobacter* species. These included gram-positive bacteria, one of which was closely related to *Arthrobacter ilicis*. It accumulated uranium intracellularly as precipitates associated with polyphosphate granules. The authors interpreted this as a biochemical detoxification mechanism. However, *D. radiodurans* was vulnerable to uranium chemical toxicity under the same conditions in this particular environment (Suzuki and Banfield 2004). In fact, certain experiments have shown that an *Arthrobacter* strain (G975) is the fastest growing and most uranium-tolerant strain, which removes up to 90% of the uranium from growth media. Various *Arthrobacter* strains, including this one, have been isolated from Hanford site soil (Katsenovich et al. 2013).

Element-specific concentration ratios (CR) are used to model the impact of radionuclides like uranium on the environment. These CR values decrease significantly as the corresponding soil concentrations increase. Interestingly, CR values can differ significantly among some soil and plant types (Sheppard and Evenden 1988). Moreover, various plant species exhibit substantial differences in the soil-plant transfer factor for certain radionuclides. For example, Indian mustard has the smallest root/shoot (R/S) ratio for both radium-226 and thorium-232, while clover has the smallest R/S ratio for uranium-238, when grown on soils from southeastern China contaminated with uranium mine tailings. Chinese mustard also has a low uranium-238 uptake (Chen et al. 2005).

The biological effects of environmental radionuclides in humans via inhalation (particularly depleted uranium), skin contact, and potential ingestion via the food chain including drinking water (for natural uranium) have been noted. The effects of DU have been described as part of the Gulf War and Balkan syndromes, and both comprise a complex set of seemingly unrelated symptoms. Some of these include incapacitating fatigue, musculoskeletal and joint pains, headaches, neuropsychiatric disorders, confusion, visual problems, change of gait, loss of memory, lymphadenopathies, respiratory impairment, impotence, and urinary tract morphological and functional alterations (Durakovic 2001, 2003). In addition, there have been reports of an increased incidence of birth defects in Iraq (Alaani et al. 2011). The overall incidence of breast and lung cancer, leukemia, and lymphoma has doubled or tripled in certain areas of Iraq contaminated with DU during the Gulf War (Fathi et al. 2013; Busby et al. 2010). An association with lung cancer has also been found in uranium miners (Grosche et al. 2006). It is strongly linked to exposure with radon that is one of the decay products of natural uranium. Miners are at risk as a result of exposure to radioactive alpha, beta, and gamma emissions from uranium, which may act as a carcinogen or cocarcinogen, as well (National Academies Press 2012).

Thus, soil remediation, plant selection and phytoremediation, and human detoxification are the main issues to be considered in relation to environmental contamination with uranium.

17.2 Soil Remediation and Soil Conservation

The remediation of uranium and other radionuclide-contaminated soils is a complex and expensive undertaking that depends on a variety of factors. These include the quantitative and qualitative aspects of the contaminant present, the structure and dynamics of the indigenous microbial community, and the geological and chemical conditions at the contaminated site. Bioremediation with microorganisms appears to be one of the most cost-effective and ecologically appealing strategies available. Bacteria, specifically, utilize various mechanisms to sequester uranium including biosorption at the cell surface, intracellular accumulation, bioprecipitation or biomineralization, and redox transformations (oxidation/reduction) (Merroun and Selenska-Pobell 2008).

In order to analyze native microbial populations in any given contaminated environment, PCR-based cloning techniques can be useful for assessing their density and structure. Using this technology, it has been determined that microbial diversity decreases with increasing uranium-associated nitrate and aluminum groundwater contamination when compared with control samples and favors bacteria metabolically adapted to the particular contaminant that is most abundant (Fields et al. 2005). Nitrate-reducing bacteria including phylotypes related to *Proteobacteria* (*Alpha*-, *Beta*-, *Delta*-, and *Gammaproteobacteria*), *Bacteroidetes*, *Actinobacteria*, *Firmicutes*, and *Planctomycetes* have been found to be the most metabolically active microbial communities in uranium-contaminated subsurface sediments using DNA/RNA amplification techniques (Akob et al. 2007). Sulfate-reducing bacteria such as *Desulfobacterales* and *Desulfovibrionales*, which can cause the reductive transformation of metals like uranium into more insoluble forms or form metal sulfides, have also been detected in a former uranium-mining area (Sitte et al. 2010).

Moreover, various *Arthrobacter* strains including *Arthrobacter* strain (G975) have been isolated from Hanford site soil and may be metabolically suited for use in the remediation of such uranium-contaminated sites. In fact, in laboratory tests, *Arthrobacter* strain (G975) is the fastest growing and most uranium-tolerant strain, which removes up to 90% of the uranium from growth media. In other experiments, *Bacillus* and *Rahnella* bacterial strains exhibiting phosphatase-positive phenotypes indicative of constitutive phosphatase activity were found to liberate sufficient phosphate to precipitate 73% and 95% of total soluble uranium added as uranyl acetate, respectively (Martinez et al. 2007). Another potential methodology for the remediation of uranium-contaminated soils involves a three-step process including extraction with citric acid, biodegradation of several metal-citrate complexes with *Pseudomonas fluorescens* resulting in the bioprecipitation and recovery of other toxic metals, and photodegradation of uranyl citrate upon exposure to light since it is resistant to biodegradation (Francis and Dodge 1998). This technique might be well suited to a site with mixed contamination. Another approach exploits the decrease in solubility following the reduction of U (VI) to U (IV), which produces the insoluble mineral uraninite. A variety of bacteria exhibit this reductive capacity via uranium reductases and may be suitable for bioremediation (Wall and Krumholz 2006). In fact, through a concerted program of biostimulation of bioreducing bacteria, organisms belonging to genera known as U (VI) reducers including *Desulfovibrio*, *Geobacter*, *Anaeromyxobacter*, *Desulfosporosinus*, and *Acidovorax* spp. were detected in sediments from a highly uranium-contaminated site (Cardenas et al. 2008).

At the same time, microorganisms are difficult to control and can mutate relatively easily. There is also a possibility that uranium may be mutagenic in bacteria according to the Ames test (Hudcová et al. 2013). Conceivably, a monumental health hazard could be created by adopting such a bioremediation practice at contaminated sites over too wide an area, especially if these bacteria were to outcompete plants, animals, and humans. Currently, a vast swathe of Eurasia may be contaminated with uranium or its decay products due to wars and nuclear accidents, so a degree of caution should be exercised in making recommendations. Thus, bioremediation with microorganisms may be a strategy best reserved for specialized use for large areas of soil and water contaminated with low concentrations of uranium and is not appropriate for every situation (Li and Zhang 2012).

Soil remediation should aim to remove the contaminant from the substratum or reduce the risk posed by the contaminant by diluting exposure. Although there are many remediation techniques available for contaminated soils, relatively few are applicable to heavy metal contamination. In some cases, a combination of more than one technique may be applied effectively. For example, soil amendments may be combined with phytostabilization, which involves the selection of tolerant plant species and genotypes, to decrease soil metal bioavailability (Vassilev et al. 2004). In practice, an integrated set of modern remediation methods should be utilized to remove uranium from contaminated environments thoroughly and efficiently with the objective of avoiding the production of secondary pollution and causing least disturbance to the ecosystem (Li and Zhang 2012).

Some of the conventional methods that have been used in the past for the treatment of uranium-contaminated soils include removal of a top soil layer or soil excavation and transfer to designated repositories, encapsulation, size separation, soil washing, leaching with chelating agents, electrokinetics, and ion exchange (Dushenkov 2003). Such methods result in the generation of solid and liquid wastes and can damage the local ecosystem by reducing soil quality. Soil components like soil microorganisms and nutrients that are essential for plant growth may also be removed in the process (Malaviya and Singh 2012).

Soil amendments convert soluble and pre-existing high-soluble solid phase forms to more stable solid phases resulting in reduced bioavailability and plant toxicity of heavy metals. Commonly used amendments include liming agents, phosphates, metal oxyhydroxides, and organic compounds such as sludge or compost. Synthetics like ammonium thiocyanate and natural zeolites have also yielded promising results (Prasad and Freitas 2003). A combination of metal immobilization and subsequent phytostabilization, which involves the use of metal-tolerant plants and/or plants tolerant to the growing conditions of a given site, can be effective. For example, grasses and fast-growing plants that are easy to care for can provide complete surface coverage and a large network of shallow roots to stabilize soil and take up soil water.

The main objectives for successful heavy metal remediation are to change the trace element speciation of the soil, to stabilize the vegetation cover and limit trace element uptake by crops, to reduce direct exposure of living organisms, and to enhance biodiversity. However, there is no standardized system for classifying the effectiveness of soil treatment, and the remedial strategy must be evaluated on a case to case basis (Vassilev et al. 2004).

Phosphate fertilizers are another major source of uranium contamination of agricultural land largely due to impurities in the phosphate rock used for fertilizer manufacture. Fertilizers contaminated with uranium (U) can significantly elevate the uranium content of fertilized soils. For example, in Germany, the cumulative use of phosphate fertilizers from 1951 to 2011 has led to an average distribution of approximately 1 kg of uranium per hectare (Schnug and Lottermoser 2013). Several German mineral waters also have a tendency toward higher than average uranium concentrations. This could be due to leaching from natural deposits, or it may be related to the extensive use of phosphate fertilizers in the region (Schnug et al. 2005). As a result, the Federal Ministry of Food, Agriculture and Consumer Protection asked its scientific advisory panel (UBA) on soil conservation matters to make recommendations in this regard. Labelling for phosphate fertilizers containing 20 mg or more of uranium per kilogram of phosphate and a limit value of 50 mg per kg of phosphate is now recommended (European Commission Brussels 2016). Thus, imposing such a regulatory control may serve to dilute pre-existing uranium contamination and limit future contamination of agricultural soil and groundwater proving to be an important initiative in environmental conservation. If labelling is not available, in certain soils, it is possible to estimate the average concentration of uranium in superphosphate fertilizers between samplings using this formula (Taylor 2007):

$$c = iAdp/f$$

where

c = concentration of U in fertilizer in μg g^{-1}
i = average annual increase in U in soil in μg g^{-1}
A = area of 1 ha in m
d = sample depth in m
ρ = bulk density in total m^3
f = annual fertilizer application to 1 ha in kg

17.3 Plant Selection and Phytoremediation

Uranium is known to be highly toxic to plants. Plants experience oxidative stress in response to heavy metals resulting in cellular damage and disruption of cellular homeostasis similar to animals and humans. As a result, they have evolved detoxification mechanisms mainly involving chelation and subcellular compartmentalization to minimize the toxic effects of heavy metal exposure. Phytochelatins (PCs) are an important class of heavy metal chelators in plants which are synthesized from reduced glutathione (GSH) in a transpeptidation reaction catalyzed by the enzyme phytochelatin synthase (PCS). So, glutathione availability is essential for heavy metal detoxification in plants (Singh et al. 2011). Organic acids also play a key role in the metal tolerance of plants. In a process of metal detoxification, organic acids can form complexes with metals (Prasad and Freitas 2003).

Phytoremediation is the use of green plants to remove pollutants from the environment or to deactivate them and is based on exploiting the metabolic diversity of plants. This technology has been applied to a number of radionuclide-contaminated sites around the world. There are four main aspects of radionuclide phytoremediation including phytoextraction, rhizofiltration, phytovolatilization, and phytostabilization. Phytoextraction involves using high-biomass radionuclide-accumulating plants to transport and concentrate radionuclides from the soil into shoots above the ground; rhizofiltration involves using plant roots to precipitate and concentrate radionuclides from polluted effluents; phytovolatilization involves using plants to extract volatile radionuclides from soil and volatilize them from the foliage; and phytostabilization involves using plants to stabilize radionuclides in soils by metabolizing them (Dushenkov 2003). However, more basic research is required in this emerging new field to maximize the metabolic differences between detoxification pathways in plants and to understand the complex interactions of metals, soil, plant roots, and soil microbes such as bacteria and mycorrhizal and non-mycorrhizal fungi for decontaminating metalliferous substrates in the environment. There are significant interactions between metals and mycorrhizal fungi and macrofungi that may result in the amelioration of plant phytotoxicity and metal accumulation. Some of the mechanisms of metal fungitoxicity include extracellular precipitation and complexation, binding to cell walls, metal-binding proteins and peptides, vacuolar compartmentation, and metal transformations such as organometalloid synthesis (Gadd 1993).

Currently, up to 400 plants that hyperaccumulate metals have been reported. Most notably different genera of Brassicaceae are known to accumulate metals, such as Indian mustard, which is effective in the removal of heavy metals including cadmium, chromium, and lead. Sunflower has been found to be effective for the removal of uranium, cesium, and other radionuclides in hydroponic solutions. Plants can be raised hydroponically for transplantation into polluted water or soil where they can absorb and concentrate metals in unexposed parts like their roots for rhizofiltration. As the plant roots become saturated with metal contaminants, roots or whole plants are harvested for disposal. Alternatively, synthetic cross-linked polyacrylates or hydrogels can protect plant roots from heavy metal toxicity by preventing the entry of toxic metals into the roots during phytoextraction. Then, after sufficient plant growth and metal accumulation, the above-ground portions of the plant are harvested and removed resulting in permanent removal of metals from the site. Certain aquatic plant species also have the ability to remove heavy metals from water including water hyacinth, pennywort, and duckweed (Prasad and Freitas 2003).

It should be noted that phytostabilization is not a technology for cleaning up contaminated soil but a management strategy for stabilizing or inactivating potentially toxic trace elements, possibly as a follow-up to rhizofiltration and/or phytoextraction. Thus, along with soil amendments, the plant cover can provide pollution control and stability to the soil. Phytostabilization may be adapted to different metal contaminants and soil types by the careful selection of plants displaying maximum uptake of the contaminant in most abundance and adopting specific cropping schemes (Vassilev et al. 2004).

Plants that have been used in the phytoremediation of uranium-contaminated soils include Chinese cabbage, Swiss chard, and maize among various other species (Malaviya and Singh 2012). Uranium accumulates mainly in the roots of plant species, and uranium remediation of soils by plants can be largely influenced by soil type. For example, plant performance is affected by uranium contamination rates, particularly in calcareous soils. Plant roots grown in soils with high carbonate-uranium content accumulate the most uranium, while the lowest plant accumulation occurs in clayey acid soils with high iron, manganese, and organic uranium content (Shahandeh and Hossner 2002).

Conversely, crops grown on partially remediated sites should be simultaneously selected for minimum radionuclide uptake since vegetable and fruit plants can accumulate uranium. As an example, Indian mustard displays low accumulation for both radium-226 and thorium-232, while clover has low accumulation of uranium-238, when grown on soils from southeastern China contaminated with uranium mine tailings. Chinese mustard also has a low uranium-238 uptake (Chen et al. 2005). Therefore, Chinese mustard and clover would be two good crop choices in a uranium-contaminated area following a degree of remediation to reduce uranium levels to nontoxic levels according to international standards such as those currently being set in Germany. Similarly, Indian mustard would be suitable for soils contaminated with uranium decay products, radium, and thorium.

Quite recently, a very interesting use has been made of transgenic plants, which detoxify/accumulate cadmium, lead, mercury, arsenic, and selenium, for the purposes of phytoremediation (Eapen and D'Souza 2004). Similarly, specific plants could be genetically engineered to selectively uptake uranium and its decay products like radium and thorium in order to remediate polluted environments.

17.4 Human Cancer Risk and the Potential for Uranium Detoxification

17.4.1 Biochemical Detoxification of Uranium

No unique biochemical pathways have been elucidated for uranium detoxification in mammals, and the existing data is limited. In rats, hydrogen sulfide has been demonstrated to attenuate uranium-induced acute nephrotoxicity through oxidative stress and the inflammatory response via the Nrf2 and NF-κB pathways. Sodium hydrosulfide administration restores glutathione levels and antioxidant enzyme activities like superoxide dismutase, catalase, glutathione peroxidase, and glutathione S-transferase in rat kidneys (Zheng et al. 2015). These results may indicate that uranium toxicity is ameliorated in mammals through the induction of antioxidant defense following an initial reduction in activity similar to that observed in certain fish species (Barillet et al. 2007). Interestingly, different patterns of antioxidant activity have been observed for depleted uranium (DU) and enriched uranium (EU) in the cerebral cortex of rats. Lipid peroxidation increased significantly after EU exposure, but not after DU exposure. The gene expression or activity of superoxide dismutase (SOD), catalase (CAT), and glutathione peroxidase (GPx) increased significantly after chronic DU exposure in drinking water, while oral EU administration induced a decrease in these antioxidant enzymes (Lestaevel et al. 2009). Experiments have also shown that uranium can alter xenobiotic-metabolizing enzymes such as *CYP3A* (cytochrome P450, family 3, subfamily A) in the rat liver or kidneys after either acute or chronic exposure causing high gene expression for several months (Gueguen et al. 2014). In addition, uranium administration has been associated in several studies with decreases in creatinine clearance congruent with an increase in the plasma concentration of creatinine and blood ureic nitrogen indicating a reduction in glomerular filtration rate (GFR). Moreover, it has been demonstrated that several species undergo some degree of renal damage in a dose-dependent and route-independent manner following uranium exposure (Vicente-Vicente et al. 2010).

In humans, nephrotoxicity has been studied the most extensively. Acute uranium intoxication appears to affect the reabsorption of filtered solutes and the excretion of other solutes leading to kidney impairment of varying degrees depending on the initial dose (Vicente-Vicente et al. 2010). Reduced creatinine clearance, increased serum nonprotein nitrogen, and a higher rate of uranium excretion have been observed in humans. Increases in the urinary excretion of proteins (Friberg et al. 1986; Lu and Zhao 1990), amino acids (Lu and Zhao 1990), and urinary catalase (Bassett et al. 1948; Friberg et al. 1986) have also been noted. Thus, in both humans and animals, acute intoxication with uranium leads to nephrotoxicity in a dose-dependent manner suggesting that the kidneys are a major site for biochemical detoxification. However, the potential role of route of exposure remains largely unexplored for these findings in humans.

17.4.2 Biochemical Mechanisms of Uranium Toxicity

Uranium (U) mainly exists in the U (+IV) or U (+VI) oxidation states in water systems depending on their reduction-oxidation potential. Uranium is present in the soil primarily in the U (+VI) state as the uranyl cation. This is the most mobile form of uranium, and it is highly soluble in this state, while U (+IV) is much less soluble. In both soil and water, uranium can form complexes with sulfate, phosphate, carbonate, and hydroxide. These complexes also increase the total solubility of uranium. However, the linear uranyl oxycation and carbonate complexes of uranium are the dominant species in solution. In body fluids, uranium is dissolved as the uranyl ion, an ionic form that may react with biological molecules (Shahandeh and Hossner 2002). Absorption of DU into the blood and deposition in tissues and organs depend mostly on particle size and solubility of the uranium-containing particle. Solubility determines the rate of absorption from the lungs. Soluble chemical forms of uranium are absorbed within days, while insoluble forms can take from months to years. Insoluble particles deposited in the lung and lymph nodes and retained for extended periods of time are more likely to be associated with the radiation effects of uranium, while more soluble forms are associated with toxic chemical effects. In the blood, uranium can form a complex with bicarbonate and bind to plasma proteins or erythrocytes, and up to 90% of the dissolved uranium is excreted in urine within the first few days of exposure. The remaining 10% is deposited in the bones, kidneys, and other organs from where it is mobilized over a longer period of time (Malaviya and Singh 2012). Long-lived isotopic U is not well absorbed from the gastrointestinal (GI) tract in humans, and uranium oxide absorption is estimated at 1–2% with the bulk passing through the GI tract in feces (Wrenn et al. 1985; Lawrence et al. 2014). However, in rats, U-233 absorption from the GI tract takes place in the small intestine, probably via a transcellular pathway (Dublineau et al. 2005). The chemical speciation of uranium may play a role in this process.

It appears that different isotopic compositions of uranium like many other heavy metals can penetrate to the subcellular level resulting in bioaccumulation and initiation of oxidative stress in zebrafish tissues (Barillet et al. 2007). At subtoxic concentrations (>100 μm), depleted uranium precipitates mainly in the nucleus of the human kidney, liver, and neuronal cell lines (Rouas et al. 2010). Chemically, uranium can activate oxygen species in the course of redox reactions via the redox chemistry of transition metals (Yazzie et al. 2003) and enhance free radical production via the ionization phenomenon induced by alpha particle emissions to produce DNA damage in the form of double-strand breaks (Miller et al. 2002a; National Academies Press 2012). Although uranium can emit alpha, beta, and gamma radiation, alpha particle emissions are of the greatest relevance in relation to depleted uranium (Hon et al. 2015). Thus, uranium is capable of initiating chemotoxicity and radiotoxicity (Ng et al. 2015) via the generation of free oxygen species and possibly via other more direct mechanisms in cells. Both biochemical pathways can stimulate cell death or apoptosis, which has been linked to carcinogenesis in various cancer models. The loss of equilibrium between cell proliferation and cell death in a tissue may play a crucial role in tumor formation (Siddoo-Atwal 2009; Siddoo-Atwal 2017a).

17.4.3 Cancer Risk Assessment

Although the cancer data on depleted uranium alone is still sparse, it is possible to analyze the combined carcinogenic potential of uranium isotopes based on the existing data according to a new method of cancer risk assessment that relies on epidemiology, animal data, and cell studies. Recently, it has become apparent that the pathogenesis of cancer is closely connected with aberrantly regulated apoptotic cell death and the resulting deregulation of cell proliferation (Denmeade AND Isaacs 1996; Martin 2006). In fact, the initiation of uncontrolled apoptosis in a tissue can serve as the trigger for carcinogenesis (Siddoo-Atwal 2017b, 2019).

A distinct biological effect on wildlife has been observed for various isotopes of uranium. An irregular distribution of U-234 and U-238 has been found in the tissues of marine birds from the Polish area of the southern Baltic Sea. The highest accumulation occurs in the liver and other organs along with the feathers, while the smallest accumulation occurs in the skin and muscles with apparent interspecies differences (Boryko et al. 2010). Depleted uranium can cause adverse reproductive effects in terrestrial animals, and the likelihood of this occurring can range from 0.1% to 44% between various species (Fan et al. 2005).

Epidemiologically, there have been reports of an increased incidence of birth defects and, possibly, infant mortality in Iraq since the advent of the Gulf War (Alaani et al. 2011). In addition, an increased incidence of congenital anomalies and perinatal mortality has been reported in Kuwaiti newborns, and birth defects have been reported in the newborns of Gulf War veterans (Araneta et al. 1997, 2000, 2003; Doyle et al. 2004). A statistically significant relationship was also found between birth defects and the mother's proximity to uranium tailings among Native American Navajo in a uranium-mining area of the United States (Shields et al. 1992).

Some studies with US Gulf War veterans have reported cancers of the upper aerodigestive tract, lymphoma, and leukemia as health consequences of DU exposure (Bertell 1999; McDiarmid et al. 2000; Cannova 1998). WHO statistics for the years 1998–2000 suggest an elevation in lymphoma and leukemia incidence for the Eastern Mediterranean region affected by the Gulf War (Shawky 2003). The overall incidence of breast and lung cancer, leukemia, and lymphoma has doubled or tripled in certain areas of Iraq contaminated with DU during the Gulf War (Fathi et al. 2013; Busby et al. 2010). Interestingly, according to WHO, there also appears to be an elevated incidence in cancers of the trachea, bronchi, and lungs in Europe during the years 1998–2000 possibly due to other regional wars (Shawky 2003). DU weapons were deployed in Bosnia and Serbia, as well.

In addition, an association with lung cancer has been found in uranium miners (Grosche et al. 2006). It is strongly linked to exposure with radon, which is one of the decay products of natural uranium. Miners are also at risk as a result of exposure to radioactive *alpha*, *beta*, and *gamma* emissions from uranium, which may act as a carcinogen or cocarcinogen in this model (National Academies Press 2012).

Moreover, in South Carolina, it has been found that regions with elevated groundwater uranium and more groundwater use may have an increased incidence of certain cancers, including kidney, breast, colorectal, and prostate (Wagner et al. 2011).

In animal studies, reproductive toxicity and teratogenicity have been observed in mice following maternal uranium exposure at different gestation periods (Domingo 2001). Like other heavy metals, depleted uranium in drinking water has estrogenic activity and acts like estrogen in the reproductive tract of female mice leading to an increased risk of fertility problems (Raymond-Whish et al. 2007). In addition, DU exposure of wild-type and metallothionein-deficient mice has been shown to result in ROS production and cell apoptosis in the kidney. The MT-deficient mice were more susceptible to both effects suggesting a protective role for metallothionein in DU nephrotoxicity (Hao et al. 2015). Implantation of DU metal fragments in rats led to the development of soft tissue sarcomas (Hahn 2002).

In cell studies, depleted uranium-uranyl chloride has been found to induce apoptosis in mouse J774 macrophages (Kalinich et al. 2002). In rat lung epithelial cells, uranium induces significant oxidative stress followed by a decrease in antioxidant potential of the cells and a decreased proliferation rate (Periyakaruppan et al. 2007). Mitochondrial membrane collapse, mitochondrial swelling, and cytochrome c release associated with apoptosis have been observed following uranyl acetate treatment of isolated rat kidney mitochondria (Shaki et al. 2012). Studies with human osteosarcoma cells (HOS) demonstrate that DU exposure results in genomic instability manifested as delayed reproductive death and micronuclei formation (Miller et al. 2003). DU exposure also causes neoplastic transformation of immortalized HOS cells (Miller et al. 2002b). Particulate DU is cytotoxic and clastogenic in human bronchial fibroblasts (Wise et al. 2007). In human kidney cells (HK-2), pre-treatment with zinc significantly inhibits DU-induced apoptosis. It does this by reducing reactive oxygen species (ROS) production in cells and increasing catalase (CAT) and glutathione (GSH) concentrations, suppressing the DU-induced soluble FAS receptor and FAS ligand overexpression, suppressing caspase activation, and suppressing the release of cytochrome c and apoptosis-inducing factor (AIF) translocation from mitochondria to cytoplasm and nucleus, respectively. AIF is involved in the caspase-independent apoptotic pathway and can act directly to cause chromatin condensation and DNA breaks. In fact, all these events inhibited by zinc are linked to biochemical pathways for triggering cellular apoptosis via caspase-dependent or caspase-independent mechanisms (Hao et al. 2014).

17.4.4 Heavy Metal Chelation Therapy

In the 1940s, animal studies found that citrate salts like sodium citrate administered intravenously or orally provided significant protection from uranium poisoning. Subsequently, it was reported that uranium can be precipitated with citric acid by forming a metal-citrate complex (uranyl citrate) under certain chemical conditions. This may still be the most practical treatment to employ in conflict areas where DU weapons have been used and metal chelation therapy is not available. Citric acid and/or citrate consumption could be recommended to individuals exposed to the easily inhaled uranium oxide aerosols resulting from DU penetrators since it seems to be effective even when taken orally (Lawrence et al. 2014).

Later, metal chelation agents were tested for their capacity to increase urinary excretion of the uranyl ion and included tiron, gallic acid, diethylenetriaminepentaacetic acid (DTPA), p-aminosalicylic acid, sodium citrate, EDTA, 5-aminosalicylic acid, and ethylenebis(oxyethylenenitrilo)tetraacetic acid (EGTA). These eight substances were tested as antidotes for acute uranium poisoning in mice and significantly increased survival rate (Ortega et al. 1989).

EDTA and DMSA have both been used as chelation agents for uranium and other heavy metal detoxification in humans. EDTA (ethylenediaminetetraacetic acid) is a synthetic amino acid food preservative that binds to various heavy metals in the body and removes them through the urine. It is administered intravenously and has been in use for nearly 50 years to treat heavy metal toxicity (Dean 2018a). Na-Ca-EDTA chelates are rapidly excreted and, generally, cause greater losses of essential minerals from the body than DMSA. EDTA has been found to be most effective in removal of lead, cadmium, and mercury (Sears 2013). DMSA (meso-2,-3-dimercaptosuccinic acid) is a sulfhydryl-containing, water-soluble, nontoxic, orally administered metal chelation agent. It has been used to treat heavy metal toxicity, particularly lead and mercury poisoning, since the 1950s. The low toxicity and efficacy of DMSA makes it the primary metal chelator of choice for removal of mercury and other heavy metals. It is also very safe and causes few side-effects. The DMSA-metal conjugates are expelled through the urine (Dean 2018b). Oral absorption of DMSA is approximately 20% with most DMSA in the blood plasma being bound to protein, mainly albumin (95%). 10–25% of orally administered DMSA is excreted in the urine, mostly within 24 h of administration. Excretion of trace elements including zinc, iron, calcium, and magnesium is much less than with Na-Ca-EDTA. Copper may be an exception (Sears 2013).

However, the individual efficacies of EDTA and DMSA for uranium detoxification are not widely documented since they have not been specifically engineered as uranium chelators. Nevertheless, an approximation of their relative efficiencies may be gained by considering the results of a case study (*in press*) in which 90 ug of uranium was completely cleared from a subject's body after two metal chelation treatments with EDTA (1900 mg) and DMSA (500 mg). According to the WHO, approximately 90 μg of uranium exist in the human body on average from normal intakes of water, food, and air (https://www.who.int/ionizing_radiation/pub_meet/

en/DU_Eng.pdf?ua=1). So, these results are not staggering, especially since the average body burden of uranium was found to be 100 times greater than the normal range in urine samples collected from a civilian population in Eastern Afghanistan following the use of depleted uranium bombs by Allied Forces in 2002 (Durakovic 2005). In some cases, uranium concentrations up to 200 times higher than in the control population were detected in affected districts (Durakovic 2003).

Moreover, metal chelation therapy is a lengthy, unpleasant, and costly treatment for patients to undergo. It is inefficient and depletes the body of essential micronutrients. Therefore, it is necessary to develop more efficient chelating agents that are specifically engineered for targeting uranium. Some interesting preliminary experiments in DU-poisoned human renal proximal tubule epithelium (HK-2) cells with Cu^{2+}-imprinted chitooligosaccharides have shown that these molecules can selectively chelate DU outside cells and reduce cellular DU accumulation in a dose-dependent manner. They also significantly reduce loss of cell viability induced by DU by preventing its cellular internalization. In addition, treatment with these chitooligosaccharides appears to increase the activity of antioxidant enzymes and to reduce membrane damage and DNA damage caused by DU oxidant injury as compared to a control drug (Zhang et al. 2011).

17.4.5 Zinc Supplementation and Metallothionein

Metallothioneins (MTs) are small, cysteine-rich proteins that bind heavy metals and participate in an array of protective stress responses (Ruttkay-Nedecky et al. 2013). Inside the cell, the harmful effects of free radicals are balanced by the antioxidant action of antioxidant enzymes and nonenzymatic antioxidants. MTs are found in bacteria, plants, invertebrates, and vertebrates. There are four main mammalian MT isoforms (MT-1–MT-4) with distinct roles in different tissues. Aerobic organisms are susceptible to damage by reactive oxygen species (ROS) and reactive nitrogen species (RNS). MT protects cells from exposure to various free radical species like the hydroxyl, peroxyl, alkoxyl, and superoxide anion radical and the nitric oxide and nitric dioxide radicals, which react readily with sulfhydryl groups. MT is also important for the regulation of zinc levels and the distribution of this metal in the extracellular space. Since zinc cannot pass easily through membranes, zinc-transporting proteins, ZIPs (Zrt-Irt-like protein or zinc iron permease), and ZnTs (zinc transporters) help to facilitate this process. The presence of zinc(II) within the cell causes an increase in the major zinc-binding protein metallothionein, and it binds to MTs forming a thermodynamically stable complex. MT can be activated by various stimuli including heavy metal ions, cytokines, growth factors, and oxidative stress within the cell. Cells that display high MT production are resistant to heavy metal toxicity by cadmium, whereas cell lines that cannot synthesize MTs are sensitive to the toxic effects of cadmium (Ruttkay-Nedecky et al. 2013).

One fascinating study with rats indicated that pre-treatment with zinc has protective effects against depleted uranium toxicity. Posttreatment with zinc also had a protective effect but to a lesser extent. In fact, the pre-treated group had a 60% higher rate of survival than the untreated group. Blood urea nitrogen, creatinine, and urine *N*-acetyl-β-D-glucosaminidase concentrations were significantly decreased. The gene expression levels of metallothionein (MT) in kidney tissues were significantly increased, and catalase levels were increased, as well. Thus, the protective mechanism that alleviated DU toxicity appeared to be related to the induction of MT synthesis and the enhancement of antioxidant function (Hao et al. 2012). It is quite possible that this result could extend to humans given that these multifunctional proteins are so well-conserved from nematodes to mammals (Isani and Carpene 2014) and zinc fortification may be worth investigating.

In rats, rice fortified with zinc oxide or zinc carbonate is a feasible vehicle for zinc absorption, although zinc oxide displays lower bioavailability than zinc carbonate (Lucia et al. 2014). In young adults, zinc absorption from supplemental zinc citrate is comparable with that from zinc gluconate but higher than from zinc oxide (Wegmuller et al. 2014).

17.5 Conclusions

The biochemical detoxification of uranium in humans is still not completely understood. It appears to activate various antioxidant enzyme activities in mammals including superoxide dismutase, catalase, and glutathione enzymes, although, the results between treatment with depleted uranium and enriched uranium are not always comparable and may vary between tissues. Nevertheless, it is known that depleted uranium/natural uranium chemotoxicity and radiotoxicity can occur through the generation of free radical species and initiation of DNA damage (double-strand breaks) within the cell resulting in apoptosis. Apoptotic deregulation is increasingly viewed as a major mechanism for carcinogenesis.

Epidemiological data suggests that uranium is implicated in the etiology of lymphoma, leukemia, renal, and respiratory tract cancers. Animal studies indicate reproductive toxicity and teratogenicity as a result of uranium exposure. DU displays estrogenic activity leading to an increased risk of fertility problems in mice, and it can cause cell apoptosis in mouse kidney. Cell studies show that depleted uranium-uranyl chloride causes apoptosis in mouse macrophages and uranium promotes oxidative stress in rat lung epithelial cells which can also induce apoptosis. Events associated with apoptosis have been observed following uranyl acetate treatment of isolated rat kidney mitochondria. In human osteosarcoma cells, DU exposure results in genomic instability, and, in human kidney cells, DU induces apoptosis. DU is cytotoxic and clastogenic in human bronchial fibroblasts. Therefore, according to the new approach to cancer risk assessment, it is highly likely that certain uranium isotopes including those of depleted uranium are implicated in respiratory disorders and lung cancer in humans. DU is also nephrotoxic in nature resulting in various renal effects such as chronic kidney disease, and it has the potential for causing renal cancer via the initiation of apoptosis. Additionally, it is probable that uranium can cause lymphoma and leukemia. Apoptotic assays should be carried out in human lymphocytes and other white blood cells to further test the carcinogenic potential of isotopic uranium in these tissues.

Metal chelation therapy and citrate and zinc supplementation may represent possible methods of uranium detoxification. There is no published data relating to the effectiveness or viability of these treatments in patients with 100 or 200 times the normal body burden of uranium. Since metal chelation depletes the body of micronutrients, which need to be replenished, and only removes a limited quantity of toxic elements, which can be redistributed from less accessible to more accessible body compartments following therapy, such a course of treatment might extend over a period of several years. Therefore, a more practical solution like the development of uranium-specific chelating agents is necessary.

Traditional methods of soil remediation are costly and labor-intensive (a typical project can cost more than $200–300 billion), while phytoremediation represents a newer, more economic, and environmentally friendly approach to cleaning up heavy metal and radionuclide contamination. On the downside, the latter has low extraction efficiency, generates large quantities of contaminated biomass, and requires long periods to complete the decontamination process. The aim of soil amendments is mainly metal immobilization and/or inactivation, while the main aim of phytoremediation is metal hyperaccumulation in various plant parts, and a combination of these two methods can be effective in the remediation of contaminated agricultural soils. However, few reports of the application of such combined techniques to cases of uranium contamination appear in the scientific literature suggesting that it is a relatively new field of research. Therefore, it is not possible to judge their efficacy for uranium removal as yet. Other biotechnological approaches that are currently being tested and may be applied to this type of pollution in the future include biomineralization (mineral synthesis by living organisms or biomaterials), biosorption (via dead microbial and renewable agricultural biomass), dendroremediation (growing trees in polluted soils), biostimulation (stimulating living microbial populations adapted to accumulating the pollutant), mycoremediation (stimulating living fungal/mycelial ultrafiltration), cyanoremediation (stimulating algal mass for remediation), and genoremediation (stimulating specific genes for remediation through selection or genetic engineering of transgenic plants) (Mani and Kumar 2014).

Uranium contamination due to the continued use of depleted uranium weapons and nuclear accidents has potentially far-reaching consequences for humans and wildlife populations. In fact, DU could prove to be an unprecedented ecological hazard. Moreover, the remediation of uranium and other related radionuclide contamination is an extremely complex and expensive biological and environmental problem. Currently, there is limited data available on effective remediation methods for the removal of uranium in contaminated areas. Furthermore, remediation measures are of indeterminate efficiency as the ratio of uranium contamination to soil, uptake by plant species, and the total body burden in humans increases. As a primary health and ecological measure, it is suggested that an immediate global ban should be placed on the use and manufacture of DU weapons. Violators of the ban should be made to bear all remediation and public health costs.

Acknowledgments I am grateful for discussions with a bright, young geologist, Talya Grewal, that helped to shape this chapter.

References

Akob DM, Mills HJ, Kostka JE (2007) Metabolically active microbial communities in uranium-contaminated subsurface sediments. FEMS Microbiol Ecol 59:95–107

Alaani S, Savabieasfahani M, Tafash M, Manduca P (2011) Four polygamous families with congenital birth defects from Fallujah, Iraq. Int J Environ Res Public Health 8:89–96

Araneta MRG, Moore CA, Olney RS, Edmonds LD, Karcher JA, McDonough C, Hiliopoulos KM, Schlangen KM, Gray GC (1997) Goldenhar syndrome among infants born in military hospitals to Gulf War veterans. Teratology 56:244–251

Araneta MRG, Destiche DA, Schlangen KM, Merz RD, Forrester MB, Gray GC (2000) Birth defects prevalence among infants of Gulf war veterans born in Hawaii, 1989–1993. Teratology 62:195–204

Araneta MR, Schlangen KM, Edmonds LD, Destiche DA, Merz RD, Hobbs CA, Flood TJ, Harris JA, Krishnamurti D, Gray GC (2003) Prevalence of birth defects among infants of Gulf War veterans in Arkansas, Arizona, California, Georgia, Hawaii, and Iowa, 1989–1993. Birth Defects Res A Clin Mol Teratol 67:246–260

Barillet S, Adam C, Palluel O, Devaux A (2007) Bioaccumulation, oxidative stress, and neurotoxicity in *Danio rerio* exposed to different isotopic compositions of uranium. Environ Toxicol Chem 26:497–505

Bassett SH, Frenkel A, Cedars N, Van Alstine H, Waterhouse C, Cusson K (1948) The excretion of hexavalent uranium following intravenous administration. II. Studies on human subjects. The University of Rochester, Rochester, URB37, Washington, DC

Bertell R (1999, May) Gulf war veterans and depleted uranium. Hague peace conference

Bleise A, Danesi PR, Burkart W (2003) Properties, use and health effects of depleted uranium (DU): a general overview. J Environ Radioact 64:93–112

Boryko A, Skwarzec B, Fabisiak J (2010) Bioaccumulation of uranium ^{234}U and ^{238}U in marine birds. J Radioanal Nucl Chem 284:165–172

Briner WE (2006) The evolution of depleted uranium as an environmental risk factor: lessons from other metals. Int J Environ Res Public Health 3:129–135

Busby C, Hamdan M, Ariabi E (2010) Cancer, infant mortality and birth sex-ratio in Fallujah, Iraq 2005–2009. Int J Environ Res Public Health 7:2828–2837

Cannova JV (1998) Multiple giant cell tumors with Gulf War syndrome. Mil Med 163:184–185

Cardenas E, Wu W-M, Leigh MB, Carley J, Carroll S, Gentry T, Luo J, Watson D, Gu B, Ginder-Vogel M, Kitanidis PK, Jardine PM, Zhou J, Criddle CS, Marsh TL, Tiedje JM (2008) Microbial communities in contaminated sediments, associated with bioremediation of uranium to submicromolar levels. Appl Environ Microbiol 74:3718–3729

Chen SB, Zhu YG, Hu QH (2005) Soil to plant transfer of ^{238}U, ^{226}Ra and ^{232}Th on a uranium mining-impacted soil from southeastern China. J Environ Radioact 82:223–236

Dean W (2018a) Chelation with EDTA: a food additive with many remarkable properties. Anti-aging and life extension medicine

Dean W (2018b) DMSA and detoxifying heavy metals. Anti-aging and life extension medicine

Denmeade SR, Isaacs JT (1996) Programmed cell death (Apoptosis) and cancer chemotherapy. Cancer Control 3:303–309

Domingo JL (2001) Reproductive and developmental toxicity of natural and depleted uranium: a review. Reprod Toxicol 15:603–609
Doyle P, Maconochie N, Davies G, Maconochie I, Pelerin M, Prior S, Lewis S (2004) Miscarriage, stillbirth and congenital malformation in the offspring of UK veterans of the first Gulf war. Int J Epidemiol 33:74–86
Dublineau I, Grison S, Baudelin C, Dudoignon N, Souidi M, Marquette C, Paquet F, Aigueperse J, Gourmelon P (2005) Absorption of uranium through the entire gastrointestinal tract of the rat. Int J Radiat Biol 81:473–482
Durakovic A (2001) On depleted uranium: Gulf war and Balkan syndrome. Croat Medl J 42:130–134
Durakovic A (2003) Undiagnosed illnesses and radioactive warfare. Croat Medl J 44:520–532
Durakovic A (2005) The quantitative analysis of uranium isotopes in the urine of the civilian population of eastern Afghanistan after operation enduring freedom. Mil Med 170:277–284
Dushenkov S (2003) Trends in phytoremediation of radionuclides. Plant Soil 249(1):167–175
Eapen S, D'Souza SF (2004) Prospects of genetic engineering of plants for phytoremediation of toxic metals. Biotechnol Adv 23:97–114
Echevarria G, Sheppard MI, Morel J-L (2001) Effect of pH on the sorption of uranium in soils. J Environ Radioact 53:257–264
European Commission Brussels, 17.3.2016 SWD (2016) SWD 2016_64_F1_IMPACT_ASSESMENT_EN_V6_P2_839082.pdf
Fan M, Thongsri T, Axe L, Tyson TA (2005) Using a probabilistic approach in an ecological risk assessment simulation tool: test case for depleted uranium (DU). Chemosphere 60:111–125
Fathi RY, Al-Salih HS, Matti LY, Godbold DL (2013) Environmental pollution by depleted uranium in Iraq with special reference to Mosul and possible effects on cancer and birth defect rates. Med Confl Surviv 29:7–25
Fields MW, Yan T, Rhee S-K, Carroll SL, Jardine PM, Watson DB, Criddle CS, Zhou J (2005) Impacts on microbial communities and cultivable isolates from groundwater contaminated with high levels of nitric acid–uranium waste. FEMS Microbiol Ecol 53:417–428
Francis AJ, Dodge CJ (1998) Remediation of soils and wastes contaminated with uranium and toxic metals. Environ Sci Technol 32:3993–3998
Friberg L, Nordberg GF, Vouk VBGFBN, Nordberg M, Friberg L (1986) General aspects, handbook on the toxicology of metals, vol 1. Elsevier, Amsterdam
Gadd GM (1993) Interactions of fungi with toxic metals. New Phytol 124:25–60
Gadd GM (2010) Metals, minerals and microbes: geomicrobiology and bioremediation. Microbiology 156:609–643. https://doi.org/10.1099/mic.0.037143-0
Grosche B, Kreuzer M, Kreisheimer M, Schnelzer M, Tschense A (2006) Lung cancer risk among German male uranium miners: a cohort study, 1946–1998. Br J Cancer 95:1280–1287
Gueguen Y, Rouas C, Monin A, Manens L, Stefni J, Delissen O, Grison S, Dublineau I (2014) Molecular, cellular, and tissue impact of depleted uranium on xenobiotic-metabolizing enzymes. Arch Toxicol 88:227–239
Hahn FF, Guilmette RA, Hoover MD (2002) Implanted depleted uranium fragments cause soft tissue sarcomas in the muscles of rats. Environ Health Perspect 110:51–59
Hao Y, Ren J, Liu J, Luo S, Ma T, Li R, Su Y (2012) The protective role of zinc against acute toxicity of depleted uranium in rats. Basic Clin Pharmacol Toxicol 111:402–410
Hao Y, Ren J, Liu C, Li H, Liu J, Yang Z, Li R, Su Y (2014) Zinc protects human kidney cells from depleted uranium-induced apoptosis. Basic Clin Pharmacol Toxicol 114:271–280
Hao Y, Huang J, Gu Y, Liu C, Li H, Liu J, Ren J, Yang Z, Peng S, Wang W, Li R (2015) Metallothionein deficiency aggravates depleted uranium-induced nephrotoxicity. Toxicol Appl Pharmacol 287:306–315
Hon Z, Osterreicher J, Navratil L (2015) Depleted uranium and its effects on humans. Sustainability 7:4063–4077

Hudcova H, Badurova M, Rozkosny M, Sova J, Funkova R, Svobodova J (2013) Quality and mutagenicity of water and sediment of the streams impacted by the former uranium mine area Olší–Drahonín (Czech Republic). J Environ Radioact 116:159–165

Isani G, Carpene E (2014) Metallothioneins, unconventional proteins from unconventional animals: a long journey from nematodes to mammals. Biomolecules 4:435–457

Kalinich F, Ramakrishnan N, Villa V, McClain DE (2002) Depleted uranium–uranyl chloride induces apoptosis in mouse J774 macrophages. Toxicology 179:105–114

Katsenovich Y, Carvajal D, Guduru R, Lagos L, Li C-Z (2013) Assessment of the resistance to uranium (VI) exposure by *Arthrobacter* sp. isolated from Hanford site soil. Geomicrobiol J 30:120–130

Lawrence GD, Patel KS, Nusbaum A (2014) Uranium toxicity and chelation therapy. Pure Appl Chem 86(7). https://doi.org/10.1515/pac-2014-0109

Lestaevel P, Romero E, Dhieux B, Bensoussan H, Berradi H, Dublineau I, Voisin P, Gourmelon P (2009) Different pattern of brain pro-/anti-oxidant activity between depleted and enriched uranium in chronically exposed rats. Toxicology 258:1–9

Li J, Zhang Y (2012) Remediation technology for the uranium contaminated environment: a review. Procedia Environ Sci 13:1609–1615

Lu S, Zhao FY (1990) Nephrotoxic limit and annual limit on intake for natural U. Health Phys 58:619–623

Lucia CMD, Santos LLM, da Cruz Rodrigues KC, da Cruz Rodrigues VC, Martino HSD, Sant'Ana HMP (2014) Bioavailability of zinc in wistar rats fed with rice fortified with zinc oxide. Nutrients 6:2279–2289

Malaviya P, Singh A (2012) Phytoremediation strategies for uranium-contaminated environments: a review. Crit Rev Environ Sci Technol 42:2575–2647

Mani D, Kumar C (2014) Biotechnological advances in bioremediation of heavy metals contaminated ecosystems: an overview with special reference to phytoremediation. Int J Environ Sci Technol 11:843–872

Martin KR (2006) Targeting apoptosis with dietary bioactive agents. Exp Biol Med 231:117–129

Martinez RJ, Beazley MJ, Taillefert M, Arakaki AK, Skolnick J, Sobecky PA (2007) Aerobic uranium (VI) bioprecipitation by metal-resistant bacteria isolated from radionuclide- and metal-contaminated subsurface soils. Environ Microbiol 9:3122–3133

McDiarmid MA, Keogh JP, Hooper FJ, McPhaul K, Squibb K, DiPino R et al (2000) Health effects of depleted uranium on exposed Gulf War veterans. Environ Res 82:168–180

Merroun ML, Selenska-Pobell S (2008) Bacterial interactions with uranium: an environmental perspective. J Contam Hydrol 102:285–295

Middle East Eye -#SyriaWar - US Pentagon confirms it used depleted uranium in Syria; www.middleeasteye.net

Miller AC, Stewart M, Brooks K, Shi L, Page N (2002a) Depleted uranium-catalyzed oxidative DNA damage: absence of significant alpha particle decay. J Inorg Biochem 91:246–252

Miller AC, Xu J, Stewart M, Brooks K, Hodge S, Shi L, Page N, McClain D (2002b) Observation of radiation-specific damage in human cells exposed to depleted uranium: dicentric frequency and neoplastic transformation as endpoints. Radiat Prot Dosim 99:275–278

Miller AC, Brooks K, Stewart M, Anderson B, Shi L, McClain D, Page N (2003) Genomic instability in human osteoblast cells after exposure to depleted uranium: delayed lethality and micronuclei formation. J Environ Radioact 64:247–259

National Academies Press (2012) Uranium mining in Virginia: scientific, technical, environmental, human health and safety, and regulatory aspects of uranium mining and processing in Virginia. Chapter 5: Potential human health effects of uranium mining, processing, and reclamation. https://www.ncbi.nlm.nih.gov/books/NBK201047/

Ng CY, Pereir S, Cheng SH, Adam-Guillermin C, Garnier-Laplace J, Yu KN (2015) Combined effects of depleted uranium and ionising radiation on zebrafish embryos. Radiat Prot Dosim 167:311–315

Obralic N, Gavrankapetanović F, Dizdarević Z, Duric O, Sisic F, Selak I, Balta S, Nakas B (2004) Regional comparison of cancer incidence. Radiol Oncol 38:145–151

Ortega A, Domingo JL, Gomez M, Corbella J (1989) Treatment of Experimental Acute Uranium Poisoning by Chelating Agents. Pharmacol Toxicol 64:247–251

Periyakaruppan A, Kumar F, Sarkar S, Sharma CS, Ramesh GT (2007) Uranium induces oxidative stress in lung epithelial cells. Arch Toxicol 81:389–395

Prasad MNV, Freitas HMO (2003) Metal hyperaccumulation in plants - biodiversity prospecting for phytoremediation technology. Electron J Biotechnol 6. https://doi.org/10.2225/vol6-issue3-fulltext-6

Priest ND (2001) Toxicity of depleted uranium. Lancet 357:244–246

Raymond-Whish S, Mayer LP, O'Neal T, Martinez A, Sellers MA, Christian PJ, Marion SL, Begay C, Proper CR, Hoyer PB, Dyer CD (2007) Drinking water with uranium below the U.S. EPA water standard causes estrogen receptor–dependent responses in female mice. Environ Health Perspect 115:1711–1716

Rouas C, Bensoussan H, Suhard D, Tessier C, Grandcolas L, Rebiere F, Dublineau I, Taouis M, Pallardy M, Lestaevel P, Gueguen Y (2010) Distribution of soluble uranium in the nuclear cell compartment at subtoxic concentrations. Chem Res Toxicol 23:1883–1889

Ruttkay-Nedecky B, Nejdl L, Gumulrc J, Zitka O, Masarik M, Eckschlager T, Stiborova M, Adam V, Kizek R (2013) The role of metallothionein in oxidative stress. Int J Mol Sci 14:6044–6066

Salbu B, Krekling T (1998) Characterisation of radioactive particles in the environment. Analyst 123:843–850

Schnug E, Lottermoser BG (2013) Fertilizer-derived uranium and its threat to human health. Environ Sci Technol 47:2433–2434

Schnug E, Steckel H, Haneklaus S (2005) Contribution of uranium in drinking waters to the daily uranium intake of humans - a case study from Northern Germany. Landbauforschung Völkenrode 55:227–236

Sears ME (2013) Chelation: harnessing and enhancing heavy metal detoxification – a review. Sci World J 2013:219840. https://doi.org/10.1155/2013/219840

Shahandeh H, Hossner LR (2002) Role of soil properties in phytoaccumulation of uranium. Water Air Soil Pollut 141:165–180

Shaki F, Hosseini M-J, Ghazi-Khansari M, Pourahmad J (2012) Toxicity of depleted uranium on isolated rat kidney mitochondria. Biochim Biophys Acta 1820(12):1940–1950. – [BBA] general subjects

Shawky S (2003) Causes of death in the Eastern Mediterranean Region during the years 1998–2000. Saudi Med J 24:380–387

Sheppard SC, Evenden WG (1988) Critical compilation and review of plant/soil concentration ratios for uranium, thorium and lead. J Environ Radioact 8:255–285

Shields LM, Wiese WH, Skipper BJ, Charley B, Benally L (1992) Navajo birth outcomes in the Shiprock uranium mining area. Health Phys 63:542–551

Siddoo-Atwal C (2009) AT, apoptosis, and cancer: a viewpoint. Indian J Ecol 36:103–110

Siddoo-Atwal C (2017a) Heavy metal carcinogenesis: a possible mechanistic role for apoptosis. Vegetos 30(Special). https://doi.org/10.5958/2229-4473.2017.00046.5

Siddoo-Atwal C (2017b) A New Approach to Cancer Risk Assessment: An Overview. Publishing, Lambert Academic

Siddoo-Atwal C (2019) An approach to cancer risk assessment and carcinogenic potential for three classes of agricultural pesticide. In: Natural resource management: ecological perspectives. Springer, Cham, pp 109–132

Singh R, Gautam N, Mishra A, Gupta R (2011) Heavy metals and living systems: An overview. Indian J Pharm 43:246–253

Sitte J, Akob DM, Kaufmann C, Finster K, Banerjee D, Burkhardt E-M, Kostka JE, Scheinost AC, Buchel G, Kusel K (2010) Microbial links between sulfate reduction and metal retention in uranium- and heavy metal-contaminated soil. Appl Environ Microbiol 76:3143–3152

Suzuki Y, Banfield JF (2004) Resistance to, and accumulation of, uranium by bacteria from a uranium-contaminated site. Geomicrobiol J 21:113–121

Taylor MD (2007) Accumulation of uranium in soils from impurities in phosphate fertilisers. Landbauforschung Völkenrode 2:133–139

Thoennessen M (2016) The discovery of isotopes – a complete compilation. Springer, Cham

UBA website – website of the German Crop Protection Association (Fertilizer)

Vassilev A, Schwitzguebel J-P, Thewys T, van der Lelie D, Vangronsveld J (2004) The use of plants for remediation of metal-contaminated soils. Sci World J 4:9–34

Vicente-Vicente L, Quiros Y, Pérez-Barriocanal F, López-Novoa JM, López-Hernández FJ, Morales AI (2010) Nephrotoxicity of uranium: pathophysiological, diagnostic and therapeutic perspectives. Toxicol Sci 118:324–347

Wagner SE, Burch JB, Bottai M, Puett R, Porter D, Bolick-Aldrich TT, Wilkerson RC, Vena JE, Hebert JR (2011) Groundwater uranium and cancer incidence in South Carolina. Cancer Causes Control 22:41–50

Wall JD, Krumholz LR (2006) Uranium reduction. Annu Rev Microbiol 60:149–166

Wegmuller R, Tay F, Zeder C, Brnic M, Hurrell RF (2014) Zinc absorption by young adults from supplemental zinc citrate is comparable with that from zinc gluconate and higher than from zinc oxide. J Nutr 144:132–136

WHO – Depleted uranium sources, exposure, and health effects (online monograph)

Wise SS, Thompson WD, Aboueissa A-M, Mason MD, Wise JP (2007) Particulate depleted uranium is cytotoxic and clastogenic to human lung cells. Chem Res Toxicol 20:815–820

World Nuclear Association Website (updated 2017) What is Uranium? How does it work?

Wrenn ME, Durbin PW, Howard B, Lipsztein J, Rundo J, Still ET, Willis DL (1985) Metabolism of ingested U and Ra. Health Phys 48:601–633

Yazzie M, Gamble SL, Civitello ER, Stearns DM (2003) Uranyl acetate causes DNA single strand breaks in vitro in the presence of ascorbate (vitamin C). Chem Res Toxicol 16:524–530

Zhang X-F, Ding C-J, Liu H, Liu L-H, Zhao C (2011) Protective effects of ion-imprinted chitooligosaccharides as uranium-specific chelating agents against the cytotoxicity of depleted uranium in human kidney cells. Toxicology 286:75–84

Zheng J, Zhao T, Yuan Y, Hu N, Tang X (2015) Hydrogen sulfide (H_2S) attenuates uranium-induced acute nephrotoxicity through oxidative stress and inflammatory response via Nrf2-NF-κ B pathways. Chem Biol Interact 242:353–362

Chapter 18
An Overview of Effective Concentration of Industrial Effluent for Improving Crop Production and Its Effect on Micro-Biodiversity Zone of Soil

Sangeeta, Gita Rani, and Rani Devi

Abstract Soil and water are integral parts of the ecosystem. These are used as resources for agriculture. In recent years, most of the water and soil have become polluted by sewage, industrial waste/effluents, and a wide range of synthetic chemicals. Our planet Earth is now overburdened with the toxic substances. Industrialization has been proven as a significant milestone in the development of human civilization. But, at the same time, it has loaded the Earth with industrial wastes including toxic solids, liquids, and gaseous discharges. With the mushrooming growth pattern of industries, all important components of ecosystems particularly soil, water, air, vegetation, and all others are being affected adversely. Management of these natural resources is very important for sustainable development of living beings on Earth. Industrializations are not only land area intensive but also lead to other serious environmental humiliation (Azumi DS, Bichi MH, J Appl Sci Environ Sanit 5:23–29, 2010). Now the challenge before us in this present scenario is the careful disposal of effluent so that adverse impact on soil fertility, plant growth, and health of animals and human beings may be reduced. Thus, there is an urgent need to have innovative strategies for wiser management of effluents and land resources for its optimum and appropriate utilization. In the present study, efforts have been made for handling sugar industry along with use of its effluent for improving wheat crop and production.

Sangeeta · R. Devi (✉)
Department of Energy and Environmental Sciences, Chaudhary Devi Lal University, Sirsa, Haryana, India

G. Rani
Department of Chemistry, Chaudhary Devi Lal University, Sirsa, Haryana, India

A. Varma et al. (eds.), *Plant Microbe Interface*,
https://doi.org/10.1007/978-3-030-19831-2_18

18.1 Introduction: Sugarcane Industry

Cultivation of sugarcane in India dates back to the Vedic period. Various historical and mythological evidences confirm that the original home of sugarcane production and sugar manufacturing is India. It is cheaper to get sugar from sugarcane than sugar beet. Sugarcane is cultivated in tropical climates. Globally, India is the largest producer of sugar (Solomon 2008). India stands as a major sugar-producing nation in the world, having 579 sugar mills and 319 distilleries (Patil and Gholey 2010), and is contributing around 18–20% of the world's cane sugar production and sugar potential in different states of India. Sugar industry is the backbone of all food industries, and sugar produced in it is the important ingredient for food preparation at home, functions, and parties and in marriages. It is one of the most important agro-based industries in India and is the second largest in the country, next to textiles. But a considerable amount of effluent is released during sugarcane crushing which is spilled into nearby water bodies and is causing a serious problem to water bodies and land/soil in its vicinity which is resulting in environmental degradation and affecting the crop quality as well as yield (Maliwal et al. 2004). It is also generating a huge amount of solid waste as bagasse. To handle this problem, the industry is using bagasse to produce ethanol, which can be used as an eco-friendly and renewable energy source by blending with petrol (Satheeskumar and Selvaraj 2007). The production of ethanol gives the boost to agriculture sector and reduces environmental pollution. The Government of India is trying to introduce supply of ethanol-doped petrol in the country.

18.2 Sugar Industry Effluents

All the industries consume huge quantity of water and throw back almost 70–80% of it as effluents which contain highly toxic materials in dissolved or suspended form. The demand for water is increasing due to population growth and urbanization, hence making this resource very scarce (CPCB 2009). Main wastewater pollutants are organic components (proteins, fats, and carbohydrates), inorganic components (alkalies, acids, inorganic salts, and other chemicals), and physical factors such as turbidity, color, temperature of effluent, radioactivity, etc. Presence of these organic compounds imparts a high biochemical oxygen demand (BOD) and chemical oxygen demand (COD) load to the liquid waste, and their level below a certain limit becomes a danger to aquatic life (Sharma and Habib 1997; Singh et al. 2005; Pandey et al. 2008; Chandra et al. 2009).

Application of sewage and industrial wastewater on agricultural land (irrigation) became an alternative water supply to crops as well as an alternative waste disposal method. Sugar industry wastes constitute a number of physicochemical effluents of suspended and dissolved solids with the high amount of BOD, COD, chlorides, sulfates, calcium, magnesium, nitrates, etc. The high contents of organic matter in stillage cause surface water pollution. Organic matters cause oxygen depletion in surface waters by biodegradation. Stillage has high concentration of potassium which can cause pollution of the soil. When released into the environment without proper treatment, effluents of sugar industry produce unpleasant color and odor. These effluents also alter the physicochemical characteristics of the receiving aquatic bodies and affect aquatic flora and fauna (Baruah et al. 1993). The polluted soil becomes unsuitable for further cultivation. So, some treatments should be given to sugar mill effluent-polluted soils before they are used for crop cultivation (Roy et al. 2007).

18.3 Wheat Cultivation

Wheat is the staple food of millions of Indians, particularly in North India, and the 2nd most important food grain of India. India is the fourth largest producer of wheat in the world and accounts for 8.7% of the world's total production of wheat. It is a rabi crop sown in the beginning of winter and is harvested in the beginning of summer.

The **Meham Co-Operative Sugar Mills Ltd., Rohtak, Haryana**, chosen for this study, is located at 28°59′49.2″N 76°14′30.1″E. The study area falls in the eastern zone, which covers around 49% of the area of the state. This zone is also called wet zone.

18.4 Microbial Species in Soil

18.4.1 Identification and Characterization of Fungal Isolates

Identification and characterization of fungal isolates in the soil affected with SME (sugar mill effluent) were carried out on the basis of colony growth (diameter), presence or absence of aerial mycelium, colony color, presence of furrows in the medium, pigment production, spore morphology, etc. The fungal species (lactophenol cotton blue preparation) were identified by microscopic analysis using taxonomic guides and standard procedure (Barnett and Hunter 1972; Ellis 1976; Domsch et al. 1980; Nelson et al. 1983; Gilman 1998).

The major cultural features used for identification included color of the colony, growth pattern, mycelial structure, spore-bearing structure, and spore morphology. The identified species are the following.

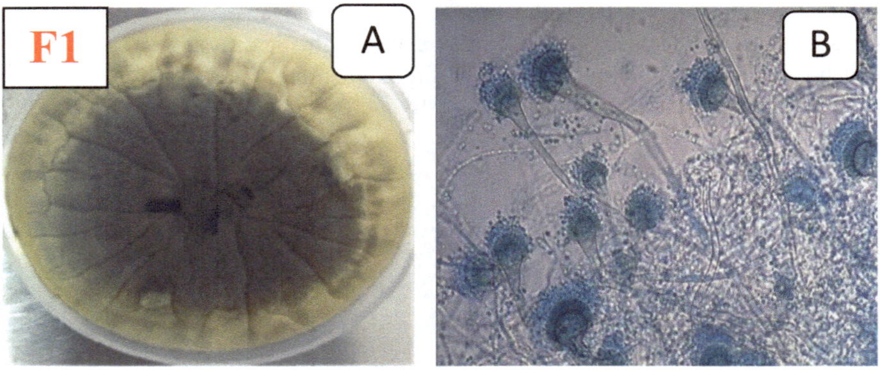

Fig. 18.1 First identified isolated fungus *Aspergillus* sp. F1 (**a**) showing growth on agar plates and (**b**) morphological observations under the microscope (100×)

18.4.1.1 First Identified Fungus F1

The first identified fungus strain was *Aspergillus* sp. F1 (Fig. 18.1a, b), which was identified on the basis of colony morphology on PDA plates and microscopic features. *Aspergillus* sp. is a filamentous fungus (mold) commonly reported from indoor environment that is used in different value added properties like food and enzyme industries. Various species of *Aspergillus* are used in the industrial preparation of citric acid, gluconic acid, and many other products of commercial importance. *Aspergillus* sp. can be isolated from many different ecological habitats such as soil, plant debris, and rotting fruits. This fungus is characterized by a round vesicle with extending conidial chains, which appear as white and fluffy strands on the substrate that the fungus inhabits. Taxonomic as well as identification points are given below.

Taxonomic Classification

- Kingdom: Fungi
- Division: *Ascomycota*
- Class: *Eurotiomycetes*
- Order: *Eurotiales*
- Family: *Trichocomaceae*
- Genus: *Aspergillus*

Identification Features

- Colonies grow very quickly.
- Colony produced is with yellow to white hyphae, turning black with the formation of conidia. Hyphae are septate.
- Conidiophores are long and globose at the tip.
- Spores are globose with conspicuous ridges or spines and not arranged in rows.

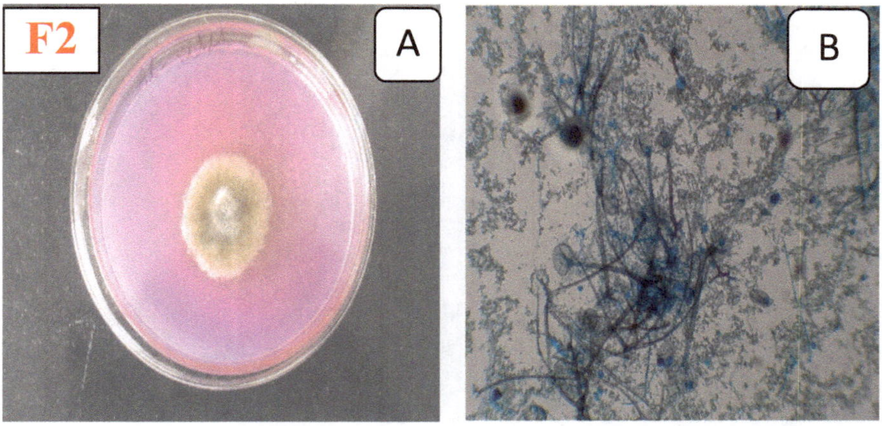

Fig. 18.2 Second and third identified isolated fungus *Rhizopus* sp. F2 (**a**) showing growth on agar plates and (**b**) morphological observations under the microscope (100×)

- Colony diameter was found to be 10–12 mm in size after 4 days of incubation on PDA plates.
- Conidia slightly roughened or finely echinulate.

18.4.1.2 Second Identified Fungus F2

The second identified fungus was *Rhizopus* sp. F2 (Fig. 18.2a, b), which is a cosmopolitan filamentous fungus frequently isolated from soil, decaying fruits and vegetables, animal feces, and old bread. Aside from being known as common contaminants, *Rhizopus* species are also occasional causes of serious, and often fatal, infections in humans. Certain species are plant pathogens as well.

Taxonomic Classification

- Kingdom: Fungi
- Phylum: *Zygomycota*
- Order: *Mucorales*
- Family: *Mucoraceae*
- Genus: *Rhizopus*

Identification Features

- Colonies spread rapidly at 25 °C, about 5–8 mm high.
- Apophysis, rhizoids, and stolons are present.
- Mycelium shows white in color at starting stage, and after that becomes dark brown or blackish brown to black in color.

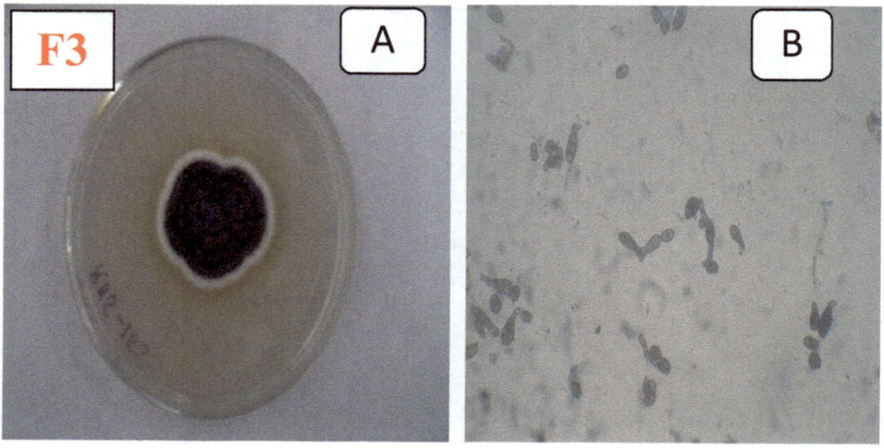

Fig. 18.3 Third identified isolated fungus *Alternaria* sp. F3 (**a**) showing growth on agar plates and (**b**) morphological observations under the microscope (100×)

18.4.1.3 Third Identified Fungus F3

Alternaria sp. F3 (Fig. 18.3a, b) is an ascomycetous fungus and is known as a major plant pathogen causing diseases in 380 host species of plant. They are also common allergens in humans, growing indoors and causing hay fever or hypersensitivity reactions. At least 20% of agricultural spoilage is caused by *Alternaria* species; the most severe losses may reach up to 80% of yield, though many human health disorders can be caused by these fungi, which grow on skin and mucous membranes, including on the eyeballs and within the respiratory tract. It is an opportunistic pathogen on numerous hosts causing leaf spots, rots, and blights on many plant parts. As a result, this pathogen propagates itself via asexual spores called conidia.
Taxonomic Classification

- Kingdom: Fungi
- Phylum: *Ascomycota*
- Class: *Dothideomycetes*
- Order: *Pleosporales*
- Family: *Pleosoporaceae*
- Genus: *Alternaria*

Identification Features

- Colonies are slow growing with whitish margin and gray to light brown in color.
- Conidiophores arise singly or in groups, usually simple, erect, straight, or curved.
- Conidiophores are occasionally geniculate, more or less cylindrical but often slightly rounded at the base.

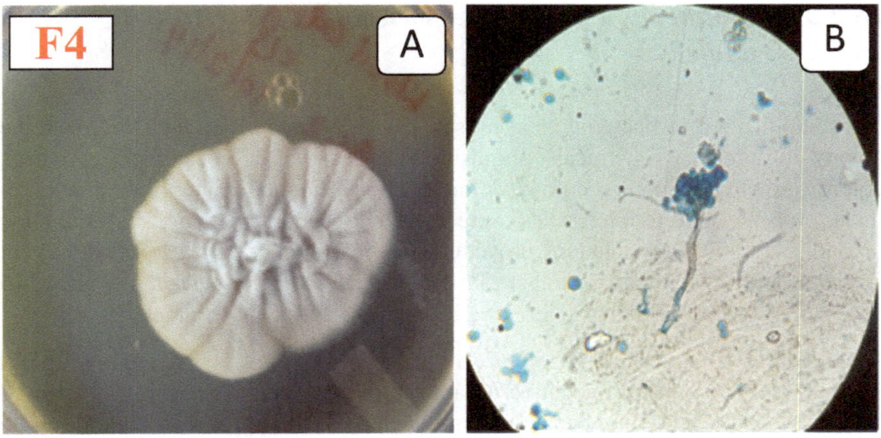

Fig. 18.4 Unidentified fungus F4 (**a**) showing growth on agar plates and (**b**) morphological observations under the microscope (100×)

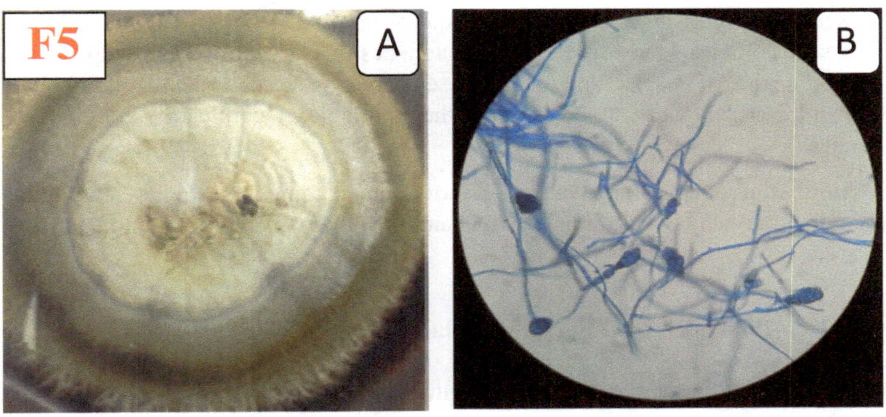

Fig. 18.5 Unidentified fungus F5 (**a**) showing growth on agar plates and (**b**) morphological observations under the microscope (100×)

- Conidia are pale brown to light brown, obclavate to obpyriform or ellipsoid, short conical beak at the tip, or beakless. Conidia are septate.

18.4.1.4 Unidentified Fungal Isolates F4 and F5

Two unidentified fungal isolates F4 and F5 (Figs. 18.4a, b and 18.5a, b) were obtained.

18.4.2 Identification and Characterization of Bacterial Isolates

The identification of bacterial isolates is based on many factors, including cell and colony morphology, chemical composition of cell walls, biochemical activities, colony growth (diameter), colony color, pigment production, nutritional requirements, etc. The bacterial species were identified by microscopic analysis using taxonomic guides and standard procedure (Barnett and Hunter 1972; Ellis 1976; Domsch et al. 1980; Nelson et al. 1983; Gilman 1998). From the samples only two bacterial isolates were identified on the basis of the following characteristics.

Gram Staining An initial step in identifying a bacterial species is determining if it is Gram-positive or Gram-negative. Gram staining is one of the most widely used tools in the identification of bacteria. Gram-negative cells appear pink after the Gram staining procedure, which enables comparison between those cells that decolorize with ethanol and those which do not.

Morphological Characteristics After the Gram staining procedure, microorganisms are also classified according to colony morphology and cell morphology. Bacterial colonies have different characteristics like size, form or shape, edge, texture, degree of opacity, and color. These characteristics describe the morphology of a single colony and may be useful in the preliminary identification of a bacterial species.

Identification of Samples On the basis of Gram staining, two isolates were found including one which was Gram-positive and the other which was Gram-negative.

18.4.2.1 First Identified Bacterial Isolate B1

The first isolated bacterial strain was identified as *Bacillus* sp. B1 (Fig. 18.6a, b).

Taxonomic Classification

- Domain: Bacteria
- Division: *Firmicutes*
- Class: *Bacilli*
- Order: *Bacillales*
- Family: *Bacillaceae*
- Genus: *Bacillus*

Identification Features

- Gram staining: Positive bacteria showing purple color cells after staining
- Shape: Rod-shaped cells

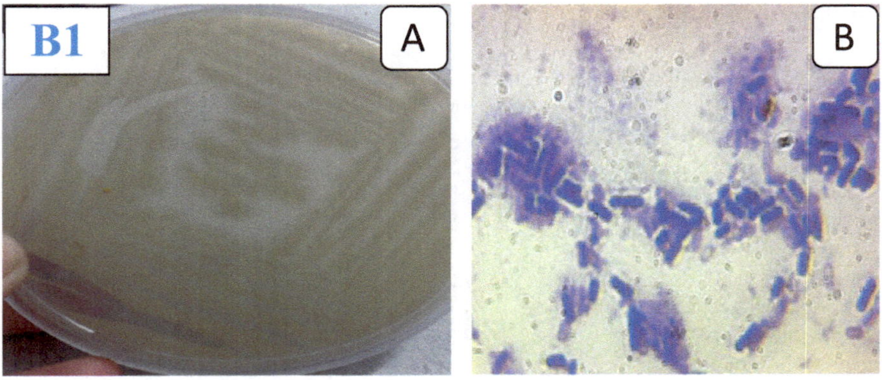

Fig. 18.6 First identified bacterial isolate *Bacillus* sp. B1, (**a**) showing growth on agar plates and (**b**) morphological features under the microscope (100×)

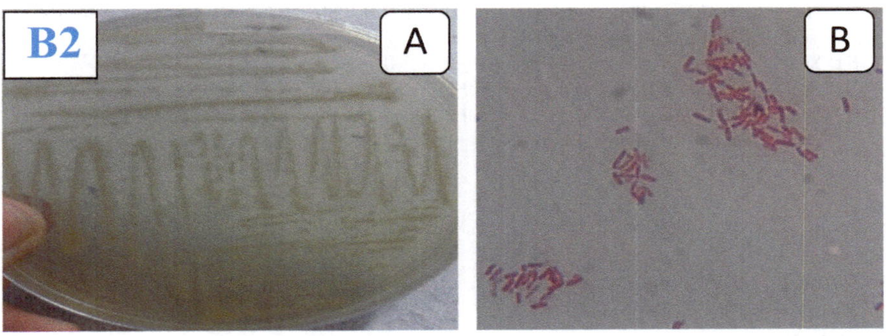

Fig. 18.7 Second identified isolated bacterial strain *Escherichia coli B*2 (**a**) showing growth on agar plates and (**b**) morphological features under the microscope (100×)

18.4.2.2 Second Identified Bacterial Isolate B2

The second identified bacterial strain was *Escherichia coli* (Fig. 18.7a, b). These are Gram-negative and rod-shaped bacteria.

Taxonomic Classification

- Domain: Bacteria
- Phylum: *Proteobacteria*
- Class: *Gammaproteobacteria*
- Order: *Enterobacteriales*
- Family: *Enterobacteriaceae*
- Genus: *Escherichia*

Box 18.1 Dilution-concentration pattern of SME with water

Water	Blank water (1: 0)	Wastewater with dilution (1:3)	Wastewater with dilution (1:1)	Wastewater wastewater (0:1)
Soil				
Soil irrigated with blank water	S1 W1 (100%)	S1 W2	S1 W3	S1 W4
Soil irrigated with wastewater with dilution	S2 W1	S2 W2	S2 W3	S2 W4
Soil irrigated with wastewater with dilution	S3 W1	S3 W2	S3 W3	S3 W4
Soil irrigated with undiluted wastewater	S4 W1	S4 W2	S4 W3	S4 W4 (100%)

Factor water was in the ratio given below:
- W1: Irrigation with fresh water or BWW containing 100% blank water (1: 0)
- W2: Irrigation with mixture containing 75% BWW and 25% wastewater (dilution ratio 1:3)
- W3: Irrigation with mixture containing 50% BWW and 50% wastewater (dilution ratio 1:1)
- W4: Irrigation with undiluted wastewater containing 0% blank water (0:1)

Factor soil was in the ratio given below:
- S1: Soil of fields containing 100% blank water (1: 0)
- S2: Soil watered with BWW and sugar mill effluents (3: 1)
- S3: Soil watered with BWW and sugar mill effluents (1: 1)
- S4: Soil watered with sugar mill effluents containing 0% blank water (0:1)

Identification Features

- Gram-negative bacteria.
- Slow-growing colonies.
- Conidiophores arising singly or in groups.
- Colonies are growing in cottony appearance.
- Colonies are light brown in color.

Isolated Microorganisms
Three fungal (*Aspergillus* sp., *Alternaria* sp., *Rhizopus* sp.) and two bacterial (*Bacillus* sp. and *Escherichia coli*) cultures have been identified on the basis of colony morphology observed on agar plates followed by microscopic features in both effluent and affected soils. However, two fungal isolates were not identified on the basis of morphological features. Moreover, all these microorganisms can be further identified on the basis of biochemical and molecular features.

18.5 Sugar Mill Effluent and Wheat Growth

Effect of effluent on wheat germination and its growth were carried out according to the following dilution-concentration pattern of SME (Box 18.1).

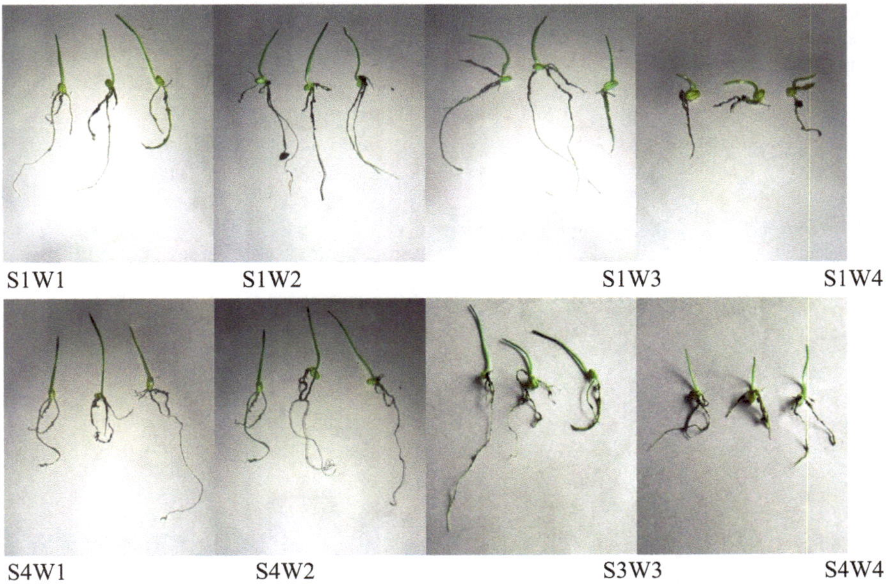

Fig. 18.8 Germination pattern of wheat plant irrigated with different concentrations (0%, 25%, 50%, and 100%) of sugar mill effluents

18.5.1 Germination of Seeds

Wheat seeds were shown in experimental pots. The pattern of germination was noted after 10 days by carefully pulling out the seedlings from each pot, and the number of roots was also counted in the seedlings. Seedling pattern of root growth and shoot growth is shown for various samples at different concentrations of sugar mill effluents (Fig. 18.8).

Wheat growth had been assessed before harvesting and after harvesting. Parameters analyzed before harvesting included germination percentage, speed of germination, peak value, vigor index, number of roots in seedling, root and shoot lengths, and fresh and total dry masses of wheat seedlings and after harvesting included plant height, length of the spike, number of spikes per plant and spikelet per spike, grain yield, straw yield, and biological yield.

18.5.2 Germination Percentage

With increase in effluent concentration, there is reduction in germination percentage of the wheat seeds but good germination rate at 50% concentration of effluents even better than with control (Table 18.1).

Table 18.1 Germination parameters of various wheat samples with different concentrations (0%, 25%, 50%, and 100%) of sugar mill effluents (mean ± SD of three values)

Samples	Germination (%)	Speed of germination
S1 W1	84.7 ± 3.85	7.21 ± 0.85
S1 W2	85.43 ± 3.85	7.58 ± 0.75
S1 W3	93.66 ± 6.67	8.21 ± 0.57
S1 W4	21.10 ± 3.85	2.16 ± 0.35
S2 W1	84.86 ± 0.00	7.98 ± 0.45
S2 W2	89.91 ± 6.67	8.54 ± 0.52
S2 W3	94.44 ± 3.85	9.21 ± 0.63
S2 W4	24.576 ± 3.85	2.11 ± 0.28
S3 W1	87.77 ± 3.85	7.59 ± 0.42
S3 W2	89.2 ± 3.85	7.23 ± 0.42
S3 W3	95.44 ± 3.85	9.34 ± 0.27
S3 W4	23.66 ± 0.00	1.89 ± 0.72
S4 W1	21.66 ± 3.85	1.21 ± 0.46
S4 W2	20.98 ± 3.85	1.29 ± 0.28
S4 W3	11.1 ± 3.85	1.01 ± 0.52
S4 W4	10.99 ± 3.85	1.11 ± 0.41

18.5.3 Speed of Germination

A comparative account of speed of germination of various wheat samples containing different ratios of sugar mill effluents (Fig. 18.8) shows clearly that speed of germination was maximum for samples S3 W3 and S2 W3, i.e., effluent concentration is beneficial up to value of 50%, and after this concentration, there is reduction in germination speed of the wheat crop.

The comparative seed germination (Fig. 18.9) represents maximum speed in green lines and minimum indicated by red lines showing inhibitory effect of effluents. Samples with 100% concentration of SME showed almost negligible germination marked by red line, and green line represents highest value of these parameters indicating usefulness of diluted SME in agriculture for better yield of crops. The seeds germinated were noted on each and every day after sowing, and then the percentage of germination was calculated. Although the seeds germinated in the samples containing S4 W4 showed least germination even at a later stage as compared to the pots irrigated with diluted effluents and containing soil prepared by irrigating with diluted effluents, still the number of seeds germinated was counted on each and every day and represented in tabular form.

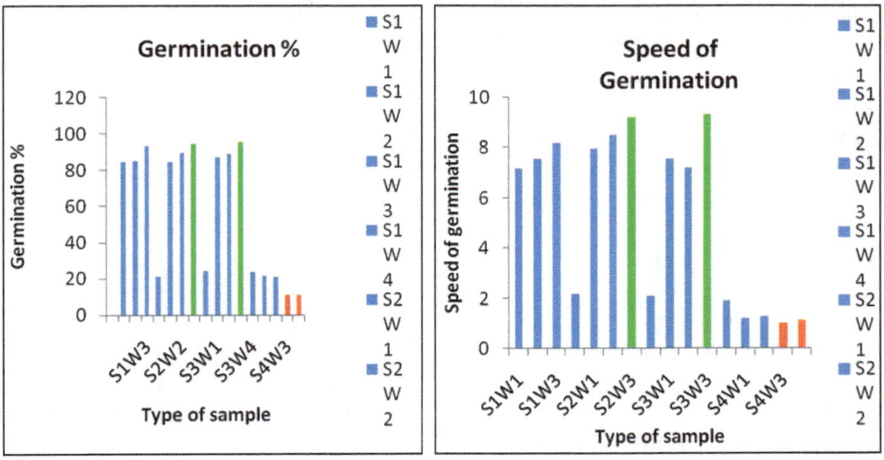

Fig. 18.9 Graph showing percentage of seed germination and speed/rate (%) of germination of various samples of wheat under different treatment conditions

18.5.4 Morphological Parameters of Wheat

18.5.4.1 Root Length (RL) and Shoot Length (SL)

The root and shoot lengths of seedlings were determined by using a scale of 10 days after sowing (DAS). The sum total of both (root and shoot) gives us the length of seedlings. The values of RL and SL of all the samples of different effluent concentrations (Tables 18.2, 18.3, and 18.4; Figs. 18.10, 18.11, and 18.12) showed clearly that the values are lowest for S4 W3 and S4 W4 and highest for S2 W3, S3 W2, and S3 W3. This is also in conformity with earlier findings that low concentration of effluents has positive effect on crops.

18.5.4.2 Number of Roots

A mature wheat plant has two distinct root types. The seminal roots are the first root types and nodal roots appear thereafter. The seminal roots are formed from the seed and nodal roots from the nodes and are related with the growth of tillers. The number of roots in various samples of different effluent concentrations is compared (Table 18.2, Fig. 18.10). It had been found that the number of roots of seedlings was maximum for S3 W3 and S2 W3 and minimum for S4 W4 and S4 W3, respectively, when compared with different treatment conditions.

Table 18.2 Shoot length, root length, number of roots, and vigor index of various samples of different effluent concentrations (mean ± SD of six values)

Samples	Shoot length (cm)	Root length (cm)	Number of roots	Vigor index
S1 W1	7.92 ± 0.3	6.65 ± 0.086	3.33 ± 0.57	12,340.7 ± 487.14
S1 W2	7.81 ± 0.15	6.51 ± 0.04	4.33 ± 0.57	12,233.5 ± 475.33
S1 W3	8.11 ± 0.19	7.11 ± 0.05	4.33 ± 0.57	14,255.0 ± 890.66
S1 W4	3.52 ± 0.079	3.08 ± 0.038	1.66 ± 0.5	1392.6 ± 44.9
S2 W1	8.41 ± 0.1	7.12 ± 0.029	4 ± 0.5	13,178.7 ± 580
S2 W2	8.45 ± 0.12	7.1 ± 0.033	5.33 ± 0.57	13,981.0 ± 914.77
S2 W3	8.92 ± 0.06	7.25 ± 0.034	5.66 ± 0.57	15,270.9 ± 534.41
S2 W4	2.79 ± 0.11	1.43 ± 0.025	2.33 ± 0.5	1037.1 ± 47.48
S3 W1	8.5 ± 0.074	7.28 ± 0.038	4.66 ± 0.57	13,850.1 ± 529.24
S3 W2	8.69 ± 0.07	8.47 ± 0.043	5.33 ± 0.57	15,306.7 ± 537.68
S3 W3	8.99 ± 0.041	8.92 ± 0.033	5.66 ± 0.57	17,093.3 ± 553.95
S3 W4	3.82 ± 0.065	2.08 ± 0.042	1.33 ± 0.57	1395.9 ± 123.12
S4 W1	2.21 ± 0.047	1.86 ± 0.058	1.33 ± 0.57	881.5 ± 78.47
S4 W2	1.39 ± 0.05	1.91 ± 0.033	1.66 ± 0.57	692.3 ± 84.76
S4 W3	1.28 ± 0.02	1.42 ± 0.035	1.33 ± 0.5	299.7 ± 66.85
S4 W4	1.18 ± 0.039	1.26 ± 0.02	0.66 ± 0.57	268.1 ± 22.88

Table 18.3 Fresh weight, dry weight, and moisture content of wheat samples under various treatment conditions (mean ± SD of six values)

Samples	Fresh weight (g)	Dry weight (g)	Moisture content (%)
S1 W1	0.82 ± 0.024	0.19 ± 0.02	0.63 ± 1.77
S1 W2	0.8 ± 0.021	0.15 ± 0.12	0.65 ± 1.23
S1 W3	0.93 ± 0.024	0.2 ± 0.012	0.73 ± 1.26
S1 W4	0.47 ± 0.033	0.09 ± 0.011	0.38 ± 1.19
S2 W1	0.72 ± 0.02	0.21 ± 0.014	0.51 ± 1.77
S2 W2	0.79 ± 0.014	0.15 ± 0.0136	0.64 ± 1.74
S2 W3	0.81 ± 0.015	0.17 ± 0.0123	0.64 ± 1.29
S2 W4	0.46 ± 0.014	0.09 ± 0.008	0.37 ± 1.42
S3 W1	0.95 ± 0.02	0.25 ± 0.011	0.7 ± 1.57
S3 W2	0.99 ± 0.017	0.3 ± 0.01	0.69 ± 1.098
S3 W3	1.05 ± 0.035	0.32 ± 0.011	0.73 ± 1.59
S3 W4	0.45 ± 0.014	0.15 ± 0.012	0.3 ± 2.66
S4 W1	0.4 ± 0.028	0.14 ± 0.007	0.26 ± 0.86
S4 W2	0.36 ± 0.035	0.11 ± 0.012	0.25 ± 1.83
S4 W3	0.21 ± 0.021	0.05 ± 0.01	0.16 ± 1.25
S4 W4	0.19 ± 0.024	0.02 ± 0.02	0.17 ± 1.86

18.5.4.3 Vigor Index

The seedling length and vigor index values of various samples of wheat are determined by the product of seeding length in mm (sum of root length and shoot

Table 18.4 Leaf length, leaf width, and leaf area of leaves of various samples under different treatment conditions after 30, 60, and 90 DAS (mean ± SD of 10 values)

	30 DAS			60 DAS			90 DAS		
Samples	Leaf length	Leaf width	Leaf area	Leaf length	Leaf width	Leaf area	Leaf length	Leaf width	Leaf area
S1 W1	9.5 ± 0.67	0.5 ± 0.019	3.56 ± 0.23	14.3 ± 0.37	1.2 ± 0.022	12.87 ± 0.45	18.5 ± 0.67	1.7 ± 0.02	23.58 ± 1.67
S1 W2	9.32 ± 0.19	0.5 ± 0.02	3.49 ± 0.41	13.12 ± 0.13	0.98 ± 0.009	9.64 ± 0.43	16 ± 0.22	1.5 ± 0.023	18 ± 1.11
S1 W3	8.66 ± 0.15	0.53 ± 0.01	3.44 ± 0.34	13.06 ± 0.12	1.15 ± 0.012	11.26 ± 0.43	15.66 ± 0.17	1.73 ± 0.019	20.33 ± 1.17
S1 W4	7.63 ± 0.13	0.32 ± 0.007	1.83 ± 0.26	11.33 ± 0.18	0.91 ± 0.01	7.73 ± 0.34	13.83 ± 0.12	1.22 ± 0.032	12.66 ± 1.34
S2 W1	8.5 ± 0.21	0.35 ± 0.02	2.23 ± 0.33	11.95 ± 0.22	0.92 ± 0.008	8.24 ± 0.23	15.5 ± 0.3	1.34 ± 0.12	15.58 ± 1.11
S2 W2	8.66 ± 0.16	0.42 ± 0.001	2.72 ± 0.28	14.66 ± 0.26	1.05 ± 0.02	11.54 ± 0.54	19.66 ± 0.3	1.42 ± 0.021	20.94 ± 1.34
S2 W3	9.33 ± 0.11	0.61 ± 0.02	4.26 ± 0.36	13.53 ± 0.21	1.42 ± 0.01	14.41 ± 0.52	16.33 ± 0.31	1.67 ± 0.16	20.45 ± 1.42
S2 W4	6.93 ± 0.04	0.38 ± 0.001	1.97 ± 0.12	10.63 ± 0.03	0.81 ± 0.01	6.46 ± 0.34	13.83 ± 0.06	1.11 ± 0.012	15.35 ± 1.65
S3 W1	8.66 ± 0.13	0.38 ± 0.005	2.46 ± 0.17	11.76 ± 0.14	0.88 ± 0.002	7.76 ± 0.26	15.66 ± 0.23	1.38 ± 0.025	21.62 ± 1.26
S3 W2	9.5 ± 0.11	0.56 ± 0.001	3.99 ± 0.46	14.5 ± 0.25	1.18 ± 0.04	12.83 ± 0.48	18.5 ± 0.16	1.76 ± 0.031	32.56 ± 1.34
S3 W3	9.83 ± 0.12	0.59 ± 0.001	4.34 ± 0.51	13.53 ± 0.2	1.16 ± 0.001	11.77 ± 0.45	16.83 ± 0.32	1.79 ± 0.01	30.13 ± 1.31
S3 W4	6.66 ± 0.13	0.46 ± 0.015	2.3 ± 0.14	11.06 ± 0.12	0.89 ± 0.003	7.38 ± 0.33	15.66 ± 0.23	1.16 ± 0.15	18.17 ± 1.42
S4 W1	8.83 ± 0.17	0.43 ± 0.013	2.84 ± 0.16	11.88 ± 0.16	0.76 ± 0.015	6.77 ± 0.22	16.83 ± 27	1.23 ± 0.03	20.70 ± 1.26
S4 W2	8.83 ± 0.14	0.48 ± 0.002	3.17 ± 0.31	15.33 ± 0.24	0.91 ± 0.013	10.46 ± 0.31	18.83 ± 0.27	1.52 ± 0.02	28.62 ± 1.37
S4 W3	9.16 ± 0.12	0.57 ± 0.008	3.19 ± 0.17	13.86 ± 0.11`	1.03 ± 0.002	10.7 ± 0.22	17.16 ± 0.12	1.67 ± 0.018	28.66 ± 1.15
S4 W4	6.33 ± 0.14	0.31 ± 0.001	1.47 ± 0.11	9.96 ± 0.15	0.69 ± 0.008	5.15 ± 0.34	12.33 ± 0.34	1.1 ± 0.011	13.56 ± 1.82

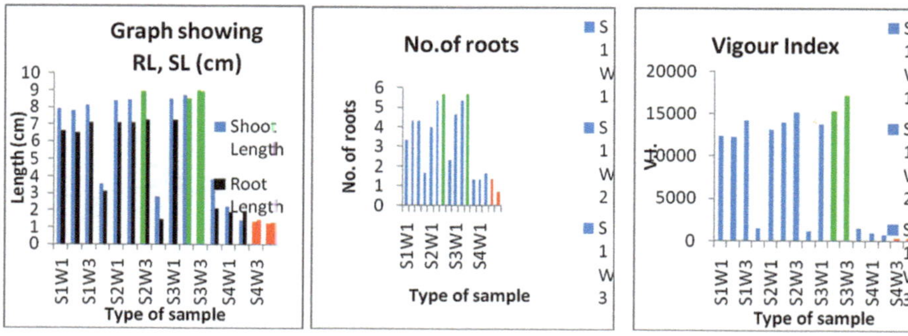

Fig. 18.10 Comparative analysis of SL, RL, no. of roots, and VI of various wheat samples with different treatments after 10 DAS

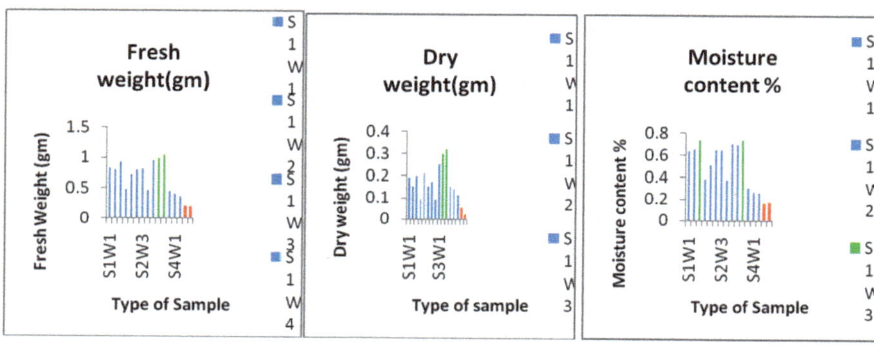

Fig. 18.11 Comparative analysis of fresh weight, dry weight, and moisture content of wheat samples under various treatment conditions after 10 DAS

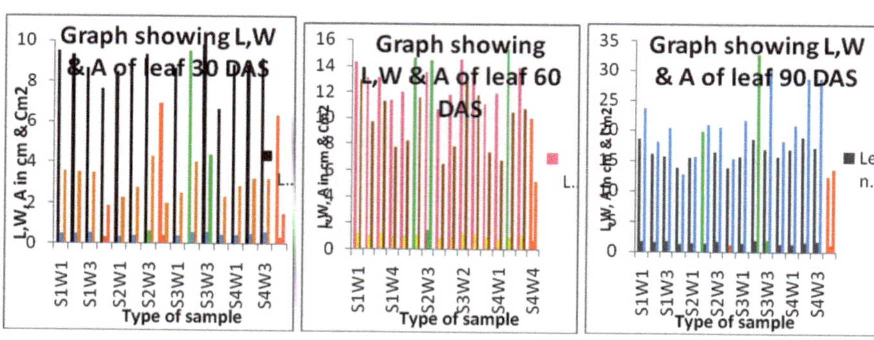

Fig. 18.12 Comparative analysis of length, width, and area of leaf of various wheat samples with different treatments after 30, 60, and 90 DAS

length) with germination percentage (Fig. 18.10). It is clearly shown that this value is lowest for S4 W3 and S4 W4 and highest for S3 W2 and S3 W3 which shows that effluent with low concentration up to 50% is suitable for good vigor crops.

18.5.4.4 Fresh Weight, Dry Weight, and Moisture Content

The fresh and total dry weights of wheat seedlings were determined after 10 days of the experiment (Table 18.3, Fig. 18.11). The plants were washed thoroughly with distilled water and were dried for 2 h under natural conditions in an open roof. The fresh weights were taken, and then the plants were packed in paper envelopes and oven-dried for 36 h at 70 °C, and the dry weight of each plant was also recorded. It showed clearly that fresh and dry weights are lowest for S4 W3 and S4 W4 and highest for S3 W2 and S3 W3. The difference of both of these, i.e., fresh weights and dry weights of the seedlings, gives the moisture content of the seedlings. Thus, it is also lowest for S4 W3 and S4 W4 but highest for S1 W3 and S3 W3 when compared with different treatment conditions which clearly shows that effluent with low concentration contributes significantly in the weight of wheat plant but high concentration of SME has negative effect on weight and moisture content of wheat.

18.5.4.5 Total Leaf Area

The leaf area was determined and was calculated on 30, 60, and 90 DAS of the experiment (Fig. 18.12). It is clearly observed (Fig. 18.12) that higher concentration of sugar mill effluents causes a decrease in leaf length, width, and area after 30 DAS, 60 DAS, and 90 DAS, whereas low concentration has stimulating effect on these factors. So it is quite clear that using low concentration of effluents in crops for irrigation purpose is beneficial for good growth of plant leaf and also increases photosynthesis rate and hence crop production. Positive effect on crops is indicated by green line and negative effect by red line (Fig. 18.12). S4 W4 and S2 W4 generally are having the least values, and S2 W2/S3 W2 are having the highest growth of leaf length, width, and area after 30 DAS, 60 DAS, and 90 DAS.

18.6 Wheat Seed and Straw Yield After Harvesting

When all the spikes became straw color, then harvesting was carried out, collected properly in separate bundles, and labeled correctly. Before drying in sunlight, plant height and number of tillers per plant for each pot were recorded. Height of the plant was recorded from ground level to the tip of the longest spike. Length of the spike was also noted in the same manner (Table 18.5).

Table 18.5 Harvested plant height and length of the spikes (cm) of various wheat samples with different treatments (mean ± SD of six values)

Samples	Plant height (cm)	Length of spike (cm)
S1 W1	73.84 ± 0.85	10.23 ± 0.24
S1 W2	70.12 ± 1.04	11.15 ± 0.38
S1 W3	68.32 ± 2.61	11.25 ± 0.58
S1 W4	34.42 ± 1.58	4.58 ± 0.38
S2 W1	70.35 ± 1.23	8.78 ± 0.36
S2 W2	77.36 ± 1.36	9.73 ± 0.27
S2 W3	85.36 ± 1.39	13.59 ± 0.4
S2 W4	29.59 ± 1.24	5.57 ± 0.36
S3 W1	70.75 ± 1.55	8.36 ± 0.36
S3 W2	66.15 ± 1.49	9.37 ± 0.15
S3 W3	88.62 ± 1.13	12.55 ± 0.39
S3 W4	26.28 ± 1.39	8.15 ± 0.47
S4 W1	30.79 ± 1.78	3.67 ± 0.2
S4 W2	32.52 ± 1.26	2.89 ± 0.26
S4 W3	32.12 ± 1.35	2.32 ± 0.31
S4 W4	23.12 ± 0.99	1.56 ± 0.43

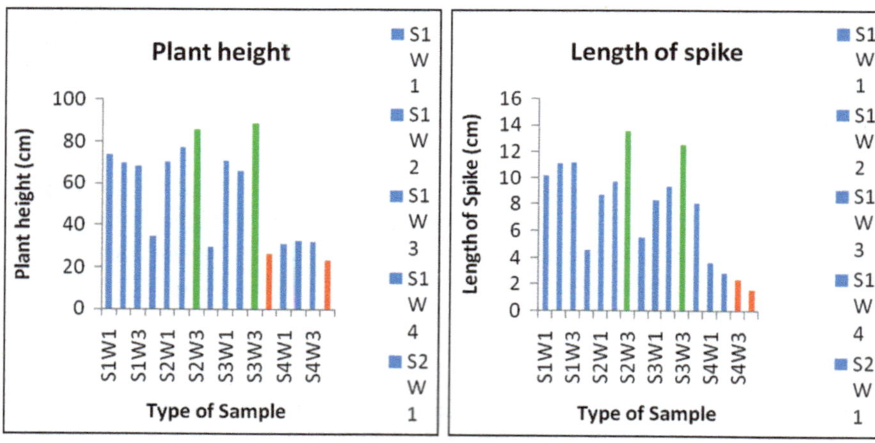

Fig. 18.13 Comparative analysis of plant height and length of spikes of various wheat samples with different treatments after harvesting

18.6.1 Plant Height and Length of the Spike

A comparative analysis (Fig. 18.13) showed that the plant height and length of spikes are lowest for S4 W3 and S4 W4 and highest for S3 W2 and S3 W3 which clearly shows that effluent with low concentration (up to 50%) is suitable for better growth over that of plant grown under control condition. The same kinds of observations were also made by Suresh et al. (2014) and Srivastava et al. (2015).

Table 18.6 Number of tillers or spikes/plant, spikelets/spike, and grains/spike wheat samples under various treatment conditions (mean ± SD of six values)

Samples	No. of tillers/plant	Spikelets/spike	Grains/spike
S1 W1	4.66 ± 0.82	15.5 ± 1.05	35.16 ± 2.79
S1 W2	4.83 ± 0.75	15 ± 0.89	34.5 ± 3.27
S1 W3	4.93 ± 0.52	17.66 ± 0.52	36.83 ± 2.63
S1 W4	2.5 ± 0.55	7.83 ± 0.75	15.66 ± 2.73
S2 W1	3.833 ± 0.75	15.5 ± 1.37	34.66 ± 3.5
S2 W2	4.66 ± 0.82	15.66 ± 1.03	32.5 ± 3.62
S2 W3	4.98 ± 0.98	17.33 ± 0.82	35.66 ± 3.26
S2 W4	2.16 ± 0.75	7.83 ± 1.17	13.33 ± 2.34
S3 W1	4.5 ± 0.54	15.66 ± 1.63	34.83 ± 2.93
S3 W2	4.66 ± 0.52	16.5 ± 1.05	36.66 ± 2.58
S3 W3	5.16 ± 0.51	16.83 ± 1.17	38 ± 2.6
S3 W4	2.3 ± 0.84	8.66 ± 0.82	15 ± 2.36
S4 W1	1.83 ± 0.75	6.83 ± 1.17	11.66 ± 1.75
S4 W2	1.23 ± 0.75	5.83 ± 1.17	8.83 ± 2.86
S4 W3	1.23 ± 0.98	5.76 ± 0.75	8.59 ± 1.67
S4 W4	1.166 ± 0.75	5.33 ± 0.82	8.34 ± 2.1

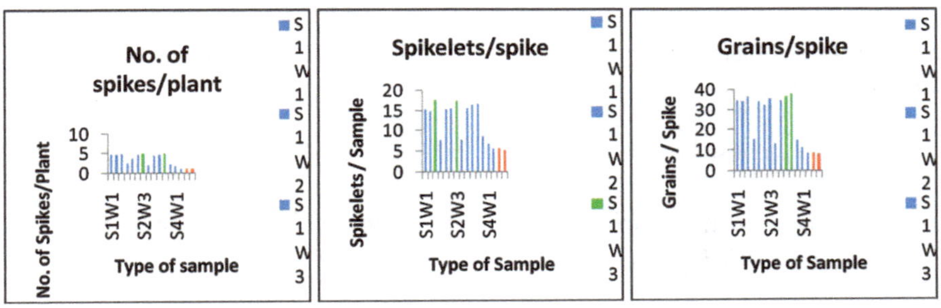

Fig. 18.14 Comparative analysis of number of spikes/plant, spikelets/spike, and grains/spike of various wheat samples with different treatments

18.6.2 Number of Tillers or Spikes/Plant, Spikelets/Spike, and Grains/Plant

The number of tillers/spikes per plant, spikelets/spike, and grains/spike had been analyzed (Table 18.6) of various wheat samples with different treatments of effluent. The comparative analysis of number of tillers/spikes per plant, spikelets/spike, and grains/spike (Fig. 18.14) shows that a number of tillers/spikes per plant, spikelets/spike, and grains/spike were maximum for samples S3 W3 and S2 W3, a number of spikelets/spike were maximum for samples S1 W3 and S2 W3, and a number grains/spike were maximum for samples S3 W3 and S3 W2 indicating that some quantity of effluent concentration (up to value of 50%) is beneficial and concentration higher

Table 18.7 Grain yield/pot, straw yield/pot, and biological yield/pot of wheat samples under various treatment conditions (mean ± SD of six values)

Samples	Grain yield/pot	Straw yield/pot	Biological yield/pot
S1 W1	95.88 ± 2.42	185.68 ± 12.31	281.56 ± 14.53
S1 W2	94.26 ± 2.97	168.22 ± 4.235	262.48 ± 7.21
S1 W3	99.67 ± 2.93	177.08 ± 8.84	276.76 ± 6.03
S1 W4	38.313 ± 3.86	90.29 ± 5.00	128.60 ± 4.09
S2 W1	99.09 ± 2.23	182.81 ± 7.37	281.90 ± 6.41
S2 W2	101.66 ± 3.01	213.44 ± 8.01	311.71 ± 10.73
S2 W3	105.303 ± 4.62	98.366 ± 4.01	318.74 ± 12.42
S2 W4	33.503 ± 3.95	93.43 ± 5.28	126.93 ± 6.48
S3 W1	95.96 ± 2.11	189.97 ± 6.91	285.94 ± 5.68
S3 W2	94.81 ± 1.78	190.28 ± 9.75	285.1 ± 10.05
S3 W3	102.006 ± 1.72	210.0433333 ± 7.74	295.76 ± 5.65
S3 W4	52.53 ± 1.11	193.76 ± 4.54	150.89 ± 3.39
S4 W1	65.716 ± 2.92	99.073 ± 1.55	164.79 ± 3.04
S4 W2	65.99 ± 1.14	91.9 ± 2.63	157.89 ± 2.88
S4 W3	68.13 ± 2.55	97.42 ± 4.82	165.5 ± 8.19
S4 W4	22.71 ± 2.41	86.61 ± 4.15	109.3 ± 5.94

than this leads to a reduction in the number of tillers/spikes per plant, spikelets/spike, and grains/spike.

18.6.3 Yield and Yield Components (Grain Yield, Straw Yield, and Biological Yield)

Grain yield, straw yield and total biological yield in grams per pot (Table 18.7) along with values of standard deviations of various wheat samples with different treatments. The grain yield per 10 plants was measured and converted into final grain yield per replicate. Similar to the grain yield, the straw yield was also determined after drying of grains in the sun and converted into yield per replicate. An average value of the pot of each replicate was noted along with values of their standard deviations. Both grain yield and straw yield collectively make biological yield, and it is also calculated per replicate, and an average value of the pot of each replicate was noted along with values of their standard deviations. It was observed that S3 W3 and S2 W2 are best conditions for grain yield and S3 W3 and S2 W2 for straw yield. The comparative analysis of grain yield, straw yield, and biological yield (Fig. 18.15) shows that some quantity of effluent concentration (up to value of 50%) is beneficial and concentration higher than this leads to reduction of wheat yield.

As mentioned in other parameters, also diluted effluents result in increase in crop yield, while increase in concentration of effluents causes reduction in yield of crops.

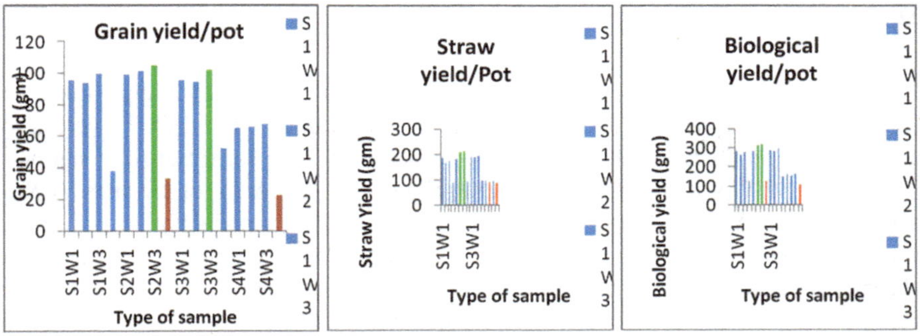

Fig. 18.15 Comparative analysis of grain yield/pot, straw yield/pot, and biological yield/pot of various wheat samples with different treatments

Grain yield is highest for S2 W3 and S3 W3, even higher than the normal field soil, whereas S2 W4 and S4 W4 have lowest yield due to contamination of soil. Similarly, S4 W2 and S4 W4 are on the lowest side, and S2 W2 and S2 W3 have the highest straw yield. These results indicate that high concentration of nutrients in the effluents cause increase in the yield of grain and straw as well. S4 W4 and S2 W4 samples have the lowest values of biological yield due to contamination of soil resulting in reduced growth, whereas S2 W2 and S2 W3 have shown the highest yield.

Variation in germination, shoot length, root length, and yield with variation in concentration of SME that was utilized for treatment of seeds clearly revealed that effluents exhibit profound effect on the abovementioned physiological parameters. Observations made from the experiments conducted indicate clearly that with gradual increase in concentration of the effluents (50–100%), a gradual decrease in germination rate, RL, SL, leaf area, and yield of crop was observed. Among various concentrations of effluents which were utilized during the study, 25 and 50% concentration of effluents was found to be most effective in increasing the germination rate and other parameters in wheat plant.

Highest germination speed was observed in SME concentrations of 25 and 50%. However, germination percentage decreases to 80–65% in wheat when seed was treated with higher concentration of effluents (more than 50%) (Pandey et al. 2008). The growth and germination percentage of seed inhibited at higher concentration of effluents may be due to osmotic pressure of high dose, which make imbibitions more difficult and reduce oxygen uptake by seedling (Khatoon et al. 2010), while diluting the effluent enhances the plant activities by providing required amount of nutrients present.

18.7 Conclusion

It is concluded that almost all the anions and cations are much higher than the permissible limits in SME. Considerable high values of BOD and COD were also observed. A number of tillers/plant and grains/plant were maximum for samples S3 W3 and S2 W3, a number of spikelets/spike were maximum for samples S1 W3 and S2 W3, S3 W3 and S2 W2 are the best conditions for grain yield and straw yield, and biological yield was best under S2 W2 and S2 W3 conditions. It has been observed that germination, harvesting, and postharvesting parameters were best for S2 W2, S2 W3, or S1 W3. Hence, some quantity of effluent concentration (up to value of 50%) is beneficial for crop growth as tests for wheat growth and concentration higher than this lead to reduction in vegetative growth as well as crop yield. Numbers of bacterial and fungal species are less in SME as compared to normal soil in this study, and it may be because of the negative impact of contaminants of SME on micro-biodiversity of soil.

References

Barnett HL, Hunter BB (1972) Illustrated genera of imperfect fungi, 3rd edn. Burgess Publishing, Minneapolis, MN, p 273

Baruah AK, Sharma RN, Borah GC (1993) Impact of sugar mill and distillery effluents on water quality of river Gelabil Assam. Indian J Environ Health 35:288–293

Chandra R, Bharagava RN, Yadav S, Mohan B (2009) Accumulation and distribution of toxic metals in wheat (*Triticum aestivum*) and Indian mustard (*Brassica juncea*) irrigated with distillery and tannery effluents. J Hazard Mater 169:1514–1521

CPCB (2009) Water quality status of Yamuna river, assessment and development of river basin series: ADSORBS/41/2009. Published by Central Pollution Control Board (CPCB), Delhi

Domsch KH, Games W, Anderson TH (1980) Compendium of soil fungi. Academic, London

Ellis MB (1976) More dematiaceous hyphomycetes. Common Wealth Mycological Institute, Kew, London

Gilman JC (1998) A manual of soil fungi. Biotech Books, New Delhi

Khatoon T, Hussain K, Majeed A, Nawaz K, Nisar MF (2010) Morphological variations in maize (*Zea mays* L.) under different levels of NaCl at germinating stage. World Appl Sci J 8:1294–1129

Maliwal GL, Patel KP, Patel KC, Patel MN (2004) Pollution studies on sugar mill effluent, physico-chemical properties and toxic metals. Pollut Res 14:231–238

Nelson PE, Toussoun TA, Marasas WFO (1983) *Fusarium* species: an illustrated manual for identification. Pennsylvania State University Press\University Park, Pennsylvania

Pandey SN, Nautiyal BD, Sharma CP (2008) Pollution level in distillery effluent and its phytotoxic effect on seed germination and early growth of maize and rice. J Environ Biol 29(2):267–270

Patil V, Gholey V (2010) Comparative toxicological studies of distillery effluent treatment such as UASB reactor followed by on oxidizer unit using *Cyprinus carpio* fish. Environ Technol 31 (2):185–191

Roy P, Ratna JP, Joshi AP (2007) Effect of sugar factory effluent on some physico-chemical properties of soils-a case study. J Environ Sci Eng 49(4):277–282

Satheeskumar L, Selvaraj V (2007) Sugar industry need of the hour. Econ Chall 10(37):61–63

Sharma BK, Habib I (1997) Irrigational impact of rubber factory effluent on elemental bioaccumulation and metabolic concentration in component parts of *Hordeum vulgare* var. J Indian Bot Soc 76:5–10

Singh DK, Kumar D, Singh VP (2005) Phytoremediation of contaminated soil. J Environ Biol 5:15–91

Solomon SK (2008) Environmental pollution and its management in sugar industry in India: an appraisal. Sugar Technol 7(1):77–81

Srivastava S, Chopra AK, Kumar V, Sehgal D (2015) Agrofertigational response of sugar mill effluent and synthetic fertilizer (DAP) on the agronomy of crop *V. unguiculata* L. walp in two seasons. Res J Agric Environ Sci 2(3):5–17

Suresh B, Abraham K, Damodharam T (2014) Effect of sugar industry effluent on changes of growth and biochemical contents of *Capsicum annuum* L. Adv Appl Sci Res 5(5):305–309

FSC
www.fsc.org
MIX
Papier | Fördert gute Waldnutzung
FSC® C083411

Kolibri 360 GmbH
Ferdinand-Jühlke-Straße 7
99095 Erfurt, Deutschland
produktsicherheit@kolibri360.de